Advances in Enzymology

Advances in Enzymology

Editor: Joshua Kelly

CALLISTO REFERENCE

www.callistoreference.com

Callisto Reference,
118-35 Queens Blvd., Suite 400,
Forest Hills, NY 11375, USA

Visit us on the World Wide Web at:
www.callistoreference.com

ISBN: 978-1-64116-254-8 (Hardback)

Cataloging-in-Publication Data

Advances in enzymology / edited by Joshua Kelly.
    p. cm.
Includes bibliographical references and index.
ISBN 978-1-64116-254-8
1. Enzymology. 2. Enzymes. 3. Biochemistry. I. Kelly, Joshua.
QP601 .A38 2020
574.192 5--dc23

Table of Contents

Preface

Enzymes are biological catalysts that are required for the sustenance of life. Nearly all metabolic processes occurring in the cell require enzyme catalysis for sustaining life. Enzymes are generally proteins, although some are catalytic RNA molecules. The difference between enzymes and any other catalysts is their high degree of specificity. Such specificity arises from their unique 3D structures. The activity of enzymes can be affected by other molecules. There are some therapeutic drugs and poisons that may inhibit enzyme activity. Enzymes are also used commercially for the synthesis of antibiotics. The activity of enzymes needs to be closely regulated for homeostasis. The overproduction, underproduction, mutation or deletion of any enzyme can lead to genetic diseases, such as phenylketonuria, Tay–Sachs disease, pseudocholinesterase deficiency, etc. This book aims to shed light on some of the unexplored aspects of enzymology and the recent researches in this field. The various studies that are constantly contributing towards advancing technologies and evolution of this field are examined in detail. Coherent flow of topics, student-friendly language and extensive use of examples make this book an invaluable source of knowledge.

After months of intensive research and writing, this book is the end result of all who devoted their time and efforts in the initiation and progress of this book. It will surely be a source of reference in enhancing the required knowledge of the new developments in the area. During the course of developing this book, certain measures such as accuracy, authenticity and research focused analytical studies were given preference in order to produce a comprehensive book in the area of study.

This book would not have been possible without the efforts of the authors and the publisher. I extend my sincere thanks to them. Secondly, I express my gratitude to my family and well-wishers. And most importantly, I thank my students for constantly expressing their willingness and curiosity in enhancing their knowledge in the field, which encourages me to take up further research projects for the advancement of the area.

Editor

Agroindustrial Wastes as Alternative for Lipase Production by *Candida viswanathii* under Solid-State Cultivation: Purification, Biochemical Properties, and its Potential for Poultry Fat Hydrolysis

Alex Fernando de Almeida,[1] **Kleydiane Braga Dias,**[1] **Ana Carolina Cerri da Silva,**[2] **César Rafael Fanchini Terrasan,**[3] **Sâmia Maria Tauk-Tornisielo,**[2] **and Eleonora Cano Carmona**[3]

[1]*Bioprocess Engineering and Biotechnology, Federal University of Tocantins (UFT), Rua Badejós, Chácaras 69/72, Zona Rural, 77402-970 Gurupi, TO, Brazil*
[2]*Environmental Studies Center (CEA), Universidade Estadual Paulista (UNESP), Avenida 24-A, 1515 Bela Vista, 13506-900 Rio Claro, SP, Brazil*
[3]*Biochemistry and Microbiology Department, Bioscience Institute (IB), Universidade Estadual Paulista (UNESP), Avenida 24-A, 1515 Bela Vista, 13506-900 Rio Claro, SP, Brazil*

Correspondence should be addressed to Alex Fernando de Almeida; alexfernando@uft.edu.br

Academic Editor: Sunney I. Chan

The aims of this work were to establish improved conditions for lipase production by *Candida viswanathii* using agroindustrial wastes in solid-state cultivation and to purify and evaluate the application of this enzyme for poultry fat hydrolysis. Mixed wheat bran plus spent barley grain (1 : 1, w/w) supplemented with 25.0% (w/w) olive oil increased the lipase production to 322.4%, compared to the initial conditions. When olive oil was replaced by poultry fat, the highest lipase production found at 40% (w/w) was 31.43 U/gds. By selecting, yeast extract supplementation (3.5%, w/w), cultivation temperature (30°C), and substrate moisture (40%, w/v), lipase production reached 157.33 U/gds. Lipase was purified by hydrophobic interaction chromatography, presenting a molecular weight of 18.5 kDa as determined by SDS-PAGE. The crude and purified enzyme showed optimum activity at pH 5.0 and 50°C and at pH 5.5 and 45°C, respectively. The estimated half-life at 50°C was of 23.5 h for crude lipase and 6.7 h at 40°C for purified lipase. Lipase presented high activity and stability in many organic solvents. Poultry fat hydrolysis was maximum at pH 4.0, reaching initial hydrolysis rate of 33.17 mmol/L/min. Thus, *C. viswanathii* lipase can be successfully produced by an economic and sustainable process and advantageously applied for poultry fat hydrolysis without an additional acidification step to recover the released fatty acids.

1. Introduction

Lipase is an enzyme that catalyzes the hydrolysis and synthesis of esters formed by the linkage of glycerol and long-chain fatty acids. Owing to properties such as catalytic activity over a wide range of temperature and pH, substrate specificity, and enantioselectivity, lipase presents important industrial application [1]. The numerous applications are also due to other catalyzed reactions that differ from their natural physiological reaction. This versatility makes it the enzyme of choice for application in food, detergent, pharmaceutical, leather, textile, cosmetic, and paper industries [2].

Lipase can be produced in submerged and solid-state cultivation by many microorganisms such as bacteria, yeast, and filamentous fungi. The utilization of solid-state cultivation (SSC) for enzyme production requires previous evaluation of important aspects, such as selection of a suitable microorganism and substrate, optimization of process parameters, and isolation and further purification of the product [3]. SSC is defined as a process in which the substrate itself acts as carbon/energy source in the absence or near-absence of free water, employing natural or inert substrates used as solid support [4]. SSC is of special economic interest for developing countries such as Brazil, with abundance of biomass and agroindustrial wastes, which can be used as inexpensive raw materials [5]. Brazil is known for its great renewable resources such as agricultural and forestry wastes such as sugarcane bagasse, rice straw, wheat straw, oat hulls, and wood chips. The production of organic residues is about 597 million tons/year. Proper use of these wastes helps to minimize environmental and energy problems; furthermore, they can be used to obtain products with important applications in the pharmaceutical and food industry [6].

Several agroindustrial wastes have been evaluated for lipase production by microorganisms. Sugarcane bagasse was used for lipase production by *Yarrowia lipolytica* [7] and *Rhizopus homothallicus* [8], *Jatropha* cake by *Pseudomonas aeruginosa* [9], rice bran by *Aspergillus niger* [10], and babassu cake by *Penicillium simplicissimum* [11]. SSC is advantageous for lipase production due to higher activity levels, increased productivity, the extracellular nature of the produced enzyme, and increased stability to pH and temperature. In addition, it allows the construction of more compact reactors with less energy requirements, causing less damage to the environment [12]. Ideally, the solid materials should act as physical support, source of nutrients, and also appropriate inducer for enzyme production; however it is difficult to obtain all these features from a single substrate. This bottleneck can be overcome by using a combination of different substrates [13]. For example, SSC using mixed wheat bran and gingelly oil cake substrates increased the lipase production to 36.0% by *A. niger* MTCC 2594 and maximum activity corresponded to 384.3 U/grams of dry substrate (gds) after cultivation at 30°C for 72 h.

Many lipase types have been purified and biochemically characterized because their properties are very important for industrial applications. Microbial lipase usually presents high thermostability and pH stability, solvent tolerance, and high specificity for hydrolysis of long-chain unsaturated fatty acids [14, 15].

Poultry fat is a low cost feedstock, which could be incorporated in delicatessen meats and has a substantial nutritional value due to its high content of unsaturated fatty acids, especially monounsaturated ones, such as oleic acid (45–50%) [16]. Production of fatty acids by hydrolysis of natural oils and fats has been exploited using renewable raw materials; and it is considered an important industrial operation, since 1.6 million tons of fatty acids is produced worldwide every year by this process [17]. The products, fatty acids, and glycerol are basic materials for a wide range of applications. Fatty acids are used as a feedstock for the production of oleochemicals such as fatty alcohols, fatty amines, and fatty esters. These oleochemicals are used as lubricant grease, antiblock agent, plasticizers, emulsifiers, and also ingredient in the manufacture of soaps, detergents, and animal feed [18].

Candida viswanathii was isolated from wastewater of a Brazilian oil refinery (Replan/Petrobras, Paulínia, São Paulo, Brazil) and selected as the best lipase producer among 19 filamentous fungi and yeasts. In this study, *C. viswanathii* growing at high fat and hydrocarbon concentrations was observed. Lipase production was induced by long-chain fatty acids and triacylglycerol and the enzyme was effective in hydrolyzing triolein and olive oil [19]. In further study, the best parameters for lipase production by this yeast were observed in liquid Vogel's medium with 1.5% (w/v) olive oil as carbon source and 0.1% (w/v) soybean lecithin as emulsifier under agitation of 210 rpm, at 27.5°C and pH 6.0 [20]. The biochemical characterization revealed that the crude lipase produced in liquid medium was quite different from those usually produced by other *Candida* species. Optimal activity at acid pH of 3.5 suggests a new lipolytic enzyme for this genus and for yeasts in general. In addition, the crude lipase presented high stability in acid condition and up to 40–45°C, remaining active in the presence of organic solvents such as dimethyl sulfoxide (DMSO) and methanol and gum Arabic emulsifier. The aim of this study was to evaluate the lipase production by *C. viswanathii* in solid-state cultivation using agroindustrial wastes and crude poultry fat as substrate. Purification and biochemical characterization of the crude and purified lipase were carried out, and also poultry fat was further hydrolyzed by both enzyme forms.

2. Materials and Methods

2.1. Microorganism and Preinoculum. *C. viswanathii* strain is available in the Culture Collection of the Environmental Studies Center, CEA/UNESP, Brazil. *C. viswanathii* was cultivated on malt extract agar (MEA) for 3 days, at 28°C, for inoculum preparation. Liquid medium was prepared using Vogel medium [21], with 1.5% (w/v) olive oil and 0.2% (w/v) yeast extract as single carbon and nitrogen sources, respectively, according to previous established conditions [19]. Erlenmeyer flasks (125 mL) containing 25 mL of medium were inoculated with 1.0 mL of cells suspension (1×10^7 cells/mL) and incubated at 28°C, 210 rpm for 24 h. Five milliliters of this suspension was used for inoculating the substrates for solid-state cultures.

2.2. Solid-State Cultivation. Cultures were performed in Erlenmeyer flasks (250 mL) containing 10 g of wheat bran (WB), cassava peel, barley spent grain (BSG), sugarcane bagasse, or citrus pulp (CP), which were previously washed with distilled water, dried until constant weight, and sieved (18 mesh). Nonsieved BSG and CP were incorporated into wheat bran to improve aeration of the substrates, constituting mixed substrate cultivation. Olive oil (25%, w/w) was used as initial carbon/inducer source to the substrates. Modified Vogel salts solution [19] prepared without nitrogen source was used to provide 50% (w/v) initial moisture. Flasks containing moisturized and supplemented substrates were autoclaved at 121°C, for 20 min. Cultures were carried out at 28°C for 5 days.

2.3. Enzyme Extraction. After cultivation 100 mL of distilled water was added to each flask and the mixture was incubated on a rotary shaker (250 rpm, 4°C) for 60 min. Then, the suspension was filtrated through a double layer gauze cloth and centrifuged (8000 ×g, 20 min, 4°C). The clear supernatant was used as source of crude extracellular lipase.

2.4. Lipase Activity. Lipase activity was assayed with p-nitrophenyl palmitate (p-NPP) as substrate [20]. p-NPP was dissolved in 0.5 mL of dimethyl sulfoxide and then diluted to 50 mM with 50 mM sodium phosphate buffer pH 7.0 containing 0.5% (w/v) Triton X-100. The hydrolysis of p-NPP was determined discontinuously at 37°C by measuring the released p-nitrophenolate (p-NP). After 5 min preincubation of 0.9 mL of the substrate, adding 0.1 mL of appropriately diluted enzyme sample started the reaction. The reaction was stopped at different intervals by heat shock (90°C, 1 min), followed by addition of 1 mL of saturated sodium tetraborate solution. The absorbance was measured at 405 nm (molar extinction coefficient for p-NP: $1.8 \times 10^4 \, M^{-1} \, cm^{-1}$). Controls were prepared without enzyme. One unit of enzyme activity was defined as the amount of enzyme that releases 1 μmol of p-NP per min.

2.5. Protein Analysis. Protein was determined with Coomassie blue G-250 [22], using bovine serum albumin as standard.

2.6. Parametric Optimization of Lipase Production

2.6.1. Carbon Supplementation. The substrates were supplemented with 25% (w/w) natural triacylglycerols, palm oil, soybean oil, corn oil, canola oil, sunflower oil, linseed oil, and babassu oil, or with wastes such as poultry fat, beef tallow, lard, and cooking oil. Poultry fat used for inducing lipase production was evaluated at 5, 10, 15, 20, 25, 30, 35, 40, 45, and 50% (w/w).

2.6.2. Nitrogen Supplementation. The medium was supplemented with 5% (w/w) corn steep liquor, yeast extract, soy protein, whey powder, and cotton protein. Yeast extract was evaluated at 1.0, 1.5, 2.0, 2.5, 3.0, 3.5, 4.0, 4.5, and 5.0% (w/w).

2.6.3. Temperature and Moisture. The effect of temperature on lipase production was verified by carrying out cultivation at 15, 20, 25, 30, 35, and 40°C.

The effect of initial moisture content of the cultures was evaluated by adding Vogel salts [19] without nitrogen sources in order to provide 20, 30, 40, 50, 60, and 70% (w/w) initial moisture.

2.7. Enzyme Purification. The crude extract was previously dialyzed against 0.02 M ammonium acetate buffer pH 6.9 (8 h, 3 changes, 4°C). The dialyzed extract was applied to a hydrophobic Octyl Sepharose column (HiPrep™ 16/10 Octyl Sepharose FF fast flow, GE Healthcare) previously equilibrated in the same buffer, at 2 mL/min flow rate. The column was washed with 50 mL of the same buffer and 3.0 mL fractions were collected. Elution of bound proteins was performed with 100 mL of a 0.0 to 1.0% (w/v) Triton X-100 linear gradient prepared in the same buffer. Fractions with

lipase activity were pooled and sample purity was evaluated by SDS-PAGE. All purification procedures were carried out at 4°C.

2.8. Enzyme Characterization

2.8.1. SDS-PAGE. The purified enzyme was previously treated with Calbiosorb™ adsorbent resin (Calbiochem®, San Diego, USA) to remove Triton X-100. The resin was equilibrated in 0.05 M ammonium acetate buffer pH 6.9 and loaded with the purified enzyme. Samples were incubated at 10°C for 45 min under slow stirring and then centrifuged (8500 ×g, 4°C, 20 min). The supernatant containing enzyme was submitted to electrophoresis. Even after treatment, residual Triton X-100 was still detected in the sample by reading absorbance at 280 nm.

SDS-PAGE was performed using 10% (w/v) polyacrylamide gels according to Hames [23]. Samples were previously treated with 8 M urea according to Lesuisse et al. [24]. Resolved protein bands were visualized after staining with 0.1% (w/v) Coomassie brilliant blue R-250 in methanol, acetic acid, and distilled water (4 : 1 : 5, v/v/v). This method was used to determine the molecular mass (MW) of the purified enzyme using appropriate standards.

2.8.2. Optimum pH and pH Lipase Stability. Enzyme activity was measured at 37°C in different pH values using 0.05 M glycine-HCl buffer pH from 2.0 to 3.0 and McIlvaine buffer pH from 3.0 to 8.0. Stability to pH was carried out with the same buffers, except in pH from 8.6 to 10.0 in which 0.05 M glycine-NaOH buffer was used. Enzyme samples were (1 : 2, v/v) diluted in each buffer and incubated for 24 h at 10°C.

2.8.3. Optimum Temperature and Thermal Stability. The optimum temperature was determined by measuring enzyme activity in temperatures from 20 to 70°C, in McIlvaine buffer pH 5.0. For thermal stability, the enzyme was incubated at 40, 45, 50, 55, and 60°C in McIlvaine buffer pH 5.0 in the absence of substrate, and the residual activity was determined in McIlvaine buffer pH 5.0 at 50°C.

2.8.4. Effect of Organic Solvents. The effect of organic solvents on activity and stability of crude and purified lipase was evaluated using 10% (v/v) glycerol, DMSO, propylene glycol, methanol, acetonitrile, ethanol, acetone, 1-propanol, 2-propanol, n-butanol, toluene, xylol, n-hexane, and isooctane. The effect of organic solvents on the activity was verified by adding each solvent into the enzymatic reactions. Stability experiments were carried out in sealed flasks shaken at 200 rpm, for 2 h at 30°C. Residual activities were determined in McIlvaine buffer pH 5.0 at 50°C and expressed in relation to the control without any substance.

2.8.5. Specificity for Substrate. Specificity was verified using 0.5 mM p-nitrophenyl acetate, p-nitrophenyl butyrate, p-nitrophenyl octanoate, p-nitrophenyl decanoate, p-nitrophenyl laurate, p-nitrophenyl myristate, p-nitrophenyl palmitate, and p-nitrophenyl stearate by performing enzyme assays in

TABLE 1: Lipase production by *C. viswanathii* with agroindustrial wastes in solid-state cultivation.

Substrate	Nonsupplemented		Supplemented with olive oil	
	Lipase activity (U/gds)	Specific activity (U/mg prot)	Lipase activity (U/gds)	Specific activity (U/mg prot)
Wheat bran	0.94 ± 0.10	0.03 ± 0.00	5.79 ± 0.87	3.50 ± 0.20
Barley spent grain	0.40 ± 0.01	0.04 ± 0.00	0.32 ± 0.04	0.37 ± 0.00
Cassava peel	ND	ND	0.17 ± 0.01	1.25 ± 0.10
Sugarcane bagasse	0.10 ± 0.01	0.95 ± 0.00	0.12 ± 0.01	1.04 ± 0.10
Citrus pulp	0.11 ± 0.02	0.01 ± 0.00	0.30 ± 0.04	0.11 ± 0.00
WB + BSG (3 : 1)	0.67 ± 0.03	0.20 ± 0.00	15.82 ± 1.17	6.72 ± 0.70
WB + BSG (3 : 2)	0.88 ± 0.07	0.28 ± 0.00	17.46 ± 1.16	7.94 ± 0.91
WB + BSB (1 : 1)	0.62 ± 0.02	0.33 ± 0.00	18.65 ± 1.77	8.85 ± 0.83
WB + CP (3 : 1)	0.20 ± 0.02	0.04 ± 0.00	9.25 ± 0.90	2.11 ± 0.11
WB + CP (3 : 2)	0.45 ± 0.03	0.17 ± 0.00	14.32 ± 0.78	6.35 ± 0.52
WB + CP (1 : 1)	0.45 ± 0.03	0.18 ± 0.00	10.10 ± 0.55	5.65 ± 0.43

Cultures were carried out for 5 days, at 28°C. Substrates were supplemented with 50% Vogel salts solution without nitrogen source. Substrates with olive oil were supplemented with 25% (w/w) of olive oil. WB: wheat bran, BSG: barley spent grain, and CP: citrus pulp.

McIlvaine buffer pH 5.0 at 45°C and pH 5.5 at 50°C for the crude and purified lipase, respectively.

2.8.6. Kinetic Parameters. The activity of purified lipase was assayed with p-nitrophenyl palmitate from 0.0 to 1.0 mM. The Michaelis-Menten constant (K_m) and maximum reaction velocity (V_{max}) were estimated from Lineweaver-Burk plot [25].

2.8.7. Hydrolysis of Poultry Fat. Hydrolysis of poultry fat was developed at 50°C by titration of released fatty acids. The oils (10%, w/v) were emulsified in McIlvaine buffer pH 4.0, 6.0, and 8.0, containing 0.5% (w/v) Triton X-100. The reaction was started by adding 1 mL of enzyme to 5 mL of this emulsion, and then it was maintained for 96 h at 200 rpm orbital agitation. The reaction was interrupted by adding 16 mL of an acetone : ethanol solution (1 : 1, v/v) to the mixture. The released fatty acids were titrated to pH 11 with a 0.05 M NaOH solution. From these values, the degree of hydrolysis was calculated according to [18]

$$X = \frac{W(V - V_0)M}{10mf_0},$$ (1)

where X is degree of hydrolysis (%); W is the mean molecular weight of the fatty acids; V is the volume (mL) of NaOH solution used for titration of the sample; V_0 is the volume (mL) of NaOH solution used for titration of the control; M is the molarity of NaOH solution; m is the weight of the sample (mL); and f_0 is weight fraction of oil in the start of reaction.

The initial rate of reaction was calculated using the following equation:

$$r_0 = \frac{10^4 S_0}{W}\left(\frac{dX}{dt}\right)_{t=0},$$ (2)

where r_0 is initial rate of hydrolysis (mmol/L/min); S_0 is initial concentration of oil (g/L); $(dX/dt)_{t=0}$ is slope of the

degree of hydrolysis (X) versus time curve at t_0. The average molecular weight of fatty acids in poultry fat was calculated as 277.3 kg/kmol.

3. Results

3.1. Selection of Agroindustrial Substrates and Olive Oil Supplementation. Solid-state cultivation of *C. viswanathii* yeast on agroindustrial wastes was evaluated for lipase production using wheat bran, barley spent grain, cassava peel, sugarcane bagasse, and citrus pulp, assessed in the absence and in the presence of olive oil as inducer (Table 1). When the substrates were individually evaluated without olive oil supplementation, the highest level of enzyme activity was observed with wheat bran; intermediate level was observed with barley spent grain and low or absent enzyme activity was observed with citrus pulp, sugarcane bagasse, and cassava peel. The low levels of lipase observed in the medium without olive oil indicated that this yeast used soluble nutrients for microbial growth and the matrixes as support. Mixtures of wheat bran plus barley spent grain and wheat bran plus citrus pulp were evaluated at different proportions since the presence of nonsieved barley spent grain and citrus pulp could increase the aeration of the system and consequently increase fungal growth and enzyme production. The use of mixed substrates slightly changed enzyme production, considering the mixture of wheat bran plus barley spent grain at 3 : 2 proportion (0.88 U/gds), although this activity still was lower than that observed in cultivation only with wheat bran (0.94 U/gds). In other mixtures, the lipase activity was still lower when compared to wheat bran. The initial supplementation of the substrates with olive oil improved lipase production and the highest activity was also verified with wheat bran (5.79 U/gds and 3.50 U/mg of protein). The use of mixed substrates and olive oil supplementation also increased the activity and the highest production was observed with wheat bran plus barley spent grain at 1 : 1 proportion (18.65 U/gds). In this case, the

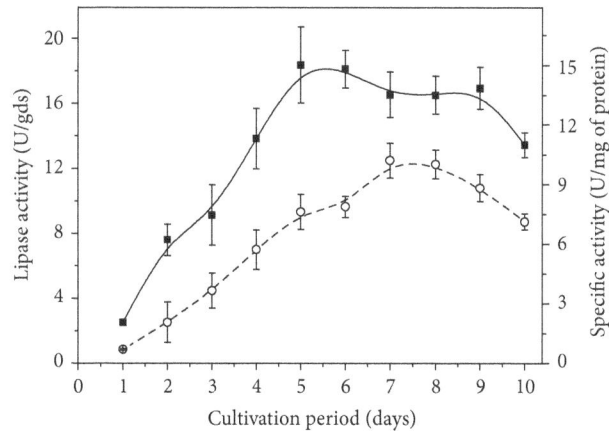

FIGURE 1: Time course of lipase production by *C. viswanathii* in solid-state cultivation using wheat bran plus barley spent grain 1:1 (w/w) supplemented with 25% (w/w) olive oil. Cultures were carried out without nitrogen source supplementation and 50% (w/v) initial moisture provided by Vogel salts solution at 28°C for 120 hours. ■: lipase production (U/gds); ○: specific activity (U/mg of protein).

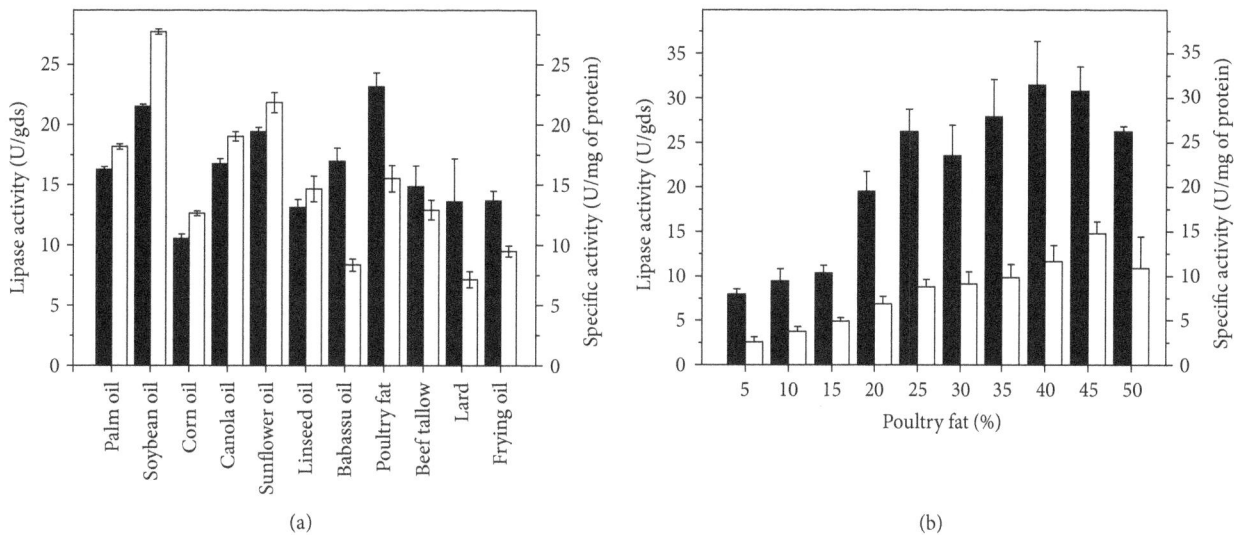

(a)

(b)

FIGURE 2: Effects of different triacylglycerol sources (a) and poultry fat concentration (b) on lipase production by *C. viswanathii* in solid-state cultivation. Cultures were carried out with wheat bran plus barley spent grain supplemented with 25% (w/w) of each triacylglycerol source without nitrogen source supplementation and 50% (w/v) initial moisture provided by Vogel salts solution, at 28°C for 5 days (a). Cultures were carried out in the same conditions with only poultry fat (b). ■: lipase production (U/gds); □: specific activity (U/mg of protein).

activity was 3.2-fold higher than that verified with olive oil supplemented wheat bran. The use of olive oil supplemented wheat bran plus citrus pulp improved enzyme production especially at 3:2 proportion (14.32 U/gds and 6.35 U/mg of protein), but to a lower level than that with wheat bran plus barley spent grain.

Lipase production by *C. viswanathii* using wheat bran plus barley spent grain 1:1 (w/w) supplemented with 25% (w/w) olive oil and 50% (v/w) initial moisture was assayed for 10 days (Figure 1). Lipase activity increased up to the 5th day (18.36 U/gds) and slightly decreased up to the 9th day. The highest specific activity (10.20 U/mg of protein) was observed after 7 days of cultivation.

3.1.1. Effect of Triacylglycerols on Lipase Production. Triacylglycerols are important inducers of lipase production and in this sense palm oil, crude babassu oil, crude linseed oil, canola oil, sunflower oil, corn oil, soybean oil, and also renewable and low cost sources such as poultry fat, lard, beef tallow, and frying oil were evaluated (Figure 2(a)). *C. viswanathii* strain was able to produce high levels of lipase in medium with all triacylglycerols. Poultry fat induced the highest lipase production (24.20 U/gds), followed by soybean oil (21.49 U/gds) and sunflower oil (19.43 U/gds). The highest specific activity (26.00 U/mg of protein) was detected with soybean oil. Due to these results, lipase production was evaluated at different concentrations of poultry fat (Figure 2(b)). Lipase

TABLE 2: Lipase production by *C. viswanathii* in solid-state cultivation with different nitrogen sources.

Nitrogen sources (5% w/w)	Lipase activity (U/gds)	Specific activity (U/mg prot)
Control	31.43 ± 4.91	11.64 ± 1.81
Yeast extract	119.91 ± 11.68	65.50 ± 6.38
Corn steep liquor	57.96 ± 5.20	49.32 ± 5.35
Whey powder	61.00 ± 4.20	93.53 ± 6.44
Soybean meal	40.65 ± 4.80	24.12 ± 4.03
Soy protein	84.60 ± 4.41	76.22 ± 7.62
Cotton seed protein	57.95 ± 2.83	66.84 ± 3.72

Cultures were carried out on wheat bran plus barley spent grain (1 : 1, w/w), 50% moisture, and 40% (w/w) poultry fat for 5 days, at 28°C. Control was carried out in the absence of nitrogen source.

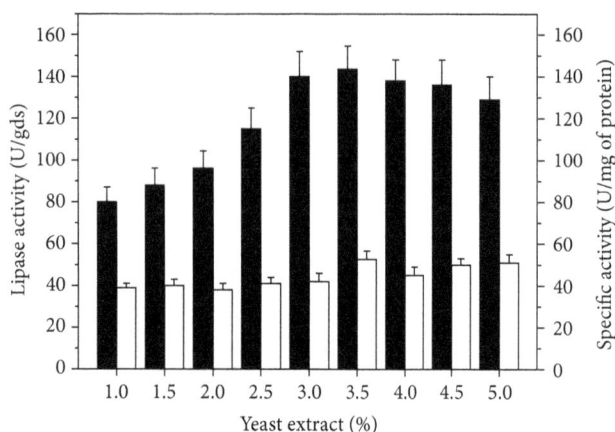

FIGURE 3: Effect of yeast extract concentration on lipase production by *C. viswanathii* in solid-state cultivation. Cultures were carried out with wheat bran plus barley spent grain supplemented with 25% (w/w) of each triacylglycerol source and 50% (w/v) initial moisture provided by Vogel salts solution, at 28°C for 5 days. ■: lipase activity (U/gds); □: specific activity (U/mg of protein).

TABLE 3: Effect of temperature and moisture on lipase production by *C. viswanathii* in solid state cultivation.

Parameter	Lipase activity (U/gds)	Specific activity (U/mg prot)
Temperature (°C)		
15	26.10 ± 4.42	13.31 ± 1.92
20	56.05 ± 6.07	22.97 ± 2.43
25	77.71 ± 3.36	28.59 ± 2.71
30	143.36 ± 9.65	65.76 ± 3.18
35	129.11 ± 8.15	43.19 ± 2.98
40	25.80 ± 6.17	10.05 ± 1.56
Moisture (%)		
20	98.70 ± 4.87	78.89 ± 3.47
30	147.55 ± 8.43	111.32 ± 4.55
40	157.33 ± 9.25	136.20 ± 5.74
50	143.36 ± 7.47	113.32 ± 6.40
60	137.56 ± 6.34	74.56 ± 5.70
70	92.43 ± 5.78	60.80 ± 4.85

Cultures were carried out using wheat bran plus barley spent grain (1 : 1, w/w), 40% (w/w) poultry fat, and 3.5% (w/w) yeast extract for 5 days (above). Cultures were carried out in the same conditions and at 30°C (below).

cific activity (65.8 U/mg of protein) at 30°C (Table 3). Above this temperature there was a reduction in enzyme production to 129.1 and 25.8 U/gds at 35 and 40°C, respectively.

The effect of substrate moisture on lipase production revealed optimal production with 40% (v/w) initial moisture (157.3 U/gds), in which the highest specific activity (136.2 U/mg of protein) was also observed. Cultures with 30 or 50% initial moisture resulted in intermediate lipase production of 147.5 and 143.4 U/gds, respectively.

3.2. Enzyme Purification. In a previous study, the crude extract obtained in cultivation under optimal conditions was subjected to ammonium sulfate precipitation. In this step, aggregation of proteins was observed obtaining inconsistent results after four attempts of precipitation using different salt concentrations (data not shown). Then, the crude extract was used to select a resin for hydrophobic chromatography. It was observed that the lipase adsorbed in phenyl and Octyl Sepharose without ammonium sulfate; then hydrophobic chromatography was carried out using Octyl Sepharose column without previous salt equilibrium. The crude extract was subjected to dialysis against 0.02 M ammonium acetate buffer pH 6.9 and applied to hydrophobic column equilibrated in the same buffer (Figure 4). Under these conditions, two peaks with lipase activity were observed. The first peak corresponded to nonbound proteins eluted in the initial fractions and the second one to proteins eluted with 1.0% (w/w) Triton X-100 gradient. Fractions with lipase activity in the second peak were pooled and subjected to SDS-PAGE, which showed electrophoretic homogeneity of one 18.5 kDa band (Figure 5). The enzyme was 47.0-fold purified with 133.6 U/mg of protein and the process presented 84.5% yield (Table 4).

production increased by increasing poultry fat concentration up to 40% (w/w) corresponding to 31.43 U/gds. Above this concentration enzyme production slowly decreased.

3.1.2. Effect of Nitrogen Sources on Lipase Production. Supplementation with different nitrogen sources improved lipase production by *C. viswanathii* (Table 2). Among them, yeast extract was the best source improving 3.8-fold the lipase production (119.91 U/gds) and also increasing the specific activity (65.50 U/mg of protein). The highest specific activity was verified with whey powder corresponding to 93.53 U/mg of protein.

The effect of different yeast extract concentrations (Figure 3) showed that lipase production increased by increasing nitrogen source concentration, reaching the highest activity (143.60 U/gds) with 3.5% (w/w) yeast extract and specific activity of 52.85 U/mg of protein.

3.1.3. Effect of Temperature and Moisture on Lipase Production. Cultivation in temperatures from 15 to 40°C showed the highest lipase production of 143.4 U/gds and the highest spe-

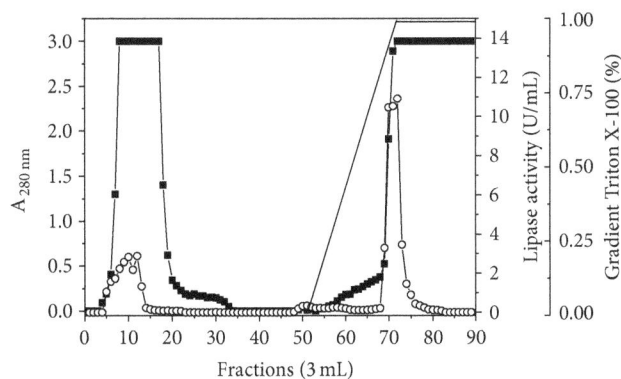

FIGURE 4: Profile of hydrophobic interaction chromatography of *C. viswanathii* lipase produced in solid-state cultivation under the best conditions for enzyme production. Chromatograph conditions: 0.02 M ammonium acetate buffer pH 6.9; 2.0 mL/min flow rate; 3.0 mL fractions, at 4°C; elution with Triton X-100 0–1.0% (w/v).

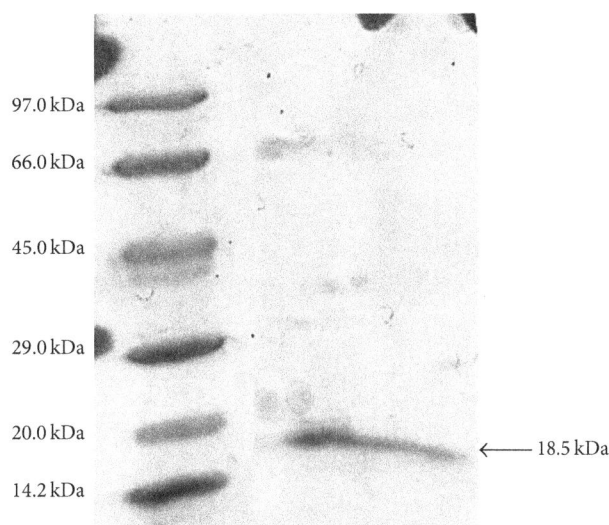

FIGURE 5: SDS-PAGE of purified lipase from *C. viswanathii* produced in solid-state cultivation. Column 1: standards: phosphorylase b (97 kDa), albumin bovine serum (66 kDa), ovalbumin (45 kDa), carbonic anhydrase (29 kDa), trypsin inhibitor (20 kDa), and α-lactalbumin (14.2 kDa). Column 2: purified lipase.

3.3. Biochemical Characterization

3.3.1. Effect of pH and Temperature on Activity and Stability. The effect of pH on the activity of crude and purified lipase was determined from 2.0 to 9.0 (Figure 6(a)). Optimal activity was observed at pH 5.0–5.5 for crude lipase and pH 6.0 for purified lipase. Crude lipase activity was higher in the pH range of 4.5 and 6.0 (92.0 and 94.0%, resp.), and purified lipase presented high activity at pH 5.0 and 6.5 (~90.0–92.0%). Enzyme activity decreased to lower levels from pH 2.0 to 3.5 and 6.5 to 9.0, presenting only 17.6 and 22.9% of activity at pH 9.0 for crude and purified lipase, respectively. Stability of the enzymes was determined after 24 h incubation in buffers of different pH at 40°C (Figure 6(b)). The crude lipase retained approximately 100% of its activity in the pH range from 3.0 to 8.0 and more than 60% at pH 2.0, 2.5, 8.5, and 9.0. However, at pH 9.5 the residual activity was 32.0%, and at pH 10.0, the enzyme completely

lost its activity. The purified lipase retained approximately 85% of its activity in the pH range from 4.0 to 5.5. The stability of the enzyme decreased in pH from 6.5 to 10.0 and below 3.5.

The effect of temperature on lipase activity was evaluated from 20 to 70°C (Figure 6(c)). The optimal crude lipase activity was observed at 50°C (100%), and high activities were also observed at 45 (90.5%) and 55°C (86.0%). The activity was about 50% reduced at 35 and 60°C; and low activity was observed at 70°C. The optimal purified lipase activity was 45°C followed by a decrease up to 70°C (19% of activity). Thermal stability of the lipase was evaluated by the incubation of the enzyme without substrate in a buffered medium at pH 5.0 for crude lipase and at pH 6.0 for purified lipase, at different temperatures. The crude *C. viswanathii* lipase retained 97.0% of activity after 24 h at 45°C. The estimated $T_{1/2}$ of the enzyme were 23.5, 1.67, and 0.25 h at 50, 55, and 60°C, respectively (Table 5). Purified lipase presented $T_{1/2}$ of

TABLE 4: Purification of *C. viswanathii* lipase produced in solid-state cultivation.

Purification step	Enzyme activity (U)	Protein total (mg)	Specific activity (U/mg of protein)	Enrichment	Yield (%)
Crude extract	550.02	193.63	2.84	1.00	100.00
Octyl Sepharose 4CL	465.02	3.48	133.63	47.05	84.55

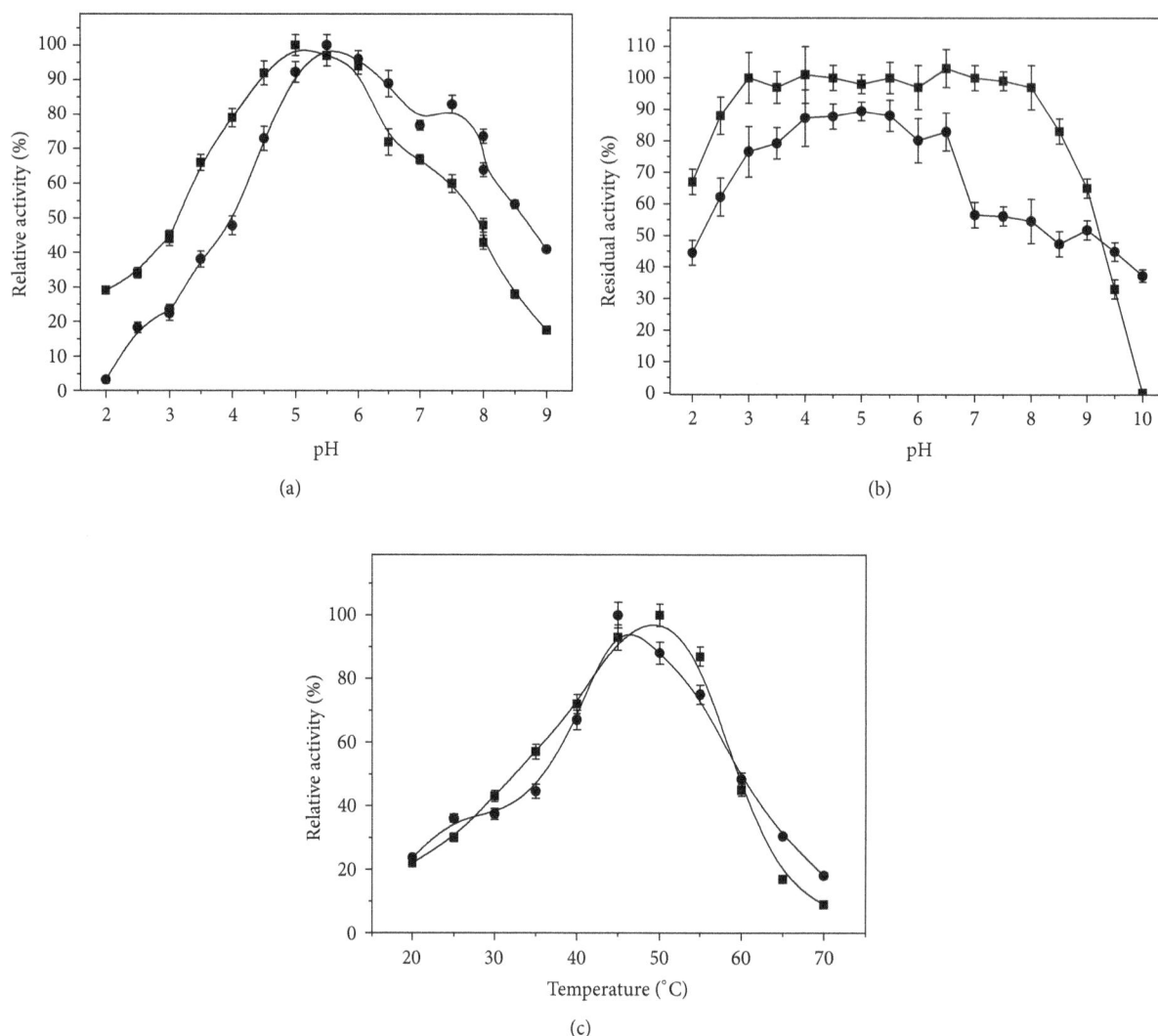

(a)

(b)

(c)

FIGURE 6: Optimum pH (a) and pH stability (b) and optimum temperature (c) and thermal stability (d) of crude and purified *C. viswanathii* lipase. Assay conditions: 0.05 M glycine-HCl buffer pH from 2.0 to 3.0, McIlvaine buffer pH from 3.0 to 8.0, and 0.05 M glycine-NaOH pH from 8.6 to 10 at 37°C (a); the crude enzyme was incubated in the same buffers for 24 h at 10°C and lipase activity was assayed in McIlvaine buffer pH 5.0 for crude enzyme and 5.5 for purified enzyme at 37°C (b); lipase activity assays were assayed in McIlvaine buffer pH 5.0 for crude enzyme and 5.5 for purified enzyme (c); ■: crude enzyme; ●: purified enzyme.

6.7, 4.2, and 0.9 h at 40, 45, and 50°C, respectively. At 55 and 60°C the half-lives of purified enzyme were 0.3 h.

3.3.2. Effect of Organic Solvents. The effect of organic solvents on lipase activity is shown in Table 6. The activity of the crude lipase increased in the presence of glycerol (135%), DMSO, propylene glycol (~111%), ethanol (106%), n-hexane,

and methanol (~105%). In media containing other organic solvents the activity remained high as in 1-propanol (98%), n-butanol and toluene (~91%), acetone and isooctane (~90%), acetonitrile (86%), and 2-propanol (78%). Purified lipase maintained its resistance to organic solvents since the activity was also increased by DMSO (113%), 1-propanol (111%), glycerol (109%), ethanol (~107%), and methanol

TABLE 5: Half-lives of crude and purified lipase from *C. viswanathii* produced in solid-state cultivation.

Temperature (°C)	Half-life (h)	
	Crude lipase	Purified lipase
40	n.d.	6.7
45	n.d.	4.2
50	23.5	0.9
55	1.67	0.3
60	0.25	0.3

n.d.: not detected after 24 hours of incubation.

TABLE 6: Effect of organic solvents on crude lipase stability produced by *C. viswanathii* under solid-state cultivation.

Organic solvent	log P	Crude enzyme		Purified enzyme	
		Relative activity (%)	Stability (%)	Relative activity (%)	Stability (%)
Control	—	100.0 ± 1.2	100.0 ± 2.0	100.0 ± 2.4	100.0 ± 3.2
Glycerol	−1.67	135.4 ± 4.2	99.5 ± 2.6	109.4 ± 5.7	60.2 ± 2.0
DMSO	−1.38	111.7 ± 4.0	99.8 ± 3.0	113.5 ± 2.5	59.5 ± 3.4
Propylene glycol	−0.92	111.5 ± 3.1	88.7 ± 3.5	97.5 ± 7.7	60.1 ± 3.4
Methanol	−0.76	105.1 ± 3.3	84.6 ± 1.9	103.2 ± 7.5	52.9 ± 6.1
Acetonitrile	−0.40	86.2 ± 1.3	68.1 ± 3.9	88.6 ± 8.2	58.6 ± 2.8
Ethanol	−0.24	106.2 ± 2.2	90.5 ± 4.8	106.8 ± 6.1	94.5 ± 2.1
Acetone	−0.23	90.5 ± 2.2	66.7 ± 2.4	85.8 ± 6.8	69.2 ± 1.8
1-Propanol	0.07	97.9 ± 2.5	72.4 ± 1.3	111.5 ± 9.9	66.9 ± 1.8
2-Propanol	0.25	78.3 ± 2.6	56.0 ± 2.4	83.0 ± 4.3	59.2 ± 1.4
n-Butanol	0.80	92.0 ± 4.0	12.1 ± 2.6	84.1 ± 9.9	58.3 ± 3.5
Toluene	2.50	91.1 ± 3.3	69.8 ± 1.9	87.2 ± 5.3	25.0 ± 2.6
Xylol	3.15	65.9 ± 1.4	51.5 ± 1.3	82.0 ± 7.4	39.1 ± 4.2
n-Hexane	3.50	105.9 ± 2.5	98.0 ± 4.7	98.5 ± 8.6	65.2 ± 3.1
Isooctane	4.51	90.4 ± 2.4	95.1 ± 3.4	91.4 ± 7.8	61.8 ± 2.0

Assay conditions: for stability assays crude and purified lipase were incubated at 30°C at 200 rpm during 2 h and the activity was assayed with p-NPP using McIlvaine buffer pH 5.0, at 50°C. log P: logarithm of the partition coefficient (P) in octanol/water two-phase system indicates the solvents hydrophobicity. DMSO: dimethyl sulfoxide.

(103%) and retained high activity levels in n-hexane, propylene glycol, isooctane, acetonitrile, toluene, acetone, n-butanol, 2-propanol, and xylol.

The stability of crude lipase after 2 h incubation at 30°C was high in media containing DMSO, glycerol, n-hexane, isooctane and ethanol (more than 90%), and propylene glycol (~89%). Intermediate stability was observed with methanol, 1-propanol, acetonitrile, toluene, acetone, 2-propanol, and xylol. Butanol sharply decreased the lipase activity. The purified lipase presented high stability in ethanol (94%). Intermediate values were observed with acetone, 1-propanol, n-hexane, isooctane, glycerol, propylene glycol, DMSO, 2-propanol, acetonitrile, n-butanol, and methanol (69–53%, resp.). Lower stability was observed with xylol (39%) and toluene (25%).

3.3.3. Kinetic Parameters. Substrate hydrolysis reactions were performed for purified lipase with p-NPP (0.0 to 1.0 mM) to determine K_m and V_{max}. From these values the turnover number (k_{cat}) and the catalytic efficiency (k_{cat}/K_m) for purified enzyme were calculated. Purified lipase showed K_m

0.12 mM, V_{max} 18.3 μmol/mL·min, k_{cat} 45.8 s^{-1}, and k_{cat}/K_m of 3.8 × 10^5 M^{-1} s^{-1}.

3.3.4. Substrate Specificity. Hydrolytic activity of crude and purified lipase was evaluated on p-nitrophenyl ester substrates (Figure 7). The highest activity of crude and purified lipase was observed with p-nitrophenyl palmitate. The activity of crude lipase increased using esters from decanoate, laurate, and myristate to palmitate; the low activities were observed with acetate, butyrate, and caproate. Activity of purified lipase was high using esters of fatty acids from C8 to C16 and low levels were observed for those of acetate and butyrate. Activity on p-nitrophenyl stearate of crude lipase was lower than on p-nitrophenyl decanoate and for purified enzyme it was lower than on p-nitrophenyl caproate.

3.4. Hydrolytic Activity on Triacylglycerols. The hydrolysis of poultry fat and initial hydrolysis rate in different pH values are shown in Figure 8 and Table 7. Higher hydrolytic activity was observed with the crude enzyme at pH 4.0 up to 72 h, reaching 40% hydrolysis (110.9 mmol of fatty acids

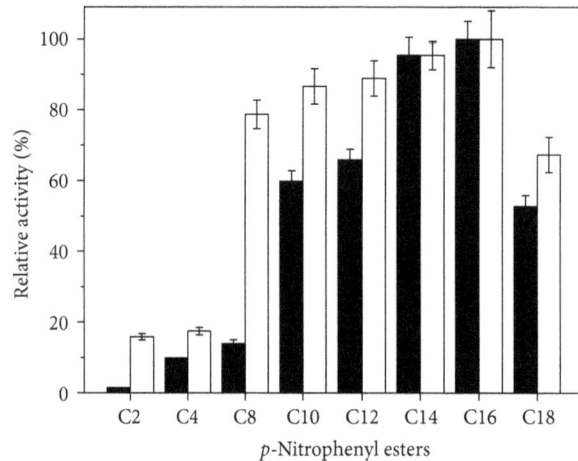

FIGURE 7: Activity of crude and purified *C. viswanathii* lipase on *p*-nitrophenyl esters. Activity was determined in McIlvaine buffer pH 5.0 at 50°C for crude enzyme and 5.5 at 45°C for purified enzyme. ■: crude lipase; □: purified lipase. C2 acetate, C4 butyrate, C8 caproate, C10 decanoate, C12 laurate, C14 myristate, C16 palmitate, and C18 stearate.

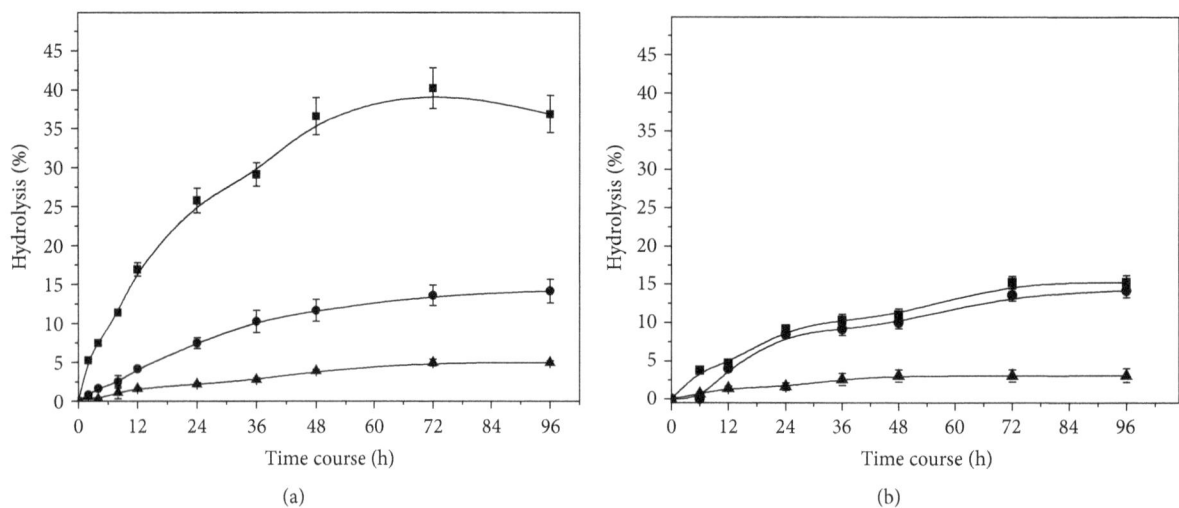

FIGURE 8: Poultry fat hydrolysis profiles by crude (a) and purified (b) *C. viswanathii* lipase. Assay conditions: hydrolysis was carried out using enzyme concentration 10 U/mL, $S_0 = 100$ g·L^{-1}, 200 rpm at 40°C. ■: pH 4.0; ●: pH 6.0; ▲: pH 8.0.

released) and initial reaction rate of 33.17 mmol/L·min. The slope degree of hydrolysis versus time at $t = 0$ $(dX/dt)_{t=0}$ was 0.0092 and the correlation coefficient (r) was 0.905. At pH 6.0 and 8.0 the hydrolytic activity sharply decreased reaching maximal hydrolysis activity of 14.1 and 5.0% at 72 h, respectively. Initial reaction rates under these conditions were 18.75 and 3.60 mmol/L·min, respectively. Purified enzyme presented low hydrolytic activity compared to the crude enzyme. Maximal hydrolysis was also observed at pH 4.0 with 15.2% and initial rate of 7.20 mmol/L·min and $r^2 = 0.845$. At pH 6.0 the hydrolytic activity was 14.3% with initial reaction rate of 6.50 mmol/L·min; at pH 8.0 the lowest rate (1.91 mmol/L·min) was observed, reaching only 3.0% of this activity.

4. Discussion

Solid-state cultivation for lipase production using yeast cells is a nonconventional practice since the access to nutrients is a limiting factor for this group of microorganisms due to the absence of well-developed hyphae and enzymes production capable of releasing soluble monomers or dimers for microbial nutrition and growth. Substrates such as wheat bran, barley spent grain, cassava peel, sugarcane bagasse, and citrus pulp used for *C. viswanathii* cultivation without or with olive oil supplementation showed wheat bran as the most promising substrate in promoting lipase production. Wheat bran contains soluble sugars (arabinose, glucose, and xylose), protein, and starch providing amino acids, nitrogen,

TABLE 7: Parameters of poultry fat hydrolysis using crude *C. viswanathii* lipase produced in solid-state cultivation.

pH	r_0 (mmol·L^{-1}·min^{-1})	r^2
Crude enzyme		
4.0	33.17	0.905
6.0	18.75	0.812
8.0	3.60	0.838
Purified enzyme		
4.0	7.20	0.720
6.0	6.50	0.733
8.0	1.91	0.754

and carbon at adequate proportion to support the production of microbial enzymes, and it is commonly considered an excellent substrate for solid-state cultivation due to its heat dissipation, improved air circulation, and loose particle binding [26, 27].

Mixed substrates formulation of wheat bran plus barley spent grain or wheat bran plus citrus pulp supplemented with olive oil increased the enzyme production by 322 and 174%, respectively, in relation to individual wheat bran, which may be related to an increase in spatial distribution or/and larger surface contact between yeast and substrate during microbial colonization. Benjamin and Pandey [28] observed the highest lipase production by *C. rugosa* using a mixture of fine wheat bran, coarse wheat bran, and thick coconut pie. Growth and yield performances can be improved in relation to traditional submerged systems. Moreover, mixture of different substrates shows better performance by providing a more suitable environment for microbial growth. Edwinoliver et al. [13] developed a mixed substrate containing wheat raw, wheat bran, and coconut oil cake for lipase production by *Aspergillus niger*. A 35% increased yield was observed in comparison to the results from individual substrates. The authors attributed these results to the synergistic effect among the three substrates.

Supplementation of substrates with renewable triacylglycerol such as 25% poultry fat instead of olive oil increased the lipase production to 132%, and at 40% it was increased to 171.2%. This is an important characteristic since olive oil is the most usual inducer in many submerged processes due the high level of oleic acid in its fatty acids composition [29, 30]. Poultry fat is a byproduct of the slaughterhouse with low added value with no application to human nutrition. In addition, it has a high amount of unsaturated fatty acids, such as oleic acid, approximately at 50% [31]. In this point, lipase production in solid-state cultivation by *C. viswanathii* becomes much more attractive since the poultry fat becomes semisolid at room temperature, which can cause problems during submerged cultivation, such as formation of solid aggregates, preventing its use by the microorganism and consequently decreasing lipase production.

Although wheat bran and barley spent grain can provide high quantity of protein, the supplementation with organic nitrogen sources may be better accessed resulting in higher enzyme production. Yeast extract supplementation increased

lipase production to 381.5% and when it was employed at 3.5% (w/w) enzyme production increased to 456.9%. Nitrogen sources play an important role in the synthesis of lipase; organic nitrogen sources supply cells with growth factors and amino acids, which are required for cell metabolism and enzyme synthesis [32]. Sun and Xu [33] showed that many organic nitrogen sources, except urea, enhanced lipase production by *Rhizopus chinensis*. Imandi et al. [34] verified lipase production by *Yarrowia lipolytica* using mustard oil cake as substrate was enhanced by supplementation with urea, yeast extract, peptone, and malt extract.

Temperature and moisture are environmental factors that affect microbial growth and enzyme production. Optimum temperature for lipase production by *C. viswanathii* was similar to those observed for other filamentous fungi and yeasts [35–37]. Temperature control is critical for process scaling-up, which cannot be observed in flasks cultivation containing a few grams of substrates, but this is an important factor on a larger scale. The mixture of substrates results in a synergistic effect on enzymes production and prevents the formation of compact mass that can aid in the dissipation of heat generated during microbial growth [3, 38]. Čertik et al. [39] related that many wastes used in solid-state cultivation caused agglomeration of substrate particles creating a more compact mass, which, in turn, interfere with microbial respiration and negatively affect substrate utilization. On the other hand, the addition of nonsieved barley spent grain to the substrates increased product accumulation and also removed heat generated during microbial growth.

C. viswanathii produced high level of lipase using 40% initial moisture of the substrates, decreasing after that. Higher moisture content may cause a decrease in substrate porosity, thereby decreasing gas exchange. Lower moisture content can promote low microbial growth and lower degree of expansion of the substrate also decreasing enzyme production [3, 9, 35]. Treichel et al. [30] found that lipase production by *C. rugosa* in solid-state fermentation was optimal using substrates with 70% initial moisture. Similarly, other yeast strains and filamentous fungi also produced high lipase levels with 70% initial moisture [33, 37, 40].

The lipase produced by solid-state fermentation under previous established conditions was purified using hydrophobic interaction chromatography. The procedure was simple involving only one chromatographic step, which did not require ammonium sulfate. Similarly, Bastida et al. [41] related that adsorption in 10 mM phosphate is more than 6-fold faster than immobilization in the presence of 1 M ammonium sulfate. This behavior is different from standard hydrophobic purification protocols in which adsorption rates and process yield strongly increase at high ammonium sulfate concentration. Lipase presents interfacial activation phenomenon in the presence of drops of natural substrates, allowing the adsorption of these enzymes to the hydrophobic interface via very hydrophobic area formed by the internal face of the lid and/or surroundings of the active center [42]. In this sense, by the use of hydrophobic supports somehow resembling the surface of natural substrates and very low ionic strength, lipase becomes selectively adsorbed on these supports via an affinity-like strategy [43].

The purified lipase presented molecular mass of 18.5 kDa, V_{max} of 18.3 μmol/min·mL, K_m 0.12 mM, k_{cat} 45.8 s^{-1}, and k_{cat}/K_m 3.8 × 10^5 M^{-1} s^{-1}. Microbial lipase usually presents MW between 20 and 90 kDa [44]. Lipase with lower MW was observed in *bacteria* such as *Bacillus licheniformis* (19.2 kDa), *Bacillus subtilis*, and *Bacillus pumilus* (19.3 kDa) [24, 45, 46]. Besides, Brush et al. [47] observed two lipase isoforms from *Ophiostoma piliferum*. The major and the minor lipase from this fungus were copurified by hydrophobic interaction chromatography on Octyl Sepharose followed by ion exchange chromatography on Q Sepharose. The major lipase presented MW of approximately 60 kDa and the minor lipase MW of 5 kDa.

Biochemical characterization of crude and purified lipase is important for further industrial applications in hydrolysis or synthesis reactions. Crude lipase from *C. viswanathii* presented optimum activity at pH 5.0 and maintained above 90% of activity in pH range from 4.5 to 6.0. Optimum temperature was observed at 50°C and the enzyme retained 97% of its activity at 45°C after 24 h of incubation. High stability was observed in the pH range from 3.0 to 8.0. Crude lipase from *C. viswanathii* produced in submerged cultivation presents optimum activity at pH 3.5 and high stability in the pH range from 3.5 to 4.5 and optimum temperature at 40°C, retaining 98% of its activity in this temperature after 24 h incubation [20]. These results show that the cultivation of *C. viswanathii* in solid-state cultivation or submerged cultivation produces enzymes with distinct biochemical properties. Yang and Wang [48] related that the protease and amylase produced by *Streptomyces rimosus* in solid-state cultivation are more stable than those produced in submerged cultivation and the first ones can be temporarily stored without significant loss of activity. These results can be related to the fact that the metabolism exhibited by microorganisms in solid-state cultivation is different from that in submerged cultivation, and the influx of nutrients and efflux of waste materials need to be carried out based on these metabolic parameters [49]. The purified lipase from *C. viswanathii* presented different optimum pH and temperature for activity. Purified lipase was stable in acid pH, but above pH 6.0 the enzyme showed decrease in the activity. The dependence of enzyme activity on pH is a consequence of the amphoteric properties of proteins. Different ionizable groups with different pKa values are present on the surface of the protein molecules and surface charge distribution on the enzyme molecules varies with the pH on the environment. These fluctuations in charges may affect the enzyme activity either by changing the structure or by changing the charge of a residue important for substrate binding or catalysis [50]. The removal of proteases and protein fragments frequently present in the crude extract, which in some cases can be considered enzyme-stabilizing substances, can make the enzyme more susceptible to pH changes [51].

The crude lipase was clearly more thermally stable than the purified lipase. Half-lives of the purified lipase were 26.1- and 5.6-fold decreased at 50 and 55°C, respectively. This finding is also observed for the crude and partially purified pectinolytic enzymes from *A. niger* [52]. According to these authors, the thermal stability can change due to the lack of (i) interaction effects among enzyme components; (ii) other proteins besides those components secreted by the organism; and/or (iii) a combination of these two factors.

Lipase stability in organic solvents is an essential prerequisite for its application in organic synthesis, since synthetic reactions with enzymes are often performed in organic media to displace the thermodynamic equilibrium towards synthesis [53, 54]. Solvents were listed according to their hydrophobicity (Log P) which ranged from −1.67 to +4.51; negative values indicate that the solvent is water soluble, whereas positive values indicate they are insoluble, causing separation of the aqueous from the organic phase [55]. Lipase has different sensitivity to solvents, but in general, water-miscible polar solvents are more destabilizing than water-immiscible solvents [56]. Nonpolar solvents probably promote changes in the equilibrium between the open and closed conformation of lipase and change substrate solubility and reaction products, while polar solvents are more destabilizing to the protein structure by removing solvation water [57, 58]. Nevertheless, no correlation between Log P values of solvents and enzyme stability was observed for crude and purified *C. viswanathii* lipase. Similarly, the lipase from *Rhizopus homothallicus* var. *rhizopodiformis* and *A. niger* MYA 135 also shows no correlation between stability and Log P values [59, 60]. The high activity and stability of *C. viswanathii* lipase observed in the majority of organic solvents indicate wide application of this enzyme for structured lipid production, biodiesel production, esters synthesis for food industries, and hydrolysis of natural triacylglycerols.

Substrate specificity is important for many industrial applications in food industry and biodiesel production. The crude and purified lipase from *C. viswanathii* showed preference for esters hydrolysis of long-chain fatty acids suggesting that this lipase is true lipase. Fojan et al. [61] related that esterases preferentially break ester bonds of shorter chain fatty acids, while lipase displays much broader substrate range than the esterases. The physical state of the substrate is a probable contributing factor towards the substrate specificity. Long-chain fatty acids are typically insoluble or at least poorly soluble. Thus, the lipase has to be capable of identifying insoluble or strongly aggregated substrates. Since lipase is active towards aggregated substrates, lipase activity is directly correlated with the total substrate area and not with the substrate concentration [61, 62].

Hydrolysis of triacylglycerol is an important industrial operation; the products, fatty acids and glycerol, are basic raw materials with wide range of applications. The fatty acids are used as feedstock for the production of oleochemicals such as fatty alcohols, fatty amines, and fatty esters. In this study, poultry fat was subjected to hydrolysis using crude and purified lipase produced by *C. viswanathii* in three different pH values. Under these conditions, it was observed that the poultry fat hydrolysis at pH 4.0 was 2.84- and 8.0-fold higher than at pH 6.0 and 8.0, respectively. This might be explained by changes in amino acids ionization in the enzyme, which alters the ionic and hydrogen bonds that determine the tridimensional structure of proteins. Purified lipase presented low hydrolytic activity on poultry fat compared to crude lipase. This fact may be related to the absence of ions stabilizing the enzyme structure during the formation of fatty acid-lipase complex. Bengtsson and Olivecrona [63] related that the formation of this complex is considered to be the major factor in product inhibition during triacylglycerol hydrolysis. Bengtsson and Olivecrona [64] reported that the cations of inorganic salts form salts with fatty acids and thus remove them from the oil-water interface. As a result, the availability of the interfacial area to the lipase increases, fatty acid-lipase complex formation remains low, and hydrolysis increases.

The alkaline hydrolysis is the most important current route used for triacylglycerol hydrolysis and it requires acidification of the formed soaps to obtain fatty acids [16]. Energy cost associated with this procedure can be prohibitive turning enzymatic hydrolysis more advantageous. Our results indicate that *C. viswanathii* lipase produced under solid-state cultivation presents optimum activity in acid medium and can be applied for poultry fat hydrolysis under these conditions without acidification of the formed soap to obtain fatty acids.

5. Concluding Remarks

This study demonstrated that the lipase from *C. viswanathii* was efficiently produced by nonconventional solid-state cultivation with mixed substrates. High lipase production was obtained from cultivation with wheat bran and spent barley grain at 1:1 (w/w) proportion supplemented with 40% poultry fat and 3.5% yeast extract. Low cost poultry fat can substitute expensive olive oil also resulting in high enzyme production. The establishment of physical parameters as temperature and moisture increased fivefold the lipase production. The lipase from *C. viswanathii* was purified to homogeneity electrophoretic using only hydrophobic interaction chromatography, and SDS-PAGE shows MW of 18.5 kDa. Crude and purified *C. viswanathii* lipase showed optimal activity at pH 5.0 and 50°C and at 5.5 and 45°C, respectively; the high stability was observed at pH from 3.0 to 8.0 and at 45°C. The crude and purified enzymes were highly active in the presence of many organic solvents and also presented prolonged stability in a variety of polar and nonpolar solvents, suggesting application of this enzyme in esterification and transesterification reactions. Finally, the crude lipase efficiently hydrolyzed poultry fat in acid condition, indicating that it can be used for this purpose, reducing the costs with the acidification step required to recover the produced fatty acids.

Competing Interests

The authors declare that they have no competing interests.

Acknowledgments

The authors would like to thank the National Council of Technological and Scientific Development (CNPq) for the scholarship awarded to the first author and financial support (Process no. 455754/2014-4), CAPES for the scholarship awarded to K. B. Dias, and UNESP for payment of this article publication.

References

[1] P. Priji, K. N. Unni, S. Sajith, P. Binod, and S. Benjamin, "Production, optimization, and partial purification of lipase from *Pseudomonas* sp. strain BUP6, a novel rumen bacterium characterized from Malabari goat," *Biotechnology and Applied Biochemistry*, vol. 62, no. 1, pp. 71–78, 2015.

[2] M. Kapoor and M. N. Gupta, "Lipase promiscuity and its biochemical applications," *Process Biochemistry*, vol. 47, no. 4, pp. 555–569, 2012.

[3] A. Pandey, "Solid-state fermentation," *Biochemical Engineering Journal*, vol. 13, no. 2-3, pp. 81–84, 2003.

[4] A. Pandey, C. R. Soccol, and D. Mitchell, "New developments in solid state fermentation: I-bioprocesses and products," *Process Biochemistry*, vol. 35, no. 10, pp. 1153–1169, 2000.

[5] S. R. Couto and M. Á. Sanromán, "Application of solid-state fermentation to food industry—a review," *Journal of Food Engineering*, vol. 76, no. 3, pp. 291–302, 2006.

[6] V. Ferreira-Leitão, L. M. F. Gottschalk, M. A. Ferrara, A. L. Nepomuceno, H. B. C. Molinari, and E. P. S. Bon, "Biomass residues in Brazil: availability and potential uses," *Waste and Biomass Valorization*, vol. 1, no. 1, pp. 65–76, 2010.

[7] I. S. Babu and G. H. Rao, "Optimization of process parameters for the production of lipase in submerged fermentation by *Yarrowia lipolytica* NCIM 3589," *Research Journal of Microbiology*, vol. 2, no. 1, pp. 88–93, 2007.

[8] J. A. Rodriguez, J. C. Mateos, J. Nungaray et al., "Improving lipase production by nutrient source modification using *Rhizopus homothallicus* cultured in solid state fermentation," *Process Biochemistry*, vol. 41, no. 11, pp. 2264–2269, 2006.

[9] N. Mahanta, A. Gupta, and S. K. Khare, "Production of protease and lipase by solvent tolerant *Pseudomonas aeruginosa* PseA in solid-state fermentation using *Jatropha curcas* seed cake as substrate," *Bioresource Technology*, vol. 99, no. 6, pp. 1729–1735, 2008.

[10] M. N. Hosseinpour, G. D. Najafpour, H. Younesi, and M. Khorrami, "Submerged cultures studies for lipase production by *Aspergillus niger* NCIM 584 on soya flour," *Middle-East Journal of Scientific Research*, vol. 7, no. 3, pp. 362–366, 2011.

[11] E. D. C. Cavalcanti, M. L. E. Gutarra, D. M. G. Freire, L. R. Castilho, and G. L. Sant'Anna Jr., "Lipase production by solid-state fermentation in fixed-bed bioreactors," *Brazilian Archives of Biology and Technology*, vol. 48, pp. 79–84, 2005.

[12] L. V. Rodríguez-Durán, B. Valdivia-Urdiales, J. C. Contreras-Esquivel, R. Rodríguez-Herrera, and C. N. Aguilar, "Novel strategies for upstream and downstream processing of tannin acyl hydrolase," *Enzyme Research*, vol. 2011, Article ID 823619, 20 pages, 2011.

[13] N. G. Edwinoliver, K. Thirunavukarasu, R. B. Naidu, M. K. Gowthaman, T. N. Kambe, and N. R. Kamini, "Scale up of a novel tri-substrate fermentation for enhanced production of *Aspergillus niger* lipase for tallow hydrolysis," *Bioresource Technology*, vol. 101, no. 17, pp. 6791–6796, 2010.

[14] F. Hasan, A. A. Shah, S. Javed, and A. Hameed, "Enzymes used in detergents: lipases," *African Journal of Biotechnology*, vol. 9, no. 31, pp. 4836–4844, 2010.

[15] P. Fickers, A. Marty, and J. M. Nicaud, "The lipases from *Yarrowia lipolytica*: genetics, production, regulation, biochemical characterization and biotechnological applications," *Biotechnology Advances*, vol. 29, no. 6, pp. 632–644, 2011.

[16] K.-T. Lee and T. A. Foglia, "Fractionation of chicken fat tria-cylglycerols: synthesis of structured lipids with immobilized lipases," *Journal of Food Science*, vol. 65, no. 5, pp. 826–831, 2000.

[17] V. R. Murty, J. Bhat, and P. K. A. Muniswaran, "Hydrolysis of oils by using immobilized lipase enzyme: a review," *Biotechnology and Bioprocess Engineering*, vol. 7, no. 2, pp. 57–66, 2002.

[18] I. M. Noor, M. Hasan, and K. B. Ramachandran, "Effect of operating variables on the hydrolysis rate of palm oil by lipase," *Process Biochemistry*, vol. 39, no. 1, pp. 13–20, 2003.

[19] A. F. De Almeida, S. M. Taulk-Tornisielo, and E. C. Carmona, "Influence of carbon and nitrogen sources on lipase production by a newly isolated *Candida viswanathii* strain," *Annals of Microbiology*, vol. 63, no. 4, pp. 1225–1234, 2013.

[20] A. F. de Almeida, S. M. Tauk-Tornisielo, and E. C. Carmona, "Acid lipase from *Candida viswanathii*: production, biochemical properties, and potential application," *BioMed Research International*, vol. 2013, Article ID 435818, 10 pages, 2013.

[21] H. J. Vogel, "A convenient growth medium for *Neurospora crassa* (medium N)," *Microbiology Genetics Bulletin*, vol. 13, pp. 42–43, 1956.

[22] J. J. Sedmak and S. E. Grossberg, "A rapid, sensitive, and versatile assay for protein using Coomassie brilliant blue G250," *Analytical Biochemistry*, vol. 79, no. 1-2, pp. 544–552, 1977.

[23] B. D. Hames, "An introduction to polyacrylamide gel electrophoresis," in *Gel Electrophoresis of Proteins: A Practical Approach*, B. D. Hames and D. Rickwood, Eds., pp. 1–91, IRL Press, Oxford, UK, 6th edition, 1987.

[24] E. Lesuisse, K. Schanck, and C. Colson, "Purification and preliminary characterization of the extracellular lipase of Bacillus subtilis 168, an extremely basic pH-tolerant enzyme," *European Journal of Biochemistry*, vol. 216, no. 1, pp. 155–160, 1993.

[25] H. Lineweaver and D. Burk, "The determination of enzyme dissociation constants," *Journal of the American Chemical Society*, vol. 56, no. 3, pp. 658–666, 1934.

[26] K. Brijwani, H. S. Oberoi, and P. V. Vadlani, "Production of a cellulolytic enzyme system in mixed-culture solid-state fermentation of soybean hulls supplemented with wheat bran," *Process Biochemistry*, vol. 45, no. 1, pp. 120–128, 2010.

[27] M. M. Javed, S. Zahoor, S. Shafaat et al., "Wheat bran as a brown gold: nutritious value and its biotechnological applications," *African Journal of Microbiology Research*, vol. 6, no. 4, pp. 724–733, 2012.

[28] S. Benjamin and A. Pandey, "Mixed-solid substrate fermentation. A novel process for enhanced lipase production by *Candida rugosa*," *Acta Biotechnologica*, vol. 18, no. 4, pp. 315–324, 1998.

[29] N. Pogori, A. Cheikhyoussef, Y. Xu, and D. Wang, "Production and biochemical characterization of an extracellular lipase from *Rhizopus chinensis* CCTCC M201021," *Biotechnology*, vol. 7, no. 4, pp. 710–717, 2008.

[30] H. Treichel, D. de Oliveira, M. A. Mazutti, M. Di Luccio, and J. V. A. Oliveira, "A review on microbial lipases production," *Food and Bioprocess Technology*, vol. 3, no. 2, pp. 182–196, 2010.

[31] K.-T. Lee and T. A. Foglia, "Synthesis, purification, and characterization of structured lipids produced from chicken fat," *Journal of the American Oil Chemists' Society*, vol. 77, no. 10, pp. 1027–1034, 2000.

[32] T. Tan, M. Zhang, J. Xu, and J. Zhang, "Optimization of culture conditions and properties of lipase from *Penicillium camembertii Thom* PG-3," *Process Biochemistry*, vol. 39, no. 11, pp. 1495–1502, 2004.

[33] S. Y. Sun and Y. Xu, "Solid-state fermentation for 'whole-cell synthetic lipase' production from *Rhizopus chinensis* and identification of the functional enzyme," *Process Biochemistry*, vol. 43, no. 2, pp. 219–224, 2008.

[34] S. B. Imandi, S. K. Karanam, and H. R. Garapati, "Use of Plackett-Burman design for rapid screening of nitrogen and carbon sources for the production of lipase in solid state fermentation by *Yarrowia lipolytica* from mustard oil cake (*Brassica napus*)," *Brazilian Journal of Microbiology*, vol. 44, no. 3, pp. 915–921, 2013.

[35] N. D. Mahadik, U. S. Puntambekar, K. B. Bastawde, J. M. Khire, and D. V. Gokhale, "Production of acidic lipase by *Aspergillus niger* in solid state fermentation," *Process Biochemistry*, vol. 38, no. 5, pp. 715–721, 2002.

[36] A. Domínguez, F. J. Deive, M. A. Sanromán, and M. A. Longo, "Effect of lipids and surfactants on extracellular lipase production by *Yarrowia lipolytica*," *Journal of Chemical Technology and Biotechnology*, vol. 78, no. 11, pp. 1166–1170, 2003.

[37] M. L. E. Gutarra, E. D. C. Cavalcanti, L. R. Castilho, D. M. G. Freire, and G. L. Sant'Anna Jr., "Lipase production by solid-state fermentation: cultivation conditions and operation of tray and packed-bed bioreactors," *Applied Biochemistry and Biotechnology*, vol. 121, no. 1–3, pp. 105–116, 2005.

[38] Y. S. P. Rahardjo, S. Sie, F. J. Weber, J. Tramper, and A. Rinzema, "Effect of low oxygen concentrations on growth and α-amylase production of *Aspergillus oryzae* in model solid-state fermentation systems," *Biomolecular Engineering*, vol. 21, no. 6, pp. 163–172, 2005.

[39] M. Čertik, L. Slávikková, S. Masrnová, and J. Šajbidor, "Enhancement of nutritional value of cereals with γ-linolenic acid by fungal solid-state fermentations," *Food Technology and Biotechnology*, vol. 44, no. 1, pp. 75–82, 2006.

[40] S. B. Imandi, S. K. Karanam, and H. R. Garapati, "Optimization of media constituents for the production of lipase in solid state fermentation by *Yarrowia lipolytica* from palm Kernal cake (*Elaeis guineensis*)," *Advances in Bioscience and Biotechnology*, vol. 1, no. 2, pp. 115–121, 2010.

[41] A. Bastida, P. Sabuquillo, P. Armisen, R. Fernández-Lafuente, J. Huguet, and J. M. Guisán, "A single step purification, immobilization, and hyperactivation of lipases via interfacial adsorption on strongly hydrophobic supports," *Biotechnology and Bioengineering*, vol. 58, no. 5, pp. 486–493, 1998.

[42] L. Sarda and P. Desnuelle, "Action de la lipase pancréatique sur les esters en émulsion," *Biochimica et Biophysica Acta*, vol. 30, no. 3, pp. 513–521, 1958.

[43] J. M. Palomo, G. Muñoz, G. Fernández-Lorente, C. Mateo, R. Fernández-Lafuente, and J. M. Guisán, "Interfacial adsorption of lipases on very hydrophobic support (octadecyl–Sepabeads): immobilization, hyperactivation and stabilization of the open form of lipases," *Journal of Molecular Catalysis B: Enzymatic*, vol. 19-20, pp. 279–286, 2002.

[44] R. Sharma, Y. Chisti, and U. C. Banerjee, "Production, purification, characterization, and applications of lipases," *Biotechnology Advances*, vol. 19, no. 8, pp. 627–662, 2001.

[45] M. B. Nthangeni, H.-G. Patterton, A. Van Tonder, W. P. Vergeer, and D. Litthauer, "Over-expression and properties of a purified recombinant *Bacillus licheniformis* lipase: a comparative report on *Bacillus* lipases," *Enzyme and Microbial Technology*, vol. 28, no. 7-8, pp. 705–712, 2001.

[46] B. Möller, R. Vetter, D. Wilke, and B. Foullois, "Alkaline *Bacillus* lipases, coding DNA sequences therefore and Bacilli which produce the lipase," Patent No. WO 91/16422, 1991.

[47] T. S. Brush, R. Chapman, R. Kurzman, and D. P. Williams, "Purification and characterization of extracellular lipases from *Ophiostoma piliferum*," *Bioorganic and Medicinal Chemistry*, vol. 7, no. 10, pp. 2131–2138, 1999.

[48] S.-S. Yang and J.-Y. Wang, "Protease and amylase production of *Streptomyces rimosus* in submerged and solid state cultivations," *Botanical Bulletin of Academia Sinica*, vol. 40, no. 4, pp. 259–265, 1999.

[49] R. Subramaniyam and R. Vimala, "Solid state and submerged fermentation for the production of bioactive substances: a comparative study," *International Journal of Science and Nature*, vol. 3, no. 3, pp. 480–486, 2012.

[50] R. Kapilan and V. Arasaratnam, "Comparison of the kinetic properties of crude and purified xylanase from *Bacillus pumilus* with commercial xylanase from *Aspergillus niger*," *Vingnanam Journal of Science*, vol. 10, no. 1, pp. 1–6, 2012.

[51] A. R. C. Braga, A. P. Manera, J. D. C. Ores, L. Sala, F. Maugeri, and S. J. Kalil, "Kinetics and thermal properties of crude and purified β-galactosidase with potential for the production of galactooligosaccharides," *Food Technology and Biotechnology*, vol. 51, no. 1, pp. 45–52, 2013.

[52] G. S. N. Naidu and T. Panda, "Studies on pH and thermal deactivation of pectolytic enzymes from *Aspergillus niger*," *Biochemical Engineering Journal*, vol. 16, no. 1, pp. 57–67, 2003.

[53] N. Doukyu and H. Ogino, "Organic solvent-tolerant enzymes," *Biochemical Engineering Journal*, vol. 48, no. 3, pp. 270–282, 2010.

[54] D. S. Dheeman, S. Antony-Babu, J. M. Frías, and G. T. M. Henehan, "Purification and characterization of an extracellular lipase from a novel strain *Penicillium* sp. DS-39 (DSM 23773)," *Journal of Molecular Catalysis B: Enzymatic*, vol. 72, no. 3-4, pp. 256–262, 2011.

[55] J. Sangster, "Octanol–water partition coefficients of simple organic compounds," *Journal of Physical and Chemical Reference Data*, vol. 18, no. 3, pp. 1111–1229, 1989.

[56] N. Nawani, N. S. Dosanjh, and J. Kaur, "A novel thermostable lipase from a thermophilic *Bacillus* sp.: characterization and esterification studies," *Biotechnology Letters*, vol. 20, no. 10, pp. 997–1000, 1998.

[57] L. D. Castro-Ochoa, C. Rodríguez-Gómez, G. Valerio-Alfaro, and R. O. Ros, "Screening, purification and characterization of the thermoalkalophilic lipase produced by Bacillus thermoleovorans CCR11," *Enzyme and Microbial Technology*, vol. 37, no. 6, pp. 648–654, 2005.

[58] M. Guncheva and D. Zhiryakova, "Catalytic properties and potential applications of *Bacillus* lipases," *Journal of Molecular Catalysis B: Enzymatic*, vol. 68, no. 1, pp. 1–21, 2011.

[59] C. M. Romero, L. M. Pera, F. Loto, C. Vallejos, G. Castro, and M. D. Baigori, "Purification of an organic solvent-tolerant lipase from *Aspergillus niger* MYA 135 and its application in ester synthesis," *Biocatalysis and Agricultural Biotechnology*, vol. 1, no. 1, pp. 25–31, 2012.

[60] B. Hernández-Rodríguez, J. Córdova, E. Bárzana, and E. Favela-Torres, "Effects of organic solvents on activity and stability of lipases produced by thermotolerant fungi in solid-state fermentation," *Journal of Molecular Catalysis B: Enzymatic*, vol. 61, no. 3-4, pp. 136–142, 2009.

[61] P. Fojan, P. H. Jonson, M. T. N. Petersen, and S. B. Petersen, "What distinguishes an esterase from a lipase: a novel structural approach," *Biochimie*, vol. 82, no. 11, pp. 1033–1041, 2000.

[62] R. Verger, "'Interfacial activation' of lipases: facts and artifacts," *Trends in Biotechnology*, vol. 15, no. 1, pp. 32–38, 1997.

[63] D. Goswami, S. De, and J. K. Basu, "Effects of process variables and additives on mustard oil hydrolysis by porcine pancreas lipase," *Brazilian Journal of Chemical Engineering*, vol. 29, no. 3, pp. 449–460, 2012.

[64] G. Bengtsson and T. Olivecrona, "Lipoprotein lipase: mechanism of product inhibition," *European Journal of Biochemistry*, vol. 106, no. 2, pp. 557–562, 1980.

2

Optimization of Xylanase Production from *Aspergillus foetidus* in Soybean Residue

Luana Cunha [iD],[1] **Raquel Martarello,**[1]
Paula Monteiro de Souza [iD],[1] **Marcela Medeiros de Freitas** [iD],[1]
Kleber Vanio Gomes Barros [iD],[1] **Edivaldo Ximenes Ferreira Filho** [iD],[2]
Mauricio Homem-de-Mello [iD],[1] **and Pérola Oliveira Magalhães** [iD][1]

[1]*Laboratory of Natural Products, School of Health Sciences, University of Brasília, Asa Norte, 70910900 Brasília, DF, Brazil*
[2]*Laboratory of Enzymology, Department of Cell Biology, University of Brasília, Asa Norte, 70910900 Brasília, DF, Brazil*

Correspondence should be addressed to Pérola Oliveira Magalhães; perolamagalhaes@unb.br

Academic Editor: Hartmut Kuhn

Enzymatic hydrolysis is an important but expensive step in the process to obtain enzyme derived products. Thus, the production of efficient enzymes is of great interest for this biotechnological application. The production of xylanase by *Aspergillus foetidus* in soybean residues was optimized using 2×2^3 factorial designs. The experimental data was fitted into a polynomial model for xylanase activity. Statistical analyses of the results showed that variables pH and the interaction of pH and temperature had influenced the production of xylanase, with the best xylanase production level (13.98 U/mL) occurring at fermentation for 168 hours, pH 7.0, 28°C, and 120 rpm.

1. Introduction

Xylanases (EC 3.2.1.8) are found in both fungi and bacteria. They randomly catalyze the endohydrolysis of $1,4$-β-D-xylosidic linkages in xylan [1]. According to the CAZy database (http://www.cazy.org), xylanases are classified under glycosyl hydrolase (GH) families 5, 7, 8, 9, 10, 11, 12, 16, 26, 30, 43, 44, 51, and 62 based on their amino acid sequences and structures [2].

Family 10 consists of endo-$1,4$-β-xylanases, endo-$1,3$-β-xylanases, and cellobiohydrolases. The major enzymes of this family are endo-$1,4$-β-xylanases; however, substrate specificity studies revealed that these may not be entirely specific for xylan and may not be active on low molecular mass cellulose substrates. In effect, it has been found that the replacement of one or two xylose residues by glucose is normally tolerated by the xylanases of this family, with this generally resulting in a lowered catalytic efficiency [2].

In contrast to other families, family 11 is monospecific, it consists solely of xylanases. Moreover, these xylanases are "true xylanases" as they are exclusively active on D-xylose containing substrates. They have a lower catalytic versatility than family 10 xylanases and indeed the products of their action can be further hydrolyzed by the family 10 enzymes. These xylanases are characterized by a high pI, low molecular weight, and a β-sheet structure [2].

Supplementary studies (data not shown) revealed a low percentage of α-helix (~3–6%) and high percentage of β-sheets (~43–48%). This suggests xylanase from *Aspergillus foetidus* is classified as a member of family 11. These results are in accordance with literature, in which xylanases present approximately 3–5% of α-helix structures and a higher percentage of β-sheets [3, 4].

Among the many microbial sources, filamentous fungi are especially interesting as they secrete these enzymes into the medium and their xylanase levels are very much higher than those found in yeasts and bacteria [5]. Several xylanases have been reported from these fungal strains for various industrial and biotechnological applications. In order to fulfill specific industrial needs, enzymes must possess pH stability, thermostability, high specific activity, and most importantly high affinity for the substrate [6, 7].

TABLE 1: Code and level of factors chosen for the trials for xylanase FD1 and FD2.

Independent variable	Symbol	Range and level				
Xylanase FD1		−1.68	−1	0	+1	+1.68
Agitation (rpm)	X_{1a}	86	100	120	140	154
Temperature (°C)	X_{2a}	18	22	28	34	38
Ph	X_{3a}	3.6	5	7	9	10.3
Xylanase FD2						
Agitation (rpm)	X_{1b}	60	84	120	156	180
Temperature (°C)	X_{2b}	10	17	28	39	46
pH	X_{3b}	3.6	5	7	9	10.3

Xylanases are extensively used in the paper and pulp industry, as well as in baking, animal feed, biofuels production, fruit and vegetable processing, manufacture of bread, food, and drinks, textiles, xylitol production, saccharification of agricultural, and industrial and municipal wastes among other utilities [5, 8]. The successful industrial application of xylanase requires its cost-effective production in bulk quantity. The production cost can be reduced by optimizing the fermentation medium and the process, for example, using cheap agroresidue as carbon source [7, 9].

Brazil is the second biggest producer of soybean worldwide, harvesting 96.5 kton in 2016, just behind United States, with 106.9 kton in 2016 [10, 11]. Therefore, soybean residues represent the major byproduct of processing soybean industry [12], which could be used as a carbon source for the production of enzymes [13].

Thus, the experimental design statistical approach for enzyme production using a response surface methodology (RSM) is an alternative strategy to reduce the production cost. Recently, RSM has been utilized successfully to improve product yield and to reduce development time and cost of biotechnological processes [7].

In this work, we employed Central Composite Design (CCD) for the planned statistical optimization of xylanase activity of an *Aspergillus foetidus* strain isolated from Brazilian Savannah, grown in submerged fermentation, using soybean residue as substrate.

2. Materials and Methods

2.1. Soybean Residue Pretreatment.
Soybean residue was autoclaved at 121°C for 2 hours and thoroughly washed with tap water. After autoclaving, it was dried at 65°C for 48 hours and then grounded to form a relatively homogeneous blend. A fine powder was obtained and used as a substrate for xylanase production [13].

2.2. Organism and Enzyme Production.
The fungi *Aspergillus foetidus* was obtained from the microorganisms' collection of Laboratory of Enzymology from University of Brasília. The *Aspergillus foetidus* was kindly provided by Professor Dr. Edivaldo Ximenes, Depositary Microorganisms Center: Collection of microorganisms for phytopathogens and weeds control from Embrapa Genetic Resources, accredited by

Genetic Heritage Management Council by CGEN deliberation n° 67 published in the DOU in 13.09.2004, Section 1, page 53, linked to the project "Biotechnological processes in application of hollocelulases from filamentous fungi", Process n° 010237/2015-1, University of Brasília. It is maintained in potato dextrose agar (PDA) medium at −80°C.

The spore concentration was determined by counting under a microscope with a Newbauer chamber and adjusted with sterile saline solution (0.9%) to a final concentration of 1×10^7 spores/mL.

For xylanase production, an aliquot (1 mL) of spore suspension (10^7 spores/mL) was inoculated into 250 mL Erlenmeyer flasks containing 50 mL of liquid medium (0.4% peptone, 0.4% yeast extract, 0.2% KH_2PO_4, 0.8% NaH_2PO_4, 0.25% $MgSO_4$) at pH 7.0 with 2% (w/v) of soybean residue. The cultures were incubated at 28°C with constant agitation at 120 rpm for 7 days.

After the culture growth, the medium was filtered through a Büchner funnel with filter paper (Whatman n°1) and stored at −20°C. The resulting filtrate, here called crude extract, was used as a source of xylanase.

2.3. Enzyme Assay.
The xylanase activity was determined by mixing 50 μL of enzyme solution with 100 μL of birchwood xylan (10 mg/mL) in 50 mM sodium acetate buffer, pH 5.4 at 50°C for 30 min [14]. The release of the reducing sugar was measured using the DNS method [15]. The absorbance was read at 540 nm by spectrophotometry (Shimadzu UV-1800) and the xylanase activity was expressed as μmol of reducing sugar released per min per milliliter (IU/mL). Xylose was used as standard.

Protein concentration was measured by the method of Bradford, using bovine serum albumin as standard [16].

2.4. Experimental Design and Statistical Analysis.
To analyze the effects of the agitation (X_1), temperature (X_2), and pH (X_3) at enzymatic production of xylanase in medium with soybean residue, two factorial designs were employed (FD1 and FD2). For both, a 2^3 factorial design with three center points and axial points was employed (Table 1). The factors were coded to allow the analysis of variance (ANOVA) by response of enzymatic activity (Y).

Center points were defined based on previous methodology used in our laboratory (data not shown); axial, −1, and +1

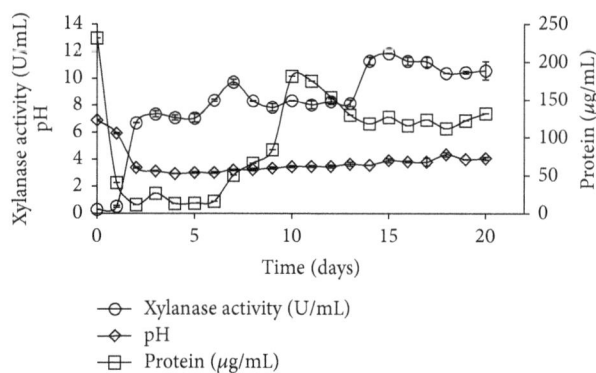

FIGURE 1: Time course of xylanase produced by A. foetidus.

points were determined in order to evaluate significant differences. FD2 was planned aiming to increase the difference observed.

The Design-Expert® software, version 9.0.6.2, was used for regression and graphical analysis of the data. Only the factors with significance higher than or equal to 5% ($p < 0.05$) were considered.

For each factorial design, 17 experiments (determination of xylanase activity) were performed and are shown in results section.

3. Results and Discussion

3.1. Enzyme Induction. To optimize induction time, *Aspergillus foetidus* xylanase specific activity was assessed every day, for 20 days (Figure 1). The specific activity was determined during this period.

The highest xylanase activity was after 15 days of culture (11.84 U/mL). On the 7th day of culture, the xylanase activity was 9.72 U/mL, and specific activity had its maximum value, which was 810.41 U/mg during the period of analysis. A specific activity of purified *Aspergillus foetidus* xylanase was 1196.53 U/mg (data not shown) [4].

The specific activity is an important parameter to assess the enzymatic activity in relation to the amount of proteins in the sample. In general, high specific activity represents the highest level of target enzyme. For this reason, the 7th day was chosen for this study.

The xylanase specific activity from *Aspergillus foetidus* cultivated with soybean residue is consistent with previous reports, which xylanases from *Aspergillus sp.* cultivated in different residues also present high activity levels, as described below.

Delabona et al. (2013) found a specific xylanolytic activity for *Aspergillus fumigatus*, 1055.6 U/g and 558.3 U/g in wheat bran and soybean, respectively, after 5 days of solid state fermentation. In the same work, he found for *Aspergillus niger*, a specific xylanolytic activity of 1285.0 U/g, 484.2 U/g, and 1050.0 U/g using as residue wheat bran, soybean, and wheat bran with sugarcane bagasse, respectively [17]. Supplementing it, Yang et al. (2015) found for *Aspergillus fumigatus* submerged liquid culture with sugarcane bagasse a xylanolytic activity of 53.1 U/mg [18]; and Farinas et al. (2010)

after 3 days of *Aspergillus niger* solid state fermentation found for xylanase activity 13.24 U/mL [19].

The amount of protein oscillated during the culture period, suggesting this result may include other enzymes besides xylanase, which are concomitantly produced and also participate in the substrate degradation process. The induction profile followed the pH variation, with maximum value of 5.92 on the 1st day of culture and minimum value of 2.94 on the 4th day. Seventh day pH was 3.20. This result indicates a xylanase production in acidic medium.

3.2. Factorial Design. The activity of the xylanase present in the crude extract produced by filamentous fungus *Aspergillus foetidus* grown on soybean residue under submerged liquid culture was assessed. Variation on agitation, temperature, and pH effects on xylanase activity were evaluated using the statistical design of experiments and RSM analysis. Table 2 presents the results of the complete factorial design for xylanase activities under the different conditions evaluated. Tables 3 and 4 exhibit the coefficients of the mathematical model and statistical parameters.

3.2.1. Xylanase FD1. In the first study, the independent variables pH, pH^2, and pH $*$ temperature (C, C^2, BC) had a significant effect on the xylanase production ($p < 0.05$). Interactions between agitation and temperature (AB) and interactions between agitation and pH (AC) had no significance ($p > 0.05$). Under the levels tested in the factorial design, the variables agitation and temperature did not interact with each other. Subsequently, the xylanase activity is not significantly modified.

Data were fitted to a quadratic model with three central points. The statistical significance of the equation was checked and the determination coefficient (R^2) was calculated to be 0.88, indicating that 88% of the variability in the response could be explained by the model. In addition, the F test (5.87 times higher than the listed F value at 90% level of confidence) was satisfactory for the prediction of the model used to describe response surface plot of the enzyme activity as a function of pH and temperature (Figure 2). Higher experimental value of enzymatic activity was found at the condition of central point, which is at 120 rpm, pH 7, and 28°C.

Other studies of xylanase report optimum pH, temperature, and agitation at specific values. De Souza Moreira et al. (2013) found an optimum pH and temperature of pH 6.0, 50°C at 120 rpm and pH 5.0, 45°C at 120 rpm for xylanases produced by *A. terreus* under submerged fermentation. Ang et al. (2013) found a maximum activity at 60°C and optimum pH of 4.0 for xylanase produced by *A. fumigatus* under solid state fermentation (SSF) [20]. The advantage of using the statistical methodology was the definition of an optimum temperature and pH range, rather than a specific value, allowing more flexibility during process development [19].

The lack-of-fit test did not result in a significant p value, indicating that the model is sufficiently accurate to predict the factors responses within the ranges studied. The "lack-of-fit F-value" of 1.38 implies that the lack of fit is not significant

TABLE 2: Results obtained for 2^3 factorial design with parameters: values of pH, temperature, and agitation.

Run order	Coded levels			Xylanase FD1 (U/mL)	Xylanase FD2 (U/mL)
	X_1	X_2	X_3		
1	−1	−1	−1	8.965	8.423
2	+1	−1	−1	8.525	8.730
3	−1	+1	−1	9.955	11.243
4	+1	+1	−1	10.603	8.512
5	−1	−1	+1	8.979	5.737
6	+1	−1	+1	8.378	3.004
7	−1	+1	+1	0.996	0.806
8	+1	+1	+1	4.783	0.674
9	0	0	0	10.811	13.982
10	0	0	0	13.880	11.506
11	0	0	0	11.709	10.915
12	0	0	−1.68	3.864	3.976
13	0	0	+1.68	1.407	1.358
14	0	−1.68	0	8.571	0.705
15	0	+1.68	0	8.131	1.016
16	−1.68	0	0	7.731	9.480
17	+1.68	0	0	8.447	8.349

TABLE 3: Analysis of variance (ANOVA) for the model regression 1.

Source	SS	DF	MS	F-value	p value
Model [1]	167.53	9	18.61	5.87	0.0147
A, agitation	1.55	1	1.55	0.49	0.5073
B, Temp.	6.27	1	6.27	1.98	0.2027
C, pH	26.56	1	26.56	8.37	0.0232
AB	3.75	1	3.75	1.18	0.3130
AC	1.11	1	1.11	0.35	0.5730
BC	26.81	1	26.81	8.45	0.0227
A^2	12.81	1	12.81	4.04	0.0844
B^2	10.68	1	10.68	3.37	0.1092
C^2	101.05	1	101.05	31.86	0.0008
Residual	22.20	7	3.17		
Lack of fit	17.22	5	3.44	1.38	0.4700
Pure error	4.98	2	2.49		
Total	189.73	16			

[1] Model regression is xylanase activity $= -81.111 + 0.431 * A + 1.784 * B + 12.937 * C + 0.006 * A * B + 0.009 * A * C - 0.153 * B * C - 0.003 * A^2 - 0.027 * B^2 - 0.748 * C^2$; $R^2 = 0.8830$; SS, sum of squares; DF, degrees of freedom; MS, mean square. Significance level = 95%.

relative to the pure error. There is a 47% ($p = 0.47$) chance that a "lack of-fit F-value" this large could occur due to noise.

The "Pred R-Squared" of 0.242 is not as close to the "Adj R-Squared" of 0.732 as one might normally expect (difference is more than 0.2). This may indicate a large block effect or a possible problem with the model. "Adeq Precision" measures the signal-to-noise ratio. A ratio of 7.91 indicates an adequate signal (a ratio greater than 4 is desirable). This model can be used to navigate the design space.

The response equation obtained is the following: $Y_1 = -81.111 + 0.431 * A + 1.784 * B + 12.937 * C + 0.006 * A * B + 0.009 * A * C - 0.153 * B * C - 0.003 * A^2 - 0.027 * B^2 - 0.748 * C^2$, where Y_1 is the predicted xylanase activity in U/mL; A, B, and

C are the coded variables of agitation, temperature, and pH, respectively. The equation in terms of actual factors can be used to make predictions about the response for given levels of each factor.

The negative effect of the factors means that an increase in one of them will reduce the enzymatic activity. For xylanase activity, pH showed a positive effect while temperature * pH and pH^2 showed a negative effect, within the range evaluated. Furthermore, the significance of the interaction effect between pH and temperature revealed synergistic effect of these variables; that is, the variables pH and temperature well-adjusted could modify the xylanase activity. The pH effect was higher than the temperature effect, as can be verified for

TABLE 4: Analysis of variance (ANOVA) for the model regression 2.

Source	SS	DF	MS	F-value	p value
Model [2]	274.80	9	30.53	4.39	0.0319
A, agitation	3.79	1	3.79	0.55	0.4844
B, Temp.	1.25	1	1.25	0.18	0.6839
C, pH	70.78	1	70.78	10.19	0.0152
AB	0.024	1	0.024	0.007	0.9549
AC	0.024	1	0.024	0.007	0.9545
BC	12.16	1	12.16	1.75	0.2274
A^2	4.67	1	4.67	0.67	0.4393
B^2	137.41	1	137.41	19.78	0.0030
C^2	91.73	1	91.73	13.20	0.0084
Residual	48.63	7	6.95		
Lack of fit	43.34	5	8.67	3.27	0.2504
Pure error	5.30	2	2.65		
Total	323.43	16			

[2]Model regression is xylanase activity = $-54.322 + 0.113 * A + 1.997 * B + 10.506 * C + 0.000 * A * B - 0.001 * A * C - 0.0560 * B * C - 0.000 * A^2 - 0.028 * B^2 - 0.713 * C^2$; $R^2 = 0.849$; SS, sum of squares; DF, degrees of freedom; MS, mean square. Significance level = 95%.

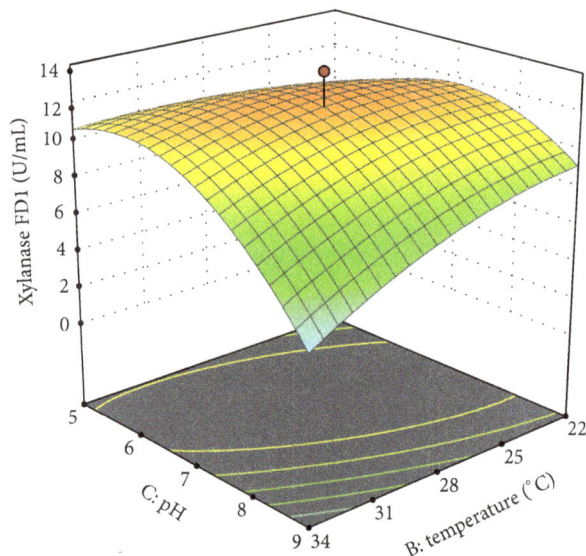

FIGURE 2: Response surface for xylanase FD1 as a function of the variables temperature and pH.

the coefficient values listed in Table 3. A similar result was found by Farinas et al. (2010) and Singh et al. (2009) on optimization of parameters for cellulase and xylanase from *Aspergillus niger* and cellulase from *Aspergillus heteromorphus*, respectively. The authors found that the change in temperature was less important than changes in pH, within the range evaluated. As the pH varies, the charge of the substrate and ionic components of the substrate changes, affecting the activity of the enzymes [19, 21].

3.2.2. Xylanase FD2. In the second study of xylanase, the range of agitation, pH, and temperature were expanded in relation to the range of the previous study, in order to evaluate a possible positive effect in the xylanase activity.

The computed F-value of 4.39 implies the model is significant at a high confidence level (Table 4). The probability p value was also relatively low ($p < 0.05$), indicating the significance of the model. The coefficient of variation ($R^2 = 0.84$) indicates a high correlation between the experimentally observed and predicted values and indicates the degree of precision with which the treatments are compared.

The "lack-of-fit F-value" of 3.27 implies the lack of fit is not significantly relative to the pure error. There is a 25.04% ($p = 0.2504$) chance that a "lack-of-fit F-value" this large could occur due to noise. Additionally, "Adeq Precision" of 5.88 indicates an adequate signal. This model can be used to represent the design space.

The independent variable pH, pH^2, and $temperature^2$ (C, C^2, B^2) had a significant effect on the xylanase production

Design-Expert software
Factor coding: actual

Xylanase FD2 (U/mL)
● Design points above predicted value
○ Design points below predicted value

0.674 ▮▮▮ 13.982

$X1 = B$: temperature
$X2 = C$: pH

Actual factor
A: agitation = 120

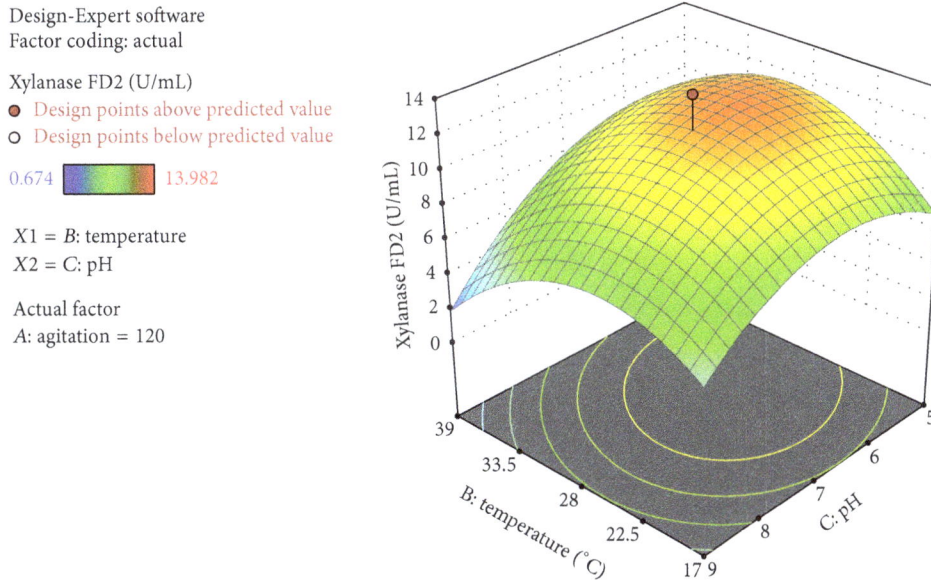

FIGURE 3: Response surface for xylanase FD2 as a function of the variables temperature and pH.

(Table 4). The other correlations had no significance ($p > 0.05$).

The response equation obtained for the second model was as follows: $Y_2 = -54.322 + 0.113 * A + 1.997 * B + 10.506 * C + 0.000 * A * B - 0.001 * A * C - 0.056 * B * C - 0.000 * A^2 - 0.028 * B^2 - 0.713 * C^2$, where Y_2 is the predicted xylanase activity and A, B, and C are the coded variables of agitation, temperature, and pH, respectively. The equation in terms of actual factors can be used to make predictions about the response for given levels of each factor.

To investigate the effects of pH and pH $*$ temperature on xylanase production, the three-dimensional contour plot was used to assess the effects of pH and temperature on xlylanase production (Figure 3). Highest xylanase activity in this second experiment was 13.98 U/mL (120 rpm, 28°C, pH 7.0). This activity was similar to the one achieved in the first study, which shows that the increase in the interval between levels of the variable did not influence the optimal condition observed in the first design. The increase of the interval, however, was able to demonstrate that the pH was the variable with greater influence in the tested design.

Xylanase activities for the first and second models were 12.028 U/mL and 11.989 U/mL, respectively, for the optimal conditions of the model (120 rpm, 28°C, and pH 7). There were very small differences in the predicted response and the activity observed during the experiments, confirming that the results are in accordance of the RSM study. The xylanase from A. foetidus is more effective in comparison to xylanase from A. niger, A. flavus, and A. fumigatus as can be seen when you compare the obtained results to some data from the literature.

Guimaraes et al. (2013) found a xylanolitic activity for Aspergillus niger and Aspergillus flavus of 10.50 and 11.92 U/mL, respectively, using as residue wheat bran 0.5% and corncob 0.5%. Supplementing it, Naseeb et al. (2015) found for A. fumigatus pea peel substrate under Solid State Fermetation (SSF) 13.97 U/mL [22, 23].

4. Conclusions

In this study, we investigated the feasibility of filamentous fungus Aspergillus foetidus to produce high level of xylanase enzyme in a liquid medium. This study highlighted a newly isolated strain A. foetidus which could produce xylanase in soybean residue medium, which is cheap and abundant. The best conditions of xylanase production were pH 7.0, 120 rpm, and 28°C (168 hours). In future studies, statistical optimization of medium, physical factors, and scaling up studies in bioreactor should be used as an alternative to contribute toward the economics of biotechnological processes.

Acknowledgments

The present work was carried out with the financial support of Foundation for Research Support of Federal District (FAPDF, Grant no. 193.000.484/2011) and the National Counsel of Technological and Scientific Development (CNPq, Grant no. 564208/2010-8).

References

[1] G. Paës, J.-G. Berrin, and J. Beaugrand, "GH11 xylanases: structure/function/properties relationships and applications," Biotechnology Advances, vol. 30, no. 3, pp. 564–592, 2012.

[2] T. Collins, C. Gerday, and G. Feller, "Xylanases, xylanase families and extremophilic xylanases," FEMS Microbiology Reviews, vol. 29, no. 1, pp. 3–23, 2005.

[3] J. C. Hurlbert and J. F. Preston III, "Functional characterization of a novel xylanase from a corn strain of Erwinia chrysanthemi," Journal of Bacteriology, vol. 183, no. 6, pp. 2093–2100, 2001.

[4] L. L. D. Cunha, Purificação e Caracterização Bioquímica e Biofísica de Xilanase de Aspergillus foetidus, Brasília: Universidade de Brasília, 2016.

[5] M. L. T. M. Polizeli, A. C. S. Rizzatti, R. Monti, H. F. Terenzi, J. A. Jorge, and D. S. Amorim, "Xylanases from fungi: properties and industrial applications," *Applied Microbiology and Biotechnology*, vol. 67, no. 5, pp. 577–591, 2005.

[6] P. Chutani and K. K. Sharma, "Biochemical evaluation of xylanases from various filamentous fungi and their application for the deinking of ozone treated newspaper pulp," *Carbohydrate Polymers*, vol. 127, pp. 54–63, 2015.

[7] S. P. Saha and S. Ghosh, "Optimization of xylanase production by Penicillium citrinum xym2 and application in saccharification of agro-residues," *Biocatalysis and Agricultural Biotechnology*, vol. 3, no. 4, pp. 188–196, 2014.

[8] J. X. Heck, S. H. Flôres, P. F. Hertz, and M. A. Z. Ayub, "Statistical optimization of thermo-tolerant xylanase activity from Amazon isolated *Bacillus circulans* on solid-state cultivation," *Bioresource Technology*, vol. 97, no. 15, pp. 1902–1906, 2006.

[9] A. Pal and F. Khanum, "Production and extraction optimization of xylanase from *Aspergillus niger* DFR-5 through solid-state-fermentation," *Bioresource Technology*, vol. 101, no. 19, pp. 7563–7569, 2010.

[10] S. Embrapa, "Soja em n·meros," *Accessed*, pp. 30-11, 2015.

[11] Conab, "Levantamentos de Safra," *Accessed*, pp. 11–30, 2015.

[12] L. R. De Souza Moreira, M. De Carvalho Campos, P. H. V. M. De Siqueira et al., "Two β-xylanases from Aspergillus terreus: Characterization and influence of phenolic compounds on xylanase activity," *Fungal Genetics and Biology*, vol. 60, pp. 46–52, 2013.

[13] L. R. D. S. Moreira, G. V. Ferreira, S. S. T. Santos, A. P. S. Ribeiro, F. G. Siqueira, and E. X. F. Filho, "The hydrolysis of agro-industrial residues by holocellulose-degrading enzymes," *Brazilian Journal of Microbiology*, vol. 43, no. 2, pp. 498–505, 2012.

[14] R. Garcia Medeiros, R. Hanada, and E. X. F. Filho, "Production of xylan-degrading enzymes from Amazon forest fungal species," *International Biodeterioration & Biodegradation*, vol. 52, no. 2, pp. 97–100, 2003.

[15] G. L. Miller, "Use of dinitrosalicylic acid reagent for determination of reducing sugar," *Analytical Chemistry*, vol. 31, no. 3, pp. 426–428, 1959.

[16] M. M. Bradford, "A rapid and sensitive method for the quantitation of microgram quantities of protein utilizing the principle of protein dye binding," *Analytical Biochemistry*, vol. 72, no. 1-2, pp. 248–254, 1976.

[17] P. D. S. Delabona, R. D. P. B. Pirota, C. A. Codima, C. R. Tremacoldi, A. Rodrigues, and C. S. Farinas, "Effect of initial moisture content on two Amazon rainforest *Aspergillus strains* cultivated on agro-industrial residues: biomass-degrading enzymes production and characterization," *Industrial Crops and Products*, vol. 42, no. 1, pp. 236–242, 2013.

[18] Q. Yang, Y. Gao, Y. Huang et al., "Identification of three important amino acid residues of xylanase AfxynA from Aspergillus fumigatus for enzyme activity and formation of xylobiose as the major product," *Process Biochemistry*, vol. 50, no. 4, pp. 571–581, 2015.

[19] C. S. Farinas, M. M. Loyo, A. Baraldo, P. W. Tardioli, V. B. Neto, and S. Couri, "Finding stable cellulase and xylanase: evaluation of the synergistic effect of pH and temperature," *New Biotechnology*, vol. 27, no. 6, pp. 810–815, 2010.

[20] S. K. Ang, E. M. Shaza, Y. A. Adibah, A. A. Suraini, and M. S. Madihah, "Production of cellulases and xylanase by Aspergillus fumigatus SK1 using untreated oil palm trunk through solid state fermentation," *Process Biochemistry*, vol. 48, no. 9, pp. 1293–1302, 2013.

[21] R. Singh, R. Kumar, K. Bishnoi, and N. R. Bishnoi, "Optimization of synergistic parameters for thermostable cellulase activity of Aspergillus heteromorphus using response surface methodology," *Biochemical Engineering Journal*, vol. 48, no. 1, pp. 28–35, 2009.

[22] N. C. D. A. Guimaraes, M. Sorgatto, S. D. C. Peixoto-Nogueira et al., "Bioprocess and biotechnology: effect of xylanase from *Aspergillus niger* and *Aspergillus flavus* on pulp biobleaching and enzyme production using agroindustrial residues as substract," *SpringerPlus*, vol. 2, no. 1, article 380, pp. 1–7, 2013.

[23] S. Naseeb, M. Sohail, A. Ahmad, and S. A. Khan, "Production of xylanases and cellulases by Aspergillus fumigatus MS16 using crude lignocellulosic substrates," *Pakistan Journal of Botany*, vol. 47, no. 2, pp. 779–784, 2015.

3

The Importance of Surface-Binding Site towards Starch-Adsorptivity Level in α-Amylase

Umi Baroroh,[1] Muhammad Yusuf,[2,3] Saadah Diana Rachman,[2] Safri Ishmayana,[2] Mas Rizky A. A. Syamsunarno,[1,3] Jutti Levita,[1,4] and Toto Subroto[2,3]

[1] *Master of Biotechnology Program, Postgraduate School, Universitas Padjadjaran, Jl. Dipati Ukur 35, Bandung, West Java, Indonesia*
[2] *Department of Chemistry, Faculty of Mathematics and Natural Sciences, Universitas Padjadjaran,*
  Jl. Raya Bandung-Sumedang Km 21, Jatinangor, Sumedang, West Java 45363, Indonesia
[3] *Research Center of Molecular Biotechnology and Bioinformatics, Universitas Padjadjaran, Jl. Singaperbangsa 2,*
  Bandung, West Java 40133, Indonesia
[4] *Department of Pharmacology and Clinical Pharmacy, Faculty of Pharmacy, Universitas Padjadjaran,*
  Jl. Raya Bandung-Sumedang Km 21, Jatinangor, Sumedang, West Java 45363, Indonesia

Correspondence should be addressed to Toto Subroto; t.subroto@unpad.ac.id

Academic Editor: Denise Freire

Starch is a polymeric carbohydrate composed of glucose. As a source of energy, starch can be degraded by various amylolytic enzymes, including α-amylase. In a large-scale industry, starch processing cost is still expensive due to the requirement of high temperature during the gelatinization step. Therefore, α-amylase with raw starch digesting ability could decrease the energy cost by avoiding the high gelatinization temperature. It is known that the carbohydrate-binding module (CBM) and the surface-binding site (SBS) of α-amylase could facilitate the substrate binding to the enzyme's active site to enhance the starch digestion. These sites are a noncatalytic module, which could interact with a lengthy substrate such as insoluble starch. The major interaction between these sites and the substrate is the CH/pi-stacking interaction with the glucose ring. Several mutation studies on the *Halothermothrix orenii*, SusG *Bacteroides thetaiotamicron*, *Barley*, *Aspergillus niger*, and *Saccharomycopsis fibuligera* α-amylases have revealed that the stacking interaction through the aromatic residues at the SBS is essential to the starch adsorption. In this review, the SBS in various α-amylases is also presented. Therefore, based on the structural point of view, SBS is suggested as an essential site in α-amylase to increase its catalytic activity, especially towards the insoluble starch.

1. Introduction

Starch is the most abundant form of storage of many economically important crops such as wheat, rice, maize, tapioca, and potato [1, 2]. Starch-containing crop is an essential constituent of the human diet, and a large proportion of the food consumed by the world's population originates from them. Starch is harvested and used as its original form or chemically or enzymatically processed into a variety of different products, for example, starch hydrolysates, glucose syrups, fructose, starch or maltodextrin derivatives, or cyclodextrins [1].

Degradation of starch into a variety of different products is performed by amylolytic enzymes, such as α-amylase, glucoamylase, β-amylase, isoamylase, pullulanase, exo-1,4-α-D-glucanase, α-D-glycosidase, and cyclomaltodextrin-D-glucotransferase [3].

The amylases are multidomain proteins. Interestingly, about 10% of amylases contain a distinct noncatalytic module that is known to facilitate binding and degradation of raw starch [4]. Initially, only two types of starch-binding domains (SBDs) were recognized: either very frequent C-terminal SBD or very scarcely occurring N-terminal SBD [5]. However, sometimes the substrate also binds to one or more surface

TABLE 1: CBM classification based on ligand specificity (http://www.cazypedia.org, taken from Barchiesi et al. [21]).

Ligand	CBM family
Cellulose	CBM1, CBM2, CBM3, CBM4, CBM6, CBM8, CBM9, CBM10, CBM16, CBM17, CBM28, CBM30, CBM37, CBM44, CBM46, CBM49, CBM59, CBM63, CBM64, CBM65, CBM73, CBM76, CBM78, CBM80, CBM81
Xylan	CBM2, CBM4, CBM6, CBM9, CBM13, CBM15, CBM22, CBM31, CBM35, CBM36, CBM37, CBM44, CBM54, CBM59, CBM60, CBM64, CBM72
Plant cell wall, other (e.g., β-glucans, porphyrans, pectins, mannans, gluco- and galacturonans)	CBM4, CBM6, CBM11, CBM13, CBM16, CBM22, CBM23, CBM27, CBM28, CBM29, CBM32, CBM35, CBM39, CBM42, CBM43, CBM52, CBM56, CBM59, CBM61, CBM62, CBM67
Chitin	CBM1, CBM2, CBM5, CBM6, CBM12, CBM13, CBM14, 16 CBM18, CBM19, CBM50, CBM54, CBM55, CBM73
α-Glucans (starch/glycogen, mutant)	CBM20, CBM21, CBM25, CBM26, CBM34, CBM41, CBM45, CBM48, CBM53, CBM58, CBM68, CBM69, CBM74
Mammalian glycans	CBM32, CBM40, CBM47, CBM51, CBM57
Other	Bacterial cell wall sugar: CBM35, CBM39, CBM50 Fructans: CBM38, CBM66 Yeast cell wall glucans: CBM54

regions called surface-binding site (SBS) [6]. In starch-based industry, α-amylase is used to break down the starch granules, which are densely packed in a polycrystalline state by inter- and intramolecular bonds. Starch granules are insoluble in cold water and often resistant to chemicals and enzymes [7]. A gelatinization step at a high temperature (105°C) would help to open the crystalline structure of starch. Hence it is easier to be digested by the enzyme [8]. Nevertheless, this process requires high energy, thus resulting in high cost of production [9]. Therefore, starch processing in lower temperature is more preferred [8, 10, 11]. The ability of the amylolytic enzyme to hydrolyze the raw starch was related to the level of starch-adsorptivity properties [11].

Amylolytic enzymes with raw starch digesting ability may contain SBD and/or SBS. Hence, in this review, we focus on the importance of starch-binding particularly SBSs. From a structural point of view, there are five examples of α-amylases, with or without SBS, which can be used to review the following aspects: (1) the most significant factor in starch-binding, (2) the type of interactions that influence the binding of these proteins to the substrate in the noncatalytic module, and (3) the reason of low substrate adsorptivity to the protein despite having high amylolytic activities.

2. Carbohydrate-Binding Module in Amylolytic Enzymes

In general, carbohydrate-active enzymes that degrade or modify polysaccharides bind to the substrate on the carbohydrate-binding site situated outside of the active-site area. These additional binding sites can be found on the carbohydrate-binding modules (CBMs) or the surface-binding sites (SBSs) [18].

Cellulose-binding domain (CBD) was originally defined as noncatalytic polysaccharide-recognizing module of glycoside hydrolases (GHs). This module binds ligand such

as cellulose and the other carbohydrates. Afterward, the term of carbohydrate-binding module (CBM) was used to reflect the diverse ligand specificity of these modules [19]. Many CBMs have been identified experimentally, and hundreds of CBMs were further identified based on the amino acid similarity [20]. There are currently 81 defined families of CBMs (http://www.cazy.org/Carbohydrate-Binding-Modules.html), and these CBMs showed substantial variation in ligand specificity (Table 1).

CBM in starch-hydrolyzing enzymes is called starch-binding domain (SBD). SBDs have been identified in α-amylase, β-amylase, maltotetraohydrolase, maltopentaohydrolase, maltogenic α-amylase, cyclodextrin glucanotransferase (CGTase), acarbose transferase, and glucoamylase [14]. The illustrative view of classical SBD architectures is shown in Figure 1.

In general, the roles of CBM in the associated catalytic modules are in the proximity effect, the targeting function, and the disruptive function. Through this sugar-binding activity, the concentrated substrate on the surface of the protein can enhance the speed of degradation of polysaccharide [20].

There are three types of CBM regarding the form of substrates, types A, B, and C (Figure 2(a)). Type A binds to the crystalline surfaces of cellulose and chitin (e.g., CBM1, CBM2, CBM3, CBM5, and CBM10 families). Their binding sites are composed of many aromatic residues, creating a flat platform to bind to the planar polycrystalline chitin or cellulose surface (Figure 2(b)). Type B, which is currently the most abundant form of CBMs, binds to the internal glycan chains (endo-type). The type B binding sites formed as extended grooves or clefts comprised binding subsites to accommodate longer sugar chains (four or more monosaccharide units), for example, CBM6, CBM36, and CBM60. Lastly, type C binds to the termini of glycans (reducing/nonreducing ends, exo-type). This site appears as a small pocket which can recognize

Aspergillus niger glucoamylase X00712gb

| Glucoamylase | CBM20 |

Rhizopus oryzae glucoamylase D00049gb

| CBM21 | Glucoamylase |

Paenibacillus polymyxa α,β-amylase P21543gb

| β-Amylase | CBM25 | CBM25 | α-Amylase | AmyC |

Lactobacillus manihotivorans α-amylase AF126051gb

| α-Amylase | AmyC | CBM26 | CBM26 | CBM26 | CBM26 |

Thermoactinomyces vulgaris α-amylase D13177gb

| CBM34 | α-Amylase | AmyC |

Thermotoga maritima pullulanase AE001821gb

| CBM41 | α-Amylase | Pullulanase |

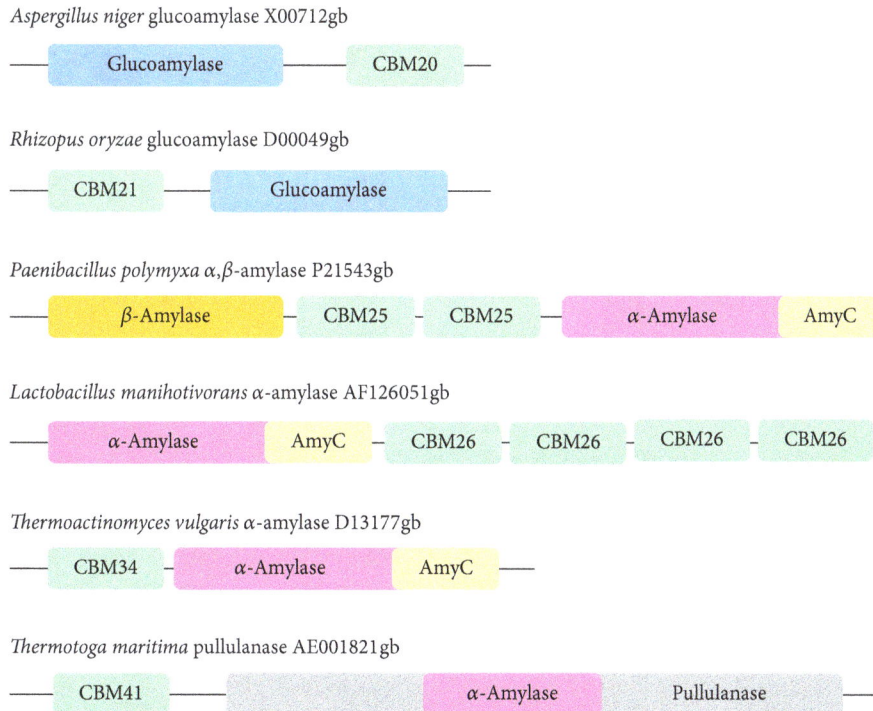

FIGURE 1: The classical architectures of starch-binding domains (SBDs), which are CBM20, CBM21, CBM25, CBM26, CBM34, and CBM41. SBDs are found at the N- or C-termini of the catalytic domain and are shown in turquoise colored boxes. The catalytic domains (CD) of glucoamylase, β-amylase, α-amylase, and pullulanase are highlighted in blue, yellow, purple, and grey colors, respectively. Accession numbers are retrieved from GenBank (adapted from Rodríguez-Sanoja et al. [12]).

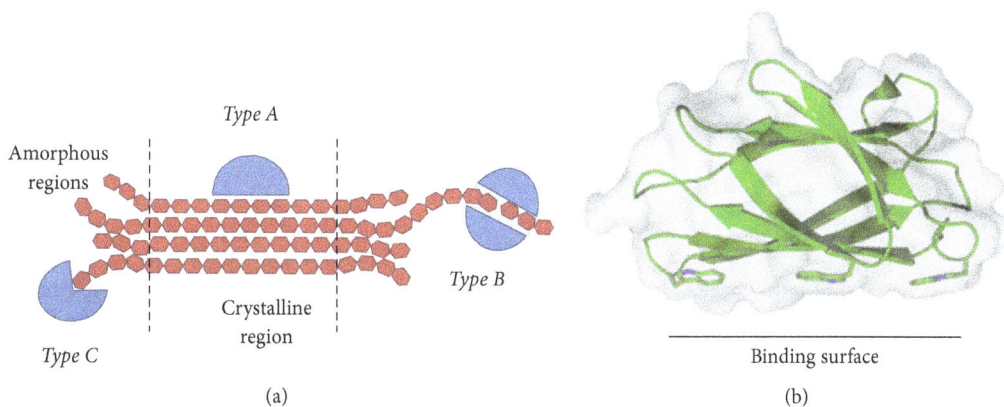

FIGURE 2: (a) Types A, B, and C of CBM bind to polysaccharides in a different region. (b) Type A of CBM2 from *Pyrococcus furiosus* (PDB ID code 2CRW [13]) shows that aromatic residues form a planar binding surface (adapted from [14]).

a short sugar ligand containing one to three monosaccharide units (e.g., CBM9, CBM13, CBM32, CBM47, CBM66, and CBM67 families) [12].

However, the noncatalytic carbohydrate-binding module does not only exist as CBM. A growing number of structural studies on various GHs have also revealed the presence of carbohydrates bound to one or more noncatalytic surface regions of the catalytic module. Carbohydrate-binding in such surface-binding sites, that is, SBSs, occurs in a fixed position relative to the catalytic site. It is different from the

noncatalytic binding in CBMs, which are usually attached to the flexible loop structure [6].

Starch granules possess crystalline and amorphous forms which are rigid and difficult to be degraded. Hence the strategy to enhance the catalytic efficiency is through the incorporation the SBSs in various enzymes. However, SBSs are restricted not only to starch-active enzymes, but also in other GHs with different specificities, belonging to several GH families, and originating from mammal, plant, archaea, fungi, and bacteria. Several functions of SBS in GHs are

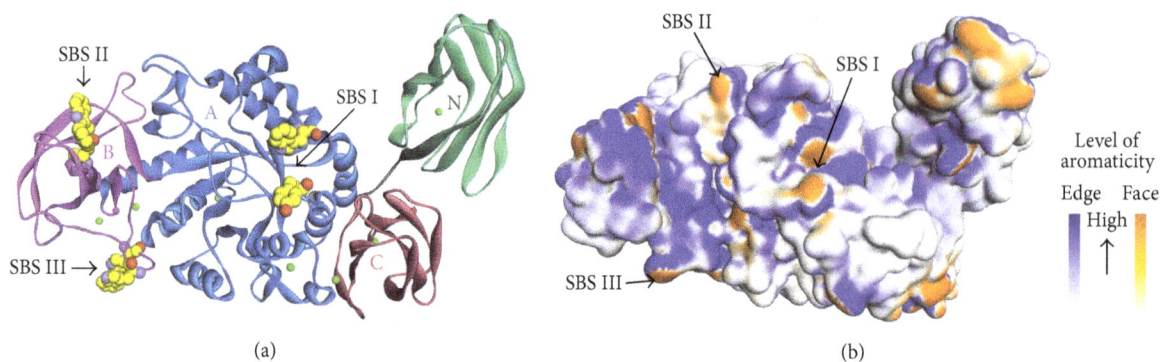

FIGURE 3: The overall structure of AmyB (PDB ID code 3BCD [15]). (a) Ribbon structure of AmyB with domains N, A, B, and C colored in green, blue, violet, and pink. Three SBSs are highlighted by a black arrow with a yellow sphere as the critical residues for binding. The eight metal ions are colored in green and purple balls. (b) Molecular surface of AmyB structure based on the aromaticity of residues. The face-side of aromatic residues forms three SBSs on the surface of AmyB.

(1) targeting towards its substrate, (2) assisting catalysis by loading substrates into the active-site pocket, (3) disrupting of the structure of substrates to facilitate catalysis, (4) keeping a substrate chain in contact with the enzyme for subsequent reactions, (5) allosteric activation of the enzyme, (6) retention and passing on the reaction products, and (7) anchoring the GH to the cell wall of the host microorganism [22–24].

CBM and SBS are crucial for starch binding. The differences between these binding sites are located on the architecture of binding. SBS is usually formed by aromatic residue on the surface of the enzyme. The importance of SBS to the starch adsorptivity in various α-amylases will be discussed below.

3. Surface-Binding Site in α-Amylase

3.1. Halothermothrix orenii α-Amylase B. Halothermothrix orenii is an anaerobic, halophilic, thermophilic, Gram-negative bacterium isolated from the sediment layer of a Tunisian salt lake in the Sahara desert. This bacterium experiences variations of salt concentration and temperature over time. The optimum pH, temperature, and salt (NaCl) concentration for the growth of *H. orenii* cells are 6.5–7.0, 60°C, and 1.7 M, respectively [25].

H. orenii produces two α-amylases, AmyA and AmyB. AmyB has an additional N-terminal domain (N domain) that forms a large groove, the N–C groove, located around 30 Å away from the active site. This N domain is important for hydrolyzing the insoluble starch by improving the binding ability of AmyB to the insoluble substrate [15].

AmyB consists of three domains, A, B, and C domain (Figure 3(a)). The A domain features the typical (β/α) 8 TIM barrel. The active site is located at the C-terminal end of the TIM barrel, composed of D350, E380, and D447 as the catalytic residues. The B domain is located between the strand $\beta3$ and the helix $\alpha3$ of the A domain. The interaction between A and B domain is also stabilized by the presence of a metal triad (Ca^{2+}–Na^{+}–Ca^{2+}). Lastly, the C domain folds as a C-terminal eight-stranded β sandwich,

following the α/β-barrel. The N domain folds into a nine-stranded immunoglobulin-like β sandwich of fibronectin III type. Although the A domain forms extensive interdomain interactions with B and C domain, it has limited interactions with the N domain [15].

Two structures of AmyB have been deposited in the Protein Data Bank. The first structure was complexed with acarbose ($AmyB_{acr}$), whereas the second one was complexed with maltoheptaose/cyclodextrin ($AmyB_{mal7-acx}$). Three SBSs were found in the crystal structures: two SBSs in the acarbose-bound complex and another SBS in the maltoheptaose/cyclodextrin-bound complex [15]. Several aromatic residues were found on the surface of this structure (Figure 3(b)).

A tetrasaccharide was present in the SBS I site of $AmyB_{acr}$ and $AmyB_{mal7-acx}$. Two aromatic residues, W488 and Y460, formed CH/pi-stacking interactions with Glc3 and Glc4, respectively. There are also ten potential hydrogen bonds, that is, E588 with O4 and O3 of Glc1, K463 with O2 of Glc2, R462 with O3 of Glc1, I459 with O3 of Glc2, S458 with O2 and O3 of Glc3, D449 with O3 of Glc3, and W488 with O6 of Glc2 (Figure 4(a)). In SBS II, a β-cyclodextrin binds to the $AmyB_{mal7-acx}$. Two tryptophans were found on this site, W287 and W260 that formed CH/pi-stacking interactions with Glc1 and Glc2, respectively. There are also four potential hydrogen bonds, W260 with O5 of Glc2, A237 with O2 of Glc3, K198 with O3 of Glc3, and M176 with O2 of Glc4 (Figure 4(b)). In SBS III, a glucose binds to the $AmyB_{acr}$. Two tryptophans, W310 and W306, formed stacking interactions with the glucose. There are also two potential hydrogen bonds, T307 with O6 of Glc1 and D311 with O5 of Glc1 (Figure 4(c)).

Interestingly, the N and C grooves also contained aromatic residues that could interact favorably with carbohydrates. The deletion of N domain decreased the starch degradation performance of AmyB as compared to the full-length sequence. This result suggests the vital role of N domain to sequester and to render the natural starch to be more accessible for further processing and hydrolysis [15].

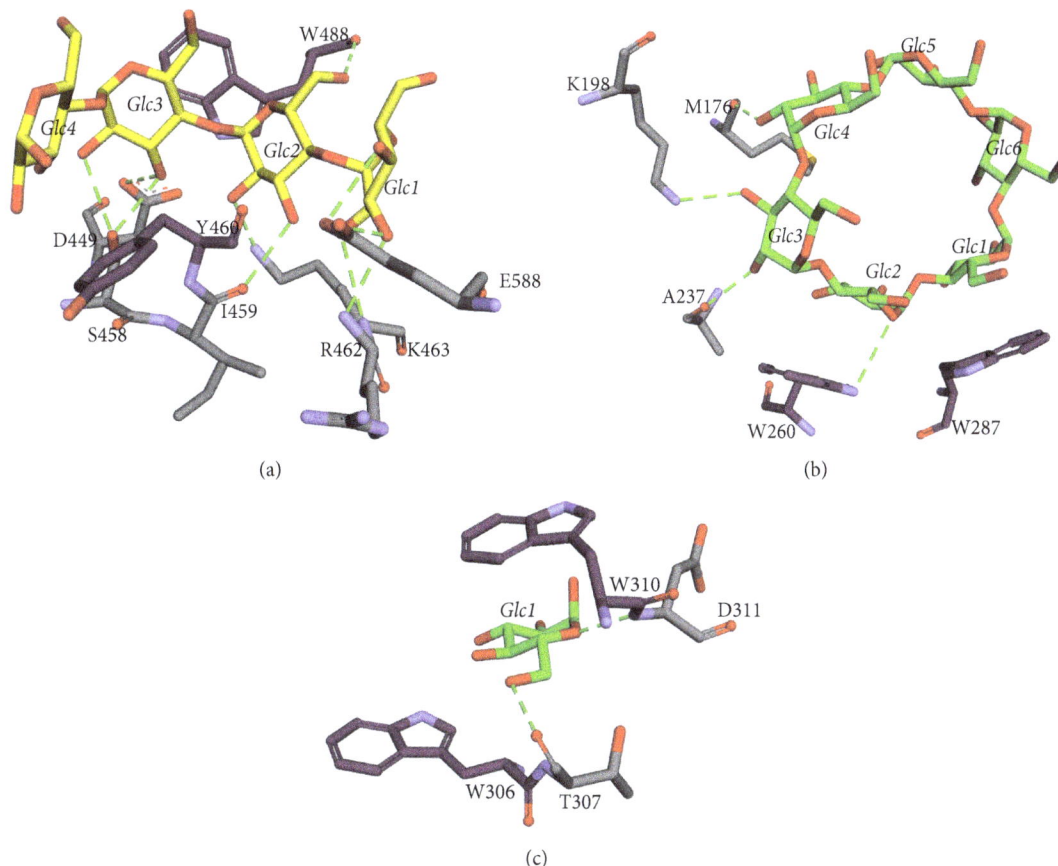

FIGURE 4: The molecular interactions around (a) SBS I, (b) SBS II, and (c) SBS III. The tetrasaccharide, β-cyclodextrin, and glucose are represented in yellow and green colored sticks, respectively. Aromatic residues and the other amino acids around the substrate that formed hydrogen bonds are shown in dark purple and grey colored sticks, respectively. A hydrogen bond is depicted in green dashed line.

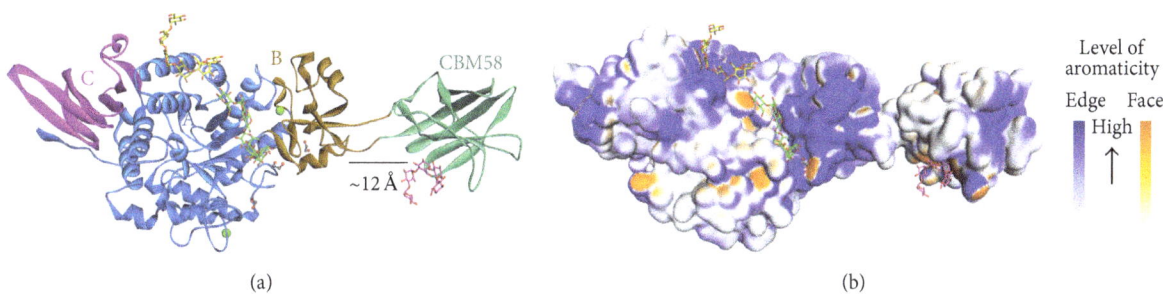

FIGURE 5: The overall structure of SusG (PDB ID code 3K8L). (a) The ribbon structure of SusG. The A, B, and C domain and CBM58 are colored in blue, brown, purple, and green, respectively. The metal ions are displayed as a green sphere, and those of ethylene glycol molecules are in grey. The maltoheptaose is represented differently based on its location at the active site, the secondary starch-binding site (SBS), and CBM58 (green, yellow, and pink colored sticks, resp.). (b) The molecular surface of SusG structure based on the edge- and face-side of aromatic residues [16].

3.2. SusG Bacteroides thetaiotaomicron α-Amylase. SusG (starch utilization system G) *Bacteroides thetaiotaomicron α-*amylase is part of a large protein complex on the outer surface of the bacterial cell. It plays a significant role in carbohydrate acquisition by the animal gut microbiota. SusG is expressed concurrently with Sus-CDEF on the outer surface of the cell and is required for cell growth on starch [16].

The structure of SusG is composed of A, B, and C domain that share structural features with the other α-amylases (Figure 5(a)). The A domain contains the catalytic site, with the B domain inserted between β3 and α3 of the A domain. The B domain contributes to the size and accessibility of the active site, whereas the C domain is a standard feature of many GH13 family enzymes. SusG displays an unusual

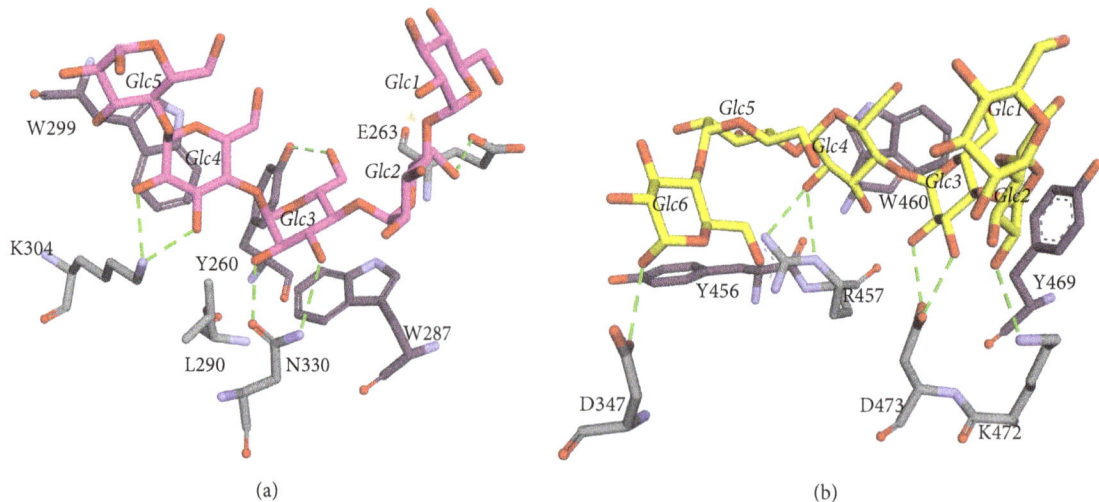

(a)

(b)

FIGURE 6: Molecular interaction around the substrate-binding site. (a) The binding of maltopentaose (pink) to the CBM58. (b) The binding of maltoheptaose (yellow) to the SBS. Aromatic residues are visualized in a darker color. Hydrogen bonds denoted by green dashed lines.

extended shape, ~12 Å in length, due to the insertion of a CBM58 that protrudes from the B domain. CBM58 makes no direct contact with the A, B, and C domain and it is linked to the core of amylase structure by two short linkers, located 12 Å away from the B domain. Naturally, these linkers are not flexible and do not directly interact with each other, either the core domains or the CBM58. They have a few potentials of interdomain water-mediated hydrogen bonds. SusG also has a secondary starch-binding site in the A domain, which is similar to the SBS [16]. Based on the aromaticity of residues on the surface of SusG, it is shown that the aromatic residues are spread around the active site and starch-binding site or SBS (Figure 5(b)).

Five glucose residues of maltoheptaose are well ordered at the CBM58. In this binding site, there are two CH/pi-stacking interactions between W287 and W299 to Glc3 and Glc4, respectively. The L290 formed hydrophobic interaction with both tryptophans. W299 has potentially formed a T-shape stacking interaction with Y260. Besides, there are also six potential hydrogen bonds: E263 with O6 of Glc2, N330 with O2 and O3 of Glc3, Y260 with O6 of Glc3, and K304 with O2 and O3 of Glc4 (Figure 6(a)). The pattern of starch binding at the CBM58 is comprised of hydrophobic interactions with the additional hydrogen bonding to the $2'$ and $3'$ hydroxyl groups of the adjacent glucose residues. This pattern is a conserved feature of many starch-binding CBMs [26]. In addition, this binding pattern is also observed in SusD [27], barley, and pancreatic α-amylases that bind raw starch on the surface of the catalytic domain [28, 29].

In addition to CBM, the SBS in SusG also has a similar characteristic. It contains tryptophan and tyrosine in the binding site. The Y469 formed CH/pi stacking with Glc2 and W460 formed stacking with Glc4. It is also noted that six potential hydrogen bonds were formed: D437 with O1 of Glc6, R457 with O2 of Glc4, D473 with O2 and O3 of Glc3, and K472 with O2 of Glc2 (Figure 6(b)).

Some mutation studies of this enzyme revealed that stacking interaction is essential to the starch-binding. The

first mutant of SusG lacking CBM58, namely, mCBM58, was generated by deleting residues 210–339 and inserting the five residues loop GSPTG, similar to that observed in the *H. orenii* amylase A, a close structural homolog of SusG without CBM58. The second mutant, namely, mSURF, was constructed by mutating the surface-binding site (W460A/Y469A/D473V) to test the importance of these residues to the starch-binding capability. The mCBM58, mSURF, and WT SusG enzymes were tested for their enzymatic activity using p-nitrophenyl-maltopentaose (PNP-G5). Their catalytic turnover rates were identical. The enzymes were then tested for their ability to degrade the soluble starch, amylopectin, pullulan, and insoluble cornstarch. For each substrate, the activity of WT SusG was used as the positive control (100%), and the mCBM58 and mSURF mutant enzymes were compared to the wild-type. The mCBM58 showed the highest activity to all substrates except for the insoluble cornstarch in which the activity was remarkably decreased up to 71%, whereas mSURF had the lowest activity for all substrates. Interestingly, its activity on the insoluble corn starch was also decreased up to 56%. Therefore, both the CBM58 and the SBS are required for the optimal degradation of insoluble corn starch [16].

3.3. Barley α-Amylase Isozyme 1. Barley α-amylase isozymes (AMY1 and AMY2) of subfamily GH13_6 [2] are among the first carbohydrate-active enzymes identified with the SBS [24, 30]. Although the SBS was first discovered in AMY2, the characterization of functional properties of these SBSs was performed on the AMY1. The reason was due to the higher yields of recombinant AMY1 produced by *Pichia pastoris,* which was about 60-fold higher than AMY2 [31]. Moreover, another preliminary work indicated that the starch binding to SBS2 in AMY2 is weaker than in AMY1. This finding was also confirmed by its crystal structure [32].

Similar to the other α-amylases, AMY1 has A, B, and C domain. The A/B domain consists of the catalytic domain, while the C domain is a common feature of many GH13 family

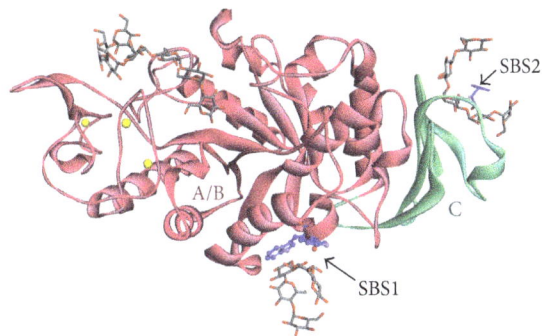

FIGURE 7: The ribbon structure of AMY1. The A/B and C domains are colored in pink and green, respectively. The metal ion is colored in yellow spheres. The maltopentaose and maltohexaose (grey sticks) bind to the SBS and active site, respectively.

enzymes. It is worth noting that CBM is not present in AMY1. However, two SBSs exist: SBS1 and SBS2 (Figure 7).

In SBS1, two aromatic residues interact with the maltopentaose, that is, W278 and W279. These tryptophans formed CH/pi-stacking interactions with Glc3 and Glc4, respectively. There are also five potential hydrogen bonds around this site, that is, Q227 with O2 and O3 of Glc4, the backbone of W278 with O6 of Glc3, and D234 with O2 and O3 of Glc3 (Figure 8(a)).

In SBS2, there is an aromatic residue which formed CH/pi-stacking interaction with maltopentaose, that is, Y380. There are also ten potential hydrogen bonds around this site, that is, V382 with O2 of Glc2, D381 with O3 of Glc2, Y380 with O2 of Glc2 and with O3 of Glc3, K375 with O2 of Glc3, D398 with O3 of Glc4, G397 with O6 of Glc4, H395 with O6 of Glc4, and T392 with O6 of Glc2 (Figure 8(b)).

SBS1 is known as starch granule binding site, and SBS2 is known as a pair of sugar tongs [18]. Nielsen and colleagues have performed the mutation of Y380A in the SBS2 [24]. As a result, its activity decreased about tenfold (Kd = 1.4 mg/mL) as compared to the wild-type AMY1. The mutant retained less than half of the activity to release the soluble reducing sugars from starch granules. Furthermore, it was noticed that these effects were more prominent for single or double SBS1 alanine mutants of W278 and W279. The complete loss of affinity for barley starch granules (Kd > 100 mg/mL) resulted when both of SBSs were modified using triple mutations W278A/W279A/Y380A. This mutant retained only 0.2% of the wild-type hydrolytic activity towards barley starch granules [24]. In contrast, both affinity and rates of hydrolysis were increased roughly tenfold when a starch-binding domain of the CBM20 family from *Aspergillus niger* glucoamylase was fused with the C-terminal of AMY1 [31].

The architecture of both SBSs corresponds to their distinct roles. A binding platform in SBS1 comprised two tryptophans, whereas the "pair of sugar tongs" in SBS2 formed by Y380 and H395, which are positioned to accommodate an individual chain of the substrate. SBS1 is suggested as the initial site for AMY1 attachment to the starch granule surface. SBS2 is a supporting site for substrate binding near the α-1,6 branch point. Thus, it feeds a linear segment of

the amylopectin into the active site, which is unable to accommodate branches near the point of hydrolysis. Once AMY1 inserts the starch granule surface, the role of SBS1 in the catalytic activity would be over. In contrast, SBS2 is continuously isolating the individual chains to be delivered to the active site [33].

3.4. Aspergillus niger α-Amylase. *Aspergillus niger* α-amylase is classified as a member of GH family 13 among the 109 GH families that are currently identified. Its sequence is 100% identical to the *A. oryzae* homolog. Its crystal structure with a resolution of 3.0 Å was reported in 1984 (PDB ID code 2TAA, [34]) and known as TAKA-amylase [17].

A. niger α-amylase in complex with maltose, the simplest substrate of this enzyme, has been published with a PDB ID code 2GVY at 1.6 Å resolution. This structure consists of four maltose molecules bound on the protein surface composed of aromatic residues (Figure 9). It is found that the two maltoses were in unusual position when compared to the acarbose in TAKA-amylase (PDB ID code 7TAA). The structure of this enzyme has a typical α-amylase structure with A, B, and C domain: A/B domain as a catalytic module and C domain as a standard feature like the other α-amylases (Figure 10). Three molecules of maltose were found in the active site in subsite −1 and −2, +1 and +2, and +4 and +5. Another maltose was found in 20 Å distance from subsite +5. This site was later known as the SBS, which is located on a loop between A and C domain. Its function is to bind the polysaccharide chain extending from the active site. The plasticity of the active-site groove in the proximity to the catalytic center might be substantial for both formations of the productive substrate-enzyme complex as well as for the release of the product from the +1 to +n subsites [17].

The M4 molecule (maltose) formed hydrophobic stacking interactions with Y382 and W385 which are located on the loop connecting the last helix of the TIM barrel and the first-strand of the C domain (Figure 11). These sites were involved in the binding of a long carbohydrate chain extending from the active site. In addition, R397 was found to stabilize the two aromatic residues with hydrophobic interaction.

3.5. Saccharomycopsis fibuligera α-Amylase. *S. fibuligera* is a food-borne yeast that is widely used in the production of rice or cassava-based fermented food [35]. The yeast, in combination with *Saccharomyces cerevisiae* or *Zymomonas mobilis*, has been used in the production of ethanol using cassava starch as the starting material [36].

One of the best strains of this yeast, *S. fibuligera* R64, produces two amylolytic enzymes: α-amylase (Sfamy) and glucoamylase (GluR) [37]. Sfamy has an optimum temperature of 50°C and is active in a broad pH range with an optimum pH of 5.0. The digestion of native Sfamy with trypsin resulted in two major fragments with apparent molecular masses of 39 kDa (p39) and 10 kDa (p10), respectively. The two fragments represent the N- and C-terminal domains of the α-amylase. According to Matsuura et al. [34], the N-terminal domain of α-amylase consists of the integrated A and B domains, in which the active site is located. The C-terminal

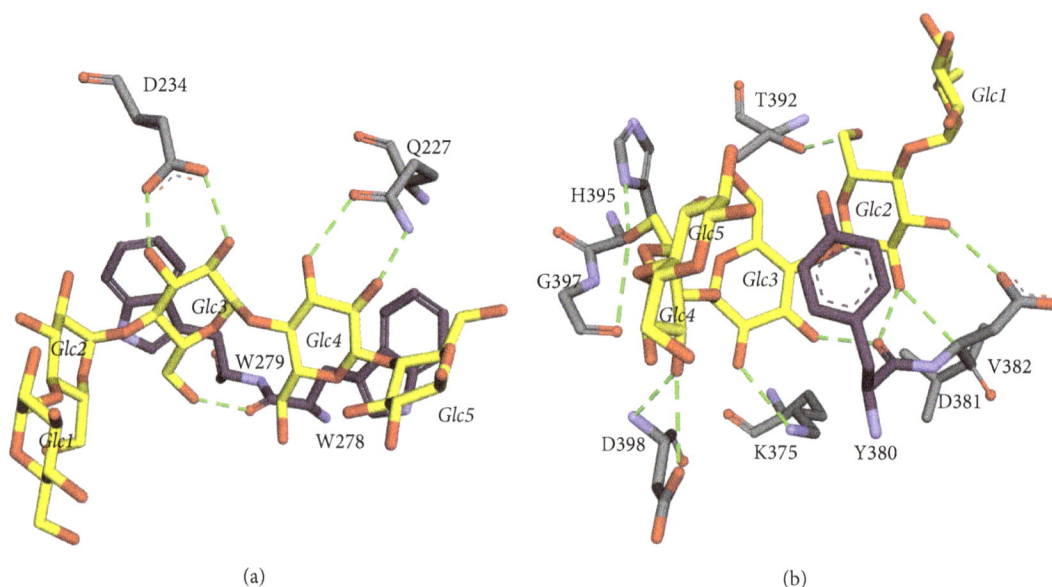

FIGURE 8: The 3D interaction of maltoheptaose (yellow) bound to the (a) SBS1 and (b) SBS2. Hydrogen bonds and aromatic residues are denoted by green dashed lines and dark violet sticks, respectively.

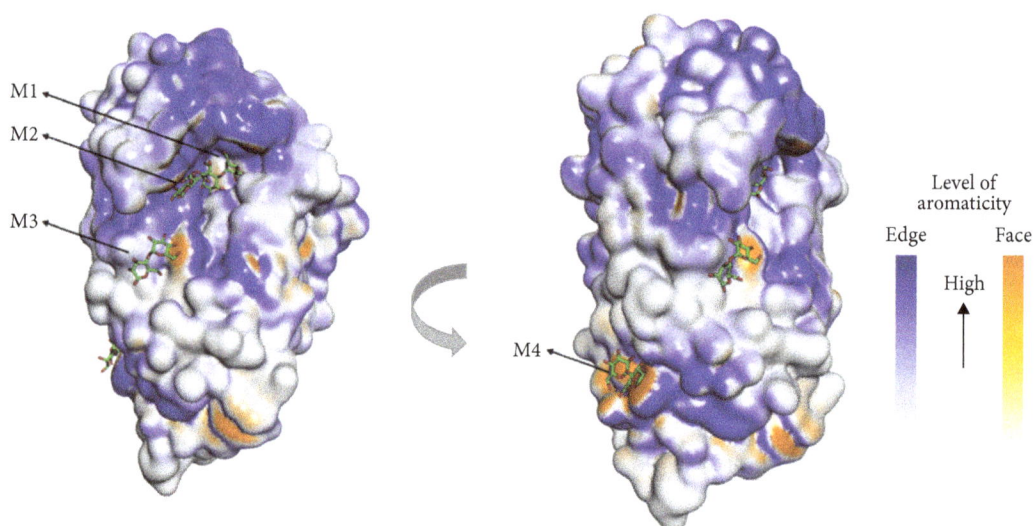

FIGURE 9: The molecular surface of *A. niger* α-amylase based on the aromaticity of amino acid. Four maltoses (substrates) bound to the surface, rich in aromatic residues.

domain consists of C domain, in which its function in Sfamy is not yet established [38].

Hasan et al. [38] reported that Sfamy has no starch binding as compared to the GluR, which has the adsorption level of 90%, 80%, 25%, and 20% to the maize, tapioca, sago, and potato starches, respectively [38].

A computational study on the differences between Sfamy and *A. niger* α-amylase was conducted as an effort to understand the low adsorptivity of Sfamy on the raw starch [39]. The sequence and homology model of Sfamy were aligned to that of *A. niger* α-amylase (PDB ID code 2GVY) [17]. The sequence of Sfamy was retrieved from NCBI with accession code HQ172905.1 [40]. As a result, these sequences

shared 54% identity and 71% homology. At the SBS region, Sfamy has two serines, while *A. niger* α-amylase has two aromatic residues (Figure 12). This difference was suggested as the reason of the low adsorptivity of Sfamy on the raw starch. Although the two serines could form hydrogen bonds with the substrate, which usually occurred in the starch-binding process, they might not be strong enough to hold the substrate on the enzyme's surface.

Furthermore, molecular dynamics simulations were performed on the structure of Sfamy and *A. niger* α-amylase to investigate their time-dependent structural behavior of substrate binding. The substrate in Sfamy was not consistently bound to the SBS region, while that in *A. niger* α-amylase

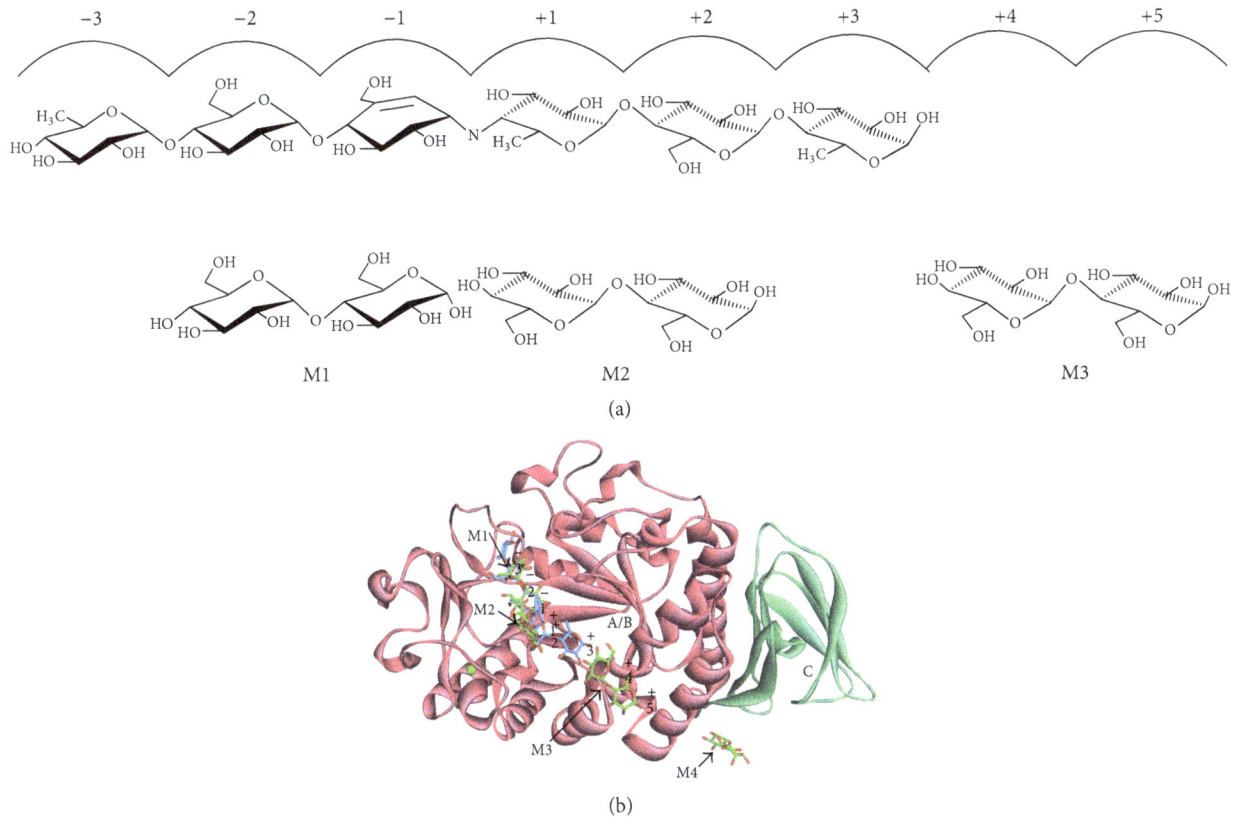

FIGURE 10: The overall structure of *A. niger* α-amylase and its ligand. (a) The comparison of subsites for acarbose and maltose binding (adapted from Vujicic-Zagar and Dijkstra [17]). (b) The structure of *A. niger* α-amylase. The A/B and C domains are colored in pink and turquoise, respectively. Acarbose and maltose are represented by blue and green colored sticks, respectively. A cofactor calcium ion is visualized by a green sphere.

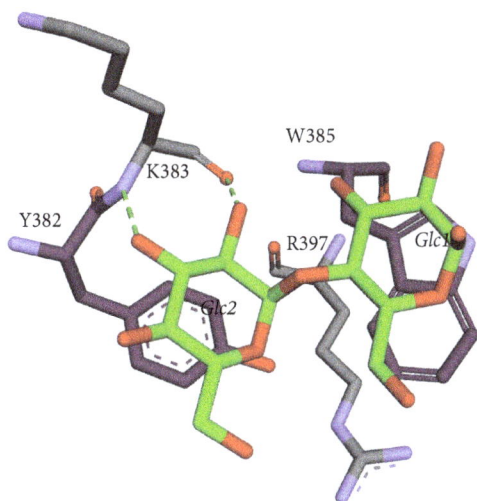

FIGURE 11: Molecular interactions around the SBS of *A. niger* α-amylase. Maltose is represented in green stick, aromatic residue in dark purple stick, and hydrogen bond in green dashed lines.

was stable over the simulation. Interestingly, a double mutant of S383Y/S386W of Sfamy showed a comparable substrate-binding activity to that of *A. niger*'s. These introduced

aromatic residues formed CH/pi-stacking interaction with the substrate [39].

In general, the interaction between CBM and carbohydrate is weak (Ka affinities in mM^{-1} to μM^{-1} range), hence making the interaction easily reversible. Once catalysis has been completed at the particular site, there is "recycling" of the appended enzyme to bind to a new region on the substrate [41]. It is suggested that the most important driving force mediating the protein-carbohydrate interactions is the position and orientation of aromatic residues within the SBS, such as tyrosine, tryptophan, or phenylalanine. These planar residues formed essential hydrophobic stacking interactions with the planar face of sugar rings. Moreover, it was noted that weak intermolecular electrostatic interactions, which occurred between CH and pi electrons in the planar ring systems, contributed around 1.5 to 2.5 kcal/mol energy to the binding reaction [42]. However, the geometric features of the interaction are not strictly unique. From the point of view of the protein structure, different architectures of the binding sites can be described, depending on the number and relative location of aromatic residues [41]. In Protein Data Bank, more than 90 of nonredundant 3D structures of CBD show carbohydrate aromatic stacking. This type of interaction has resulted in the improvement of protein modeling strategies, especially those that are of a low similarity, by introducing

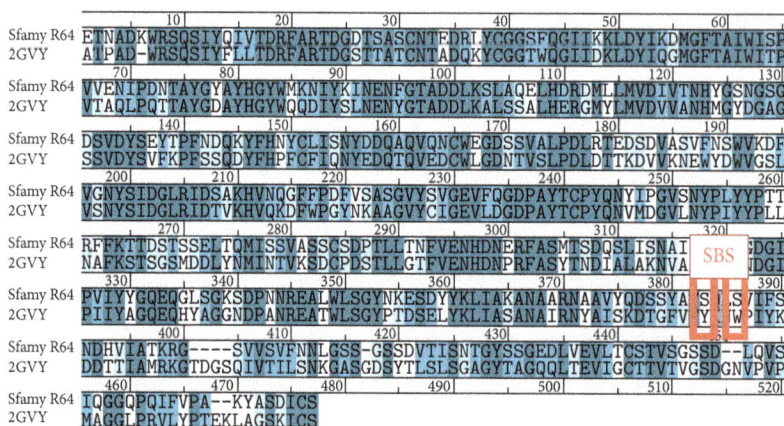

FIGURE 12: Sequence alignment between Sfamy R64 and *A. niger* α-amylase. The SBS is highlighted by red colored box.

a "hydrophilic aromatic residue" parameter as a restriction for structural modeling [43].

4. Conclusion

Starch binding in α-amylases, with or without SBS, is shown to be influenced by the presence of CH/pi-stacking interaction. This interaction occurs between aromatic residues (tyrosine, tryptophan, and sometimes phenylalanine) and the partial positively charged hydrogen atom of the substrate. These aromatic residues should have a specific topology to bind well to the substrate. Also, their conformations have to be stable (e.g., stabilized by hydrophobic interaction around aromatic residue). The CBM or SBS does not significantly influence the catalytic activities towards the short chain polysaccharides, but they are essential to hydrolyze the long or insoluble starch (raw starch). Therefore, the presence of SBS should be considered as the critical aspect of improving the starch adsorptivity of α-amylase.

Acknowledgments

This work was supported by Universitas Padjadjaran and Ministry of Research, Technology and Higher Education of the Republic of Indonesia through Internal Grant and PPTI, respectively.

References

[1] M. J. van der Maarel, B. van der Veen, J. C. Uitdehaag, H. Leemhuis, and L. Dijkhuizen, "Properties and applications of starch-converting enzymes of the α-amylase family," *Journal of Biotechnology*, vol. 94, no. 2, pp. 137–155, 2002.

[2] M. R. Stam, E. G. J. Danchin, C. Rancurel, P. M. Coutinho, and B. Henrissat, "Dividing the large glycoside hydrolase family 13 into subfamilies: towards improved functional annotations of α-amylase-related proteins," *Protein Engineering, Design and Selection*, vol. 19, no. 12, pp. 555–562, 2006.

[3] J. F. Kennedy, V. M. Cabalda, and C. A. White, "Enzymic starch utilization and genetic engineering," *Trends in Biotechnology*, vol. 6, no. 8, pp. 184–189, 1988.

[4] R. M. Gibson and B. Svensson, "Identification of tryptophanyl residues involved in binding of carbohydrate ligands to barley α-amylase 2," *Carlsberg Research Communications*, vol. 52, no. 6, pp. 373–379, 1987.

[5] S. Bozonnet, B. C. Bonsager, B. Kramhoft et al. et al., "Binding of carbohydrates and protein inhibitors to the surface of alpha-amylases," *Biologia*, vol. 60, no. 17, pp. 27–36, 2005.

[6] S. Cuyvers, E. Dornez, J. A. Delcour, and C. M. Courtin, "Occurrence and functional significance of secondary carbohydrate binding sites in glycoside hydrolases," *Critical Reviews in Biotechnology*, vol. 32, no. 2, pp. 93–107, 2012.

[7] L. M. Hamilton, C. T. Kelly, and W. M. Fogarty, "Purification and properties of the raw starch-degrading α-amylase of Bacillus sp. IMD 434," *Biotechnology Letters*, vol. 21, no. 2, pp. 111–115, 1999.

[8] P. Nahid, M. Vossoughi, R. Roostaazad, M. Ahmadi, A. Zarrabi, and S. M. Hosseini, "Production of glucoamylase by aspergillus niger under solid state fermentation," *International Journal of Engineering, Transactions B: Applications*, vol. 25, no. 1, pp. 1–7, 2012.

[9] H. Sun, P. Zhao, X. Ge et al., "Recent advances in microbial raw starch degrading enzymes," *Applied Biochemistry and Biotechnology*, vol. 160, no. 4, pp. 988–1003, 2010.

[10] N. Goyal, J. K. Gupta, and S. K. Soni, "A novel raw starch digesting thermostable α-amylase from *Bacillus* sp. I-3 and its use in the direct hydrolysis of raw potato starch," *Enzyme and Microbial Technology*, vol. 37, no. 7, pp. 723–734, 2005.

[11] R.-J. Shiau, H.-C. Hung, and C.-L. Jeang, "Improving the thermostability of raw-starch-digesting amylase from a Cytophaga sp. by site-directed mutagenesis," *Applied and Environmental Microbiology*, vol. 69, no. 4, pp. 2383–2385, 2003.

[12] R. Rodríguez-Sanoja, N. Oviedo, and S. Sánchez, "Microbial starch-binding domain," *Current Opinion in Microbiology*, vol. 8, no. 3, pp. 260–267, 2005.

[13] H. Zhang, P. F Hayashi, and S. Yokoyama, "Solution Structure of the ArfGap Domain of ADP-Ribosylation Factor GTPaseactivating Protein 3 (ArfGap 3) Structure Summary Page," 2005.

[14] H. J. Gilbert, J. P. Knox, and A. B. Boraston, "Advances in understanding the molecular basis of plant cell wall polysaccharide

recognition by carbohydrate-binding modules," *Current Opinion in Structural Biology*, vol. 23, no. 5, pp. 669–677, 2013.

[15] T.-C. Tan, B. N. Mijts, K. Swaminathan, B. K. C. Patel, and C. Divne, "Crystal structure of the polyextremophilic α-Amylase AmyB from halothermothrix orenii: details of a productive enzyme-substrate complex and an n domain with a role in binding raw starch," *Journal of Molecular Biology*, vol. 378, no. 4, pp. 850–868, 2008.

[16] N. M. Koropatkin and T. J. Smith, "SusG: a unique cell-membrane-associated α-amylase from a prominent human gut symbiont targets complex starch molecules," *Structure*, vol. 18, no. 2, pp. 200–215, 2010.

[17] A. Vujicic-Zagar and B. W. Dijkstra, "Monoclinic crystal form of aspergillus niger a -amylase in complex with maltose at 1 . 8 A resolution," *Protein Structure Communications*, vol. 62, pp. 716–721, 2006.

[18] D. Cockburn, C. Wilkens, C. Ruzanski et al., "Analysis of surface binding sites (SBSs) in carbohydrate active enzymes with focus on glycoside hydrolase families 13 and 77—a mini-review," *Biologia (Poland)*, vol. 69, no. 6, pp. 705–712, 2014.

[19] O. Shoseyov, Z. Shani, and I. Levy, "Carbohydrate binding modules: biochemical properties and novel applications," *Microbiology and Molecular Biology Reviews*, vol. 70, no. 2, pp. 283–295, 2006.

[20] A. B. Boraston, D. N. Bolam, H. J. Gilbert, and G. J. Davies, "Carbohydrate-binding modules: fine-tuning polysaccharide recognition," *Biochemical Journal*, vol. 382, no. 3, pp. 769–781, 2004.

[21] J. Barchiesi, N. Hedin, D. F. Gomez-Casati, M. A. Ballicora, and M. V. Busi, "Functional demonstrations of starch binding domains present in Ostreococcus tauri starch synthases isoforms," *BMC Research Notes*, vol. 8, no. 1, article 1598, 2015.

[22] Z. Ye, H. Miyake, M. Tatsumi, S. Nishimura, and Y. Nitta, "Two additional carbohydrate-binding sites of β-amylase from bacillus cereus var. Mycoides are involved in hydrolysis and raw starch-binding," *The Journal of Biochemistry*, vol. 135, no. 3, pp. 355–363, 2004.

[23] J. Ševčík, E. Hostinová, A. Solovicová, J. Gašperík, Z. Dauter, and K. S. Wilson, "Structure of the complex of a yeast glucoamylase with acarbose reveals the presence of a raw starch binding site on the catalytic domain," *FEBS Journal*, vol. 273, no. 10, pp. 2161–2171, 2006.

[24] M. M. Nielsen, S. Bozonnet, E.-N. Seo et al., "Two secondary carbohydrate binding sites on the surface of barley α-amylase 1 have distinct functions and display synergy in hydrolysis of starch granules," *Biochemistry*, vol. 48, no. 32, pp. 7686–7697, 2009.

[25] J.-L. Cayol, B. Ollivier, B. K. C. Patel, G. Prensier, J. Guezennec, and J. L. Garcia, "Isolation and characterization of Halothermothrix orenii gen. nov., sp. nov., a halophilic, thermophilic, fermentative, strictly anaerobic bacterium," *International Journal of Systematic Bacteriology*, vol. 44, no. 3, pp. 534–540, 1994.

[26] A. B. Boraston, M. Healey, J. Klassen, E. Ficko-Blean, A. L. Van Bueren, and V. Law, "A structural and functional analysis of α-glucan recognition by family 25 and 26 carbohydrate-binding modules reveals a conserved mode of starch recognition," *The Journal of Biological Chemistry*, vol. 281, no. 1, pp. 587–598, 2006.

[27] N. M. Koropatkin, E. C. Martens, J. I. Gordon, and T. J. Smith, "Starch catabolism by a prominent human gut symbiont is directed by the recognition of amylose helices," *Structure*, vol. 16, no. 7, pp. 1105–1115, 2008.

[28] M. Qian, R. Haser, and F. Payan, "Carbohydrate binding sites in a pancreatic α-amylase-substrate complex, derived from X-ray structure analysis at 2.1 Å resolution," *Protein Science*, vol. 4, no. 4, pp. 747–755, 1995.

[29] X. Robert, R. Haser, H. Mori, B. Svensson, and N. Aghajari, "Oligosaccharide binding to barley α-amylase 1," *The Journal of Biological Chemistry*, vol. 280, no. 38, pp. 32968–32978, 2005.

[30] S. Bozonnet, M. T. Jensen, M. M. Nielsen et al., "The 'pair of sugar tongs' site on the non-catalytic domain C of barley α-amylase participates in substrate binding and activity," *FEBS Journal*, vol. 274, no. 19, pp. 5055–5067, 2007.

[31] N. Juge, J. Nøhr, M.-F. Le Gal-Coëffet et al., "The activity of barley α-amylase on starch granules is enhanced by fusion of a starch binding domain from Aspergillus niger glucoamylase," *Biochimica et Biophysica Acta (BBA)—Proteins and Proteomics*, vol. 1764, no. 2, pp. 275–284, 2006.

[32] E.-S. Seo, C. Christiansen, M. Abou Hachem et al., "An enzyme family reunion—similarities, differences and eccentricities in actions on α-glucans," *Biologia*, vol. 63, no. 6, pp. 967–979, 2008.

[33] D. Cockburn, M. M. Nielsen, C. Christiansen et al., "Surface binding sites in amylase have distinct roles in recognition of starch structure motifs and degradation," *International Journal of Biological Macromolecules*, vol. 75, pp. 338–345, 2015.

[34] Y. Matsuura, M. Kusunoki, W. Harada, and M. Kakudo, "Structure and possible catalytic residues of Taka-amylase A," *The Journal of Biochemistry*, vol. 95, no. 3, pp. 697–702, 1984.

[35] S. Endang, K. Hidehiko, and K. Yamamoto, "Study on amyloglucosidase of a newly isolated saccharomycopsis Sp. TJ-1 from the indonesian fermented food (Tape)," *AnnalesBogorienses*, vol. 5, no. 2, pp. 77–83, 1998.

[36] Z. Chi, Z. Chi, G. Liu, F. Wang, L. Ju, and T. Zhang, "Saccharomycopsis fibuligera and its applications in biotechnology," *Biotechnology Advances*, vol. 27, no. 4, pp. 423–431, 2009.

[37] W. T. Ismaya, K. Hasan, and T. Subroto, *Chromatography as the Major Tool in the Identification and the Structure-Function Relationship Study of Amylolytic Enzymes from Saccharomycopsis Fibuligera R64*, InTech, 2012.

[38] K. Hasan, W. Tirta Ismaya, I. Kardi et al., "Proteolysis of α-amylase from Saccharomycopsis fibuligera: characterization of digestion products," *Biologia*, vol. 63, no. 6, pp. 1044–1050, 2008.

[39] M. Yusuf, U. Baroroh, K. Hasan, S. D. Rachman, S. Ishmayana, and T. Subroto, "Computational model of the effect of a surface-binding site on the saccharomycopsis fibuligera R64 α-amylase to the substrate adsorption," *Bioinformatics and Biology Insights*, vol. 11, pp. 1–8, 2017.

[40] S. Gaffar, W. T. Ismaya, K. Hasan et al. et al., "Saccharomycopsis Fibuligera Isolate R64 Alpha-Amylase (Amy) Gene, Complete Cds," 2011.

[41] J. L. Asensio, A. Ardá, F. J. Cañada, and J. Jiménez-Barbero, "Carbohydrate-aromatic interactions," *Accounts of Chemical Research*, vol. 46, no. 4, pp. 946–954, 2013.

[42] E. A. Meyer, R. K. Castellano, and F. Diederich, "Interactions with aromatic rings in chemical and biological recognition," *Angewandte Chemie International Edition*, vol. 42, no. 11, pp. 1210–1250, 2003.

[43] W.-Y. Chou, T.-W. Pai, T.-Y. Jiang, W.-I. Chou, C.-Y. Tang, and M. D.-T. Chang, "Hydrophilic aromatic residue and in silico structure for carbohydrate binding module," *PLoS ONE*, vol. 6, no. 9, Article ID e24814, 2011.

4

Highly Active and Stable Large Catalase Isolated from a Hydrocarbon Degrading *Aspergillus terreus* MTCC 6324

Preety Vatsyayan[1] and Pranab Goswami[2]

[1]*Institute of Analytical Chemistry, Chemo- and Biosensors, University of Regensburg, 93053 Regensburg, Germany*
[2]*Department of Biosciences and Bioengineering, Indian Institute of Technology Guwahati, Guwahati, Assam 781039, India*

Correspondence should be addressed to Preety Vatsyayan; preety.vatsyayan@chemie.uni-regensburg.de

Academic Editor: John David Dignam

A hydrocarbon degrading *Aspergillus terreus* MTCC 6324 produces a high level of extremely active and stable cellular large catalase (CAT) during growth on *n*-hexadecane to combat the oxidative stress caused by the hydrocarbon degrading metabolic machinery inside the cell. A 160-fold purification with specific activity of around 66×10^5 U mg^{-1} protein was achieved. The native protein molecular mass was 368 ± 5 kDa with subunit molecular mass of nearly 90 kDa, which indicates that the native CAT protein is a homotetramer. The isoelectric pH (*pI*) of the purified CAT was 4.2. BLAST aligned peptide mass fragments of CAT protein showed its highest similarity with the catalase B protein from other fungal sources. CAT was active in a broad range of pH 4 to 12 and temperature 25°C to 90°C. The catalytic efficiency (K_{cat}/K_m) of 4.7×10^8 M^{-1} s^{-1} within the studied substrate range and alkaline pH stability (half-life, $t_{1/2}$ at pH 12~15 months) of CAT are considerably higher than most of the extensively studied catalases from different sources. The storage stability ($t_{1/2}$) of CAT at physiological pH 7.5 and 4°C was nearly 30 months. The haem was identified as haem b by electrospray ionization tandem mass spectroscopy (ESI-MS/MS).

1. Introduction

The importance of catalase for different industrial applications such as biosensors [1–5], therapeutics [6], and food and textile [7] is tremendously growing as revealed from the volume of publications on these areas since recent past. Catalase (EC 1.11.1.6) is a haem-containing antioxidant enzyme known for its ability to degrade hydrogen peroxide (H_2O_2) into water and oxygen [8]. The degradation occurs in two steps: the first step involves reduction of a H_2O_2 molecule into water with the concomitant oxidation of the catalase haem Fe^{3+} to oxyferryl species (Fe^{4+}=O), while the second step involves oxidation of a second molecule of H_2O_2 into water and oxygen with the associated reduction of the oxyferryl species that regenerates the haem Fe^{3+}.

Catalases are ubiquitous homotetrameric enzymes present in archaea, bacteria, fungi, plants, and animals. They are monophyletic in origin and are grouped into three clades: clade 1 in green algae and plants; clade 2 in archaea, bacteria, and fungi; and clade 3 in archaea, bacteria, fungi, and animals

[9]. Clades 1 and 3 are composed of 55 to 69 kDa subunits, while clade 2 enzymes are formed by larger subunits of nearly 75 to 86 kDa [10]. In bacteria, the large catalases described are from *E. coli* (HPII or KatE), *Bacillus subtilis* (KatE), *Bacillus firmus* (KatE), *Mycobacterium avium* (KatE), *Pseudomonas putida* (CatC), and *Xanthomonas oryzae* (KatX) [11], whereas from fungi they are mostly reported from *Neurospora crassa* (Cat-1 and Cat-3) [12], *Penicillium vitale* [13], *Claviceps* (CPCAT1) [14], and *Aspergillus* species, namely, *A. nidulans* (CatA and CatB) [15–17], *A. fumigatus* (CatA and CatB) [18, 19], and *A. niger* (CatR) [20, 21]. The works on biochemical characterisation of large catalases from the *Aspergillus* species are mostly limited to *A. nidulans* [17], *A. fumigatus* [19], and *A. niger* [21].

Fungi, *A. terreus*, are widespread and abundant in terrestrial ecosystem due to their capability to utilise a wide variety of carbon substrates for growth and survival [22–24]. This makes them a potential target for search of novel biocatalysts (enzymes) involved in metabolic machinery of these diverse susbstrates. We already made detailed investigation

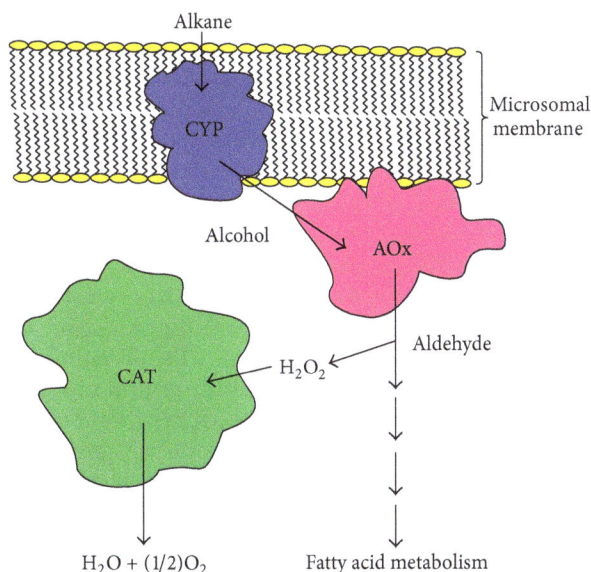

SCHEME 1: A schematic representation of metabolic machinery involved in hydrocarbon utilisation inside the cells of *A. terreus* that produces high level of H_2O_2 as byproduct which is then neutralised by CAT. CYP: cytochrome P450 monoxygenase and AOx: alcohol oxidase. (diagram not to scale).

on enzymes involved in metabolic machinery involved in the hydrocarbon utilisation in *A. terreus* MTCC 6324 isolated from oil fields of Assam (India) [23, 25–27]. The isolated enzymes cytochrome P450 monoxygenase (CYP) and alcohol oxidase (AOx) were used succesfully for their biosensor and industrial applications [28, 29]. However, as shown in Scheme 1, the metabolic utilisation of hydrocarbon produces a huge amount of H_2O_2 as biproduct inside the cells which may be of damage to cell. We expected the presence of some highly efficient H_2O_2 metabolising enzyme inside the cells to combat this oxidative stress for cell survival. This led us to identify, purifiy, and characterise a large catalse (CAT) from this fungal strain which showed excellent catalytic efficiency and stability. The work on utilisation of this CAT protein for fabrication of a highly stable and sensitive biosensor has been published [30]. In this work, we focus on the isolation, purification, characterisation, and establishment of the potentially interesting properties of this novel CAT from *A. terreus* which has lot of untapped potential for future applications.

2. Materials and Methods

2.1. Organism and Culture Conditions. The fungal strain *A. terreus* used in this investigation is a stock culture of Microbial Type Culture Collection (MTCC), Chandigarh, with accession number 6324. The culture conditions and maintenance of *A. terreus* MTCC 6324 used in this study were described elsewhere [25]. Briefly, the organism was cultivated in 500 mL Erlenmeyer flasks containing 2% *n*-hexadecane/glucose in 50 mL of basal medium [31]. The pH of the medium was adjusted to 5.8 and the flasks were incubated

at 28°C under static condition. The organism was maintained on fungal agar slants with periodic transfer to a new slant following each 15 days.

2.2. Purification of CAT. The fungal mycelia were harvested from the early stationary phase of growth (72 hours) and were disrupted at 30 kpsi using mechanical cell disruptor (Constant Systems, UK) by following the procedure described by Kumar and Goswami [25]. The disrupted cell homogenate was centrifuged successively at 10,000 ×g for 10 min and 20,000 ×g for 20 min to pellet undisrupted cell mass and light mitochondrial fractions. The supernatant thus obtained was initially adjusted to 50% $(NH_4)_2SO_4$ saturation to precipitate the contaminating proteins. After removing the precipitated protein by centrifugation at 10,000 ×g for 30 min, the supernatant was finally adjusted to 80% $(NH_4)_2SO_4$ saturation to precipitate the CAT protein and centrifuged at 10,000 ×g for 30 min. The pellet was resuspended in 50 mM sodium phosphate buffer (SPB) containing 1 M $(NH_4)_2SO_4$ and loaded on a Phenyl Sepharose 6 Fast Flow hydrophobic interaction chromatography (HIC) column (1.5 × 25 cm) connected to fast pressure liquid chromatography (FPLC) (AKTA prime plus, GE), preequilibrated with 50 mM SPB (pH 7.5) containing 1 M $(NH_4)_2SO_4$. The fractions were eluted from the column using 50 mM SPB (pH 7.5) containing a decreasing step gradient of $(NH_4)_2SO_4$ (0.75 M, 0.5 M, and 0.25 M). The final fraction was eluted with 50 mM SPB (pH 7.5). Fractions containing CAT activity were pooled, dialysed, concentrated, and then loaded on HiPrep Sephacryl S-300 HR (1.6 × 60 cm, 50 μm) size exclusion chromatography (SEC) column preequilibrated with 50 mM SPB (pH 7.5) to be eluted with the same buffer and flow rate. CAT-containing fractions were pooled and concentrated. The lyophilised purified CAT was stored at −20°C and was used for further characterisation after resuspension in 50 mM SPB (pH 7.5).

2.3. Molecular Weight Determination and Electrophoresis. Molecular weight of purified native CAT was determined by SEC (HiPrep Sephacryl S-300 HR, 1.6 × 60 cm, 50 μm) using the FPLC (AKTA prime plus, GE). The gel filtration molecular weight standards used were thyroglobulin (669 kDa), apoferritin (443 kDa), β-amylase (200 kDa), alcohol dehydrogenase (150 kDa), and carbonic anhydrase (29 kDa) (Sigma).

Native and SDS-PAGE analyses of the purified CAT protein were done following the method of Laemmli [32] using 7% and 10% separating gels, respectively, with 5% stacking gel and thickness of 0.75 mm. The gels were stained with Coomassie brilliant blue (CBB) R 250 (Merck, India). Zymogram analysis of the native PAGE was done by incubating the gel in 10 mM H_2O_2 for 10 min. The SDS-PAGE protein markers used were carbonic anhydrase (bovine erythrocyte, 29 kDa), fumarase (porcine heart, 48.5 kDa), serum albumin (bovine, 66 kDa), phosphorylase b (rabbit muscle, 97.4 kDa), and β-galactosidase (*E. coli*, 116 kDa) (Sigma). Glycostaining of the protein band in SDS-PAGE was done following the method of Segrest and Jackson [33] using periodic acid Schiff (PAS) reagent.

The *pI* of CAT was determined by isoelectric focusing following the method described elsewhere [26].

The haem from the purified CAT isolated by 2-butanone/HCl method [34] was characterised by electrospray ionization tandem mass spectroscopy (ESI-MS/MS) (Q-TOF Premier, Waters) following the method of Sana et al. [35] and Fourier transform infrared spectroscopy (FTIR) (Perkin-Elmer).

2.4. Peptide Mass Fingerprinting. The protein identification work was carried out at ProtTech, Inc. (USA) by using the nanoliquid chromatography-tandem mass spectroscopy (LC-MS/MS) peptide sequencing technology following the method described elsewhere [26].

2.5. Enzyme Assay and Kinetics. The CAT activity was assayed by the method of Beers and Sizer [36] with a partial modification. The standard reaction mixture for the assay contained 50 mM SPB (pH 7.5) and 4.9 mM H_2O_2 ($\varepsilon_{240} = 43.6\,M^{-1}\,cm^{-1}$). Initial reaction rate was determined spectrophotometrically (Cary 100Bio, Varian) by measuring the decrease of absorbance at 240 nm. One unit (U) of enzyme activity is the amount of enzyme that consumes $2\,\mu mol\,H_2O_2$ in one minute at room temperature. The substrate saturation kinetics was measured with H_2O_2 concentrations in a range of 0.05 to 75 mM and enzyme preparation bearing specific catalase activity $30.74 \times 10^5\,U\,mg^{-1}$ protein. The apparent Michaelis-Menten constant (K_m) for H_2O_2 was calculated from the Lineweaver-Burk plot.

2.6. pH and Temperature Optima and Stability of CAT. To measure pH optima, the enzyme reactions were carried out in different pH buffers, namely, trisodium citrate (pH 2.5 to 4), sodium acetate (pH 5), sodium phosphate (pH 6 to 7.5), tris (pH 8), ethanolamine (pH 9), and sodium phosphate (pH 10 to 12.5), each at a concentration of 50 mM. For pH stability measurement the enzyme samples were incubated for different time periods in the aforementioned pH buffers at 4°C and the residual CAT activity was determined at room temperature in SPB (50 mM, pH 7.5). The temperature optima were determined by carrying out the enzyme reactions at different temperatures and the CAT activity was measured spectrophotometrically. For determination of temperature stability, enzyme samples were incubated at different temperatures (4°C to 80°C) in circulating water bath (Julabo SW22, Germany) for different time periods followed by measurement of residual activity of CAT at room temperature in SPB (50 mM, pH 7.5).

The half-life ($t_{1/2}$) was calculated from the value of $0.693/k$, where "k" is the first-order rate constant. The value of "k" was calculated from the slope of a curve obtained by plotting $\ln a_0/a_t$ versus time (t), where a_0 is the initial CAT activity and a_t is the residual CAT activity obtained by incubating the CAT for different time periods at said pH or temperature.

Protein was estimated by Bradford's method using bovine serum albumin as standard [37] and carbohydrate was estimated by anthrone method using glucose as standard [38].

3. Results and Discussion

A very high level of catalase activity was detected in the vegetative cells of *A. terreus* during its growth on *n*-hexadecane (2% v/v) as the sole source of carbon for growth. A total activity of $17 \times 10^6\,U\,gm^{-1}$ dry cell mass was obtained in the cells harvested from the late exponential phase (90 h) of growth in *n*-hexadecane substrate. This activity in the corresponding glucose grown cells was $4.25 \times 10^6\,U\,gm^{-1}$ dry cell mass. The average catalase activity from late log phase to late exponential phase was $9.5 \times 10^6\,U\,gm^{-1}$ dry cell mass in *n*-hexadecane grown cells. The overall activity decreased in the stationary phase of the growth of fungus. Thus, around 4-fold increased catalase activity was recorded in the late exponential phase of hydrocarbon grown cells of *A. terreus* when compared to glucose grown cells of the same strain. Such increased level of catalase activity in the *n*-hexadecane grown cells can be linked to the metabolism of this hydrocarbon substrate in the cells of fungal mycelia (Scheme 1). The initial oxidation of *n*-hexadecane by *A. terreus* produces hexadecanol [23], which is further oxidised by a long-chain alcohol oxidase with concomitant generation of H_2O_2 as byproduct [25, 26]. It is expected that the high level of catalase produced in the cells of *A. terreus* rapidly neutralises the harmful H_2O_2 continuously generated as byproduct in the cells during metabolism of the hydrocarbon substrate. Thus, the induction of such high level of cellular large catalase (as identified by biochemical studies discussed later in this section, abbreviated as CAT) can be credited as the cellular defence mechanism against the oxidative stress of H_2O_2 generated during metabolism of *n*-hexadecane. The fact that large catalases have been reported to be less sensitive to oxidative damage by H_2O_2 also supports the given assumption [12, 21]. Although the induction of peroxide degrading enzyme genes (katG encoding catalase-peroxidase) has been reported from polycyclic aromatic hydrocarbon degrading *Mycobacterium* sp. [39], any such induction of cellular large catalase from the hydrocarbon degrading cells of fungus is not reported before. Notably, except the level of induction, other properties of the CAT from *n*-hexadecane grown cells (as described later in this section) were alike to the one produced in the glucose grown cells (results not separately shown).

CAT was isolated and purified successively by differential centrifugation, $(NH_4)_2SO_4$ precipitation, HIC, and SEC. A 160-fold purification with specific activity of $66 \times 10^5\,U\,mg^{-1}$ protein with an overall yield of 47% in comparison to the original crude extract was achieved (Table 1). The homogeneity of the purified protein was demonstrated by single protein band in native PAGE (Figure 1(a), lanes 1 and 2). The molecular weight of the CAT determined by SDS-PAGE of the purified protein sample (Figure 1(b), lanes 1 and 2) was 90±2 kDa. The native protein molecular mass determined by SEC was 368± 5 kDa (see Figure S1 of the Supplementary Material available online at http://dx.doi.org/10.1155/2016/4379403), which indicates that the native CAT protein is a homotetramer with subunit molecular mass ~90 kDa and thus belongs to the clade 2 of catalases constituting larger subunits (>80 kDa).

TABLE 1: Purification table for CAT.

Fractions	Activity ($\times 10^5$ U mL^{-1})	Protein (mg mL^{-1})	Specific activity ($\times 10^5$ U mg^{-1})	Fold purification	Yield %
crude extract	1.2 ± 0.5	2.9 ± 0.8	0.41 ± 0.10	1.0	100
sup-1[a]	1.7 ± 0.3	1.8 ± 0.5	0.94 ± 0.20	2.3	88.6
pellet-1[a]	2.7 ± 0.8	5.4 ± 1.0	0.50 ± 0.09	1.2	10.8
sup-2[b]	2.3 ± 0.4	1.4 ± 0.3	1.64 ± 0.60	4.0	75.0
pellet-2[b]	0.3 ± 0.1	7.2 ± 1.2	0.04 ± 0.01	0.1	12.4
50% pellet[c]	2.4 ± 0.7	7.2 ± 1.9	0.33 ± 0.10	0.8	10.1
80% pellet[c]	3.3 ± 1.2	1.4 ± 0.5	2.36 ± 0.50	5.8	62.5
HIC purified[d]	14.7 ± 1.4	0.5 ± 0.1	29.40 ± 1.42	71.7	54.4
SEC purified	39.6 ± 2.0	0.6 ± 0.1	66.00 ± 3.97	160.9	47.0

[a]sup-1 and pellet-1 are supernatant and pellet of 10,000 ×g centrifugation.
[b]sup-2 and pellet-2 are supernatant and pellet of 20,000 ×g centrifugation.
[c]50% and 80% pellet are $(NH_4)_2SO_4$ precipitated fractions.
[d]HIC purified fraction was eluted at 250 mM $(NH_4)_2SO_4$ concentration.
Each value represents the mean ± standard error at $p < 0.05$.

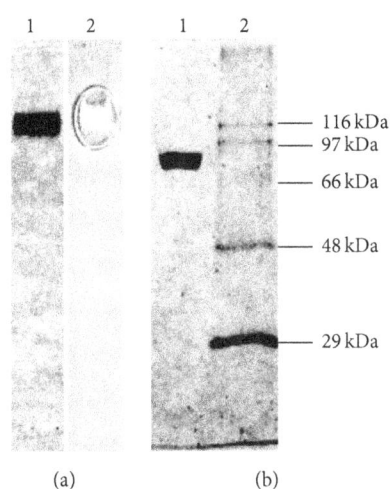

FIGURE 1: PAGE analysis of purified CAT. (a) Native PAGE: lane 1, CBB staining of CAT and lane 2, activity staining of CAT (loaded 10 μg purified protein). (b) SDS-PAGE: lane 1, CBB staining of CAT protein and lane 2, CBB staining of standard molecular markers.

The purity of the CAT was further demonstrated by the high optical purity ratio (Reinheitszahl or Rz, A_{405}/A_{280}) of 0.8 ± 0.05 (A_{405} is Soret band typical for catalases [40]) (Figure S2). A distinct but low intensity peak at around 570 nm in the same spectrum is attributed to the presence of a low concentration of inactive form of CAT [41]. The inactivation of CAT may have resulted from the addition of dithiothreitol in disrupted cell homogenate following the procedure described by Kumar and Goswami [25]. The characteristic charge-transfer band (reflecting the heme Fe$^{(III)}$ to Tyr axial ligand electronic structure in catalase) at ca. 630 nm also appears to be masked by this peak. The presence of this peak also provides the reason for Rz (A_{405}/A_{280}) of 0.8 ± 0.05 (less than 1), despite the fact that the protein was purified to

homogeneity as depicted from the fold purification (Table 1) and absence of any other contaminating band in SDS- and native PAGE. The isoelectric pH (pI) determined by isoelectric focusing of the purified CAT was found to be 4.2 ± 0.1. The CAT protein was identified as glycoprotein based on the PAS staining of the protein band in SDS-PAGE. The protein to carbohydrate ratio of the purified catalase was 3.33. The haem was isolated from the purified CAT and the molecular mass of 616 of the isolated haem was determined by ESI-MS/MS (Figure 2(a)), which is similar to the reported molecular mass of haem b [35]. The isolated haem was further characterised for its functional groups -OH, -CH$_3$, -CH$_2$-, and -C=O by FTIR spectra (Figure 2(b)). Many variants of haem, namely, haem-b, -d, and derivatives of haem have been reported in catalases. Notably, the presence of only haem d in large catalases from Aspergillus has been described so far [21]. Thus, this finding on the presence of haem b in the large catalase from A. terreus is interesting and warrants further investigation to understand the exact role of these two types of haem (b and d) in the large catalases, though haem in general is known as the catalytic centre of these redox enzymes.

BLAST alignment of peptide mass fragments from the SDS-PAGE separated protein band of the purified CAT showed its highest similarity with the catalase B precursor protein from A. terreus NIH 2624 (Accession code: XP_001216098, http://www.ncbi.nlm.nih.gov/). Figures S3 and S4 show the LC-MS/MS tandem mass spectra of proximal haem binding and tetramer interface domains of the CAT protein, respectively. The presence of fully conserved positions for tyrosine (Y) residue in the proximal haem binding domain is evident from the multiple sequence alignment of proximal haem binding residues of catalases from different source organisms (Table 2). It is known that phenolic side chain of the conserved tyrosine acts as the fifth haem iron ligand, the other four being the nitrogens of the porphyrin ring. The overall sequence of CAT was found to be conserved

TABLE 2: Multiple sequence alignment of amino acid residues of proximal haem binding domain of CAT with other known catalases.

Catalase source organism	Amino acid sequence	Accession code
Aspergillus terreus MTCC 6324	R L F S **Y** L D T Q	This work
Aspergillus terreus NIH 2624	R L F S **Y** L D T Q	XP_001216098.1
Aspergillus niger	R L F S **Y** L D T Q	XP_001388621.1
Aspergillus nidulans FGSC A4	R L F S **Y** L D T Q	XP_682608.1
Aspergillus fumigatus Af293	R L F S **Y** L D T Q	XP_748550.1
Penicillium marneffei ATCC 18224	R L F S **Y** L D T Q	XP_002153601.1
Neurospora crassa OR74A	R L F S **Y** L D T Q	XP_957826.1
Pseudomonas stutzeri A1501	R L F S **Y** A D T Q	YP_001174038.1
Escherichia coli SE11	R L F S **Y** T D T Q	YP_002293177.1
Mycobacterium vanbaalenii PYR-1	R L F S **Y** L D T Q	YP_954009.1
Candida dubliniensis CD36	R L F S **Y** A D T H	dbj|BAD77826.1
Arabidopsis thaliana	R V F S **Y** A D T Q	emb|CAA64220.1
Rattus norvegicus	R L F A **Y** P D T H	NP_036652.1
Bos taurus	R L F A **Y** P D T H	NP_001030463.1
Homo sapiens	R L F A **Y** P D T H	NP_001743.1

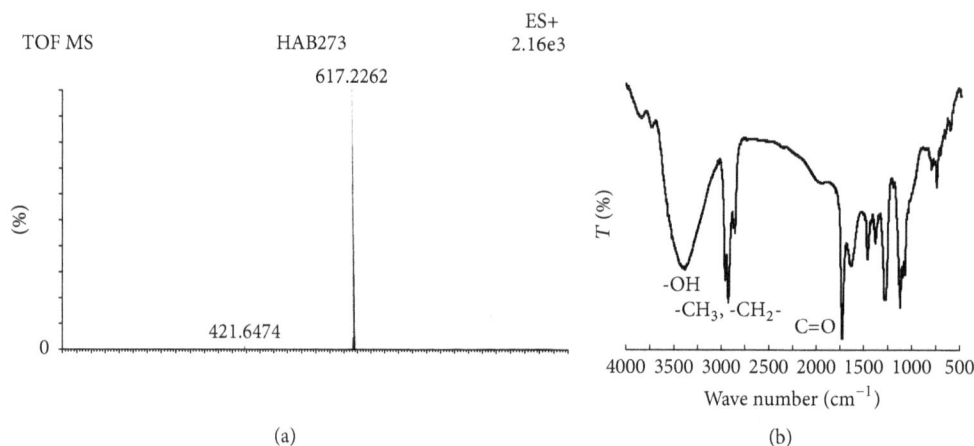

(a)   (b)

FIGURE 2: (a) MS spectrum of the extracted haem from the native purified CAT. The sample was run in the ESI positive mode. Details of the experiment and results are discussed in the paper. (b) FTIR spectra of the isolated haem. Peaks corresponding to the representative functional groups of haem are shown in the figure.

among the large catalases from bacteria and fungi while being distinct from smaller catalases from bacteria and higher organisms.

Catalases in general are not known to exhibit Michaelis-Menten kinetics over the entire substrate range (from mM to molar concentrations of H_2O_2) and they usually have the two-step nature of the catalytic reaction [21]. The large catalase from *A. terreus* showed increased catalytic activity with increasing H_2O_2 concentrations (within 0.05 to 75 mM H_2O_2). The velocity of enzymatic reaction showed a Michaelis-Menten-type dependence on substrate concentration with a saturation of catalytic activity at around 75 mM. The apparent K_m of 14.15 mM was calculated from the Lineweaver-Burk plot with a linear equation, $y = 4.265x + 0.301$, and correlation coefficient (R^2) of 0.9949. The apparent K_m of CAT from *A. terreus* correlates well with the apparent K_m (21.7 mM) of large catalase (Cat-1) of *N. crassa* at H_2O_2 concentrations below 100 mM [12]. However,

at molar concentrations of H_2O_2, a 10 times higher apparent K_m was reported for *N. crassa* Cat-1. The living organisms in nature are seldom exposed to molar concentrations of H_2O_2; thus the studies with such high concentrations of H_2O_2 were consciously avoided in this work. However, the inhibition studies of CAT and its kinetics at molar concentration of H_2O_2 are subject of future investigations. A high catalytic efficiency (K_{cat}/K_m) of 4.7×10^8 M^{-1} s^{-1} was deduced from the catalytic turnover number (K_{cat}) of 6.65×10^6 s^{-1} and apparent K_m of the CAT (for the above-mentioned range of H_2O_2 concentrations). The high catalytic efficiency reported by us is only comparable to the durable Cat-1 of *N. crassa* (4.11×10^8 M^{-1} s^{-1}, within 100 mM H_2O_2) [12], which is stated to be produced to combat the oxidative stress developed during the conidia formation. This correlates well with our previous assumption for production of high level of extremely efficient CAT in *A. terreus* to combat the oxidative stress during assimilation of long-chain hydrocarbons in the cell.

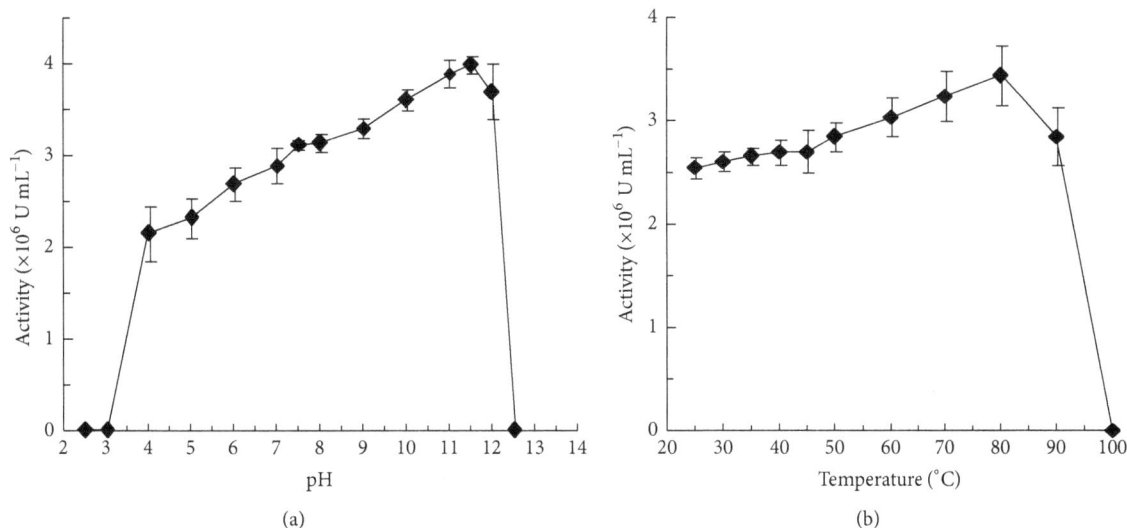

FIGURE 3: CAT activity as a function of pH (a) and temperature (b). The different pH buffers (each at a concentration of 50 mM) were trisodium citrate (pH 2.5 to 4), sodium acetate (pH 5), sodium phosphate (pH 6 to 7.5), tris (pH 8), ethanolamine (pH 9), and sodium phosphate (pH 10 to 12.5). Each datum point represents the average of the analysis of triplicate values.

The large catalase from *A. terreus* did not show any specific pH or temperature optimum for maximum activity but was active throughout the broad range of pH 4 to 12 and temperatures 25°C to 90°C (Figures 3(a) and 3(b)). However, a variation in CAT activity was observed within this range, where the activity increased with the increasing pH and temperature. The CAT activity was drastically lost outside this pH range (pH 4 to 12) and when boiled. Although CAT was active throughout the mentioned pH range, the maximum stability of CAT (measured as retention of its catalytic activity) was recorded at around physiological pH (pH 7.5) with the $t_{1/2}$ of ~30 months. CAT was also found to be exceptionally stable at alkaline pH. The $t_{1/2}$ of the CAT at pH 12 was ~15 months. The stability of CAT diminished steadily with the increasing temperature from 25°C to 80°C with $t_{1/2}$ of 54 days and 42 min at 25°C and 80°C, respectively. The maximum stability of CAT was recorded when stored at 4°C and pH 7.5, as more than 98% of initial catalase activity was retained even after one month and a $t_{1/2}$ of nearly 30 months was calculated. Although the broad pH and temperature optima and many other biochemical properties discussed before in this section such as acidic *pI*, glycoprotein nature, and binding to hydrophobic column of the isolated CAT from *A. terreus* are similar to the widely studied large catalases reported from other *Aspergillus* strains [17, 19, 21], certain properties of CAT were found to be unique and not reported to date from these sources. The catalytic efficiency and stability near extreme alkaline pH of CAT are considerably higher than most of the extensively studied catalases from different sources [17, 21, 42, 43]. The thermal stability of CAT at 80°C is considerably high [21] and is only comparable with the catalase from thermophilic origin [44], Cat-1 from *N. crassa* [12], and HPII of *E. coli* [45]. High storage stability

at 4°C and 25°C (room temperature) is also interesting feature of the isolated CAT. The stability of CAT at extreme alkaline pH (pH 12) was intriguing and drove us to investigate the reasons behind such behaviour (not part of this report). The early spectroscopic and electrochemical studies of CAT at different pH conditions (from pH 3 to 12) showed that acidic conditions induced the dissociation of haem from CAT whereas the association of haem with protein matrix was quite stable at basic pH values [46]. These findings provided some of the reasons for the higher activity and stability of CAT at alkaline pH in this study and sudden loss of activity at pH 3. Thus, besides the sequence conservation and similarities in various biochemical properties of CAT from *A. terreus* with other large catalases from bacterial and fungal sources, understanding the differences in some of the specific behaviours on the basis of structure-function relationship studies is always a possibility. Such variations in protein chemical properties such as specific activities, reaction velocities, and stability among a large number of catalases bearing extensive sequence similarities have been reported earlier also by other research groups [21], further supporting our claim for the requirement of separate studies for different catalases from diverse sources. The remarkable activity and stability of CAT under broad thermal and pH conditions are construed as attractive properties for potential applications in areas like biosensors as well as therapeutics, textile, and food industries where the enzyme is usually required to work in diverse pH and temperature conditions for long periods. The biosensing application of CAT with a considerably high sensitivity and stability is already established in one of our investigations [30], fortifying the potential of this large catalase from *A. terreus* in areas where catalytic efficiency and stability of protein are major concerns.

4. Conclusions

A considerably high level of catalase activity is reported inside the cells of *A. terreus* (originally isolated from the oil fields of Assam, India) when grown on hydrocarbon substrate. The reason for this high level of catalase activity is attributed to the necessity to neutralise the high level of H_2O_2 produced as a byproduct of the metabolic machinery for hydrocarbon assimilation inside the cells. Although the isolated large catalase from *A. terreus* shares many functional similarities with other large catalases reported from different microbial sources, its remarkable stability at alkaline pH and catalytic efficiency are intriguing. These properties establish the potential of this large catalase for biosensing and other industrial applications and also require further investigations to understand the structure-function relationship behind such specificities.

Acknowledgments

This work was carried out with the financial support from the Department of Biotechnology (DBT), Government of India (Grants nos. BT/PR8522/PID/06/346/2006 and BT/01/NE/PS/08). Dr. Preety Vatsyayan is also thankful to the Alexander von Humboldt Foundation for the postdoctoral fellowship.

References

[1] L. D. Mello and L. T. Kubota, "Biosensors as a tool for the antioxidant status evaluation," *Talanta*, vol. 72, no. 2, pp. 335–348, 2007.

[2] K. B. O'Brien, S. J. Killoran, R. D. O'Neill, and J. P. Lowry, "Development and characterization in vitro of a catalase-based biosensor for hydrogen peroxide monitoring," *Biosensors and Bioelectronics*, vol. 22, no. 12, pp. 2994–3000, 2007.

[3] B. Modrzejewska, A. J. Guwy, R. Dinsdale, and D. L. Hawkes, "Measurement of hydrogen peroxide in an advanced oxidation process using an automated biosensor," *Water Research*, vol. 41, no. 1, pp. 260–268, 2007.

[4] L. Campanella, M. P. Sammartino, M. Tomassetti, and S. Zannella, "Hydroperoxide determination by a catalase OPEE: application to the study of extra virgin olive oil rancidification process," *Sensors and Actuators, B: Chemical*, vol. 76, no. 1–3, pp. 158–165, 2001.

[5] L. Campanella, R. Roversi, M. P. Sammartino, and M. Tomassetti, "Hydrogen peroxide determination in pharmaceutical formulations and cosmetics using a new catalase biosensor," *Journal of Pharmaceutical and Biomedical Analysis*, vol. 18, no. 1-2, pp. 105–116, 1998.

[6] B. J. Day, "Catalytic antioxidants: a radical approach to new therapeutics," *Drug Discovery Today*, vol. 9, no. 13, pp. 557–566, 2004.

[7] D. Akertek and L. Tarhan, "Characteristics of immobilized catalases and their applications in pasteurization of milk," *Applied Biochemistry and Biotechnology*, vol. 50, pp. 9555–9560, 1995.

[8] M. Zámocký and F. Koller, "Understanding the structure and function of catalases: clues from molecular evolution and in vitro mutagenesis," *Progress in Biophysics and Molecular Biology*, vol. 72, no. 1, pp. 19–66, 1999.

[9] M. G. Klotz and P. C. Loewen, "The molecular evolution of catalatic hydroperoxidases: evidence for multiple lateral transfer of genes between prokaryota and from bacteria into eukaryota," *Molecular Biology and Evolution*, vol. 20, no. 7, pp. 1098–1112, 2003.

[10] A. Díaz, V.-J. Valdés, E. Rudiño-Piñera, E. Horjales, and W. Hansberg, "Structure-function relationships in fungal large-subunit catalases," *Journal of Molecular Biology*, vol. 386, no. 1, pp. 218–232, 2009.

[11] M. G. Klotz, G. R. Klassen, and P. C. Loewen, "Phylogenetic relationships among prokaryotic and eukaryotic catalases," *Molecular Biology and Evolution*, vol. 14, no. 9, pp. 951–958, 1997.

[12] A. Díaz, P. Rangel, Y. Montes De Oca, F. Lledías, and W. Hansberg, "Molecular and kinetic study of catalase-1, a durable large catalase of *Neurospora crassa*," *Free Radical Biology and Medicine*, vol. 31, no. 11, pp. 1323–1333, 2001.

[13] B. K. Vainshtein, W. R. Melik-Adamyan, V. V. Barynin et al., "Three-dimensional structure of catalase from *Penicillium vitale* at 2.0 A resolution," *Journal of Molecular Biology*, vol. 188, no. 1, pp. 49–61, 1986.

[14] V. Garre, U. Müller, and P. Tudzynski, "Cloning, characterization, and targeted disruption of *cpcat1*, coding for an in planta secreted catalase of *Claviceps purpurea*," *Molecular Plant-Microbe Interactions*, vol. 11, no. 8, pp. 772–783, 1998.

[15] R. E. Navarro, M. A. Stringer, W. Hansberg, W. E. Timberlake, and J. Aguirre, "*catA*, a new *Aspergillus nidulans* gene encoding a developmentally regulated catalase," *Current Genetics*, vol. 29, no. 4, pp. 352–359, 1996.

[16] L. Kawasaki, D. Wysong, R. Diamond, and J. Aguirre, "Two divergent catalase genes are differentially regulated during *Aspergillus nidulans* development and oxidative stress," *Journal of Bacteriology*, vol. 179, no. 10, pp. 3284–3292, 1997.

[17] J. A. Calera, J. Sánchez-Weatherby, R. López-Medrano, and F. Leal, "Distinctive properties of the catalase B of *Aspergillus nidulans*," *FEBS Letters*, vol. 475, no. 2, pp. 117–120, 2000.

[18] T. Takasuka, N. M. Sayers, M. J. Anderson, E. W. Benbow, and D. W. Denning, "*Aspergillus fumigatus* catalases: cloning of an *Aspergillus nidulans* catalase B homologue and evidence for at least three catalases," *FEMS Immunology and Medical Microbiology*, vol. 23, no. 2, pp. 125–133, 1999.

[19] R. Lopez-Medrano, M. C. Ovejero, J. A. Calera, P. Puente, and F. Leal, "An immunodominant 90-kilodalton *Aspergillus fumigatus* antigen is the subunit of a catalase," *Infection and Immunity*, vol. 63, no. 12, pp. 4774–4780, 1995.

[20] T. Fowler, M. W. Rey, P. Vaha-Vahe, S. D. Power, and R. M. Berka, "The catR gene encoding a catalase from *Aspergillus niger*: primary structure and elevated expression through increased gene copy number and use of a strong promoter," *Molecular Microbiology*, vol. 9, no. 5, pp. 989–998, 1993.

[21] J. Switala and P. C. Loewen, "Diversity of properties among catalases," *Archives of Biochemistry and Biophysics*, vol. 401, no. 2, pp. 145–154, 2002.

[22] L. N. Britton, "Microbial degradation of aliphatic hydrocarbons," in *Microbial Degradation of Organic Compounds*, D. T. Gibson, Ed., vol. 5, pp. 89–129, Marcel Dekker, New York, NY, USA, 1984.

[23] P. Vatsyayan, A. K. Kumar, P. Goswami, and P. Goswami, "Broad substrate Cytochrome P450 monooxygenase activity in the cells of *Aspergillus terreus* MTCC 6324," *Bioresource Technology*, vol. 99, no. 1, pp. 68–75, 2008.

[24] N. Dashti, H. Al-Awadhi, M. Khanafer, S. Abdelghany, and S. Radwan, "Potential of hexadecane-utilizing soil-microorganisms for growth on hexadecanol, hexadecanal and hexadecanoic acid as sole sources of carbon and energy," *Chemosphere*, vol. 70, no. 3, pp. 475–479, 2008.

[25] A. K. Kumar and P. Goswami, "Functional characterization of alcohol oxidases from *Aspergillus terreus* MTCC 6324," *Applied Microbiology and Biotechnology*, vol. 72, no. 5, pp. 906–911, 2006.

[26] A. K. Kumar and P. Goswami, "Purification and properties of a novel broad substrate specific alcohol oxidase from *Aspergillus terreus* MTCC 6324," *Biochimica et Biophysica Acta—Proteins and Proteomics*, vol. 1784, no. 11, pp. 1552–1559, 2008.

[27] M. Chakraborty, M. Goel, S. R. Chinnadayyala, U. R. Dahiya, S. S. Ghosh, and P. Goswami, "Molecular characterization and expression of a novel alcohol oxidase from *Aspergillus terreus* MTCC6324," *PLoS ONE*, vol. 9, no. 4, Article ID e95368, 2014.

[28] P. Vatsyayan, M. Chakraborty, S. Bordoloi, and P. Goswami, "Electrochemical investigations of fungal cytochrome P450," *Journal of Electroanalytical Chemistry*, vol. 662, no. 2, pp. 312–316, 2011.

[29] A. Kakoti, A. K. Kumar, and P. Goswami, "Microsome-bound alcohol oxidase catalyzed production of carbonyl compounds from alcohol substrates," *Journal of Molecular Catalysis B: Enzymatic*, vol. 78, pp. 98–104, 2012.

[30] P. Vatsyayan, S. Bordoloi, and P. Goswami, "Large catalase based bioelectrode for biosensor application," *Biophysical Chemistry*, vol. 153, no. 1, pp. 36–42, 2010.

[31] L. D. Bushnell and F. F. Haas, "The utilization of certain hydrocarbons by microorganisms," *Journal of Bacteriology*, vol. 41, no. 5, pp. 653–673, 1941.

[32] U. K. Laemmli, "Cleavage of structural proteins during the assembly of the head of bacteriophage T4," *Nature*, vol. 227, no. 5259, pp. 680–685, 1970.

[33] J. P. Segrest and R. L. Jackson, "Molecular weight determination of glycoproteins by polyacrylamide gel electrophoresis in sodium dodecyl sulfate," in *Methods in Enzymology*, vol. 28, pp. 54–63, Elsevier, Philadelphia, Pa, USA, 1972.

[34] F. W. J. Teale, "Cleavage of the haem-protein link by acid methylethylketone," *BBA—General Subjects*, vol. 35, article 543, 1959.

[35] T. R. Sana, K. Waddell, and S. M. Fischer, "A sample extraction and chromatographic strategy for increasing LC/MS detection coverage of the erythrocyte metabolome," *Journal of Chromatography B*, vol. 871, no. 2, pp. 314–321, 2008.

[36] R. F. Beers Jr. and I. W. Sizer, "A spectrophotometric method for measuring the breakdown of hydrogen peroxide by catalase," *The Journal of Biological Chemistry*, vol. 195, no. 1, pp. 133–140, 1952.

[37] M. M. Bradford, "A rapid and sensitive method for the quantitation of microgram quantities of protein utilizing the principle of protein-dye binding," *Analytical Biochemistry*, vol. 72, no. 1-2, pp. 248–254, 1976.

[38] L. C. Mokrasch, "Analysis of hexose phosphates and sugar mixtures with the anthrone reagent," *The Journal of Biological Chemistry*, vol. 208, no. 1, pp. 55–59, 1954.

[39] R.-F. Wang, D. Wennerstrom, W.-W. Cao, A. A. Khan, and C. E. Cerniglia, "Cloning, expression, and characterization of the katG gene, encoding catalase-peroxidase, from the polycyclic aromatic hydrocarbon-degrading bacterium *Mycobacterium* sp. strain PYR-1," *Applied and Environmental Microbiology*, vol. 66, no. 10, pp. 4300–4304, 2000.

[40] A. Ivancich, H. M. Jouve, B. Sartor, and J. Gaillard, "EPR investigation of-compound I in *Proteus mirabilis* and bovine liver catalases: formation of porphyrin and tyrosyl radical intermediates," *Biochemistry*, vol. 36, no. 31, pp. 9356–9364, 1997.

[41] A. Takeda, T. Miyahara, A. Hachimori, and T. Samejima, "The interactions of thiol compounds with porcine erythrocyte catalase," *Journal of Biochemistry*, vol. 87, no. 2, pp. 429–439, 1980.

[42] D. Monti, E. Baldaro, and S. Riva, "Separation and characterization of two catalase activities isolated from the yeast *Trigonopsis variabilis*," *Enzyme and Microbial Technology*, vol. 32, no. 5, pp. 596–605, 2003.

[43] J. Ogawa, W. T. Sulistyaningdyah, Q.-S. Li et al., "Two extracellular proteins with alkaline peroxidase activity, a novel cytochrome c and a catalase-peroxidase, from *Bacillus sp.* No.13," *Biochimica et Biophysica Acta—Proteins and Proteomics*, vol. 1699, no. 1-2, pp. 65–75, 2004.

[44] V. S. Thompson, K. D. Schaller, and W. A. Apel, "Purification and characterization of a novel thermo-alkali-stable catalase from *Thermus brockianus*," *Biotechnology Progress*, vol. 19, no. 4, pp. 1292–1299, 2003.

[45] J. Switala, J. O. O'Neil, and P. C. Loewen, "Catalase HPII from *Escherichia coli* exhibits enhanced resistance to denaturation," *Biochemistry*, vol. 38, no. 13, pp. 3895–3901, 1999.

[46] P. Vatsyayan and P. Goswami, "Acidic pH conditions induce dissociation of the haem from the protein and destabilise the catalase isolated from *Aspergillus terreus*," *Biotechnology Letters*, vol. 33, no. 2, pp. 347–351, 2011.

Inhibition of α-Amylases by Condensed and Hydrolysable Tannins: Focus on Kinetics and Hypoglycemic Actions

Camila Gabriel Kato,[1,2] **Geferson de Almeida Gonçalves,**[1,2] **Rosely Aparecida Peralta,**[3]
Flavio Augusto Vicente Seixas,[2] **Anacharis Babeto de Sá-Nakanishi,**[1,2] **Lívia Bracht,**[1,2]
Jurandir Fernando Comar,[1,2] **Adelar Bracht,**[1,2] **and Rosane Marina Peralta**[1,2]

[1]*Postgraduate Program of Food Science, University of Maringá, Avenida Colombo 5790, 87020900 Maringá, PR, Brazil*
[2]*Department of Biochemistry, University of Maringá, Maringá, PR, Brazil*
[3]*Department of Chemistry, Federal University of Santa Catarina, Florianópolis, SC, Brazil*

Correspondence should be addressed to Rosane Marina Peralta; rosanemperalta@gmail.com

Academic Editor: Sunney I. Chan

The aim of the present study was to compare the in vitro inhibitory effects on the salivary and pancreatic α-amylases and the in vivo hypoglycemic actions of the hydrolysable tannin from Chinese natural gall and the condensed tannin from *Acacia mearnsii*. The human salivary α-amylase was more strongly inhibited by the hydrolysable than by the condensed tannin, with the concentrations for 50% inhibition (IC$_{50}$) being 47.0 and 285.4 μM, respectively. The inhibitory capacities of both tannins on the pancreatic α-amylase were also different, with IC$_{50}$ values being 141.1 μM for the hydrolysable tannin and 248.1 μM for the condensed tannin. The kinetics of the inhibition presented complex patterns in that for both inhibitors more than one molecule can bind simultaneously to either the free enzyme of the substrate-complexed enzyme (parabolic mixed inhibition). Both tannins were able to inhibit the intestinal starch absorption. Inhibition by the hydrolysable tannin was concentration-dependent, with 53% inhibition at the dose of 58.8 μmol/kg and 88% inhibition at the dose of 294 μmol/kg. For the condensed tannin, inhibition was not substantially different for doses between 124.4 μmol/kg (49%) and 620 μmol/kg (57%). It can be concluded that both tannins, but especially the hydrolysable one, could be useful in controlling the postprandial glycemic levels in diabetes.

1. Introduction

Both the pancreatic α-amylase and the salivary α-amylase (α-1,4-glucan-4-glucanohydrolase, EC 3.2.1.1) catalyse the hydrolysis of the α-1,4-glycosidic linkages in starch, glycogen, and other oligo- and polysaccharides. The salivary amylase (HSA), the most abundant enzyme in human saliva, initiates the digestion of complex carbohydrates in the human oral cavity, where especially starch is partly digested into oligosaccharides, maltose, and glucose [1]. The process is subsequently completed by the pancreatic α-amylase. In humans, five isoenzymes of α-amylase (α-1,4-glucan-4-glucanohydrolase, EC 3.2.1) have been described. The three isoforms of salivary α-amylase and the two isoforms of pancreatic α-amylase are classified as two different families of isoenzymes. The three-dimensional structures of the

α-amylases from human pancreas and saliva and from porcine pancreas have already been determined by X-ray crystallography [2–4]. Structurally these enzymes are all very closely related. Due to its importance in several metabolic disorders including diabetes and obesity, the pancreatic α-amylase has been more extensively studied than the salivary α-amylase. In consequence, a series of pancreatic α-amylase inhibitors are available in the market, such as acarbose, voglibose, and miglitol [5–7]. The administration of these molecules can be a useful first-line treatment for diabetic patients who have a combination of slightly raised basal plasma glucose concentrations and marked postprandial hyperglycemia.

α-Amylase inhibitors help in the prevention and medical treatment of metabolic syndromes such as type 2 diabetes and obesity, in which they control the elevation of blood glucose levels by delaying and blocking postprandial carbohydrate digestion and absorption [8]. Different types of molecules were reported to possess α-amylase inhibitory activity. Among these molecules are flavonoids, polyphenolics, condensed tannins, hydrolysable tannins, terpenes, and cinnamic acid derivatives [9–12]. Tannins are naturally occurring plant polyphenols. Their main characteristic is that they bind proteins, basic compounds, pigments, large molecular weight compounds, and metallic ions and display antioxidant activities. They are amply distributed in nature and are present in fruits, teas, trees, and grasses. Hydrolysable tannins are derivatives of gallic acid (3,4,5-trihydroxybenzoic acid). Gallic acid is esterified to a core polyol, and the galloyl groups may be further esterified or oxidatively cross-linked to yield more complex hydrolysable tannins. One of the most simple and common hydrolysable tannins is the gallotannin with up to 12 esterified galloyl groups and a core glucose (Figure 1(a)) as the gallotannin from Chinese natural gallnuts [13]. Condensed tannins are oligomeric and polymeric proanthocyanidins that can possess different interflavanyl coupling and substitution patterns [13, 14]. One of the most extensively studied proanthocyanidins is that one extracted from the bark of the black wattle tree (Acacia mearnsii). It is rich in the catechin-like flavan-3-ols monomers robinetinidol and fisetinidol (Figure 1(b) [15]).

The aim of the present study is to compare the in vitro inhibitory effects of two tannins with well-known chemical structures on the salivary and pancreatic α-amylases and their putative in vivo hypoglycemic actions. The first one is the hydrolysable tannin from Chinese natural gall and the second one the condensed tannin from A. mearnsii (Figures 1(a) and 1(b)). In the in vitro experiments, especial attention has been devoted to the kinetics of the inhibition, with a detailed search for the model that best describes the mechanism of action.

2. Materials and Methods

2.1. Materials.
Porcine pancreatic α-amylase (Type VI-B), human salivary α-amylase, acarbose, and the hydrolysable tannin from Chinese natural gallnuts were purchased from Sigma-Aldrich Co. The condensed tannin from Acacia mearnsii bark was purchased from Labsynth, Brazil.

2.2. Reaction Rate Measurements.
The kinetic experiments with the porcine pancreatic α-amylase and the human salivary α-amylase were carried out at 37°C in 20 mmol/L phosphate buffer, pH 6.9, containing 6.7 mmol/L NaCl. Both temperature and pH of the assay are close to the optimum values reported in several studies. Potato starch (Sigma-Aldrich) was used as substrate. The substrate (0.05–1.0 g/100 mL) and one of the two inhibitors, A. mearnsii condensed tannin (up to 620 μM) and hydrolysable tannin (tannic acid; up to 294 μM), were mixed and the reaction was initiated by adding the enzyme. The specific activity of both enzymes was 500 units/mg protein. The amount of enzyme added to each reaction system was 1 unit. The reaction was allowed

to proceed for 5 min. The reducing sugars resulting from the starch hydrolysis were assayed by the 3,5-dinitrosalicylic acid (DNS) method, using maltose as standard [18]. The aldehyde group of reducing sugars converts 3,5-dinitrosalicylic acid to 3-amino-5-nitrosalicylic acid, which is the reduced form of DNS. The formation of 3-amino-5-nitrosalicylic acid results in a change in the amount of light absorbed at 540 nm. The absorbance measured using a spectrophotometer is directly proportional to the amount of reducing sugar. The pH of the reaction medium was tested for all situations. No changes were detected during the incubation time.

2.3. Animal Experiments.
Male healthy Wistar rats weighing 200–250 g were used in all experiments. The rats were housed, fed, and treated in accordance with the universally accepted guidelines for animal experimentation. Prior to the investigations, the animals were kept for one week under standard environmental conditions. Throughout the experimentation period, the rats were maintained in single cages and had access to standard pelleted diet and water ad libitum. Food was withdrawn 18 h before the experiments. All experiments involving rats were done in accordance with the worldwide accepted ethical guidelines for animal experimentation and previously approved by the Ethics Committee for Animal Experimentation of the University of Maringá (Protocol number 067/2014-CEUA-UEM).

2.4. Glycemic Levels of Rats after Starch Administration.
Rats were divided into 9 groups (n = 4 rats per group). To group I (positive control) commercial corn starch (1 g per kg body weight) was given intragastrically. Group II (negative control) received only tap water. Group III (positive control) received intragastrically commercial corn starch plus acarbose (50 mg/kg). Groups IV, V, and VI received intragastrically commercial corn starch plus A. mearnsii tannin doses of 100, 250, and 500 mg/kg, respectively. Finally, groups VII, VIII, and IX received intragastrically commercial corn starch plus tannic acid doses of 100, 250, and 500 mg/kg, respectively. The amounts of inhibitors given to the rats were based on literature data [19]. Fasting blood glucose levels were determined before the administration of starch and amylase inhibitors (0 time). Later evaluations of blood glucose levels took place at 15, 30, 45, and 60 min. Blood glucose from cut tail tips was determined using Accu-Chek® Active Glucose Meter.

2.5. Calculations and Statistical Criteria.
Statistical analysis of the data was done by means of the Statistica program (Statsoft, Inc., Tulsa, OK). Numerical interpolation for the determination of the half-maximal inhibitor concentrations (IC$_{50}$) was done using the Scientist Software from MicroMath Scientific Software (Salt Lake City, UT). The same program was used for fitting the rate equations to the experimental initial rates by means of an iterative nonlinear least-squares procedure. The decision about the most adequate model (equation) was based on the model selection criterion (MSC) and on the standard deviations of the optimized parameters.

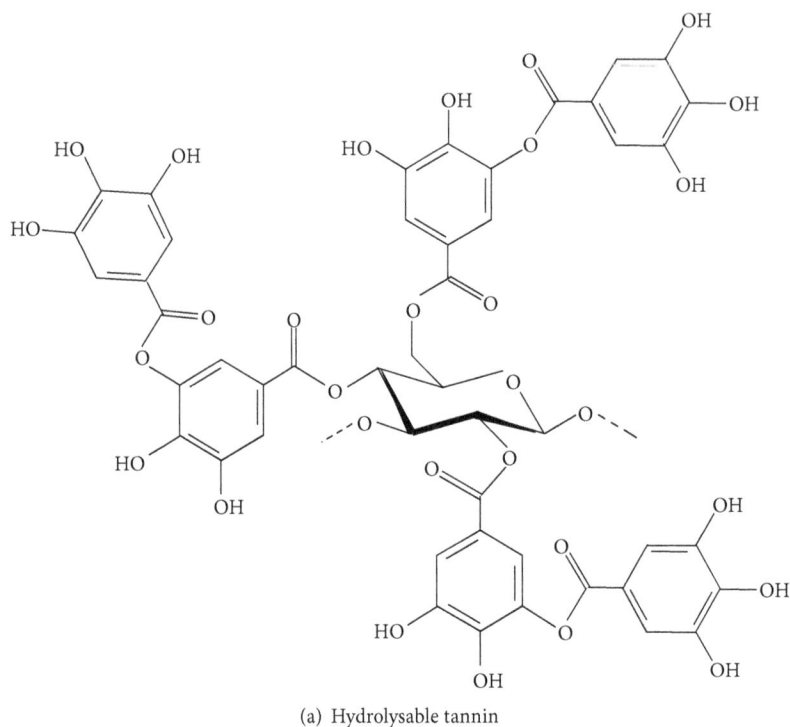

(a) Hydrolysable tannin

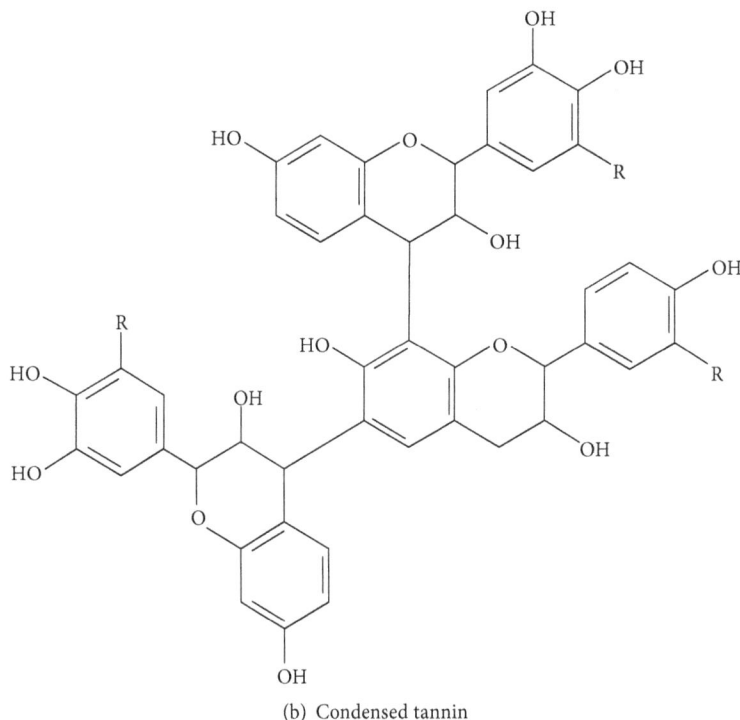

(b) Condensed tannin

FIGURE 1: Repetitive structures of the hydrolysable tannin from Chinese natural gallnuts (tannic acid; mw 1701.2 g/mol [16]) and the condensed tannin from *A. mearnsii* (mw 806.0 g/mol [17]).

The model selection criterion, which corresponds to the normalized Akaike Information Criterion [20], is defined as

$$\text{MSC} = \ln \left[\frac{\sum_{i=1}^{n} w_i \left(Y_{\text{obs}_i} - \overline{Y}_{\text{obs}} \right)^2}{\sum_{i=1}^{n} w_i \left(Y_{\text{obs}_i} - Y_{\text{cal}_i} \right)^2} \right] - \frac{2p}{n}. \quad (1)$$

Y_{obs} are the experimental reaction rates, $\overline{Y}_{\text{obs}}$ is the mean experimental reaction rate, Y_{cal} is the theoretically calculated reaction rate, w is the weight of each experimental point, n is the number of observations, and p is the number of parameters of the set of equations. In the present work, the model with the largest MSC value was considered the most

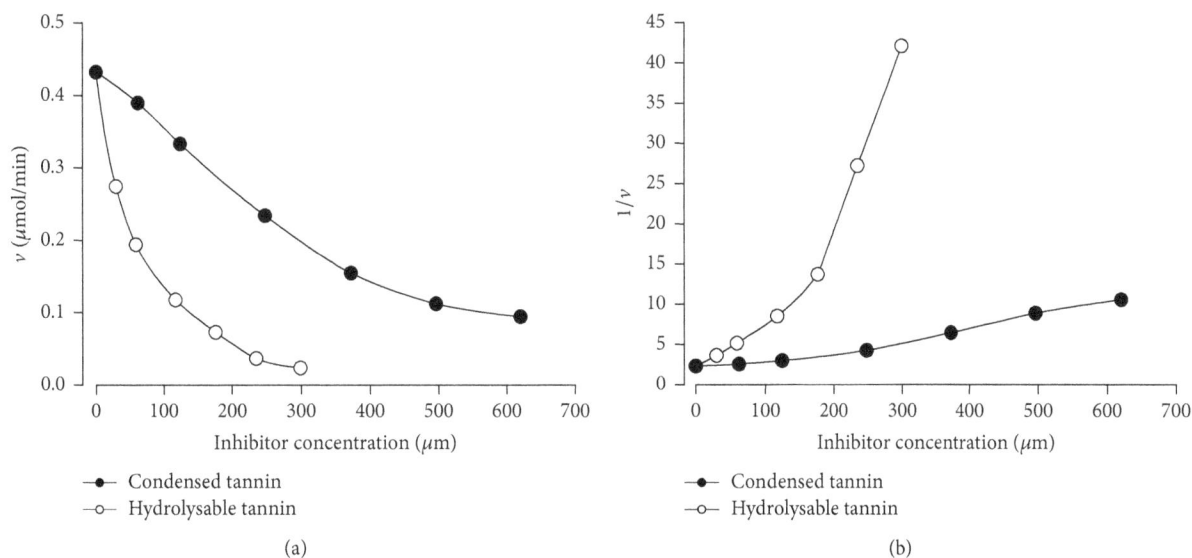

FIGURE 2: Inhibition of the human salivary α-amylase by condensed and hydrolysable tannins: concentration dependence. Initial reaction rates were measured as described in the Material and Methods. Each datum point represents the mean of four independent determinations. (a) Reaction rates (v); (b) inverse reaction rates ($1/v$).

appropriate, provided that the estimated parameters were positive. When the MSC values differed by less than 5%, the mode yielding the smallest standard deviations for the estimated parameters was considered the most appropriate one.

3. Results

3.1. Concentration Dependence of the α-Amylases Inhibition. Initially, inhibition of the activity of both α-amylases by the hydrolysable and condensed tannins was characterized in terms of the corresponding concentration dependence. For this purpose, initial rates were measured at a fixed starch concentration (1 g/100 mL) and variable tannin concentrations. Concentrations were expressed in μmol/L (μM), as the molecular weights of both tannins have already been determined [16, 17] and molar concentrations are much more informative about the number of molecules involved. The results of the experiments with the human salivary α-amylases are summarized in Figure 2. In Figure 2(a) the rates were represented against the inhibitor concentration. It is apparent that both condensed and hydrolysable tannins inhibited the enzyme with clear concentration dependence. From the graph it is also apparent that the hydrolysable tannin is a stronger inhibitor than the condensed tannin. IC$_{50}$ value (the concentration of inhibitor required to reduce the rate of the enzymatic reaction by 50%) allows a quantitative evaluation of the effectiveness of each compound: it is equal to 47 μM for the hydrolysable tannin and 285.4 μM for the condensed tannin. This means that at the starch concentration of 1 g/100 mL the hydrolysable tannin is 6 times more effective as an inhibitor of the salivary α-amylase than the condensed tannin. In Figure 2(b) the reciprocals of the reaction rates ($1/v$) were represented against the corresponding concentrations. In both cases the relationship was parabolic even

though this is less evident for the inhibition caused by the condensed tannin. This occurs because a single $1/v$ scale was used for both inhibitors and the inhibition degree with the hydrolysable tannin is much more pronounced, what causes a much more evident upward concavity.

The results of the measurements that were done with the porcine pancreatic α-amylase are shown in Figure 3. Both the hydrolysable and the condensed tannin inhibited the enzyme. The hydrolysable tannin was again more efficient, though the difference was less pronounced. In fact, IC$_{50}$ for the hydrolysable tannin was 141.1 μM and that for the condensed tannin 248.1 μM. It is also noteworthy, because it has mechanistic implications, that the $1/v$ versus concentration plots revealed parabolic relationships for both inhibitors.

3.2. Kinetics of the Human Salivary α-Amylase Inhibition by the Hydrolysable and Condensed Tannins. When investigating the kinetic mechanism of the inhibitions caused by the hydrolysable and condensed tannins it is indispensable to take into account the form of the $1/v$ versus [I] plots shown in Figure 2. The parabolic relationships reveal that more than one inhibitor molecule can bind to at least one enzyme form [21, 22]. There are several mechanistic possibilities. The best way of investigating this is to measure the reaction rates by varying simultaneously the substrate concentration and the inhibitor concentration with subsequent model analysis in order to find out the mechanism that gives the best description of the experimental data. The results of the experiments that were done with the human salivary α-amylase are shown in Figure 4. Both the hydrolysable and the condensed tannins showed saturation curves that were progressively lowered as the tannins were added at progressively increased concentrations. The saturation curves do not show any tendency of convergence at high substrate concentrations, which excludes the possibility of competitive

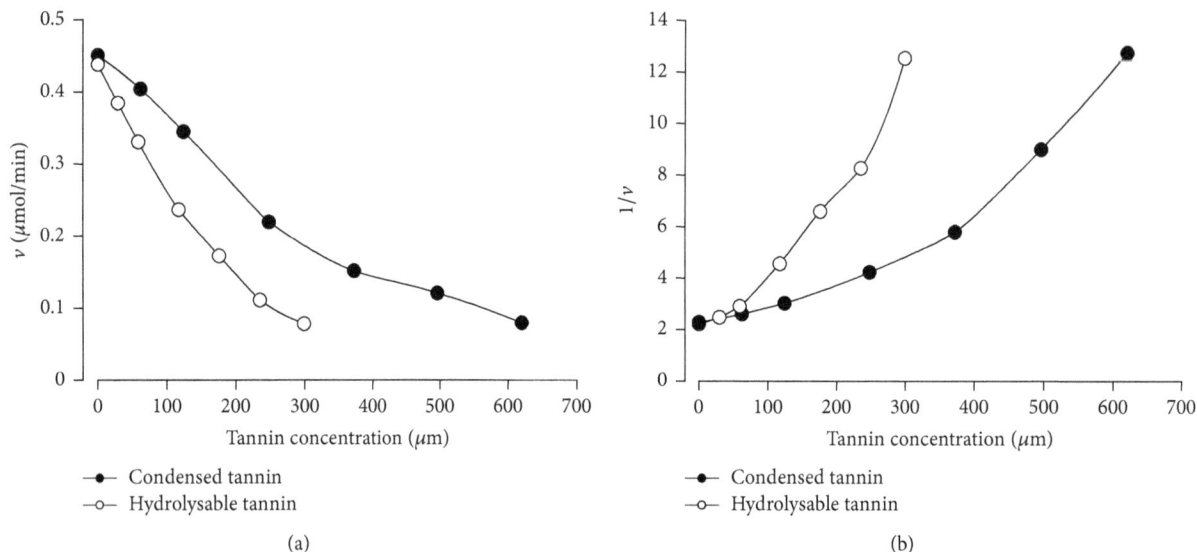

FIGURE 3: Concentration dependence of the porcine pancreatic α-amylase inhibition caused by the condensed and hydrolysable tannins: concentration dependence. Each datum point is the mean of three determinations. Reaction rates (v) and reciprocals of the reaction rates ($1/v$) were represented against the inhibitor concentrations.

inhibition [21, 22]. Likewise, there is no decrease in the inhibition degree at low substrate concentrations, which would be indicative of uncompetitive inhibition [21, 22]. Most likely, thus, mixed (competitive-noncompetitive) inhibition must be considered in addition to the probability that at least

two inhibitor molecules can bind to at least one form of the enzyme. The complete equation that applies to a mechanism in which the inhibitor binds twice and sequentially to the free enzyme (E) and to the enzyme-substrate complex (ES) is [22, 23]

$$v = \frac{V_{\max} [S]}{K_M \left(1 + [I]/K_{i1} + [I]^2/K_{i1}K_{i2}\right) + [S]\left(1 + [I]/K_{i3} + [I]^2/K_{i3}K_{i4}\right)}. \tag{2}$$

In (2), $V_{\max}$ is the maximal reaction rate, K_M the Michaelis-Menten constant, [S] the substrate concentration, and [I] the inhibitor concentration. The following inhibitory complexes are allowed: EI, EI$_2$, ESI, and ESI$_2$; K_{i1}, K_{i2}, K_{i3}, and K_{i4} are the corresponding dissociation constants of these complexes (inhibitor constants). If one of these complexes is lacking, limiting forms of (2) will apply [21–23]. It must be noted that the squared inhibitor concentration ($[I]^2$) accounts for the parabolic inhibition. Agreement between theory and experiment was tested by means of a least-squares fitting procedure. Fitting was done simultaneously with two independent variables ([S] and [I]), including the rate versus inhibitor concentration data shown in Figure 2. Attempts of fitting (2) to the set of data in Figure 4(a) (hydrolysable tannin) failed in that it was not possible to distinguish K_{i3} and K_{i4}. This means that the enzyme-substrate complex (ES) forms only one type of complex with the inhibitor, which could be either ESI or ESI$_2$. The latter implies in a simultaneous or almost simultaneous binding of two inhibitor molecules to the enzyme. After fitting the equations corresponding to 10 mechanistic possibilities, the best fit was achieved with (3), which describes the

mechanism that allows the formation of complexes EI, EI$_2$, and ESI:

$$v = \frac{V_{\max} [S]}{K_M \left(1 + [I]/K_{i1} + [I]^2/K_{i1}K_{i2}\right) + [S]\left(1 + [I]/K_{i3}\right)}. \tag{3}$$

All parameters have the meaning already described above. The continuous lines in Figure 4(a) represent the curves calculated by introducing the optimized parameters, given in the legend of Figure 4, into (3). It should be remarked that K_{i1} is much smaller than K_{i2} (26-fold). This means that the first binding of the hydrolysable tannin to the free enzyme occurs much more readily than the second one. Furthermore, K_{i3} is 46-fold higher than K_{i1}, indicating that the complex ESI forms only at relatively high concentrations of the hydrolysable tannin. The legend of Figure 4 also gives the values of the sum of squared deviations and the model selection criterion (MSC), on which the decision about the most probable mechanism was based (see Materials and Methods). It should be stressed that (3) describes quite well both v versus [S] and v versus [I] curves (Figure 4(a)). Only at the highest [I] values a small systematic deviation was found,

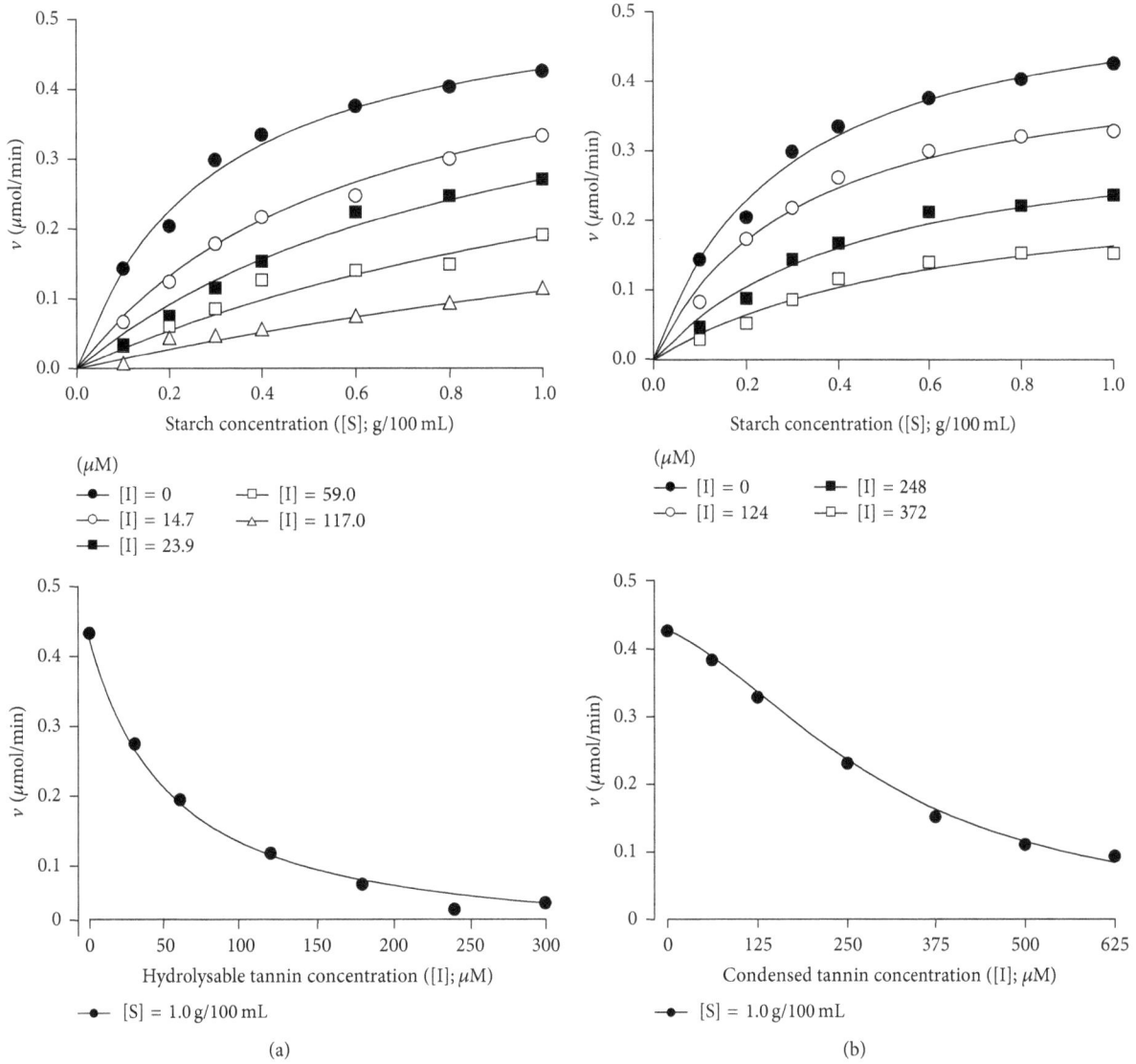

FIGURE 4: Reaction rates of the human salivary α-amylase obtained by varying simultaneously the substrate (starch) and the hydrolysable tannin (a) or condensed tannin (b) concentrations. Each datum point is the mean of four determinations. The lines running through the experimental points were calculated using optimized parameters obtained by fitting (3) ((a) hydrolysable tannin) or (4) ((b) condensed tannin) to the experimental data by means of a nonlinear least-squares procedure. Values of the optimized parameters and goodness-of-fit indicators for the hydrolysable tannin data (panel (a); equation (3)) are K_M, 0.290 ± 0.027 g/100 mL; $V_{\max}$, 0.553 ± 0.018 μmol/min; K_{i1}, 13.2 ± 1.7 μM; K_{i2}, 343.4 ± 119.5 μM; K_{i3}, 609.6 ± 1086.2 μM; sum of squared deviations, 0.00489; MSC, 4.535. For the condensed tannin data (panel (b); equation (4)) the optimized parameters are K_M, 0.281 ± 0.024 g/100 mL; $V_{\max}$, 0.548 ± 0.016 μmol/min; $K_{i1\text{-}2}$, 194.9 ± 15.6 μM; K_{i3}, 705.9 ± 192.5 μM; K_{i4}, 369.6 ± 233.6 μM; sum of squared deviations, 0.00381; MSC, 4.493.

which could be indicating the existence of a small fraction of ESI_2 complex.

Fitting of (2) to the data obtained with the condensed tannin (Figure 4(b)) was troubled by the impossibility of discriminating between K_{i1} and K_{i2}. This could be indicating that the free enzyme (E) forms only one type of complex with the inhibitor, which could be EI or EI_2. After fitting the equations corresponding to 10 mechanistic possibilities, the best fit was achieved with (4), which describes the

mechanism that allows the formation of complexes EI_2, ESI, and ESI_2:

$$v = \frac{V_{\max}[S]}{K_M\left(1 + [I]^2/(K_{i1\text{-}2})^2\right) + [S]\left(1 + [I]/K_{i3} + [I]^2/K_{i3}K_{i4}\right)}. \quad (4)$$

$K_{i1\text{-}2}$ is the dissociation constant for the complex EI_2 (formed by the reaction $E + 2I \rightarrow EI_2$) and all other parameters have the meanings already specified above. In this particular

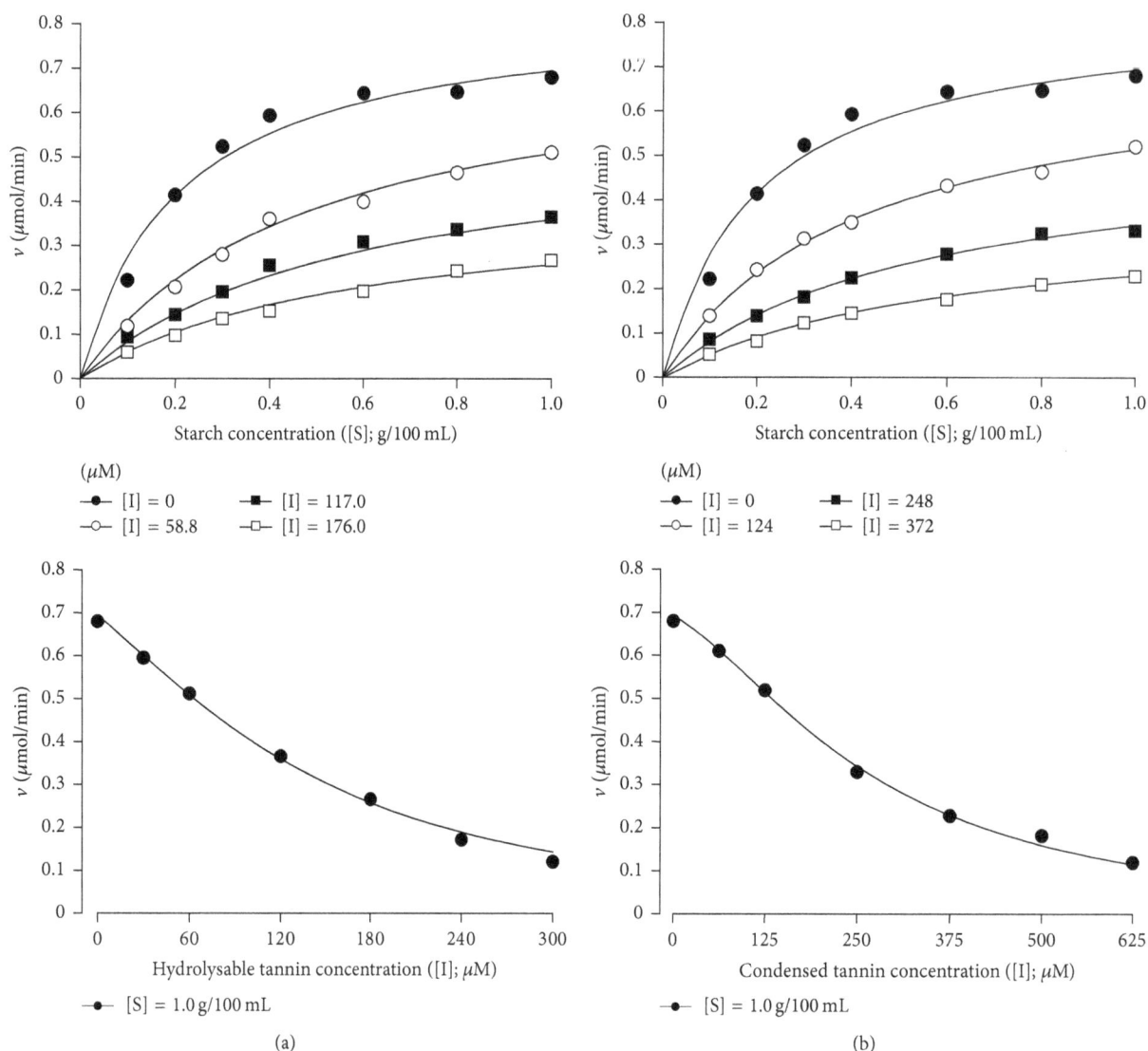

FIGURE 5: Reaction rates of the porcine pancreatic α-amylase measured by varying simultaneously the substrate (starch) and the hydrolysable tannin (a) or condensed tannin (b) concentrations. Each datum point is the mean of four determinations. The lines running through the experimental points were calculated using optimized parameters obtained by fitting (5) ((a) hydrolysable tannin) or (6) ((b) condensed tannin) to the experimental data by means of a nonlinear least-squares procedure. Values of the optimized parameters and goodness-of-fit indicators for the hydrolysable tannin data (panel (a); equation (5)) are K_M, 0.207 ± 0.017 g/100 mL; V_{max}, 0.839 ± 0.020 µmol/min; K_{i1}, 37.7 ± 3.7 µM; K_{i3-4}, 169.6 ± 10.6 µM; sum of squared deviations, 0.0101; MSC, 4.545. For the condensed tannin data (panel (b); equation (6)) the optimized parameters are K_M, 0.201 ± 0.015 g/100 mL; V_{max}, 0.832 ± 0.018 µmol/min; K_{i1}, 147.0 ± 21.5 µM; K_{i2}, 663.2 ± 345.4 µM; K_{i3-4}, 441.6 ± 37.8 µM; sum of squared deviations, 0.00786; MSC, 4.793.

case binding of two inhibitor molecules to the free enzyme occurs simultaneously or nearly so in such a way that the complex EI is practically absent. Comparison of theory and experiment in Figure 4(b) reveals a very good agreement, without systematic deviations at high inhibitor concentrations. Comparison of the numerical values of the inhibitor constants reveals stronger binding to the free enzyme (E) as K_{i1-2} is considerably smaller than K_{i3} by a factor of 3.6. Singularly, the second binding to the substrate-complexed enzyme (i.e., the formation of ESI_2) is facilitated over the first one, as K_{i4} is smaller than K_{i3}.

K_M and V_{max} values obtained when fitting (4) to the condensed tannin data and those obtained when (3) was fitted to the hydrolysable tannin data were practically the same, as given in the legend of Figure 4. This is actually expected because the data were obtained with the same enzyme, but agreement also speaks in favour of the correctness and reliability of the numerical analyses.

3.3. Kinetics of the Porcine Pancreatic α-Amylase Inhibition by the Hydrolysable and Condensed Tannins. The results of the experiments that were done with the hydrolysable tannin are shown in Figure 5(a). Simple inspection reveals many qualitative similarities to the inhibition caused by the hydrolysable tannin on the human salivary α-amylase (Figure 4(a)). In the search for the best mechanism that describes the set of data in Figure 5(a) the equations corresponding to 10 different mechanisms were fitted to the experimental data, including (2). Here again fitting of the complete (2) was unsuccessful and the best fitting was achieved with an equation that predicts the formation of complexes EI and ESI$_2$:

$$v = \frac{V_{max}\,[S]}{K_M\left(1 + [I]/K_{i1}\right) + [S]\left(1 + [I]^2/\left(K_{i3\text{-}4}\right)^2\right)}. \quad (5)$$

In (5), $K_{i3\text{-}4}$ is the dissociation constant for the complex ESI$_2$ (formed by the reaction ES + 2I $\rightarrow$ ESI$_2$) and all other symbols have the same meanings specified above. The optimized parameters are listed in the legend of Figure 5. As can be deduced from the graphs in Figure 5(a), the calculated curves agree pretty well with the experimental ones with no systematic deviations at the extremes of both substrate and inhibitor concentrations. K_{i1} is 4.5 times smaller than $K_{i3\text{-}4}$; the free enzyme, thus, binds much more strongly the hydrolysable tannin than the substrate-complexed enzyme.

The results of the kinetic investigations on the inhibition caused by the condensed tannin on the pancreatic α-amylase are shown in Figure 5(b). In this case, the best fit was found with an equation that allows the formation of complex EI in addition to the complexes EI$_2$ and ESI$_2$:

$$v$$
$$= \frac{V_{max}\,[S]}{K_M\left(1 + [I]/K_{i1} + [I]^2/K_{i1}K_{i2}\right) + [S]\left(1 + [I]^2/\left(K_{i3\text{-}4}\right)^2\right)}. \quad (6)$$

K_{i2} is the inhibitor constant for the formation of complex EI$_2$ from complex EI. Agreement between theory and experiment was as good as in the preceding analyses, as can be concluded by inspecting the graphs in Figure 5(b) and the statistical parameters in the legend of Figure 5. As in all preceding cases, formation of the EI complex is greatly favoured in comparison with the formation of all other complexes, as $K_{i3\text{-}4}$ (ESI$_2$) exceeds K_{i1} by a factor of 3.0 and K_{i2} (EI$_2$) exceeds K_{i1} by a factor of 4.5.

3.4. In Vivo Inhibition of α-Amylase. For testing in vivo the inhibition caused by both the hydrolysable and condensed tannin, starch was given to rats and the glycemic levels were followed during 60 minutes. The basis for these experiments is given by the well-established notion that hydrolysis of intragastrically administered starch is a prerequisite for the entrance of the derived glucosyl units into blood. Figure 6(a) shows the time course of the experiments that were done by administering various doses of the hydrolysable tannin. When an aqueous solution of starch was administered alone the glycemic levels raised producing a concave down curve with a peak increment of 85% at 30 minutes after administration. When water was administered the glycemic levels remained relatively constant. Administration of starch in combination with various doses of hydrolysable tannin

produced increases in the glycemic levels that were less pronounced than those found when starch was administered alone. A dose-dependent effect is apparent. In all cases, however, concave down curves were obtained. Starch plus acarbose administration, the positive control experiment, also diminished the increase in blood glucose concentration, especially during the first 30 minutes, with a peak at 45 minutes.

Figure 6(b) shows the results of the experiments done with the condensed tannin. The control curves are the same shown in Figure 6(a). Coadministration of starch and condensed tannin resulted in diminished increases in the glycemic levels. However, the fivefold increase in the administered dose (124.1 to 620.0 μmol/kg) did not result in a pronounced enhancement of the effect. This phenomenon can be best appreciated by comparing the areas under the glycemic curves in Figure 7. The areas were computed numerically and subtracted from the area under the curve obtained when water was administered alone. This area can be regarded as a measure of the extra glucose in the circulating blood during the first 60 minutes following starch and tannin administration. Figure 7 shows that the action of the hydrolysable tannin shows a well-defined dose-dependent action. The lowest dose already diminished the glycemic response by 53%; with the highest dose the diminution reached 88%. The action of the condensed tannin was similar to that of the hydrolysable tannin at the lowest dose (49%), but further increases in the administered dose were poorly effective, as the highest dose reduced the glycemic response by not more than 57%.

4. Discussion

Inhibition of the human salivary and porcine pancreatic α-amylases by both the hydrolysable and condensed tannins presents several complexities in that for both inhibitors more than one molecule can bind simultaneously to the enzymes [21, 22]. This is revealed a priori by the nonlinear $1/v$ versus [I] plots and confirmed by the numerical analysis in which attempts of fitting an equation describing linear inhibition (single binding) always produced unfavourable results. Even assuming some limited degree of heterogeneity for the preparations that were used, especially for the condensed tannin [17], it should be remarked that the phenomenon does not invalidate (2) or its limiting forms, provided that all concentrations are kept at constant ratios as it occurs when different amounts of the same preparation are added [21, 22]. In the latter case, however, the inhibition constants are no longer true dissociation constants but rather complex functions of several individual dissociation constants. They remain, notwithstanding, a measure of the potency of a given inhibitor [21–23]. Parabolic inhibition is a common phenomenon among phenolics and tannins. The inhibition of α-amylases by a *pinhão* coat tannin [24] and by the *Phaseolus* protein inhibitorα-AI [23] has been reported to be parabolic. Inhibition of the pancreatic lipase by a *pinhão* coat tannin is also of the parabolic type [25]. Furthermore, the fact that the same phenomenon occurs with a pure and well-defined substance such as acarbose, depending on the substrate [23, 26], is a proof that it is not generated by an eventual

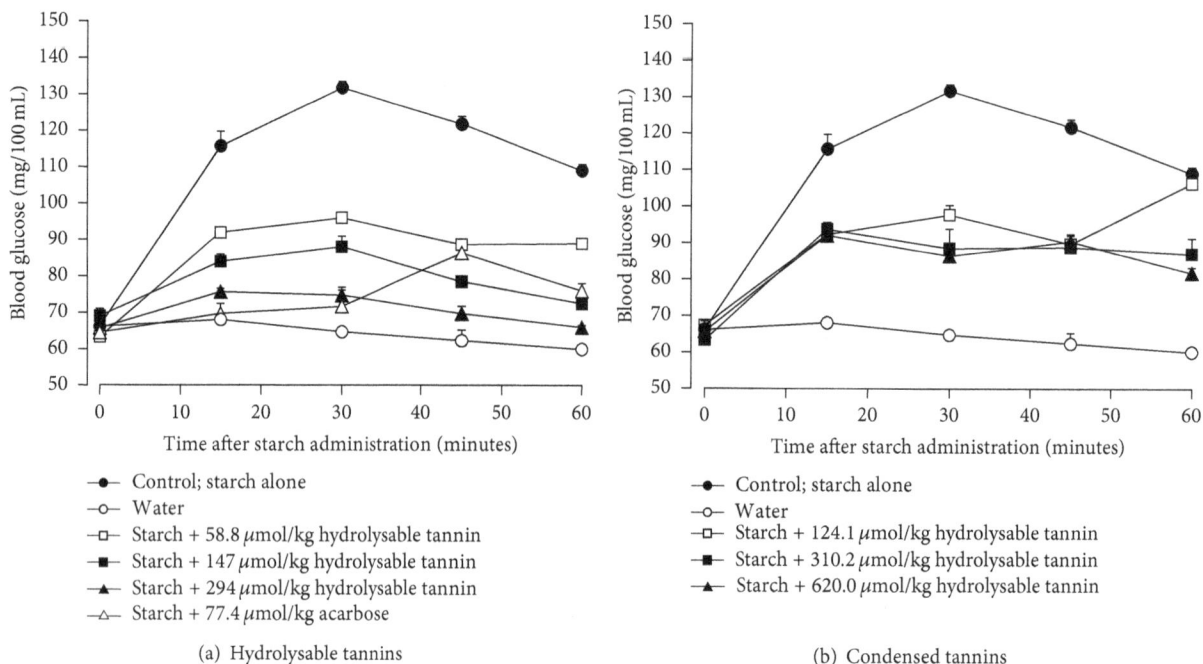

FIGURE 6: Influence of the hydrolysable tannins and acarbose administration on the glycemic levels of fasting rats during 60 min following starch administration. Blood samples from the tail vein were analyzed by means of a glucometer after intragastric starch administration (1 g per kg body weight). The hydrolysable and condensed tannins as well as acarbose were administered intragastrically at the doses given on the top. Each datum point represents the mean ± mean standard errors of three experiments. Experimental details are given in the Materials and Methods.

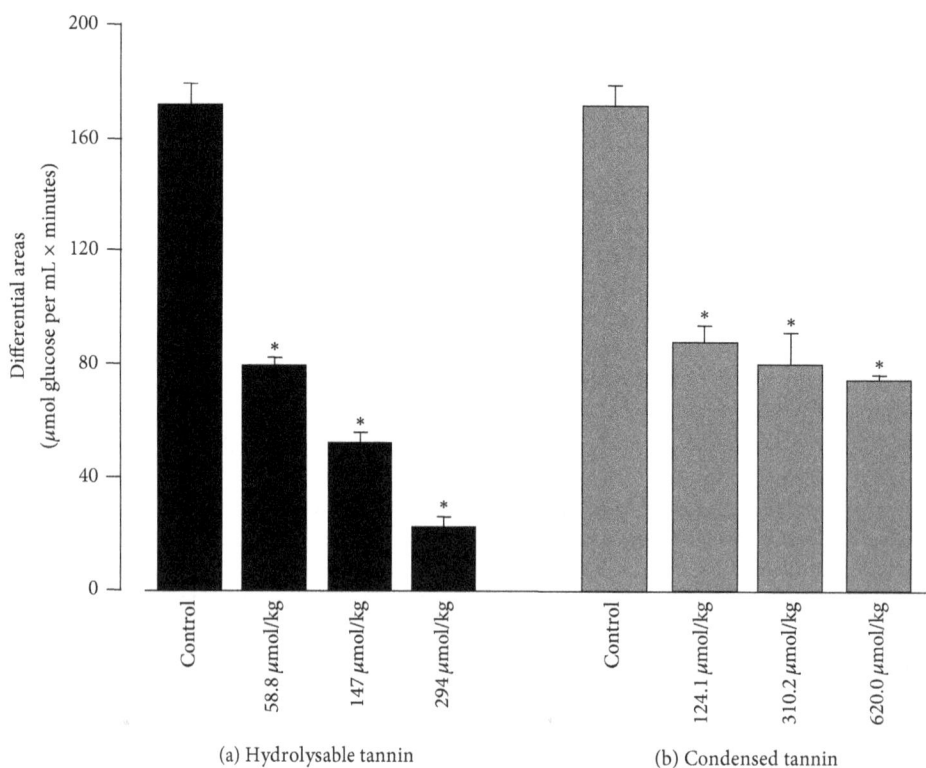

FIGURE 7: Areas between the glycemic curves after starch administration (starch alone or starch + α-amylase inhibitors) and the glycemic basal levels. The areas were determined with the corresponding data in Figures 6(a) and 6(b) using the numerical integration procedures of the Scientist Software from MicroMath Scientific Software (Salt Lake City, UT). The error terms correspond to standard errors of the means. Asterisks indicate statistical difference relative to the control experiment according to ANOVA followed by post hoc Student-Newman-Keuls testing ($p \leq 0.05$).

heterogeneity of the inhibitor. On the other hand, on some occasions the phenomenon has been neglected. For example, the inhibition of the human α-amylase by a gallotannin was analyzed as being of the linear type even though the Dixon plots ($1/v$ versus [I]) that were presented are clearly indicating parabolic inhibition [27]. It should be noted that, in the experiments in which the substrate concentration was varied, the maximal tannin concentrations were smaller than those used in v versus [I] experiments. This occurred because it is difficult to measure accurately low initial reaction rates at low substrate concentrations. Even so, the description of v versus [I] relationships by the fitted equations was very good, with minimal deviations at the highest inhibitor concentrations. For all cases analyzed in the present work inhibition was of the mixed (competitive-noncompetitive) type. This is the most frequently reported mode of inhibition [23–27]. The longan pericarp proanthocyanidins, however, have been reported as a singular case of uncompetitive inhibition on the α-amylase [28].

By comparison of our data on both tannins with those on acarbose in the literature [23, 24, 26] it is obvious that acarbose is a much better inhibitor of the pancreatic α-amylase than the tannins. The question of which tannin is more effective can be unambiguously answered in the case of the human salivary α-amylase inhibition. The hydrolysable tannin concentration for half-maximal inhibition (IC_{50}) of the salivary enzyme is considerably smaller than that of the condensed tannin ($47\,\mu M$ compared to $285\,\mu M$) and the same can be said about the tendency of binding to the free enzyme, which is 14.7 times superior in the case of the hydrolysable tannin. With respect to the pancreatic enzyme the difference is not as pronounced. In terms of their masses and at the substrate concentration of $1\,g/100\,mL$ the condensed tannin was a moderately better inhibitor because 50% inhibition occurred at the concentration of $141\,\mu M$ compared to the $248\,\mu M$ required for the same degree of inhibition by the hydrolysable tannin. However, the inhibition degree will vary with the substrate concentration and at low substrate concentrations the hydrolysable tannin will be a better inhibitor because of the smaller value of its inhibition constant K_{i1} (see legend of Figure 5).

A characteristic of the inhibitory action of both the hydrolysable and the condensed tannin is the observation that they are bound more tightly by the free enzyme (E) than by the enzyme complexed with the substrate (SE). This is revealed by the observation that for both enzymes K_{i1} (or K_{i-12}) values for the hydrolysable and the condensed tannins were always considerably smaller than K_{i3} or K_{i3-4} values. This suggests that binding of the substrate, which is a large molecule, to the enzyme promotes structural modifications that make binding of the inhibitor more difficult in the case of the tannins.

Both hydrolysable and condensed tannins are well-known protein precipitating agents. Two different models have been suggested for explaining the capability of tannins to precipitate proteins [29]. It has been proposed that nonpolar tannins, such as pentagalloyl glucose, form a hydrophobic coat around the proteins, whereas polar tannins, such as epicatechin$_{16}$ (4 $\rightarrow$ 8), form hydrogen-bonded cross-links

between the protein molecules [29]. However, precipitation of proteins requires high concentrations of tannins. In the present and other studies, inhibition of the human salivary α-amylase by both the hydrolysable and the condensed tannin occurred in the presence of relatively low tannin concentrations, implying that the inhibitory action is due to specific molecular interactions involving specific amino acid residues of the protein and well-defined structural parts of the tannins. Binding of tannins to proteins have been demonstrated by methods that do not involve kinetics and they have been investigated for several combinations of tannins and proteins, including enzymes [30–32]. It can be deduced from these studies that binding of tannins to proteins involves both hydrophilic and hydrophobic interactions. It can be nonspecific in some cases and specific with a certain degree of cooperation in others [30]. In the case of the hydrolysable and condensed tannins used in the present study, binding is certainly a complex phenomenon, as indicated by the parabolic inhibition kinetics and probably facilitated by the numerous hydroxyl groups present in these molecules (see Figure 1). These groups could be especially important for the interactions at low concentrations [31]. The higher density of these groups on the hydrolysable tannin could be an explanation for its higher degrees of inhibition at low concentrations, especially with respect to the human salivary α-amylase. Consistent with this is the fact that hydroxylation of flavonoids improves the inhibitory effect on α-amylase exerted by these compounds [33]. At high concentrations, however, random hydrophobic stacking of the planar aromatic rings may occur between tannin and protein [30]. In this context, based on data from surface plasmon resonance binding experiments and nuclear magnetic resonance analyses [34] it has been proposed that the inhibitory effect caused by pentagalloyl glucose on the human salivary α-amylase results from the interaction of aromatic rings of the former with aromatic amino acids of the protein. The role of the aromatic amino acids of the human salivary α-amylase in the pentagalloyl glucose binding was reinforced by kinetic studies with W58L and Y151M mutants of the enzyme: replacement of the aromatic amino acids in the active site by aliphatic ones decreased inhibition dramatically, what seems to be in accordance with a participation of these residues in the interaction of tannins with the human salivary α-amylase [34].

Inhibition of α-amylase (as well as α-glucosidase) resulted in delayed carbohydrate digestion and glucose absorption with attenuation of postprandial hyperglycemic excursions. The diminution of hyperglycemia in rats to which starch was given by both the hydrolysable and condensed tannin is thus an expected phenomenon. Demonstration of the phenomenon also proofs that both tannins are able to exert α-amylase inhibition under in vivo conditions and not only in the test tube. In spite of the relatively small difference in their inhibitory activities toward the enzyme, the hydrolysable tannin was more effective in lowering hyperglycemia when compared to the condensed tannin. The response practically ceased to increase with condensed tannin doses above $124.1\,\mu mol/kg$ ($100\,mg/kg$), whereas the effects of the hydrolysable tannin increased progressively until the dose of 620 μmol/kg. The reasons for this behavior are not clear. One possible reason is that the hydrolysable tannin is active on other enzymes equally involved in starch digestion as the α-glucosidases and invertases, for example, whereas the condensed tannin is inactive or less active [19, 35]. Such a phenomenon would enhance the effectiveness of the hydrolysable tannin. Another possible reason, which does not exclude the former, is that the two types of tannins could be suffering the consequence of different gastric events and movements able to affect their effectiveness as inhibitors.

Besides participating in the initial hydrolysis of starch and other carbohydrate constituents of the diet, the salivary α-amylase exerts two additional functions, namely, binding to the tooth surface and binding to oral streptococci [36–38]. All three actions contribute to the process of dental plaque and caries formation. Binding to the enzyme is likely to restrict these three activities. A number of studies have shown that tea extracts (*Camellia sinensis* (L.) Kuntze) reduce dental caries [39, 40]. Based on the α-amylase inhibitory activity of tea extracts, the hypothesis has been raised that this activity could be involved in the reduction of the cariogenicity of starch-containing foods [41, 42]. For this reason, the hydrolysable tannin, which is bound very strongly by the enzyme, as indicated by K_{i1} value of 13.2 μM, can be regarded as a useful agent for oral health.

5. Conclusion

In conclusion, both tannins are potentially useful in controlling the postprandial glycemic levels in diabetic patients, with the hydrolysable one, however, being superior. To our knowledge, the present study is the first one that presents a comparison of the effects of both types of tannins under exactly the same conditions. Clinical studies are evidently indispensable for evaluating the viability and safety of the use of preparations containing the hydrolysable or even condensed tannins, especially as food additives. With reference to the latter, more studies are also necessary to evaluate the possibility of incorporating this tannin into dental products such as dentifrices, mouthwashes, dental flosses, and chewing gums that could be helpful in the prevention of dental caries.

Acknowledgments

This work was financially supported by grants from the Conselho Nacional de Desenvolvimento Científico e Tecnológico (CNPq) and Coordenação do Aperfeiçoamento de Pessoal do Ensino Superior (CAPES).

References

[1] C. Boehlke, O. Zierau, and C. Hannig, "Salivary amylase—the enzyme of unspecialized euryphagous animals," *Archives of Oral Biology*, vol. 60, no. 8, pp. 1162–1176, 2015.

[2] G. D. Brayer, Y. Luo, and S. G. Withers, "The structure of human pancreatic α-amylase at 1.8 Å resolution and comparisons with related enzymes," *Protein Science*, vol. 4, no. 9, pp. 1730–1742, 1995.

[3] N. Ramasubbu, V. Paloth, Y. Luo, G. D. Brayer, and M. J. Levine, "Structure of human salivary α-amylase at 1.6 Å resolution: implications for its role in the oral cavity," *Acta Crystallographica Section D: Biological Crystallography*, vol. 52, no. 3, pp. 435–446, 1996.

[4] M. Qian, R. Haser, and F. Payan, "Structure and molecular model refinement of pig pancreatic α-amylase at 2.1 Å resolution," *Journal of Molecular Biology*, vol. 231, no. 3, pp. 785–799, 1993.

[5] T. Fujisawa, H. Ikegami, K. Inoue, Y. Kawabata, and T. Ogihara, "Effect of two α-glucosidase inhibitors, voglibose and acarbose, on postprandial hyperglycemia correlates with subjective abdominal symptoms," *Metabolism: Clinical and Experimental*, vol. 54, no. 3, pp. 387–390, 2005.

[6] A. S. Dabhi, N. R. Bhatt, and M. J. Shah, "Voglibose: an alpha glucosidase inhibitor," *Journal of Clinical and Diagnostic Research*, vol. 7, no. 12, pp. 3023–3027, 2013.

[7] E. Y. Lee, S. Kaneko, P. Jutabha et al., "Distinct action of the α-glucosidase inhibitor miglitol on SGLT3, enteroendocrine cells, and GLP1 secretion," *Journal of Endocrinology*, vol. 224, no. 3, pp. 205–214, 2015.

[8] S. Jayaraj, S. Suresh, and R.-K. Kadeppagari, "Amylase inhibitors and their biomedical applications," *Starch/Staerke*, vol. 65, no. 7-8, pp. 535–542, 2013.

[9] S. Devarajan and S. Venugopal, "Antioxidant and α-amylase inhibition activities of phenolic compounds in the extracts of Indian honey," *Chinese Journal of Natural Medicines*, vol. 10, no. 4, pp. 255–259, 2012.

[10] N. Ikarashi, W. Sato, T. Toda, M. Ishii, W. Ochiai, and K. Sugiyama, "Inhibitory effect of polyphenol-rich fraction from the bark of Acacia mearnsii on itching associated with allergic dermatitis," *Evidence-Based Complementary and Alternative Medicine*, vol. 2012, Article ID 120389, 9 pages, 2012.

[11] M. Miao, B. Jiang, H. Jiang, T. Zhang, and X. Li, "Interaction mechanism between green tea extract and human α-amylase for reducing starch digestion," *Food Chemistry*, vol. 186, pp. 20–25, 2015.

[12] Á. Zajácz, G. Gyémánt, N. Vittori, and L. Kandra, "Aleppo tannin: structural analysis and salivary amylase inhibition," *Carbohydrate Research*, vol. 342, no. 5, pp. 717–723, 2007.

[13] F. Melone, R. Saladino, H. Lange, and C. Crestini, "Tannin structural elucidation and quantitative 31P NMR analysis. 1. Model compounds," *Journal of Agricultural and Food Chemistry*, vol. 61, no. 39, pp. 9307–9315, 2013.

[14] H. D. Naumann, A. E. Hagerman, B. D. Lambert, J. P. Muir, L. O. Tedeschi, and M. M. Kothmann, "Molecular weight and protein-precipitating ability of condensed tannins from warm-season perennial legumes," *Journal of Plant Interactions*, vol. 9, no. 1, pp. 212–219, 2014.

[15] R. Kusano, S. Ogawa, Y. Matsuo, T. Tanaka, Y. Yazaki, and I. Kouno, "α-amylase and lipase inhibitory activity and structural characterization of acacia bark proanthocyanidins," *Journal of Natural Products*, vol. 74, no. 2, pp. 119–128, 2011.

[16] L. Costadinnova, M. Hristova, T. Kolusheva, and N. Stoilova, "Conductometric study of the acidity properties of tannic acid (Chinese Tannin)," *Journal of the University of Chemical Technology and Metallurgy*, vol. 47, no. 3, pp. 289–296, 2012.

[17] A. De Cássia Oliveira Carneiro, B. R. Vital, A. M. M. L. Carvalho, A. C. Oliveira, B. L. C. Pereira, and B. G. De Andrade, "Tannins molar mass determination using gel permeation chromatography technique," *Scientia Forestalis*, vol. 38, no. 87, pp. 419–429, 2010.

[18] G. L. Miller, "Use of dinitrosalicylic acid reagent for determination of reducing sugar," *Analytical Chemistry*, vol. 31, no. 3, pp. 426–428, 1959.

[19] N. Ikarashi, R. Takeda, K. Ito, W. Ochiai, and K. Sugiyama, "The inhibition of lipase and glucosidase activities by acacia polyphenol," *Evidence-Based Complementary and Alternative Medicine*, vol. 2011, Article ID 272075, 8 pages, 2011.

[20] H. Akaike, "A new look at the statistical model identification," *IEEE Transactions on Automatic Control*, vol. 19, pp. 716–723, 1974.

[21] W. W. Cleland, "The kinetics of enzyme-catalyzed reactions with two or more substrates or products. II. Inhibition: nomenclature and theory," *Biochimica and Biophysical Acta—Biochimica et Biophysica Acta*, vol. 67, pp. 173–187, 1963.

[22] K. M. Plowman, *Enzyme Kinetics*, McGraw-Hill Book Company, New York, NY, USA, 1972.

[23] V. Desseaux, R. Koukiekolo, Y. Moreau, M. Santimone, and G. Marchis-Mouren, "Mechanism of porcine pancreatic α-amylase: inhibition of amylose and maltopentaose hydrolysis by various inhibitors," *Biologia*, vol. 57, no. 11, pp. 163–170, 2002.

[24] S. M. da Silva, E. A. Koehnlein, A. Bracht et al., "Inhibition of salivary and pancreatic α-amylases by a pinhão coat (araucaria angustifolia) extract rich in condensed tannin," *Food Research International*, vol. 56, pp. 1–8, 2014.

[25] R. F. Oliveira, G. A. Gonçalves, F. D. Inácio et al., "Inhibition of pancreatic lipase and triacylglycerol intestinal absorption by a Pinhão coat (Araucaria angustifolia) extract rich in condensed tannin," *Nutrients*, vol. 7, no. 7, pp. 5601–5614, 2015.

[26] M. Alkazaz, V. Desseaux, G. Marchis-Mouren, F. Payan, E. Forest, and M. Santimone, "The mechanism of porcine pancreatic α-amylase. Kinetic evidence for two additional carbohydrate-binding sites," *European Journal of Biochemistry*, vol. 241, no. 3, pp. 787–796, 1996.

[27] L. Kandra, G. Gyémánt, Á. Zajácz, and G. Batta, "Inhibitory effects of tannin on human salivary α-amylase," *Biochemical and Biophysical Research Communications*, vol. 319, no. 4, pp. 1265–1271, 2004.

[28] C. Fu, X. Yang, S. Lai, C. Liu, S. Huang, and H. Yang, "Structure, antioxidant and α-amylase inhibitory activities of longan pericarp proanthocyanidins," *Journal of Functional Foods*, vol. 14, pp. 23–32, 2015.

[29] A. E. Hagerman, M. E. Rice, and N. T. Ritchard, "Mechanisms of protein precipitation for two tannins, pentagalloyl glucose and epicatechin16 (4→8) catechin (Procyanidin)," *Journal of Agricultural and Food Chemistry*, vol. 46, no. 7, pp. 2590–2595, 1998.

[30] R. A. Frazier, A. Papadopoulou, I. Mueller-Harvey, D. Kissoon, and R. J. Green, "Probing protein-tannin interactions by isothermal titration microcalorimetry," *Journal of Agricultural and Food Chemistry*, vol. 51, no. 18, pp. 5189–5195, 2003.

[31] O. Cala, N. Pinaud, C. Simon et al., "NMR and molecular modeling of wine tannins binding to saliva proteins: revisiting astringency from molecular and colloidal prospects," *FASEB Journal*, vol. 24, no. 11, pp. 4281–4290, 2010.

[32] A. Barrett, T. Ndou, C. A. Hughey et al., "Inhibition of α-amylase and glucoamylase by tannins extracted from cocoa, pomegranates, cranberries, and grapes," *Journal of Agricultural and Food Chemistry*, vol. 61, no. 7, pp. 1477–1486, 2013.

[33] J. Xiao, X. Ni, G. Kai, and X. Chen, "A review on structure-activity relationship of dietary polyphenols inhibiting α-amylase," *Critical Reviews in Food Science and Nutrition*, vol. 53, no. 5, pp. 497–506, 2013.

[34] G. Gyémánt, Á. Zajácz, B. Bécsi et al., "Evidence for pentagalloyl glucose binding to human salivary α-amylase through aromatic amino acid residues," *Biochimica et Biophysica Acta—Proteins and Proteomics*, vol. 1794, no. 2, pp. 291–296, 2009.

[35] H. Laube, "Acarbose: an update of its therapeutic use in diabetes treatment," *Clinical Drug Investigation*, vol. 22, no. 3, pp. 141–156, 2002.

[36] J. Zhang and S. Kashket, "Inhibition of salivary amylase by black and green teas and their effects on the intraoral hydrolysis of starch," *Caries Research*, vol. 32, no. 3, pp. 233–238, 1998.

[37] C. Hannig, M. Hannig, and T. Attin, "Enzymes in the acquired enamel pellicle," *European Journal of Oral Sciences*, vol. 113, no. 1, pp. 2–13, 2005.

[38] A. E. Nikitkova, E. M. Haase, and F. A. Scannapieco, "Taking the starch out of oral biofilm formation: molecular basis and functional significance of salivary α-amylase binding to oral streptococci," *Applied and Environmental Microbiology*, vol. 79, no. 2, pp. 416–423, 2013.

[39] S. Rosen, M. Elvin-Lewis, F. M. Beck, and E. X. Beck, "Anticariogenic effects of tea in rats," *Journal of Dental Research*, vol. 63, no. 5, pp. 658–660, 1984.

[40] B. Narotzki, A. Z. Reznick, D. Aizenbud, and Y. Levy, "Green tea: a promising natural product in oral health," *Archives of Oral Biology*, vol. 57, no. 5, pp. 429–435, 2012.

[41] S. Kashket and V. J. Paolino, "Inhibition of salivary amylase by water-soluble extracts of tea," *Archives of Oral Biology*, vol. 33, no. 11, pp. 845–846, 1988.

[42] S. Aizawa, H. Miyasawa-Hori, K. Nakajo et al., "Effects of α-amylase and its inhibitors on acid production from cooked starch by oral streptococci," *Caries Research*, vol. 43, no. 1, pp. 17–24, 2009.

6

A Comparative Study of New *Aspergillus* Strains for Proteolytic Enzymes Production by Solid State Fermentation

Gastón Ezequiel Ortiz, Diego Gabriel Noseda, María Clara Ponce Mora,
Matías Nicolás Recupero, Martín Blasco, and Edgardo Albertó

Instituto de Investigaciones Biotecnológicas-Instituto Tecnológico de Chascomús (IIB-INTECH),
Universidad Nacional de San Martín (UNSAM) and Consejo Nacional de Investigaciones Científicas y Técnicas (CONICET),
San Martín, 1650 Buenos Aires, Argentina

Correspondence should be addressed to Gastón Ezequiel Ortiz; gas.ortiz@gmail.com

Academic Editor: Raffaele Porta

A comparative study of the proteolytic enzymes production using twelve *Aspergillus* strains previously unused for this purpose was performed by solid state fermentation. A semiquantitative and quantitative evaluation of proteolytic activity were carried out using crude enzymatic extracts obtained from the fermentation cultures, finding seven strains with high and intermediate level of protease activity. Biochemical, thermodynamics, and kinetics features such as optimum pH and temperature values, thermal stability, activation energy (E_a), quotient energy (Q_{10}), K_m, and V_{max} were studied in four enzymatic extracts from the selected strains that showed the highest productivity. Additionally, these strains were evaluated by zymogram analysis obtaining protease profiles with a wide range of molecular weight for each sample. From these four strains with the highest productivity, the proteolytic extract of *A. sojae* ATCC 20235 was shown to be an appropriate biocatalyst for hydrolysis of casein and gelatin substrates, increasing its antioxidant activities in 35% and 125%, respectively.

1. Introduction

Proteases constitute a large and complex group of hydrolytic enzymes with important applications in medical, pharmaceutical, biotechnology, leather, detergent, and food industries [1]. They can be synthesized by plants, animals, and microorganisms constituting around 60% of the worldwide enzyme market [2]. Among such sources, microorganisms present remarkable potential for proteolytic enzymes production due to their extensive biochemical diversity and susceptibility to genetic manipulation [3]. Filamentous fungi have been utilized for the production of diverse industrial enzymes because these organisms exhibit the capacity to grow on solid substrates and secrete a wide range of hydrolyzing enzymes. Particularly, several species of *Aspergillus* have been exploited as important sources of extracellular enzymes including proteases [4]. According to the state of the art, most of the current scientific knowledge associated with the proteases production by *Aspergillus* fungus

is related to the use of *A. oryzae* and *A. niger* species (Supplementary Data 1 in Supplementary Material available online at http://dx.doi.org/10.1155/2016/3016149). Products of *Aspergillus* species such as *A. niger*, *A. sojae*, and *A. oryzae* have acquired a Generally Recognized as Safe (GRAS) status from the US Food and Drug Administration, which has approved their use in the food industry [5].

The production of proteases can be performed by solid state fermentation (SSF) and submerged fermentation (SmF) [6]. The application of SSF is of interest for fungi enzymes production due to its advantages in comparison to SmF, such as low fermentation technology, low cost, higher yields and concentration of the enzymes, and reduced waste output [7, 8]. Furthermore, inexpensive and widely available agricultural solid wastes could be used with the aim of providing nutritional and physical support throughout SSF procedures [9]. The intense demand for industrial proteolytic enzymes requires the search of new strains with high level of protease productivity in order to enhance the enzyme production capacity and their applications [2]. In such context, the aim of this work is to expand the range of species and strains of the genus *Aspergillus* suitable for the production of proteolytic enzymes under solid state fermentation for industrial use.

2. Materials and Methods

2.1. State of the Art Analysis. The state of the art for proteases production by *Aspergillus* was evaluated through a literature search in the Scopus and PubMed database using the following criteria of search: TITLE ((*Aspergillus* AND Proteases) and Production). The publications nondirectly related to the proteases production were dismissed (Supplementary Data 1).

2.2. Microbial Strains. The *Aspergillus* strains used in this work were (1) *A. terreus* ICFC 744/11; (2) *A. oryzae* NRRL 2217; (3) *A. awamori* NRRL 356; (4) *A. flavipes* NRRL 295; (5) *A. kawachii* IFO 4308; (6) *Aspergillus* sp. ICFC 7/14; (7) *A. japonicus* NRRL 1782; (8) *A. oryzae* ICFC 8/12; (9) *A. giganteus* NRRL 10; (10) *A. rhizopodus* NRRL 6136; (11) *A. sojae* NRRL 5595; and (12) *A. sojae* ATCC 20235. Such strains are conserved in the IIB-INTECH Collection of Fungal Cultures (ICFC), reference in the WDCM database: WDCM 826. All the strains were periodically propagated and maintained on potato dextrose agar slants.

2.3. Inoculum Preparation. In order to produce conidia for inoculation of the main cultures, the strains were grown on agar-plates containing sugarcane molasses (45 g/L), peptone (18 g/L), NaCl (5 g/L), KCl (0.5 g/L), $FeSO_4 \cdot 7H_2O$ (15 mg/L), KH_2PO_4 (60 mg/L), $MgSO_4$ (50 mg/L), $CuSO_4 \cdot 5H_2O$ (15 mg/L), $MnSO_4$ (15 mg/L), and agar (20 g/L). Plates were incubated at 28°C until sporulation. Conidia were harvested from the plates by the addition of 5 mL of 0.08% (w/v) Tween80. The number of conidia/mL in the conidia suspension was determined using Neubauer cell-counting chamber.

2.4. Culture Conditions. Erlenmeyer flasks (250 mL) containing 10 g of wheat bran with a homogeneous particle size of 2000 μm in average were moistened at 110% with Czapek-Dox medium corresponding to 0.96 units of water activity (Supplementary Data 2). The flasks with the sterile culture substrate (sterilization conditions, 121°C for 20 min) were inoculated with 10^6 conidia/g of dry substrate (gds) and incubated at 28°C in a moist chamber for 2, 4, and 6 days.

2.5. Enzyme Recovery. Enzymes produced were recovered by the addition of 10 mL/gds of distilled water into each culture flask and mixing in a shaker at 250 rpm, 28°C, for 30 min. Then, the mixtures were clarified by filtration through cotton followed by centrifugation at 2000 ×g, 4°C, for 20 min. The clarified supernatants were used for the following analysis.

2.6. Semiquantitative Determination of Proteolytic Activity by Agar-Plate Assay. A semiquantitative determination of proteolytic activity was carried out at different values of pH

by the agar-plate diffusion assay according to the technique described by Heerd et al. [5]. Agar-plates were prepared with 1.5% (w/v) agar and 1.5% (w/v) skim milk as substrate. Both agar and skim milk were dissolved in 0.1 M Tris-HCl buffer in order to adjust and maintain the pH between 6 and 9. Wells of 4 mm diameter were punched in the solid media and loaded with 20 μL crude enzymatic extracts. After 18 h incubation at 30°C, the milk agar-plates were stained with Coomassie Brilliant Blue G-250 for 20 min. The diameter of the halos (*D*) corresponding to the zone of milk degradation was converted to $\log_{10}$ by (1) and reported as hydrolysis index ($\log_{10} mm^2$) [5]:

$$\log_{10} \text{ adjusted zone area}$$
$$= \log_{10} \left[\left(\frac{D}{2} \right)^2 \pi - \left(\frac{4.0}{2} \right)^2 \pi \right]. \tag{1}$$

Semiquantitative determination was performed by means of 2 independent assays. Proteolytic activity values from different crude extracts were evaluated by Multifactorial ANOVA and cluster analysis using the statistical software Statgraphics Centurion XVII trial version (Supplementary Data 3).

2.7. Quantitative Analysis of Proteolytic Activity. Protease activity was quantitatively measured using azo-casein assay according to Cavello et al. [10], with modifications. Reaction mixture containing 20 μL of enzyme extract diluted in 0.1 M Tris-HCl buffer (pH: 8) (buffer T) and 50 μL of 1% (w/v) azo-casein solution in buffer T was incubated for 60 min at 37°C in thermocycler machine. Reaction was stopped with the addition of 100 μL of 10% (v/v) trichloroacetic acid. The mixture was kept at room temperature for 15 min and then centrifuged at 2000 ×g, 20°C, for 10 min. Finally, 50 μL of each sample was diluted by addition of 50 μL of 1 M NaOH solution and absorbance was measured at 415 nm with a microplate spectrophotometer. All determinations were performed in duplicate and a heat-inactivated enzyme extract was used for blank. One unit of proteolytic activity (U) was defined as the amount of enzyme that produces an increase of 0.1 units in the absorbance at 415 nm per min under test conditions. Azo-casein was synthesized as described by Riffel et al. [11].

2.8. Effect of Temperature and pH on Proteolytic Activity. The optimal pH and temperature for crude enzymatic extracts were determined employing a central composite design (CCD) with four axial points and three central points. In order to maximize the variability, the experiments were randomized and performed as shown in Supplementary Data 4. Enzymatic extracts corresponding to 2 days of fermentation were normalized to 30 U/mL and incubated at different temperature and pH values during 1 h. For this, the enzymatic extracts were diluted with 0.1 M maleate buffer or 0.1 M Tris-HCl to adjust the pH to lower or higher values than 7.0, respectively. The protease activity was determined using the azo-casein method and the second-order model represented by (2) was used to describe this response:

$$Y = \beta_0 + \sum_{i=1}^{n} \beta_i x_i + \sum_{i=1}^{n-1} \sum_{j=i+1}^{n} \beta_{ij} x_i x_j, \qquad (2)$$

where Y is the estimated response, β_0 is the constant term, i and j have values from 1 to the number of variables (n), β_i is the linear coefficient, β_{ij} is the quadratic coefficient, and x_i and x_j are the coded independent variables. The coefficient of determination R^2 and the F value from analysis of variance (ANOVA) were used to confirm the quality of the model. Relationships between the responses and variables were evaluated using Statgraphics Centurion XVII software trial version.

2.9. Enzymatic Stability. The proteolytic stability was analyzed by incubating the crude extracts under optimal pH and temperature values at different periods of time before conducting enzymatic activity determination by azo-casein method. Due to the complexity of the reaction occurring during inactivation, several equations have been proposed to model this kinetic. In this work, a first-order kinetic model (3) was selected to represent the residual enzyme activity (A/A_0) at time (t, min). The parameter k (min^{-1}) is the rate constant of the reaction under assay conditions:

$$\frac{A}{A_0} = e^{-kt}. \qquad (3)$$

2.10. SDS-PAGE and Zymogram Analysis. SDS-PAGE electrophoresis was performed according to the technique described by Laemmli [12]. The enzyme extracts were diluted in loading buffer without DDT to a final concentration of 10 U/mL. A volume of $20\,\mu$L of each sample was loaded on a 10% (v/w) separating SDS-PAGE gel by duplicate in symmetric disposition. The electrophoresis was conducted at 160 V and 4°C during 1 h. After that, the gel was cut in halves, keeping one for the zymogram and staining the other with colloidal Coomassie Brilliant Blue G-250 [13]. Zymography analysis was performed according to Cavello et al. [14] with slight modifications. Briefly, the gel was submerged in 100 mM Tris-HCl buffer (pH 8.0) (buffer T) containing 2.5% Triton X-100 during 60 min, with constant agitation, and then washed three times with buffer T. Finally, the gel was incubated with 1% (w/v) casein in buffer T at 30°C for 60 min and then stained with Coomassie Brilliant Blue R-250. The development of clear bands on the blue background of the gel indicated the presence of protease activity. The molecular weight of proteolytic enzymes was estimated through densitometry analysis using Image J software 1.44p version [15].

2.11. Determination of Activation Energy and Temperature Quotient (Q_{10}). Activation energy (E_a) values of the proteases produced by *Aspergillus* strains were calculated by incubating enzyme extracts, under optimal pH condition, with 1% azo-casein at several temperatures ranging from 30 to 50°C.

The dependence of the rate constants with temperature was assumed to follow Arrhenius Law (4) and E_a was calculated from the slope of the plot of $1000/T$ versus ln (protease activity), where $E_a = -$slope $\times R$, R (gas constant) = 8.314 (J/K·mol), and T is the absolute temperature in Kelvin (K) [16]. The temperature quotient (Q_{10}) values of *Aspergillus* proteases were determined using (5) [17]:

$$\ln (\text{Activity}) = \ln (A) - \frac{E_a}{R} \times \frac{1}{T}, \qquad (4)$$

$$Q_{10} = \text{antilog} \times e^{-(E_a \times 10)/RT^2}. \qquad (5)$$

2.12. Kinetics Parameters K_m and V_{max}. Kinetics parameters K_m and V_{max} of the proteases produced by *Aspergillus* strains were determined by incubating crude extracts, under optimal pH and temperature values, with azo-casein over the concentration range 1.0–10.0 mg/mL. Protease activity of each extract was quantitatively measured as described in Section 2.7. The Michaelis-Menten constant (K_m) and maximum velocity (V_{max}) values were calculated using the GraphPad Prism 4® trial version. The values employed in the nonlinear regression are shown in Supplementary Data 7.

2.13. Preparation of Protein Hydrolysates. For the production of protein hydrolysates, the enzyme concentration of a selected crude extract was adjusted to 400 U per gram of dry substrate (U/gds) according to its proteolytic activity established previously. Gelatin and casein substrates were suspended in buffer citrate phosphate 0.1 M to a final concentration of 1.0% (w/v). A total of 15 mL of each mixture was distributed in 50 mL Erlenmeyer flasks and incubated in a water bath shaker operating at 50 rpm. The hydrolysis was performed at the optimal temperature and pH values of the enzyme extract during 120 min. Aliquots (0.5 mL) of each mixture were taken at regular intervals and inactivated in a water bath at 100°C for 15 min. A volume of 0.1 mL was reserved to determine the degree of hydrolysis. The remaining volume (0.4 mL) was centrifuged at 12,500 ×g in Eppendorf MiniSpin® for 10 min for separating the soluble peptides from nonhydrolyzed proteins. The supernatants were collected to determine the antioxidant activities.

2.14. Determination of Antioxidant Activities. The antioxidant activity of the protein hydrolysates was determined by the DPPH radical-scavenging method as described by Bougatef et al. [18] with adaptations for microplate assay. A volume of $50\,\mu$L of the diluted hydrolysates was mixed with $50\,\mu$L of 99.5% ethanol and $30\,\mu$L of 0.02% DPPH in 99.5% ethanol. The mixtures were kept at room temperature in the dark for 60 min, and the reduction of the DPPH radical was measured at 540 nm using a microplate reader Bio-Rad Benchmark®. The DPPH radical-scavenging activity was calculated using

Radical scavening activity (%)

$$= \frac{\text{Control Abs}_{540\,\text{nm}} - \text{Sample Abs}_{540\,\text{nm}}}{\text{Control Abs}_{540\,\text{nm}}} \times 100\%. \qquad (6)$$

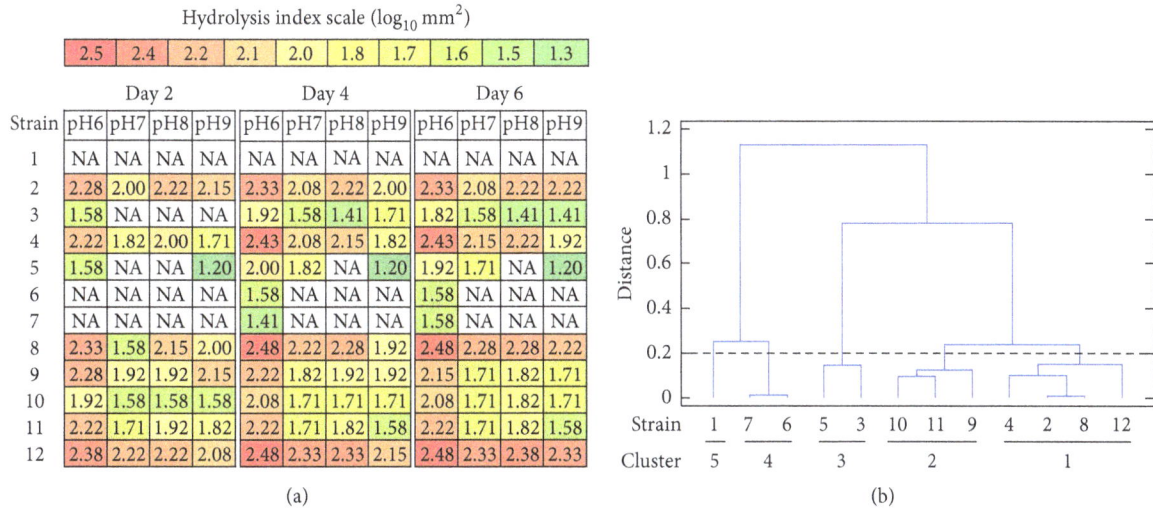

FIGURE 1: Semiquantitative analysis of proteolytic activity from *Aspergillus* spp. extracts. Protease activity was semiquantitatively analyzed for different values of pH from crude extracts of *Aspergillus* ssp. cultures obtained by solid state fermentation. (a) Proteolytic activity values expressed by the hydrolysis index ($\log_{10}$ mm^2). The values correspond to the average of two independent studies. NA: no activity. (b) Dendrogram constructed from the proteolytic activities using median method and squared Euclidean distance. Distance of 0.2 indicates the cut-line for cluster designation. For strain reference numbers see Section 2.2.

2.15. Determination of the Degree of Hydrolysis. The degree of hydrolysis of the protein hydrolysates was determined according to the method described by Adler-Nissen [19]. Mixtures of 0.5 mL of diluted hydrolysates or 0.825 mmol/L standard cysteine amino acid and 0.250 mL of 0.01% (w/v) TNBSA in 0.1 M sodium bicarbonate pH 8.5 were incubated for 30 min at 37°C. A volume of 200 μL of each mixture was taken and the absorbance measured at 415 nm in a Bio-Rad Benchmark microplate reader. Degree of hydrolysis (DH) values were calculated using

$$\text{Cysteine-NH}_2 \left(\frac{\text{meqv}}{\text{g}} \right)$$

$$= X \times V \times f$$

$$\times \frac{\text{Sample Abs}_{415\,\text{nm}} - \text{Blank Abs}_{415\,\text{nm}}}{\text{Standard Abs}_{415\,\text{nm}} - \text{Blank Abs}_{415\,\text{nm}}}$$

$$\times 0.82 \text{ meqv/L},$$

$$h = \frac{[\text{Cysteine-NH}_2] - \beta}{\alpha},$$

$$\text{DH}\,(\%) = \frac{h}{h_{\text{tot}}} \times 100\%,$$

(7)

where cysteine-NH$_2$ is milliequivalents of cysteine amine groups per gram of protein; X is mass of sample protein in gram (g); V is the reaction volume in liter (L); and f is dilution factor. The value h is the number of hydrolyzed peptide bonds and h_{tot} is the total number of peptide bonds per protein equivalent. The values α, β, and h_{tot} for casein and gelatin are $\alpha = 1.039$, $\beta = 0.383$, and $h_{\text{tot}} = 8.2$ and $\alpha = 0.796$, $\beta = 0.457$, and $h_{\text{tot}} = 11.1$, respectively [20].

3. Results and Discussion

3.1. Semiquantitative Determination of Proteolytic Activity. To evaluate the capability of *Aspergillus* strains for proteases production, crude enzymatic extracts obtained by SSF were semiquantitatively analyzed for total proteolytic activity at different pH values by agar-plate diffusion assay. In this study the diameters of proteolysis halos were adjusted as indicated in Section 2.6 to report the enzymatic activity index. Such values were employed to compare the strains through multifactorial analysis of variance (ANOVA) (Supplementary Data 3). Wheat bran was chosen as the screening substrate because it possesses a suitable carbon to nitrogen ratio (C:N) and water absorption index, necessary for an appropriate growth and protease production as was reported by Soares de Castro et al. [21, 22]. Analysis of variance revealed that most of the evaluated strains registered significantly higher proteolytic activity at days 4 and 6 of fermentation and with a pH value of 6.0 (Figure 1(a)). Furthermore, the cluster analysis (Figure 1(b)) revealed the presence of five homogeneous clusters of strains based on its enzymatic activity. Cluster 1, which included *A. sojae* ATCC 20235, *A. oryzae* ICFC 8/12, *A. oryzae* NRRL 2217, and *A. flavipes* NRRL 295, showed a high protease activity throughout the tested pH range, while cluster 2, formed by *A. giganteus* NRRL 10, *A. sojae* NRRL 5595, and *A. rhizopodus* NRRL 6136, presented intermediate values of proteolytic activity in the pH range 7.0–9.0, displaying an increase in the activity at pH 6.0. Furthermore, cluster 3 (*A. kawachii* IFO 4308 and *A. awamori* NRRL 356 strains) exhibited a reduced proteolytic activity throughout the pH range for days 4 and 6. Cluster 4 (*Aspergillus* sp. ICFC 7/14 and *A. japonicus* NRRL 1782 strains) showed very low proteolytic activity at pH 6.0 on days 4 and 6. Finally, cluster 5 did not record proteolytic activity. Based on these results it can be concluded that crude

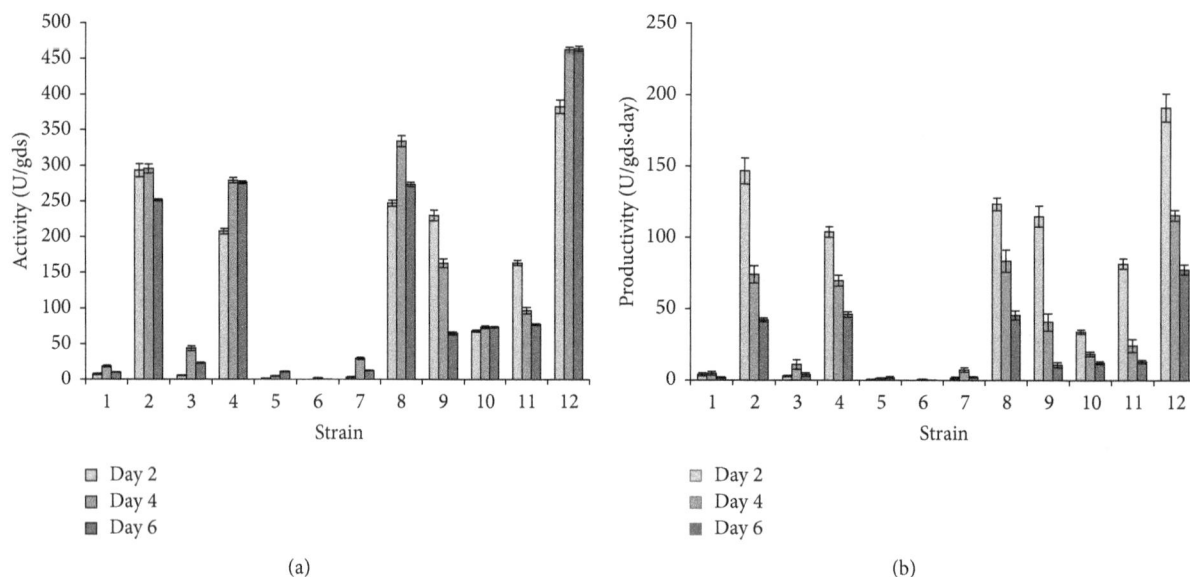

(a)

(b)

FIGURE 2: Protease production and productivity from *Aspergillus* spp. extracts. (a) Protease production expressed as proteolytic activity for 3 different periods of culture. (b) Proteolytic productivity during fermentation process. Experiments were performed in triplicate and error bars represent the standard deviation. For strain reference numbers see Section 2.2.

extracts from strains of cluster 1 provided the highest level of protease activity. It should be mentioned that the strains of such cluster, *A. flavipes* NRRL 295 and *A. sojae* ATCC 20235, do not possess scientific reports for proteolytic enzymes production. However, it is well known that *A. sojae* and *A. oryzae* are employed industrially for commercial production of proteases [2]. On the other hand, the *A. sojae* NRRL 5595, *A. giganteus* NRRL 10, and *A. rhizopodus* NRRL 6136 strains (cluster 2) provided intermediate level of proteolytic activity. It is important to highlight that the last two species were not previously reported for protease production.

3.2. Proteolytic Activity and Productivity Analysis. In order to validate the results achieved previously a quantitative study of the proteolytic activity and enzymatic productivity was conducted. As shown in Figure 2(a) the strains belonging to cluster 1 (Figure 1(b)) presented the highest level of proteolytic activity. In this cluster, the *A. sojae* ATCC 20235 and *A. flavipes* NRRL 295 showed a similar protease production pattern in which the activity increased and remained constant during the culture suggesting an adequate stability of the crude extracts, whereas *A. oryzae* ICFC 8/12 and *A. oryzae* NRRL 2217 exhibited a decrease in the proteolytic activity on day 6 of incubation, which suggested certain instability of these enzymes. Furthermore, *A. giganteus* NRRL 10, *A. rhizopodus* NRRL 6136, and *A. sojae* NRRL 5595 that formed cluster 2 (Figure 1(b)) showed a lower enzyme activity which decreased over the culture course for *A. giganteus* NRRL 10 and *A. sojae* NRRL 5595, but remained stable with a low value for *A. rhizopodus* NRRL 6136 suggesting an appropriate enzymatic stability for this strain. On the other hand, the protease production profile observed for *A. giganteus* NRRL 10, *A. rhizopodus* NRRL 6136, and *A. sojae* NRRL 5595 is similar to the production profile reported by Soares de Castro et al. for

A. oryzae LBA01 and *A. niger* LB02, two species well known for their good productivity of proteolytic enzymes [3, 23]. In addition, maximum protease productivity was registered for day 2 of fermentation for strains of clusters 1 and 2 (Figures 2(b) and 1(b)). These results confirmed that strains from cluster 1 presented the highest level of protease production and demonstrated that maximum productivity was achieved after a short fermentation time. On the other hand, and according to the results obtained with the semiquantitative analyses, crude extracts from the strains of the clusters 3, 4, and 5 showed the lowest proteolytic activity (Figures 2(a) and 1(b)) and productivity (Figure 2(b)).

3.3. Influence of Temperature and pH on Proteolytic Activity. The optimal temperature of fungal proteases ranged between 35 and 50°C with few exceptions and the optimum pH values for acid and alkaline proteolytic enzymes range between 2.0–6.0 and 8.0–11.0, respectively, while optimal pH for neutral proteases and metalloproteases ranges from 6 to 8 [2]. Therefore, in order to determine the optimum pH and temperature of proteases from the strains with higher proteolytic productivity, *A. flavipes* NRRL 295, *A. oryzae* ICFC 8/12, *A. giganteus* NRRL 10, and *A. sojae* ATCC 20235, we performed a central composite design with four axial points as was described in Section 2.8 and the results were analyzed by ANOVA to confirm the quality of each model and the adequacy of these models was validated by three verification trials (Supplementary Data 4). It should be mentioned that *A. oryzae* ICFC 8/12 was selected as a control strain for this analysis since it is a well-studied strain. As shown in Figure 3, enzymatic extracts from strains *A. oryzae* ICFC 8/12 and *A. sojae* ATCC 20235 presented maximum proteolytic activity in a pH range of 6.0 to 6.8 and a temperature range between 43 and 53°C, suggesting

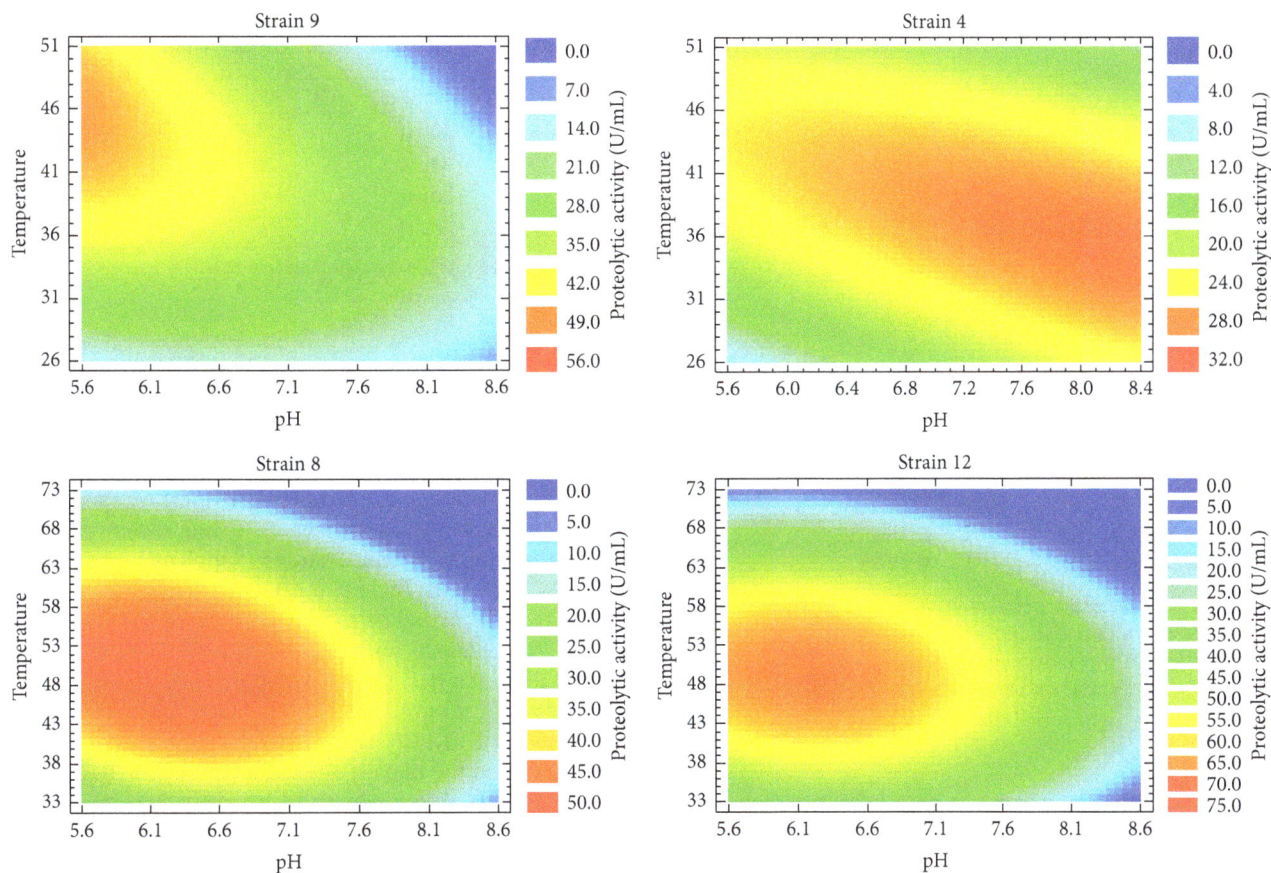

FIGURE 3: Effect of pH and temperature on proteolytic activity. Enzyme activities analyzed through central composite design. The dependence of the activity with pH and temperature is represented as a contour plot. Red and blue colors indicate the values of maximum and minimum activities, respectively. For strain reference numbers see Section 2.2.

that these extracts would be composed mainly of neutral and thermophilic proteases. In this sense, the production of neutral proteolytic enzymes had previously been reported for *A. oryzae* NRRL 2220 and *A. sydowii* [24, 25]. Furthermore, we determined that *A. flavipes* NRRL 295 had maximum activity around pH 8.4 and in the temperature range 32–43°C, suggesting the presence of alkaline and mesophilic proteases in such extract. In this respect, proteases with these biochemical features were previously described for *A. tamarii* [26]. Meanwhile, *A. giganteus* NRRL 10 exhibited maximum activity around pH 5.6 and with a temperature range 41–51°C, indicating that the extract would be mainly composed of acidic and thermophilic proteases. In this sense, the production of enzymes with similar characteristics has previously been reported for *A. oryzae* LBA01 and MTCC 5341 [23, 27]. In accordance with the results obtained from the semiquantitative analysis (Figure 1), *A. sojae* ATCC 20235, *A. oryzae* ICFC 8/12, and *A. flavipes* NRRL 295 showed proteolytic activity over the broad pH range 5.6–8.4, while *A. giganteus* NRRL 10 presented higher activity at acidic pH values.

3.4. Thermal and pH Stability of Aspergillus Proteases.
The stability of enzymes appears as an important critical aspect to conduct industrial application. Therefore, thermal and pH stability of proteolytic extracts were evaluated under the optimal conditions previously determined. Figure 4 and Table 3 show the decay curves of proteolytic activity for the evaluated enzymatic extracts. The extracts from *A. sojae* ATCC 20235 and *A. flavipes*, tested under pH 6.4 and 48°C and pH 8.4 and 36°C, respectively, presented similar stability with an inactivation constant of 0.003 min^{-1} and a half-life time of around 215 min. These enzymes resulted as more stable than other proteases previously reported for *Aspergillus* species such us *A. clavatus* strains ES1 (half-life: 30 min at 50°C) and CCT2759 (half-life: 18 min at 50°C) [28]. On the other hand, the extract from *A. giganteus*, analyzed at pH 5.6 and 48°C showed a high inactivation constant of 0.021 min^{-1} and a half-life of about 33 min reflecting poor level of stability. Proteases from *Penicillium* sp. showed similar stability properties with a half-life of 30 min at 45°C [29]. Meanwhile, *A. oryzae* ICFC 8/12 which was evaluated under pH 6.4 and 48°C exhibited an intermediate stability with an inactivation constant of 0.009 min^{-1} and a half-life time of 75 min.

3.5. Electrophoresis SDS-PAGE and Zymogram Analysis.
The molecular weights of fungal proteases are generally in the range of 20 and 50 kDa [2], with some exception such as low-molecular weight alkaline protease (6.8 kDa) from

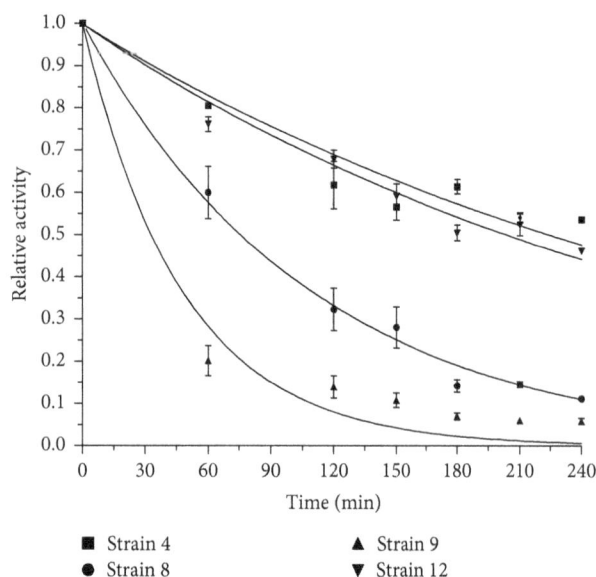

FIGURE 4: Thermal and pH stability study of protease extracts. Fraction of enzymes remaining active after incubation under optimal conditions. Three independent samples were assayed for each enzymatic extract and error bars represent the standard deviation.

TABLE 1: Activation energy (E_a) and Q_{10} for azo-casein hydrolysis of the proteolytic extracts from selected *Aspergillus strains*.

Strain	Temperature range (°C)	E_a (kJ/mol)	Q_{10}*	R^2
4	30–40	21.82 ± 1.43[†]	1.31	0.98
8	30–50	38.64 ± 3.27	1.61	0.97
9	30–46	31.66 ± 1.87	1.48	0.99
12	30–46	29.28 ± 3.35	1.44	0.97

*Q_{10} determined using the average temperature range.
[†]Std. error of linear regression coefficient.

3.6. *Activation Energy and Temperature Quotient.* The activation energies of the proteases produced by *Aspergillus* strains were determined at the temperature ranges that are indicated in Table 1. The Arrhenius plots for such proteases showed a linear variation with temperature increase, suggesting that these proteolytic enzymes have single conformations up to the transition temperatures (Supplementary Data 5) [16]. As shown in Table 1, the minimal activation energy (21.82 kJ/mol) necessary to conduct the hydrolysis of azo-casein was obtained for alkaline proteases from *A. flavipes* NRRL 295. Likewise, the intermediate energies values were obtained for acid proteases from *A. sojae* ATCC 20235 and *A. giganteus* NRRL 10. On the other hand *A. oryzae* ICFC 8/12 presented the highest activation energy (38.64 kJ/mol). In this context, Melikoglu and coworkers [16] reported an activation energy of the 36.8 kJ/mol for bread protein hydrolysis employing proteases from *A. awamori* in a temperature range of 30–55°C. This value was similar to those obtained for the strains of *A. sojae*, *A. giganteus*, and *A. oryzae*. On the other hand proteolytic extracts from *A. niger* in a range of 35–50°C exhibited activation energies values for azo-casein hydrolysis that ranged from 16.32 to 19.48 kJ/mol [17]. These values are similar to that obtained from the strain of *A. flavipes*. To study the effect of temperature on the rate of reaction, we investigated the temperature quotient (Q_{10}), that is, a measure of the rate of change of a biological or chemical system as a consequence of increasing the temperature by 10°C. The Q_{10} values of the examined extracts ranged from 1.31 to 1.61 (Table 1). Generally enzymatic reactions show Q_{10} values between 1.00 and 2.00 units and any deviation from this value is indicative of involvement of some factor other than temperature in controlling the rate of reaction. Soares de Castro and coworkers reported the Q_{10} values between 1.20 and 1.28 for azo-casein hydrolysis with temperatures between 30 and 60°C [17]. The maximum value of Q_{10} was obtained for proteases from *A. oryzae* ICFC 8/12 reflecting that for every 10°C raise in temperature the rate of reaction increased 61%.

3.7. *Kinetic Parameters K_m and V_{max}.* In order to evaluate the kinetics properties of proteolytic enzymes, the parameters V_{max}, K_m, and V_{max}/K_m were calculated for each proteases extract under optimal pH and temperature conditions (Table 2). The maximum catalysis velocity (V_{max}) is the amount of enzyme involve in the enzymatic reaction. The K_m parameter provides the affinity of enzyme for substrate: a low value of this parameter indicates a higher affinity enzyme-substrate. The ratio V_{max}/K_m is related to the specificity and

Conidiobolus coronatus [30] and high-molecular weight thiol proteinase (237 kDa) from *Humicola lanuginosa* UAMH 1623 [31]. Likewise, the diversity in molecular weights of fungal proteases has been used to differentiate the species *A. sojae* and *A. oryzae* on the basis of specific mobility of alkaline proteases in polyacrylamide gel disc electrophoresis [32]. In order to evaluate the total protein pattern and the taxonomic proteolytic enzyme profile, crude enzymatic extracts corresponding to strains *A. sojae* ATCC 20235, *A. oryzae* ICFC 8/12, *A. giganteus* NRRL 10, and *A. flavipes* NRRL 295 were analyzed under alkaline condition by polyacrylamide gel electrophoresis and zymogram. Hence, SDS-PAGE analysis (Figure 5) allowed us to distinguish a clear difference between the protein patterns of the four examined strains. Likewise, such patterns presented remarkable bands throughout the whole range of molecular weight (100–10 kDa). Furthermore, as shown in Table 4, *A. flavipes* NRRL 295 and *A. oryzae* ICFC 8/12 strains exhibited bands with high proteolytic activity in the range 20–35 kDa. These results are in accordance with previous studies that reported an alkaline serine-protease from *A. clavatus* ES1 with a molecular mass of 32 kDa and an alkaline protease from *A. terreus* with a molecular weight of 37 kDa [33, 34]. In addition, crude extracts from the strains of *A. giganteus* and *A. sojae* presented remarkable active bands with molecular masses in the range 85–95 kDa which was consistent with a high-molecular mass protease (124 kDa) from *A. fumigatus* TKU003 reported by Wang et al. [35]. Chien and coworkers reported the purification, characterization, and cloning of a leucine aminopeptidase (LAP) from *A. sojae* with an apparent molecular mass of 37 kDa [36]. In this sense, a protease with a similar relative molecular weight could be produced by *A. sojae* ATCC 20235 (Table 4).

TABLE 2: Kinetics parameters K_m and V_{Max} for proteolytic extracts from selected *Aspergillus* strains.

Strain	V_{Max} (U/mL)	V_{Max} (U/gds)	K_m (mg/mL)	V_{Max}/K_m (U/mg)	V_{Max}/K_m (U·mL/mg·gds)	R^2
12	$23.38 \pm 1.16^\dagger$	$233.8 \pm 2.1^\dagger$	$1.34 \pm 0.26^\dagger$	$17.45 \pm 0.20^\dagger$	$174.5 \pm 2.0^\dagger$	0.98
9	20.79 ± 0.98	207.9 ± 9.8	1.48 ± 0.26	14.04 ± 0.18	140.4 ± 1.8	0.98
8	15.51 ± 0.66	155.1 ± 6.6	0.86 ± 0.18	18.03 ± 0.22	180.3 ± 2.2	0.98
4	4.46 ± 0.21	44.6 ± 2.1	0.97 ± 0.21	4.59 ± 0.22	45.9 ± 2.2	0.98

†Std. error of nonlinear regression coefficient (Michaelis-Menten kinetics fit).

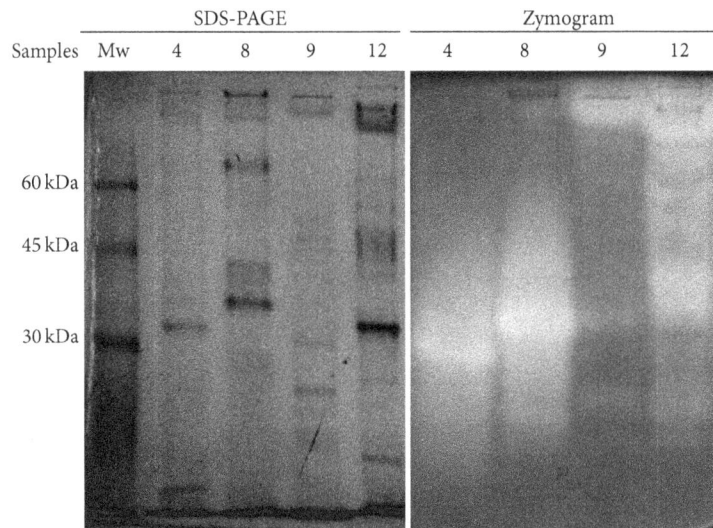

FIGURE 5: Analysis of crude extracts by SDS-PAGE electrophoresis and zymogram. Samples: Mw: protein molecular weight marker; 4: *A. flavipes* NRRL 295; 8: *A. oryzae* ICFC 8/12; 9: *A. giganteus* NRRL; 10, 12: *A. sojae* ATCC 20235.

TABLE 3: Inactivation and statistical parameters estimated using a first-order inactivation model (3). Parameter K is the inactivation rate constant and A_0 is the initial activity. For strain reference numbers see Section 2.2.

	Strain 4	Strain 8	Strain 9	Strain 12
Enz. extract				
A_0	0.998	1.000	0.999	0.999
K (min)	0.003	0.009	0.021	0.003
Half-life (min)	225.000	75.260	32.910	204.000
Std. error				
A_0	0.004	0.003	0.005	0.003
K	0.000	0.000	0.001	0.000
R^2	0.973	0.995	0.994	0.990

efficiency of enzymes: a high value of this parameter indicates a higher catalytic specificity and efficiency [37, 38].

The highest affinity for azo-casein was observed for proteolytic extract from *A. oryzae* ICFC 8/12 with a K_m value expected at 0.86 mg/mL, followed by proteases from *A. flavipes* NRRL 295, *A. sojae* ATCC 20235, and *A. giganteus* NRRL 10 (Table 2). A wide K_m value for azo-casein hydrolysis has been reported. Li and coworkers reported a K_m value of 0.96 mg/mL for an acid protease from a *A. oryzae* and *A. niger* fusant strain [4]. Furthermore, Soares de Castro et

al. reported a wide range of K_m (0.44–1.92 mg/mL) when *A. niger* was grown on different substrates [17]. Murthy and Naidu reported a higher K_m value (3 mg/mL) for an alkaline protease from *A. oryzae* growing in solid state fermentation using coffee waste as substrate [39]. The V_{max} values of the proteolytic extracts ranged between 44.6 U/gds and 233.8 U/gds (Table 2). Similar range values were reported by Soares de Castro et al. for *A. niger* grown on different substrates [17]. The highest substrate specificity and catalytic efficiency (V_{max}/K_m ratio) was achieved for proteases from *A. oryzae* ICFC 8/12 and *A. sojae* ATCC 20235 (Table 2); similar values were obtained for proteases from *A. niger* grown on wheat bran [17].

3.8. Hydrolysis Degree and Antioxidant Activity. Proteases from different *A. oryzae* strains have been used in the hydrolysis of whey protein or gelatin with the purpose of obtaining functional peptides with antioxidant activity [40, 41]. In this context and considering that *A. sojae* ATCC 20235 achieved the highest productivity and exhibited similar biochemical characteristics (Figure 3) and casein specificity (Table 2) with respect to *A. oryzae* ICFC 8/12 (control strain), we decided to explore the potentiality of the *A. sojae* ATCC 20235 proteases for the production of antioxidant peptides from the hydrolysis of casein or gelatin. As shown in Figure 6(a) the degree of hydrolysis (DH) of casein was significantly higher

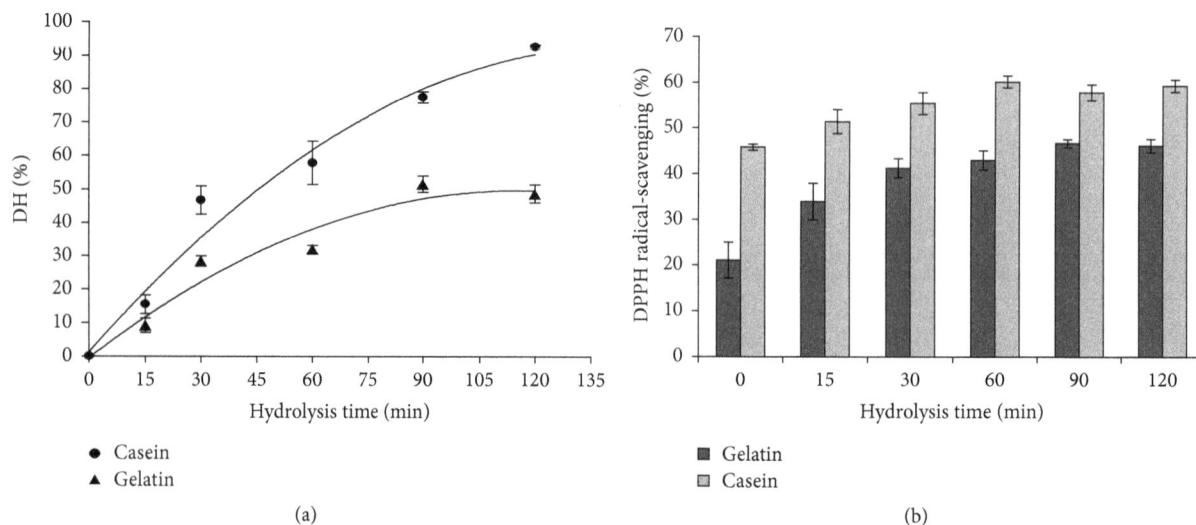

FIGURE 6: Hydrolysis and antioxidant activity assay. (a) Degree of hydrolysis (DH) for casein and gelatin substrates using proteases from *A. Sojae* ATCC 20235. (b) DPPH radical-scavenging activity using the hydrolysates from casein and gelatin. The error bars correspond to three independent assays.

TABLE 4: Estimated molecular weights corresponding to the bands with proteolytic activity for each extract.

Sample	Estimate Mw
	32
4	29
	22
	32
8	34
	26
	92
9	85
	33
	23
	88
	78
	68
	58
	51
12	40
	36
	29
	22
	19

than gelatin after fifteen minutes of hydrolysis reaction, reaching 95% of hydrolysis degree at 120 min. Similar degree of hydrolysis values has been obtained for casein hydrolysis using an immobilized *A. oryzae* protease [42].

Likewise, the antioxidant activity was higher for casein hydrolysates than for gelatin hydrolysates obtained throughout all the hydrolysis time. However the major increase rate was registered for gelatin, possibly because of the rapid increase of small peptides production that ranged between 11 and 25 kDa (Figure 6(b) and Supplementary Data 6). In this context it is important to mention that the initial higher DPPH radical-scavenging activity and the low increase rate observed for casein are due to partial hydrolysis (small peptides) present in the substrate at the initial time (Supplementary Data 6).

4. Conclusions

Based on the results obtained in this study we were able to find new *Aspergillus* species and strains with the ability of achieving high and intermediate levels of proteolytic activity. This is the first report for protease production under solid state fermentation for the strains *A. oryzae* NRRL 2217, *A. flavipes* NRRL 295, *A. oryzae* ICFC 8/12, *A. giganteus* NRRL 10, *A. rhizopodus* NRRL 6136, *A. sojae* NRRL 5595, and *A. sojae* ATCC 20235. It was determined that these strains exhibited maximum protease productivity within a short fermentation time using wheat bran as substrate, which significantly reduces the production costs. It is important to note that the crude extracts from *A. sojae* ATCC 20235, *A. oryzae* ICFC 8/12, *A. flavipes* NRRL 295, and *A. giganteus* NRRL 10, which presented the maximum levels of protease productivity, showed remarkable proteolytic activity in a wide range of pH and temperature. The extracts of *A. flavipes* NRRL 295 and *A. sojae* ATCC 20235 presented a notable stability under optimum pH and temperature values with a high half-life (240 min). Likewise, these extracts showed a minimal Q_{10} and E_a values for azo-casein hydrolysis. However, the proteolytic extract from *A. sojae* ATCC 20235 presented a high affinity for azo-casein. Finally, this extract is composed of a variety of proteolytic enzymes of different molecular weights and proved to be an appropriate biocatalyst for hydrolysis of casein and gelatin, increasing the antioxidant activities in 35% and 125%, respectively.

In conclusion, *A. sojae* ATCC 20235 is a promising strain for the production of proteolytic enzymes useful for obtaining peptides from food by-products with nutritional and medicinal relevance.

Authors' Contribution

Diego Gabriel Noseda and Edgardo Albertó equally contributed to this work as co-senior authors.

Acknowledgments

The work was supported by the PICT Start Up 2010-1312 grant from the National Agency for Science and Technology Promotion from the National Ministry of Science and Technology of Argentina (issued to Dr. E. Albertó). The authors would like to acknowledge Dr. Fernandez-Lahore (*Jacobs University Bremen*, Germany) for generously providing *A. sojae* ATCC 20235 strain. The RNNL strains employed in this work were kindly provided by ARS Culture Collection.

References

[1] V. Ramakrishna, S. Rajasekhar, and L. S. Reddy, "Identification and purification of metalloprotease from dry grass pea (*Lathyrus sativus* L.) seeds," *Applied Biochemistry and Biotechnology*, vol. 160, no. 1, pp. 63–71, 2010.

[2] N. P. Nirmal, S. Shankar, and R. S. Laxman, "Fungal proteases: an overview," *International Journal of Biotechnology & Biosciences*, vol. 1, no. 1, 2011.

[3] R. J. Soares de Castro and H. H. Sato, "Production and biochemical characterization of protease from *Aspergillus oryzae*: an evaluation of the physical-chemical parameters using agroindustrial wastes as supports," *Biocatalysis and Agricultural Biotechnology*, vol. 3, no. 3, pp. 20–25, 2014.

[4] C. Li, D. Xu, M. Zhao, L. Sun, and Y. Wang, "Production optimization, purification, and characterization of a novel acid protease from a fusant by *Aspergillus oryzae* and *Aspergillus niger*," *European Food Research and Technology*, vol. 238, no. 6, pp. 905–917, 2014.

[5] D. Heerd, S. Yegin, C. Tari, and M. Fernandez-Lahore, "Pectinase enzyme-complex production by *Aspergillus* spp. in solid-state fermentation: a comparative study," *Food and Bioproducts Processing*, vol. 90, no. 2, pp. 102–110, 2012.

[6] R. R. da Silva, T. P. de Freitas Cabral, A. Rodrigues, and H. Cabral, "Production and partial characterization of serine and metallo peptidases secreted by *Aspergillus fumigatus* Fresenius in submerged and solid state fermentation," *Brazilian Journal of Microbiology*, vol. 44, no. 1, pp. 235–243, 2013.

[7] A. Pandey, "Solid-state fermentation," *Biochemical Engineering Journal*, vol. 13, no. 2-3, pp. 81–84, 2003.

[8] A. Pandey, C. R. Soccol, and D. Mitchell, "New developments in solid state fermentation: I-bioprocesses and products," *Process Biochemistry*, vol. 35, no. 10, pp. 1153–1169, 2000.

[9] S. Y. Sun and Y. Xu, "Membrane-bound 'synthetic lipase' specifically cultured under solid-state fermentation and submerged fermentation by *Rhizopus chinensis*: a comparative investigation," *Bioresource Technology*, vol. 100, no. 3, pp. 1336–1342, 2009.

[10] I. A. Cavello, R. A. Hours, and S. F. Cavalitto, "Bioprocessing of 'hair waste' by *Paecilomyces lilacinus* as a source of a bleach-stable, alkaline, and thermostable keratinase with potential application as a laundry detergent additive: characterization and wash performance analysis," *Biotechnology Research International*, vol. 2012, Article ID 369308, 12 pages, 2012.

[11] A. Riffel, F. Lucas, P. Heeb, and A. Brandelli, "Characterization of a new keratinolytic bacterium that completely degrades native feather keratin," *Archives of Microbiology*, vol. 179, no. 4, pp. 258–265, 2003.

[12] U. K. Laemmli, "Cleavage of structural proteins during the assembly of the head of bacteriophage T4," *Nature*, vol. 227, no. 5259, pp. 680–685, 1970.

[13] V. Neuhoff, N. Arold, D. Taube, and W. Ehrhardt, "Improved staining of proteins in polyacrylamide gels including isoelectric focusing gels with clear background at nanogram sensitivity using Coomassie Brilliant Blue G-250 and R-250," *Electrophoresis*, vol. 9, no. 6, pp. 255–262, 1988.

[14] I. A. Cavello, M. Chesini, R. A. Hours, and S. F. Cavalitto, "Study of the production of alkaline keratinases in submerged cultures as an alternative for solid waste treatment generated in leather technology," *Journal of Microbiology and Biotechnology*, vol. 23, no. 7, pp. 1004–1014, 2013.

[15] C. A. Schneider, W. S. Rasband, and K. W. Eliceiri, "NIH Image to ImageJ: 25 years of image analysis," *Nature Methods*, vol. 9, no. 7, pp. 671–675, 2012.

[16] M. Melikoglu, C. S. K. Lin, and C. Webb, "Kinetic studies on the multi-enzyme solution produced via solid state fermentation of waste bread by *Aspergillus awamori*," *Biochemical Engineering Journal*, vol. 80, pp. 76–82, 2013.

[17] R. J. Soares de Castro, A. Ohara, T. G. Nishide, J. R. M. Albernaz, M. H. Soares, and H. H. Sato, "A new approach for proteases production by *Aspergillus niger* based on the kinetic and thermodynamic parameters of the enzymes obtained," *Biocatalysis and Agricultural Biotechnology*, vol. 4, no. 2, pp. 199–207, 2015.

[18] A. Bougatef, M. Hajji, R. Balti, I. Lassoued, Y. Triki-Ellouz, and M. Nasri, "Antioxidant and free radical-scavenging activities of smooth hound (*Mustelus mustelus*) muscle protein hydrolysates obtained by gastrointestinal proteases," *Food Chemistry*, vol. 114, no. 4, pp. 1198–1205, 2009.

[19] J. Adler-Nissen, "Determination of the degree of hydrolysis of food protein hydrolysates by trinitrobenzenesulfonic acid," *Journal of Agricultural and Food Chemistry*, vol. 27, no. 6, pp. 1256–1262, 1979.

[20] P. M. Nielsen, D. Petersen, and C. Dambmann, "Improved method for determining food protein degree of hydrolysis," *Journal of Food Science*, vol. 66, no. 5, pp. 642–646, 2001.

[21] R. J. Soares de Castro, T. G. Nishide, and H. H. Sato, "Production and biochemical properties of proteases secreted by *Aspergillus niger* under solid state fermentation in response to different agroindustrial substrates," *Biocatalysis and Agricultural Biotechnology*, vol. 3, no. 4, pp. 236–245, 2014.

[22] R. J. Soares de Castro and H. H. Sato, "Synergistic effects of agroindustrial wastes on simultaneous production of protease and α-amylase under solid state fermentation using a simplex

centroid mixture design," *Industrial Crops and Products*, vol. 49, pp. 813–821, 2013.

[23] R. J. Soares de Castro and H. H. Sato, "Advantages of an acid protease from *Aspergillus oryzae* over commercial preparations for production of whey protein hydrolysates with antioxidant activities," *Biocatalysis and Agricultural Biotechnology*, vol. 3, no. 3, pp. 58–65, 2014.

[24] A. Belmessikh, H. Boukhalfa, A. Mechakra-Maza, Z. Gheribi-Aoulmi, and A. Amrane, "Statistical optimization of culture medium for neutral protease production by *Aspergillus oryzae*. Comparative study between solid and submerged fermentations on tomato pomace," *Journal of the Taiwan Institute of Chemical Engineers*, vol. 44, no. 3, pp. 377–385, 2013.

[25] M. Hiyama, T. Ohmoto, M. Shinozuka, K. Ito, M. Iizuka, and N. Minamiura, "A serine proteinase of a fungus isolated from dried bonito 'Katsuobushi'," *Journal of Fermentation and Bioengineering*, vol. 80, no. 5, pp. 462–466, 1995.

[26] C. G. Boer and R. M. Peralta, "Production of extracellular protease by *Aspergillus tamarii*," *Journal of Basic Microbiology*, vol. 40, no. 2, pp. 75–81, 2000.

[27] K. S. Vishwanatha, A. G. Appu Rao, and S. A. Singh, "Characterisation of acid protease expressed from *Aspergillus oryzae* MTCC 5341," *Food Chemistry*, vol. 114, no. 2, pp. 402–407, 2009.

[28] C. R. Tremacoldi, R. Monti, H. S. Selistre-De-Araújo, and E. C. Carmona, "Purification and properties of an alkaline protease of *Aspergillus clavatus*," *World Journal of Microbiology and Biotechnology*, vol. 23, no. 2, pp. 295–299, 2007.

[29] S. Germano, A. Pandey, C. A. Osaku, S. N. Rocha, and C. R. Soccol, "Characterization and stability of proteases from *Penicillium* sp. produced by solid-state fermentation," *Enzyme and Microbial Technology*, vol. 32, no. 2, pp. 246–251, 2003.

[30] I. I. Sutar, M. C. Srinivasan, and H. G. Vartak, "A low molecular weight alkaline proteinase from *Conidiobolus coronatus*," *Biotechnology Letters*, vol. 13, no. 2, pp. 119–124, 1991.

[31] S. Shenolikar and K. J. Stevenson, "Purification and partial characterization of a thiol proteinase from the thermophilic fungus *Humicola lanuginosa*," *The Biochemical Journal*, vol. 205, no. 1, pp. 147–152, 1982.

[32] S. Nasuno, "Differentiation of *Aspergillus sojae* from *Aspergillus oryzae* by polyacrylamide gel disc electrophoresis," *Journal of General Microbiology*, vol. 71, no. 1, pp. 29–33, 1972.

[33] M. Hajji, S. Kanoun, M. Nasri, and N. Gharsallah, "Purification and characterization of an alkaline serine-protease produced by a new isolated *Aspergillus clavatus* ES1," *Process Biochemistry*, vol. 42, no. 5, pp. 791–797, 2007.

[34] S. K. Chakrabarti, N. Matsumura, and R. S. Ranu, "Purification and characterization of an extracellular alkaline serine protease from *Aspergillus terreus* (IJIRA 6.2)," *Current Microbiology*, vol. 40, no. 4, pp. 239–244, 2000.

[35] S.-L. Wang, Y.-H. Chen, C.-L. Wang, Y.-H. Yen, and M.-K. Chern, "Purification and characterization of a serine protease extracellularly produced by *Aspergillus fumigatus* in a shrimp and crab shell powder medium," *Enzyme and Microbial Technology*, vol. 36, no. 5-6, pp. 660–665, 2005.

[36] H.-C. R. Chien, L.-L. Lin, S.-H. Chao et al., "Purification, characterization, and genetic analysis of a leucine aminopeptidase from *Aspergillus sojae*," *Biochimica et Biophysica Acta (BBA)— Gene Structure and Expression*, vol. 1576, no. 1-2, pp. 119–126, 2002.

[37] X. Sun, E. Salih, F. G. Oppenheim, and E. J. Helmerhorst, "Kinetics of histatin proteolysis in whole saliva and the effect on bioactive domains with metal-binding, antifungal, and wound-healing properties," *The FASEB Journal*, vol. 23, no. 8, pp. 2691–2701, 2009.

[38] I. E. Crompton and S. G. Waley, "The determination of specificity constants in enzyme-catalysed reactions," *Biochemical Journal*, vol. 239, no. 1, pp. 221–224, 1986.

[39] P. S. Murthy and M. Naidu, "Protease production by *Aspergillus oryzae* in solid," *World Applied Science Journal*, vol. 2, no. 8, pp. 199–205, 2010.

[40] H. J. Kim, K. H. Park, J. H. Shin et al., "Antioxidant and ACE inhibiting activities of the rockfish Sebastes hubbsi skin gelatin hydrolysates produced by sequential two-step enzymatic hydrolysis," *Fisheries and Aquatic Science*, vol. 14, no. 1, pp. 1–10, 2011.

[41] R. J. Soares de Castro and H. H. Sato, "Protease from *Aspergillus oryzae*: biochemical characterization and application as a potential biocatalyst for production of protein hydrolysates with antioxidant activities," *Journal of Food Processing*, vol. 2014, Article ID 372352, 11 pages, 2014.

[42] S.-J. Ge, H. Bai, H.-S. Yuan, and L.-X. Zhang, "Continuous production of high degree casein hydrolysates by immobilized proteases in column reactor," *Journal of Biotechnology*, vol. 50, no. 2-3, pp. 161–170, 1996.

7

Production of Recombinant *Trichoderma reesei* Cellobiohydrolase II in a New Expression System based on *Wickerhamomyces anomalus*

Dennis J. Díaz-Rincón,[1] **Ivonne Duque,**[1] **Erika Osorio,**[1]
Alexander Rodríguez-López,[1,2] **Angela Espejo-Mojica,**[1] **Claudia M. Parra-Giraldo,**[3]
Raúl A. Poutou-Piñales,[4] **Carlos J. Alméciga-Díaz,**[1] **and Balkys Quevedo-Hidalgo**[4]

[1]*Institute for the Study of Inborn Errors of Metabolism, Facultad de Ciencias, Pontificia Universidad Javeriana, Bogotá, Colombia*
[2]*Departamento de Química, Facultad de Ciencias, Pontificia Universidad Javeriana, Bogotá, Colombia*
[3]*Unidad de Proteómica y Micosis Humanas, Grupo de Enfermedades Infecciosas, Departamento de Microbiología,*
 Facultad de Ciencias, Pontificia Universidad Javeriana, Bogotá, Colombia
[4]*Grupo de Biotecnología Ambiental e Industrial (GBAI), Departamento de Microbiología, Facultad de Ciencias,*
 Pontificia Universidad Javeriana, Bogotá, Colombia

Correspondence should be addressed to Carlos J. Alméciga-Díaz; cjalmeciga@javeriana.edu.co
and Balkys Quevedo-Hidalgo; bquevedo@javeriana.edu.co

Academic Editor: Sunney I. Chan

Cellulase is a family of at least three groups of enzymes that participate in the sequential hydrolysis of cellulose. Recombinant expression of cellulases might allow reducing their production times and increasing the low proteins concentrations obtained with filamentous fungi. In this study, we describe the production of *Trichoderma reesei* cellobiohydrolase II (CBHII) in a native strain of *Wickerhamomyces anomalus*. Recombinant CBHII was expressed in *W. anomalus* 54-A reaching enzyme activity values of up to 14.5 U L^{-1}. The enzyme extract showed optimum pH and temperature of 5.0–6.0 and 40°C, respectively. Enzyme kinetic parameters (K_M of 2.73 mM and Vmax of 23.1 μM min^{-1}) were between the ranges of values reported for other CBHII enzymes. Finally, the results showed that an enzymatic extract of *W. anomalus* 54-A carrying the recombinant *T. reesei* CBHII allows production of reducing sugars similar to that of a crude extract from cellulolytic fungi. These results show the first report on the use of *W. anomalus* as a host to produce recombinant proteins. In addition, recombinant *T. reesei* CBHII enzyme could potentially be used in the degradation of lignocellulosic residues to produce bioethanol, based on its pH and temperature activity profile.

1. Introduction

Cellulases are enzymes from the glycoside hydrolase family (EC 3.2.1.-) that are expressed by a broad spectrum of bacteria and fungi strains [1, 2]. Among these microorganisms, white- and brown-rot fungi are considered the major producers of cellulases, including *Trichoderma* sp., *Aspergillus* sp., *Schizophyllum commune*, and *Volvariella volvacea* [1, 2]. Cellulase is a family of at least three groups of enzymes: endo-(1,4)-β-D-glucanase (EC 3.2.1.4, EG), exo-(1,4)-β-D-glucanase (EC 3.2.1.91, CBH), and β-glucosidases (EC 3.2.1.21, BG) [3]. These

enzymes act sequentially on the cellulose hydrolysis: (1) EG randomly attacks on O-glycosidic bonds resulting in glucan chains of different lengths, (2) CBH acts on glucan chains ends to release β-cellobiose, and (3) BG hydrolyzes the β-cellobiose to produce glucose [3].

Trichoderma reesei is one of the most important cellulases producing filamentous fungi, since it contains the core enzymes necessary for the complete hydrolysis of lignocellulose material [4]. Among these enzymes, cellobiohydrolases have been shown to be one of the most important components within this process [4]. Enzymes involved in hydrolysis of cellulose polymer molecules have been recombinantly expressed [5], which might allow reducing the production times and increasing the low proteins concentrations obtained with filamentous fungi. Recombinant expression might also facilitate the scaling up of the enzyme production, as well as the implementation of Simultaneous Saccharification and Fermentation (SSF) and Separate Hydrolysis and Fermentation (SHF) processes. *Saccharomyces cerevisiae* has been used for the production of recombinant CBH from *Phanerochaete chrysosporium* [6] and *Talaromyces emersonii* [7], while *T. reesei* CBHII has been expressed in *S. cerevisiae* [8, 9], *P. pastoris* [10, 11], and *Y. lipolytica* [11].

Conventional microbial systems (e.g., *E. coli*, *S. cerevisiae*, and *P. pastoris*) have allowed the production of a long list of recombinant proteins [12]. However, there is a growing interest in the identification of new hosts that allow the production of high-quality or cost-efficient recombinant proteins [13]. Recently, a native strain of *Wickerhamomyces anomalus* was isolated from sugar cane bagasse at Puerto López, Meta, Colombia (unpublished results). *W. anomalus* is a yeast that has been isolated from different sources [14], and its reported applications include food biopreservation, bioremediation, and production of phytases and biofuels [14, 15]. However, to the best of our knowledge, there are no reports showing its use as host to produce recombinant proteins.

In this paper, we describe the production of recombinant cellobiohydrolase II (CBHII, EC 3.2.1.91) from *T. reesei* using the native strain *W. anomalus* 54-A. Furthermore, the effects of pH and temperature on enzyme activity were characterized, as well as the kinetic properties of the enzyme and its application on the hydrolysis of a lignocellulose material. Overall, the results showed the potential of *W. anomalus* 54-A as host to produce recombinant proteins and showed that recombinant *T. reesei* CBHII has similar characteristics to those reported for the wild-type enzyme.

2. Materials and Methods

2.1. Microorganism, Culture Media, and Vectors. *W. anomalus* 54-A was previously isolated from sugar cane bagasse at Puerto López, Meta, Colombia. Initially, the microorganism was identified by carbon assimilation profile using the API 20C AUX kit (bioMérieux, Marcy l'Etoile, France). Further microorganism identification was done by amplification of internal transcribed spacer (ITS), using the primers ITS1 5′-TCCGTAGGTGAACCTGCGG-3′ and ITS4 5′-TCCTCCGCTTATTGATATGC-3′ and *Pfu* DNA polymerase (Thermo Fisher Scientific Inc., San Jose, CA, US), under manufacturer instructions. Identity of *W. anomalus* 54-A was finally confirmed by MALDI-TOF analysis (Unidad de Investigación en Proteómica y Micosis Humana, Pontificia Universidad Javeriana).

Plasmid pKS2-ST (Dualsystems Biotech, Schlieren, Switzerland) was used as expression vector. In this vector, protein expression is regulated by the alcohol dehydrogenase II promoter (EC 1.1.1.1, ADH2), while the secretion signal of the *Saccharomyces cerevisiae* invertase SUC2 mediates the secretion of the recombinant protein. *Escherichia coli* DH5 was used for cloning purposes, which was

cultured in Luria-Bertani (LB) medium supplemented with $100 \, mg \, mL^{-1}$ ampicillin. Yeast cultures were performed in YPD supplemented with $500 \, mg \, mL^{-1}$ geneticin for the selection of clones and $60 \, mg \, mL^{-1}$ for protein expression cultures.

2.2. Expression Vector and Recombinant Strains. Cellobiohydrolase II gene (*cbhII*) from *T. reesei* (GenBank Number GU724763.1) was previously codon-optimized for its expression in *S. cerevisiae* by GeneArt™ (Thermo Fisher Scientific Inc.) (unpublished data). The sequence encoding for the CBHII signal peptide was identified by using SignalP [16] and removed from the gene sequence. Comparison of the codon usage tables of *S. cerevisiae* and *W. anomalus* showed a similar profile between both microorganisms (Supplementary Figure 1 in Supplementary Material available online at https://doi.org/10.1155/2017/6980565) Two codons (CGC and CGG), which encode Arg (14 residues, 3%, in CBHII) showed significant differences between both microorganisms. Nevertheless, the other two Arg codons have a comparable usage between *S. cerevisiae* and *W. anomalus*. In this sense, this codon optimized sequence was used for the CBHII expression in *W. anomalus* 54-A. Codon optimized *cbhII* gene was inserted downstream of the SUC2 secretion signal in the vector pKS2-ST, to produce pKS2-ST::CBHII (Supplementary Figure 2). The pKS2-ST::CBHII vector was used to transform competent cells of *W. anomalus* 54-A by electroporation, using the Gene Pulser electroporation system Xcell™ (Bio-Rad Laboratories) at 1400 V and 200 Ω. The *W. anomalus* 54-A clones were selected in YPD medium supplemented with $500 \, mg \, mL^{-1}$ geneticin. Clones were confirmed by PCR using the primers 5′-GAGGAGAGCATAGAAATGGGG-3′ and 5′-CAGCAGTAGCCATAGCACCA-3′, which amplify a fragment from *cbhII* gene. The PCR-positive clones were evaluated at 55 mL scale, according to vector pKS2-ST manufacturer's instructions (Dualsystems Biotech). Briefly, each clone was incubated for 10 h in 10 mL YPD, after which 15 mL of fresh YPD medium was added. After 14 h incubation, 30 mL of fresh YPD medium was added to reach a final culture volume of 55 mL. After 6 h of incubation, it was expected that the glucose was exhausted, and this time was considered as the beginning of the induction phase. 1 mL aliquots were taken every 24 h for 96 h to measure extracellular enzyme activity and cell density. All the assays were performed in triplicate at 28°C and 180 rpm. Residual glucose quantitation was carried out by DNS method [17]. Crude extracellular fractions of the clone with the highest enzyme activity were loaded and processed by SDS-PAGE.

2.3. Evaluation of Carbon Sources. The clone that showed the highest enzyme activity was used to evaluate the effect of carbon source (i.e., glucose and glycerol) on the enzyme production. Cultures were carried out in 2% (w/v) tryptone and 1% (w/v) yeast extract, supplemented with glycerol and glucose in different concentrations: 2% (w/v) glucose, 1% (w/v) glucose, 1% (w/v) glycerol, 2% (w/v) glucose, 1% (w/v) glycerol, and 2% (w/v) glycerol. All cultures were supplemented with $60 \, mg \, mL^{-1}$ geneticin.

2.4. Culture at 2.4 L Scale. The *W. anomalus* 54-A clone that showed the highest CBHII activity at 55 mL scale was scaled up to 2.4 L in a 3.7 L Bioengineering KFL2000 bioreactor. Cultures were done in modified YPD medium [2% (w/v) tryptone, 1% (w/v) yeast extract, 1% (w/v) glucose, and 1% (w/v) glycerol] supplemented with $60\,mg\,mL^{-1}$ geneticin. Briefly, 2 mL from the cell bank was inoculated into 18 mL of culture medium and incubated for 24 h at 28°C and 180 rpm. The preinoculum was used to inoculate 180 mL of fresh culture medium and incubated for 24 h at 28°C and 180 rpm. Finally, the inoculum was used to inoculate 1000 mL of fresh culture medium at the bioreactor. After 15 h culture, 1200 mL of fresh medium was added to reach a final volume of 2.4 L and cultured during additional 96 h at 28°C, 400 to 800 rpm, and pH 6.0.

2.5. Cellobiohydrolase Activity Assay. The CBHII activity was carried out as previously described [18, 19], using 5 mM p-nitrophenyl β-D-cellobioside (pNPC, Sigma-Aldrich) in sodium acetate buffer 50 mM, pH 5.0. One enzyme unit was defined as the amount of enzyme releasing 1 μmol of p-nitrophenol per min, and the activity was expressed as $U\,L^{-1}$.

2.6. Characterization of Recombinant CBHII. The CBHII enzyme extract was evaluated at different pH and temperature conditions, using pNPC. To evaluate the effect of pH and temperature on enzyme activity, the enzyme activity assay was performed at 3.0, 4.0, 5.0, 6.0, 7.0, and 8.0 ± 0.2 and at 30, 40, 50, 60, and 70°C for 1h. The reactions were stopped and read as described above. Apparent kinetic parameters (K_M and Vmax) were estimated for the crude extract by using pNPC between 0 and 6.5 mM and fitting the experimental data to a Michaelis-Menten model using the software GraphPad PRISM 6.0.

2.7. Chrysanthemum Wastes Degradation Assay. The effect of recombinant CBHII on cellulose hydrolysis was evaluated by using *Chrysanthemum* wastes, as previously described [18, 19], with modifications. *Chrysanthemum* wastes were autoclaved at 121°C for 15 min. Experiments were performed in substrate submerged cultures carried out in 100 mL flasks in a rotatory shaker at 150 rpm and 30°C, for 20 days. Flasks contained 1% *Chrysanthemum* wastes in 40 mL of the following enzyme crude extracts: (1) cellulolytic and hemicellulolytic extract from a *Penicillium* sp. culture (hereafter Ce-Hem extract), (2) concentrated recombinant CBHII extract from *W. anomalus* 54-A (hereafter rCBHII), and (3) a 1 : 1 Ce-Hem : rCBHII extracts mixture. Control culture was carried out by using 40 mL of distilled water. All cultures were carried out with 10 mM sodium azide, 0.1% tween 80, and pH 5.0. The response variable was the concentration of reducing sugars as measured by DNS method [17]. Results are reported as $g\,L^{-1}$ of reducing sugars after subtraction of the results obtained with the control cultures, and normalized by the initial unis of CBHII present in each extract. Production of Ce-Hem extract was carried out by using a *Penicillium* sp. strain (Collection of Microorganisms of Pontificia Universidad Javeriana). For this purpose, 20 mL of a 10^6 conidia mL^{-1} suspension was inoculated in 200 mL of a rice straw medium without nitrogen source and incubated for 8 days at 28°C and 120 rpm. The crude extract was obtained by centrifugation and sequential filtrations through Whatman paper Number 42 and 0.45 and 0.22 μm polyether sulphone membranes (Pall Corp, Port Washington, NY, USA). Concentrated recombinant CBHII extract from *W. anomalus* 54-A was produced by using a 2.4 L culture of the *W. anomalus* 54-A Celo-3.2 clone, as described above. Culture medium was centrifuged at $8000g$ and filtered sequentially through Whatman paper Number 42 and 0.45 and 0.22 μm polyether sulphone membranes (Pall Corp). Permeate was ultrafiltered through a 30 kDa cut-off membrane (Millipore, Billerica, MA, USA), until reaching a final volume of 20 mL. CBHII activity in both extracts was assayed as described above. The β-glucosidase (BG) activity in both extracts was assayed using 5 mM 4-nitrophenyl β-D-glucuronide (Sigma-Aldrich) as previously described [18, 19]. One BG unit was defined as the amount of enzyme releasing 1 μmol of p-nitrophenol per min, and the activity was expressed as $U\,L^{-1}$. The endoglucanase (EG) activity in both extracts was assayed using 2% carboxymethylcellulose (Sigma-Aldrich) [18, 19] and the DNS method for quantitation of reducing sugars [17]. One EG unit was defined as the amount of enzyme releasing 1 μmol of reducing sugars, equivalent to glucose, per min [20], and the activity was expressed as $U\,L^{-1}$.

2.8. Statistical Analysis. Differences between groups were tested for statistical significance by using one-way ANOVA. An error level of 5% ($p < 0.05$) was considered significant. All analyses were performed using the software GraphPad PRISM 6.0.

3. Results and Discussion

3.1. Production of Recombinant CBHII in W. anomalus 54-A. In this study, we evaluated the production of the recombinant CBHII in the native yeast strain 54-A, previously isolated from sugar cane bagasse from Puerto López, Meta, Colombia. This strain showed a carbon assimilation profile (API 20C AUX, bioMérieux) different from that observed for *S. cerevisiae* strains, and it was identified as *Pichia anomala* (recently renamed as *Wickerhamomyces anomalus* [21]) with a 94.5% ID and 0,25 *T* value. Identity of this strain was further confirmed as *W. anomalus* through ITS amplification (GenBank accession number KX676490, Supplementary Figure 3) and MALDI-TOF analysis (Supplementary Figure 4). *W. anomalus* is a yeast that have been isolated from different sources, such as food- and feed-related systems, insects, and human clinical samples. This yeast has a wide robustness to environmental stresses conditions like extreme pH or low water activity [14]. Biotechnological applications of *W. anomalus* strains include biopreservation agent due to antimicrobial activity against variety of microorganisms in fruits and cereals and production of enzymes (e.g., phytases), biofuels, and bioremediation [14, 15]. However, to the best of our knowledge, there are no reports of the use of this yeast as a host to produce recombinant proteins.

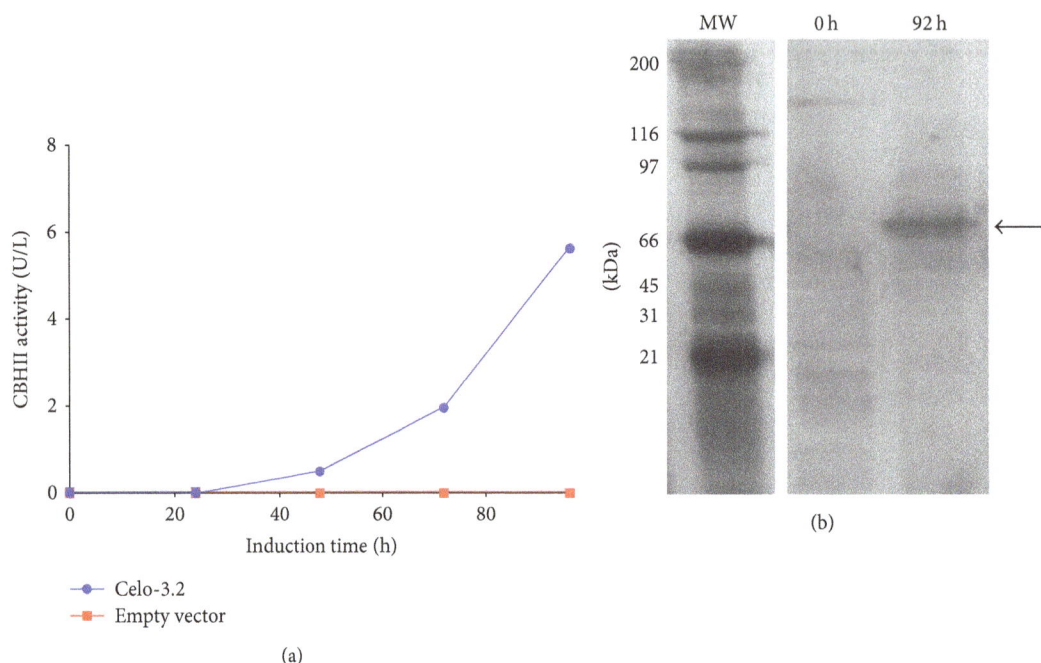

FIGURE 1: Recombinant *T. reesei* CBHII produced in *W. anomalus* 54-A. (a) PCR-positive clones of *W. anomalus* 54-A were evaluated at 55 mL scale and the activity of recombinant *T. reesei* CBHII was assayed in the extracellular fraction. Among the four PCR-positive clones, only *W. anomalus* 54-A Celo 3.2 clone showed CBHII activity. No activity was observed in *W. anomalus* 54-A transfected with the empty vector. Each clone was evaluated in triplicate. (b) Crude extracellular extracts from *W. anomalus* 54-A Celo 3.2 at 0 and 92 h were evaluated by SDS-PAGE, followed by silver staining. The arrow indicates the recombinant CBHII.

Four clones were obtained after transformation of *W. anomalus* 54-A with the vector pKS2-ST::CBHII (Celo-2.2, Celo-3.2, Celo-4.2, and Celo-5.1), which were confirmed by PCR. The *ADH2* gene promoter is repressed in the presence of glucose, since this metabolite inhibits the expression of the alcohol dehydrogenase regulator protein (Adr1) that is an activator of the *ADH2* gene. On the other hand, under glucose depletion conditions, the expression of Adr1 is increased leading to the induction of the *ADH2* gene [22]. In this sense, we first determined the residual glucose during the culture of *W. anomalus* 54-A pKS2-ST::CBHII clones, to establish the beginning of the induction phase. We observed, for all the evaluated clones, that, at the final culture step, the glucose was consumed within the first 6 h of incubation (Supplementary Figure 5). In this sense, this point was considered as the beginning of the induction phase, which agrees with previous reports of the production of recombinant proteins in yeast under the control of the *ADH2* gene promoter [22, 23]. Under these culture conditions, extracellular CBHII activity was only detected in *W. anomalus* 54-A Celo-3.2 clone, reaching a final enzyme activity of $5.7 \, U \, L^{-1}$ after 96 h of induction (Figure 1(a)). As expected, CBHII activity was not detected in *W. anomalus* 54-A transfected with the empty vector.

As was mentioned above, *T. reesei* CBHII has been expressed in *S. cerevisiae* [8, 9], *P. pastoris* [10, 11], and *Y. lipolytica* [11]. Recombinant CBHII produced in *S. cerevisiae* showed an activity of 25 S.I. $U \, mL^{-1}$ using barely β-glucan as substrate [8], while recombinant CBHII produced in *P. pastoris* GS115 showed an enzyme activity of $5.84 \, U \, mL^{-1}$

at 96 h using microcrystalline cellulose PH101 as substrate [10]. In *P. pastoris* X-33 and *Y. lipolytica* POld, extracellular recombinant CBHII activities were 0.25 and $0.36 \, U \, mL^{-1}$, respectively, using phosphoric acid swollen cellulose (PASC) prepared from Avicel PH-101 as substrate [11]. The differences in promoters, secretion signals, hosts, and enzyme activity substrates limit the comparison between these results and those obtained for recombinant CBHII produced in *W. anomalus* 54-A. However, a crude culture broth from *Penicillium* sp. containing wild-type cellulase and hemicellulase enzymes (see *Chrysanthemum* Wastes Degradation Assay) showed a CBHII activity of $8.5 \, U \, L^{-1}$ using the pNPC substrate, which was similar to that observed for the recombinant CBHII produced in *W. anomalus* 54-A. Nevertheless, these results are lower than those observed for erlenmeyer flask cultures of *Pleurotus ostreatus*, which showed an activity of $445.3 \pm 27.6 \, U \, L^{-1}$ using the pNPC substrate [18, 19].

SDS-PAGE analysis of extracellular fraction of *W. anomalus* 54-A Celo-3.2 showed that recombinant CBHII has an apparent molecular weight of about 66 kDa (Figure 1(b)). This recombinant CBHII is higher than the wild-type enzyme produced by *T. reesei* (50–58 kDa) [24]. However, recombinant CBHII produced in *W. anomalus* 54-A has a molecular weight similar to that reported for a CBHII produced in *P. pastoris* X33 and *Y. lipolytica* POld (~63–65 kDa) [11] but higher than that reported for the enzyme produced in *P. pastoris* GS115 (~58 kDa) [10]. Differences in the molecular weight between wild-type and the recombinant counterpart could be associated with the hypermannosylation observed

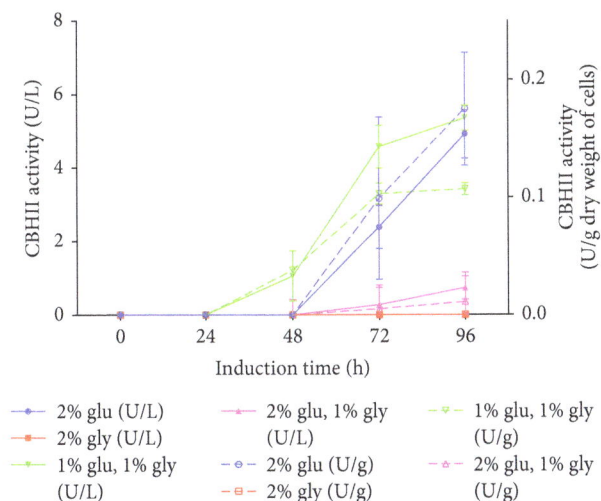

FIGURE 2: Effect of the carbon source on the production of recombinant *T. reesei* CBHII. Production of recombinant CBHII was carried out at 55 mL with 2% (w/v) glucose (circle), 2% (w/v) glycerol (square), 2% (w/v) glucose, 1% (w/v) glycerol (up triangle), and 1% (w/v) glucose, 1% (w/v) glycerol (down triangle). Enzyme activity is reported as $U L^{-1}$ (continuous line) and $U g^{-1}$ dry weight of cells (dashed line). Each experiment was done in triplicate.

FIGURE 3: Production of recombinant *T. reesei* CBHII in *W. anomalus* 54-A at 2.4 L scale. Recombinant *T. reesei* CBHII was produced at 2.4 L with 1% (w/v) glucose and 1% (w/v) glycerol. Production was followed by 117 h after the final feeding step. Dissolved oxygen (%), cell density (gL^{-1}), and extracellular enzyme activity ($U L^{-1}$) are reported.

in yeast, while the differences in the molecular weight among the recombinant CBHII could be associated with several factors of the culture conditions that affect the expression of glycosyltransferases [25].

3.2. Evaluation of Carbon Sources. To improve the production of recombinant CBHII, two glucose and glycerol concentrations were evaluated. First, 1% w/v glucose was evaluated, which is lower than the concentration suggested by the vector manufacturer (2% w/v). We hypothesize that 1% w/v glucose would be depleted faster than 2% w/v, allowing a reduction in the induction time and a longer and higher enzyme production [22]. In addition, glycerol, alone or in combination with glucose, was also evaluated as a carbon source, since it was expected that glycerol would be used for yeast growth after depletion of glucose, increasing the production of the recombinant enzyme.

Figure 2 shows the results of volumetric activity ($U L^{-1}$) and enzyme activity normalized by biomass ($U g^{-1}$ dry weight of cells). A similar activity profile was observed between volumetric and normalized enzyme activities, suggesting that the production of recombinant CBHII was associated with yeast growth. In fact, the profile of recombinant CBHII production was similar to that previously reported for the ADH2 promoter, with a complete induction in the late stationary phase (>36 h culture) [22]. The results showed that in 2% (w/v) glycerol cultures no CBHII activity was detected, suggesting that the glycerol was used for cell growth or that the ADH2 promoter needs to sense the glucose depletion to induce the gene expression. On the other hand, lower CBHII activity was observed at 2% (w/v) glucose, 1% (w/v) glycerol, cultures compared to that obtained with 2% (w/v) glucose ($p < 0.001$). Finally, the highest CBHII activity was observed

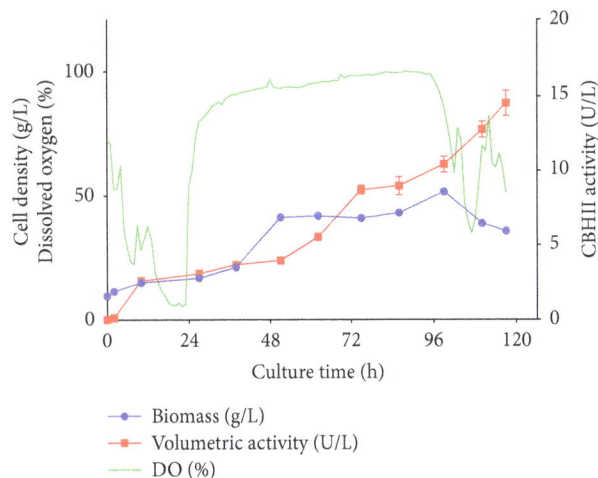

at 1% (w/v) glucose, 1% (w/v) glycerol, although this was not statistically different regarding the levels obtained with 2% (w/v) glucose ($p > 0.05$). However, 1% (w/v) glucose, 1% (w/v) glycerol, allowed an earlier detection of the enzyme activity than that observed with 2% (w/v) glucose. In this sense, 1% (w/v) glucose, 1% (w/v) glycerol, was selected to produce recombinant CBHII at 2.4 L scale. The reasons of the effect of glycerol on CBHII activity are unknown, and further investigations should be carried out to clarify this aspect. Nevertheless, it is important to mention that there is a wide variation among yeasts, and even strains, regarding glycerol metabolism (i.e., uptake and catabolism) [26]. Although the metabolic pathways and the uptake and regulation (e.g., catabolic repression) process have been widely described for *S. cerevisiae*, limited information is available for other yeasts [26]. In this sense, the effect of glycerol on the changes of enzyme activity levels might be associated with the specific glycerol metabolism of *W. anomalus* 54-A.

3.3. Production of Recombinant CBHII at 2.4 L Scale. The production of recombinant CBHII enzyme at 2.4 L scale was carried out for 117 h after the final feeding step (see Materials and Methods). Figure 3 shows the behavior of dissolved oxygen (DO) during the production of recombinant CBHII at bioreactor scale, which could be associated with the consumption of the carbon sources. The CBHII activity was detected from the first 2 h of culture (0.12 ± 0.08 $U L^{-1}$). This activity significantly increased after 10 h (2.6 ± 0.09 $U L^{-1}$) and continuously kept increasing to reach a final enzyme activity of 14.5 ± 0.86 $U L^{-1}$, which was 2.7-fold higher than that observed at 55 mL scale. These results might suggest that, under controlled culture conditions, the strict repression of the ADH2 promoter might be affected

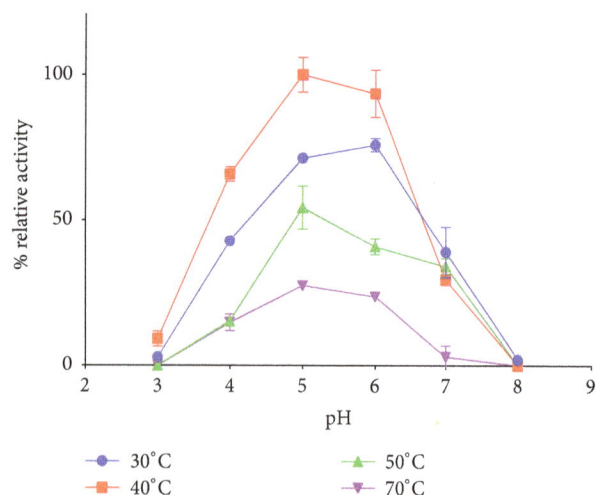

FIGURE 4: Effect of temperature and pH on reaction of recombinant *T. reesei* CBHII. The enzyme activity of recombinant *T. reesei* CBHII produced in *W. anomalus* 54-A was evaluated at different pH and temperatures reactions condition. Each experiment was carried out in triplicate. Activity is reported as % Relative Activity as compared with the highest activity obtained in the experiment (40°C, pH 5).

allowing the expression of the recombinant protein even in the presence of residual glucose.

3.4. Characterization of CBH Recombinant Extract. The effect of temperature and pH on the enzyme activity of the recombinant CBHII produced in *W. anomalus* 54-A was evaluated. In this work, crude extract rather than the purified recombinant CBHII was used for characterization, since this crude extract would be used in the production of bioethanol by the degradation of *Chrysanthemum* residues [19]. As observed in Figure 4, optimum reaction temperature for recombinant CBHII crude extract was 40°C ($p < 0.001$), while the highest enzyme activities were observed at pH 5.0 and 6.0 ± 0.2 at all evaluated temperatures ($p < 0.001$). On the other hand, at pH 3.0 and 8.0 ± 0.2, a marked decrease in the enzymatic activity was observed regardless of the reaction temperature.

Using microcrystalline cellulose PH101 as substrate, recombinant CBHII produced in *P. pastoris* showed an optimum reaction pH and temperature of 5.0 ± 0.2 and 50°C, respectively. In addition, this recombinant enzyme showed a significant loss of activity at pH of 3.0, 7.0, and 8.0 and at 30°C and temperatures above 65°C [10]. Recombinant CBHII enzyme produced in *P. pastoris* and *Y. lipolytica* showed an optimum pH between 5.0 and 6.0 and at a temperature of 60°C. Recombinant CBHII produced in *P. pastoris* and *Y. lipolytica* also showed rapid inactivation at pH and temperatures above 7.0 and 60°C, respectively, while at temperatures below 40°C, an 80% loss in the enzyme activity was observed [11]. In this sense, recombinant *T. reesei* CBHII produced in *W. anomalus* 54-A has a reaction pH profile similar to that reported for recombinant CBHII enzymes expressed in other yeast. Nevertheless, it seems that the optimum temperature of reaction differs among the recombinant CBHII enzymes, which might be associated with the substrate used for the

enzyme activity assay or the posttranslational modifications carried out for each host. It is noteworthy that recombinant CBHII produced in *W. anomalus* 54-A only showed 30% reduction in enzyme activity at 30°C. Functionality at 30°C might allow the use of this enzyme in SSF processes for the production of bioethanol, since this temperature is also the optimum growth temperature for *S. cerevisiae* that is the preferable microorganism used during the fermentation stage of bioethanol production [27].

Kinetic parameters for the crude extract were estimated using pNPC as substrate (Supplementary Figure 6). Recombinant *T. reesei* CBHII produced in *W. anomalus* 54-A showed K_M and Vmax of 2.73 mM (est. error 0.26) and 23.1 μM min^{-1} (est. error 0.89), respectively. This K_M value agrees with those reported for wild-type CBHII enzymes (EC 3.2.1.91) isolated from different microorganisms, which vary between 0.1 and 3.1 mM, depending on the source of the enzyme [28]. In the case of *T. reesei* CBHII enzymes there are no reports of kinetic parameters using pNPC substrate. Nevertheless, a wide variation in K_M values is observed (i.e., between 0.041 and 380 mM), depending on the substrate used for this estimation [28].

3.5. Chrysanthemum Wastes Degradation Assay. To evaluate the effect of recombinant CBHII on cellulose hydrolysis, *Chrysanthemum* wastes degradation assay was performed using three enzyme extracts: (1) a crude culture broth from *Penicillium* sp. containing wild-type cellulase and hemicellulase enzymes (Ce-Hem extract), (2) a concentrated recombinant CBHII extract from *W. anomalus* 54-A (rCBHII extract), and (3) a 1:1 Ce-Hem:rCBHII extracts mixture. After treatment, 0.2 ± 0.02, 0.4 ± 0.01, and 1.25 ± 0.1 g L^{-1} of reducing sugars were observed for Ce-Hem, rCBHII, and 1:1 Ce-Hem:rCBHII extracts, respectively. However, when reducing sugar concentration was normalized by the CBHII units present in each extract at the beginning of the experiment (Figure 5), it was observed that rCBHII extract had production of reducing sugars similar to that obtained with the Ce-Hem extract ($p > 0.05$). Noteworthy, the use of a 1:1 Ce-Hem:rCBHII extracts mixture allowed an increase of about 5-fold, in the reducing sugars per CBHII unit, in comparison with the results obtained by using the Ce-Hem or concentrated recombinant CBHII extracts alone, suggesting a synergistic effect of the combination of these two enzymatic extracts. To understand the reasons of this synergistic effect, we measured the activity of β-glucosidase (BG) and endoglucanase (EG) in the extracts (Figure 5). Whereas BG was detected in the three evaluated extracts, EG was only detected in Ce-Hem and Ce-Hem:rCBHII extracts. Nevertheless, there was no correlation between BG and EG activity and the increase in reducing sugars observed with the Ce-Hem:rCBHII extract. In this sense, we hypothesize that the synergistic effect observed in the Ce-Hem:rCBHII treatment could be associated with the presence of other enzymes, such as lignin peroxidase, xylanase, laccase, and manganese peroxidase [29].

Chrysanthemum wastes degradation using enzymatic extracts with native or recombinant cellulases has not been

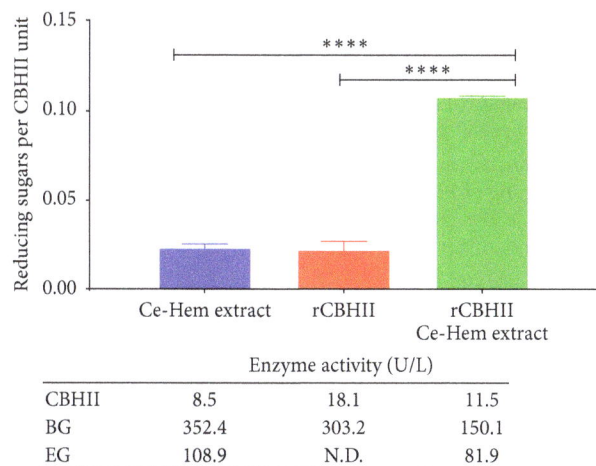

Enzyme activity (U/L)		
CBHII	8.5	18.1	11.5
BG	352.4	303.2	150.1
EG	108.9	N.D.	81.9

FIGURE 5: Chrysanthemum wastes degradation assay. The effect of recombinant CBHII on cellulose hydrolysis was evaluated by using *Chrysanthemum* wastes. Flasks that contained 1% *Chrysanthemum* wastes were incubated for 20 days at 150 rpm and 30°C with a crude culture broth from *Penicillium* sp. containing wild-type cellulase and hemicellulase enzymes (Ce-Hem extract), concentrated recombinant CBHII extract from *W. anomalus* 54-A (rCBHII), and a 1:1 Ce-Hem:rCBHII extracts mixture. Results are reported as $g\,L^{-1}$ of reducing sugars per unit of CBHII after subtraction of the results obtained with the control cultures. $* * ** = p < 0.0001$. The table shows the enzyme activities of CBHII, β-glucosidase (BG), and endoglucanase (EG) present in the enzymatic extracts at the beginning of the assay.

previously reported. For this lignocellulosic waste, degradation with *P. ostreatus* showed a production of $9.6\,g\,L^{-1}$ of reducing sugars [18]. The higher production of reducing sugars using *P. ostreatus* could be associated with the production of several lignocellulolytic enzymes by *P. ostreatus* such as laccases (up to 6 isoforms), manganese peroxidases, cellulases, and xylanases [30–32]. The presence of all these lignocellulolytic enzymes has shown an improvement in the delignification and reducing sugars production processes [29, 33]. In addition, the use of enzymatic extracts with recombinant cellulases, to degrade lignocellulosic materials, has been preceded by chemical or physical-chemical delignification treatments, such as alkaline [10] or steam [34] pretreatments. These pretreatments allowed higher reducing sugar levels than those reported in the present study (between 1.2 and $2.0\,g\,L^{-1}$), showing the importance of lignin degradation to increase biomass digestibility [29]. Nevertheless, these results showed that an enzymatic extract of *W. anomalus* 54-A carrying a recombinant *T. reesei* CBHII allows production of reducing sugars similar to that of a crude extract from cellulolytic fungi, showing the potential of *W. anomalus* 54-A as a host to produce recombinant cellulases.

In conclusion, we reported the production of the recombinant *T. reesei* CBHII in *W. anomalus* 54-A strains. The results showed production of recombinant CBHII in *W. anomalus* 54-A, with enzyme activity levels of up to $14.5\,U\,L^{-1}$. Recombinant CBHII showed optimum pH and temperature reaction of $5.0–6.0 \pm 0.2$ and 40°C, respectively,

which were similar to those reported for other recombinant *T. reesei* CBHII enzymes. *Km* of this recombinant CBHII was between the ranges of values reported for other CBHII enzymes. The results showed that an enzymatic extract of *W. anomalus* 54-A carrying the recombinant *T. reesei* CBHII allows production of reducing sugars similar to that of a crude extract from cellulolytic fungi, showing the potential of *W. anomalus* 54-A as a host to produce recombinant cellulases. Noteworthily, this is the first report on the use of *W. anomalus* as a host to produce heterologous proteins. Further studies should be focused on the consolidation of *W. anomalus* as a platform to produce recombinant proteins such as the design of specific expression vectors and the generation of auxotrophic strains.

Additional Points

One-Sentence Summary. This work represents the first report on the use of the yeast *Wickerhamomyces anomalus* as a host to produce heterologous proteins.

Acknowledgments

The authors thank Dr. Pilar Melendez from Pharmacy Department at Universidad Nacional de Colombia for the kind donation of the *W. anomalus* A-54 strain. Dennis J. Díaz-Rincón and Alexander Rodríguez-López received young researcher and doctoral scholarship, respectively, from Pontificia Universidad Javeriana. Angela Espejo-Mojica received a doctoral scholarship from the Administrative Department of Science, Technology and Innovation, Colciencias. The authors thank Nicolas Contreras for his assistance during performing experiments. This work was supported by Pontificia Universidad Javeriana [Grant ID 00005578, Producción de Etanol Mediante la Degradación de Residuos de Crisantemo Empleando Enzimas Recombinantes Producidas en *Saccharomyces cerevisiae*].

References

[1] L. Cuervo, J. L. Folch, and R. E. Quiroz, "Lignocelulosa como fuente de azúcares para la producción de etanol," *BioTecnología*, vol. 13, pp. 11–25, 2009.

[2] V. Juturu and J. C. Wu, "Microbial cellulases: Engineering, production and applications," *Renewable and Sustainable Energy Reviews*, vol. 33, pp. 188–203, 2014.

[3] R. C. Kuhad, R. Gupta, and A. Singh, "Microbial cellulases and their industrial applications," *Enzyme Research*, vol. 2011, Article ID 280696, 11 pages, 2011.

[4] J. Zhou, Y.-H. Wang, J. Chu, L.-Z. Luo, Y.-P. Zhuang, and S.-L. Zhang, "Optimization of cellulase mixture for efficient hydrolysis of steam-exploded corn stover by statistically designed experiments," *Bioresource Technology*, vol. 100, no. 2, pp. 819–825, 2009.

[5] M. Tanghe, B. Danneels, A. Camattari et al., "Recombinant Expression of Trichoderma reesei Cel61A in Pichia pastoris:

Optimizing Yield and N-terminal Processing," *Molecular Biotechnology*, vol. 57, no. 11-12, pp. 1010-1017, 2015.

[6] S. H. Petersen, W. H. Van Zyl, and I. S. Pretorius, "Development of a polysaccharide degrading strain of Saccharomyces cerevisiae," *Biotechnology Techniques*, vol. 12, no. 8, pp. 615–619, 1998.

[7] J. H. D. Van Zyl, R. Den Haan, and W. H. Van Zyl, "Overexpression of native Saccharomyces cerevisiae exocytic SNARE genes increased heterologous cellulase secretion," *Applied Microbiology and Biotechnology*, vol. 98, no. 12, pp. 5567–5578, 2014.

[8] M. Bailey, M. Siika-aho, A. Valkeajarvi, and M. Penttila, "Hydrolytic properties of two cellulases of Trichoderma reesei expressed in yeast," *Biotechnology and Applied Biochemistry*, vol. 17, no. 1, pp. 65–76, 1993.

[9] M. E. Penttilä, L. André, P. Lehtovaara, M. Bailey, T. T. Teeri, and J. K. C. Knowles, "Efficient secretion of two fungal cellobiohydrolases by Saccharomyces cerevisiae," *Gene*, vol. 63, no. 1, pp. 103–112, 1988.

[10] H. Fang and L. Xia, "Heterologous expression and production of Trichoderma reesei cellobiohydrolase II in Pichia pastoris and the application in the enzymatic hydrolysis of corn stover and rice straw," *Biomass and Bioenergy*, vol. 78, pp. 99–109, 2015.

[11] N. Boonvitthya, S. Bozonnet, V. Burapatana, M. J. O'Donohue, and W. Chulalaksananukul, "Comparison of the heterologous expression of trichoderma reesei endoglucanase II and cellobiohydrolase II in the yeasts pichia pastoris and yarrowia lipolytica," *Molecular Biotechnology*, vol. 54, no. 2, pp. 158–169, 2013.

[12] A. L. Demain and P. Vaishnav, "Production of recombinant proteins by microbes and higher organisms," *Biotechnology Advances*, vol. 27, no. 3, pp. 297–306, 2009.

[13] J. L. Corchero, B. Gasser, D. Resina et al., "Unconventional microbial systems for the cost-efficient production of high-quality protein therapeutics," *Biotechnology Advances*, vol. 31, no. 2, pp. 140–153, 2013.

[14] V. Passoth, M. Olstorpe, and J. Schnürer, "Past, present and future research directions with Pichia anomala," *Antonie van Leeuwenhoek, International Journal of General and Molecular Microbiology*, vol. 99, no. 1, pp. 121–125, 2011.

[15] L. De Vuyst, H. Harth, S. Van Kerrebroeck, and F. Leroy, "Yeast diversity of sourdoughs and associated metabolic properties and functionalities," *International Journal of Food Microbiology*, vol. 239, pp. 26–34, 2016.

[16] T. N. Petersen, S. Brunak, G. Von Heijne, and H. Nielsen, "SignalP 4.0: discriminating signal peptides from transmembrane regions," *Nature Methods*, vol. 8, no. 10, pp. 785-786, 2011.

[17] G. L. Miller, "Use of dinitrosalicylic acid reagent for determination of reducing sugar," *Analytical Chemistry*, vol. 31, no. 3, pp. 426–428, 1959.

[18] B. Quevedo-Hidalgo, P. C. Narvaéz-Rincón, A. M. Pedroza-Rodríguez, and M. E. Velásquez-Lozano, "Degradation of chrysanthemum (dendranthema grandiflora) wastes by pleurotus ostreatus for the production of reducing sugars," *Biotechnology and Bioprocess Engineering*, vol. 17, no. 5, pp. 1103–1112, 2012.

[19] B. Quevedo-Hidalgo, F. Monsalve-Marín, P. C. Narváez-Rincón, A. M. Pedroza-Rodríguez, and M. E. Velásquez-Lozano, "Ethanol production by *Saccharomyces cerevisiae* using lignocellulosic hydrolysate from *Chrysanthemum* waste degradation," *World Journal of Microbiology and Biotechnology*, vol. 29, no. 3, pp. 459–466, 2013.

[20] N. Ali, Z. Ting, H. Li et al., "Heterogeneous Expression and Functional Characterization of Cellulose-Degrading Enzymes from Aspergillus niger for Enzymatic Hydrolysis of Alkali Pretreated Bamboo Biomass," *Molecular Biotechnology*, vol. 57, no. 9, pp. 859–867, 2015.

[21] C. P. Kurtzman, "Phylogeny of the ascomycetous yeasts and the renaming of Pichia anomala to Wickerhamomyces anomalus," *Antonie van Leeuwenhoek, International Journal of General and Molecular Microbiology*, vol. 99, no. 1, pp. 13–23, 2011.

[22] K. M. Lee and N. A. DaSilva, "Evaluation of the Saccharomyces cerevisiae ADH2 promoter for protein synthesis," *Yeast*, vol. 22, no. 6, pp. 431–440, 2005.

[23] V. L. Price, W. E. Taylor, W. Clevenger, M. Worthington, and E. T. Young, "Expression of heterologous proteins in Saccharomyces cerevisiae using the ADH2 promoter," *Methods in Enzymology*, vol. 185, pp. 308–318, 1990.

[24] T. T. Teeri, P. Lehtovaara, S. Kauppinen, I. Salovuori, and J. Knowles, "Homologous domains in Trichoderma reesei cellulolytic enzymes: Gene sequence and expression of cellobiohydrolase II," *Gene*, vol. 51, no. 1, pp. 43–52, 1987.

[25] B. Laukens, C. De Visscher, and N. Callewaert, "Engineering yeast for producing human glycoproteins: Where are we now?" *Future Microbiology*, vol. 10, no. 1, pp. 21–34, 2015.

[26] M. Klein, S. Swinnen, J. M. Thevelein, and E. Nevoigt, "Glycerol metabolism and transport in yeast and fungi: established knowledge and ambiguities," *Environmental Microbiology*, vol. 19, no. 3, pp. 878–893, 2017.

[27] D. Kennes, H. N. Abubackar, M. Diaz, M. C. Veiga, and C. Kennes, "Bioethanol production from biomass: Carbohydrate vs syngas fermentation," *Journal of Chemical Technology and Biotechnology*, vol. 91, no. 2, pp. 304–317, 2016.

[28] A. Chang, I. Schomburg, S. Placzek et al., "BRENDA in 2015: Exciting developments in its 25th year of existence," *Nucleic Acids Research*, vol. 43, no. 1, pp. D439–D446, 2015.

[29] O. Sánchez, R. Sierra, and C. J. Alméciga-Díaz, "Delignification Process of Agro-Industrial Wastes an Alternative to Obtain Fermentable Carbohydrates for Producing Fuel," in *Alternative Fuel*, M. Manzanera, Ed., InTech, pp. 111–154, Rijeka, 2011.

[30] R. Castanera, G. Pérez, A. Omarini et al., "Transcriptional and enzymatic profiling of pleurotus ostreatus laccase genes in submerged and solid-state fermentation cultures," *Applied and Environmental Microbiology*, vol. 78, no. 11, pp. 4037–4045, 2012.

[31] J. M. R. da Luz, M. D. Nunes, S. A. Paes, D. P. Torres, M. D. C. S. da Silva, and M. C. M. Kasuya, "Lignocellulolytic enzyme production of Pleurotus ostreatus growth in agroindustrial wastes," *Brazilian Journal of Microbiology*, vol. 43, no. 4, pp. 1508–1515, 2012.

[32] O. S. Isikhuemhen and N. A. Mikiashvilli, "Lignocellulolytic enzyme activity, substrate utilization, and mushroom yield by *Pleurotus ostreatus* cultivated on substrate containing anaerobic digester solids," *Journal of Industrial Microbiology and Biotechnology*, vol. 36, no. 11, pp. 1353–1362, 2009.

[33] J. Plácido and S. Capareda, "Ligninolytic enzymes: a biotechnological alternative for bioethanol production," *Bioresources and Bioprocessing*, vol. 2, no. 23, pp. 1–12, 2015.

[34] S. P. Das, A. Ghosh, A. Gupta, A. Goyal, and D. Das, "Lignocellulosic fermentation of wild grass employing recombinant hydrolytic enzymes and fermentative microbes with effective bioethanol recovery," *BioMed Research International*, vol. 2013, Article ID 386063, 2013.

8

Overexpression of Soluble Recombinant Human Lysyl Oxidase by using Solubility Tags: Effects on Activity and Solubility

Madison A. Smith,[1] Jesica Gonzalez,[2] Anjum Hussain,[3] Rachel N. Oldfield,[3] Kathryn A. Johnston,[3] and Karlo M. Lopez[3]

[1]MI-SWACO, Shafter, CA 93263, USA
[2]University of California, San Francisco, San Francisco, CA 94143, USA
[3]California State University, Bakersfield, Bakersfield, CA 93311, USA

Correspondence should be addressed to Karlo M. Lopez; klopez@csub.edu

Academic Editor: Paul Engel

Lysyl oxidase is an important extracellular matrix enzyme that has not been fully characterized due to its low solubility. In order to circumvent the low solubility of this enzyme, three solubility tags (Nus-A, Thioredoxin (Trx), and Glutathione-S-Transferase (GST)) were engineered on the N-terminus of mature lysyl oxidase. Total enzyme yields were determined to be 1.5 mg for the Nus-A tagged enzyme (0.75 mg/L of media), 7.84 mg for the Trx tagged enzyme (3.92 mg/L of media), and 9.33 mg for the GST tagged enzyme (4.67 mg/L of media). Enzymatic activity was calculated to be 0.11 U/mg for the Nus-A tagged enzyme and 0.032 U/mg for the Trx tagged enzyme, and no enzymatic activity was detected for the GST tagged enzyme. All three solubility-tagged forms of the enzyme incorporated copper; however, the GST tagged enzyme appears to bind adventitious copper with greater affinity than the other two forms. The catalytic cofactor, lysyl tyrosyl quinone (LTQ), was determined to be 92% for the Nus-A and Trx tagged lysyl oxidase using the previously reported extinction coefficient of $15.4 \text{ mM}^{-1} \text{ cm}^{-1}$. No LTQ was detected for the GST tagged lysyl oxidase. Given these data, it appears that Nus-A is the most suitable tag for obtaining soluble and active recombinant lysyl oxidase from *E. coli* culture.

1. Introduction

Lysyl oxidase (LOX) is a copper-dependent amine oxidase that catalyzes a key cross-linking step in collagen and elastin [1, 2]. The resulting network of cross-linked molecules serves to provide flexibility and stability within connective tissues in the cardiovascular, respiratory, and skeletal systems of biological organisms [1, 3]. Aside from its structural role in the extracellular matrix, LOX has also been implicated as playing a role in other biological processes including morphogenesis and repair of connective tissues, developmental control, and chemotaxis [2, 4, 5].

LOX is synthesized as an N-glycosylated 50 kDa propeptide [6] that is exported to the extracellular matrix where it undergoes proteolytic cleavage by procollagen-C-proteinase (bone morphogenetic protein I) at a specific gly-asp bond, confirmed to be Gly 168 and Asp 169 in humans [7, 8]. LOX is dependent on two cofactors: a tightly bound Cu(II) atom [9], which is required for the posttranslational modification of Tyr 355 to form the second cofactor, lysyl tyrosyl quinone (LTQ) [10, 11]. LTQ is formed by the cross-linking of Tyr 355 with Lys 320 to form the catalytic site of the enzyme. Figure 1 is a schematic of the synthesis of LOX both in the intracellular space (left panel) and the extracellular matrix (right panel). It has been postulated that the copper atom is required for the catalytic activity of the enzyme [1, 9]; however, this postulate has been disputed [12].

Until recently, it has been extremely difficult to obtain large amounts of mammalian recombinant LOX due to the low solubility of the enzyme in aqueous buffers. Previous work with overexpression systems required that the enzyme be refolded [13, 14], a process that required lengthy dialysis in several different buffers. A breakthrough was achieved in 2010 when an overexpression system was developed that

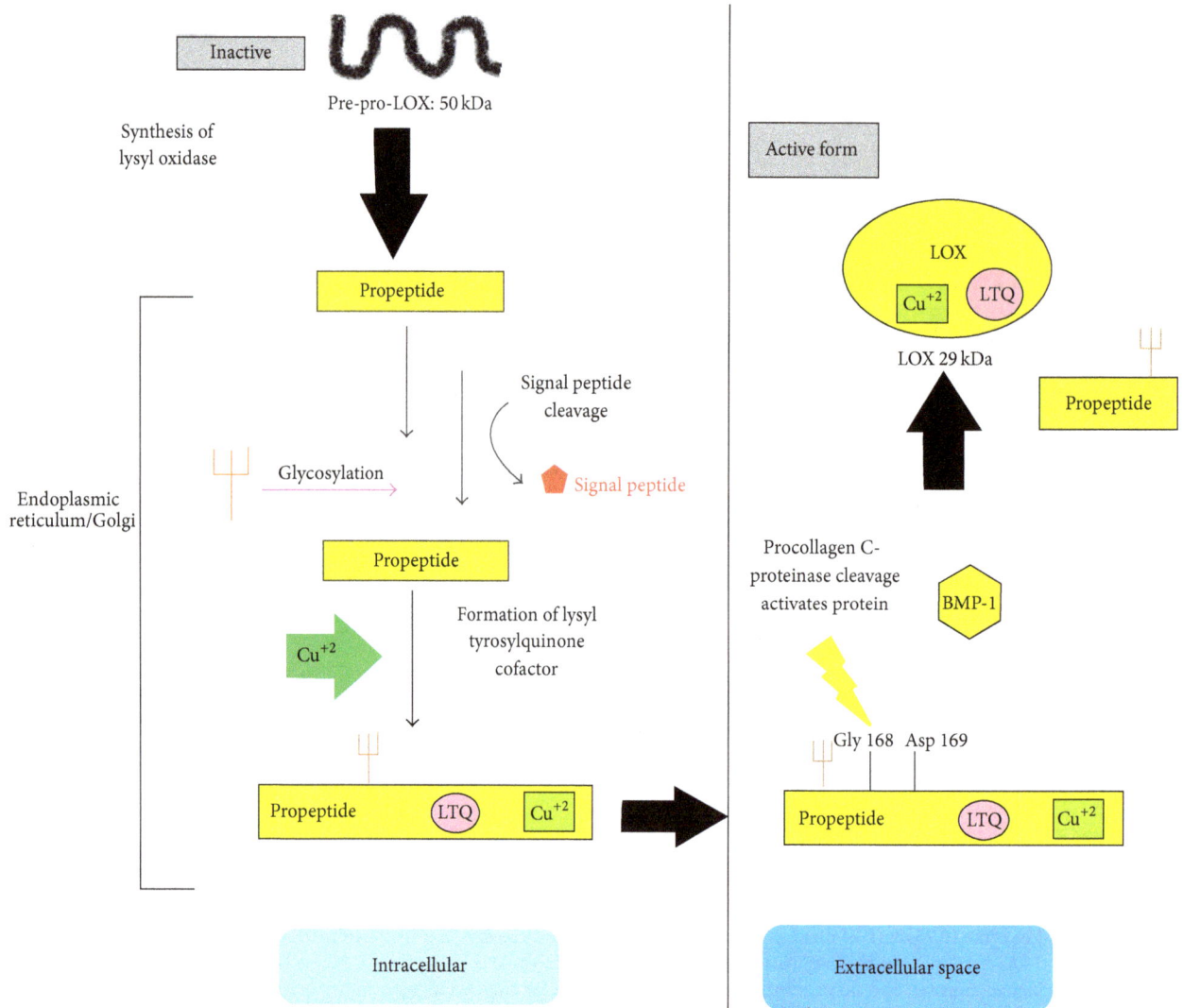

FIGURE 1: Schematic of the synthesis of active lysyl oxidase. An inactive propeptide is glycosylated and exported to the extracellular matrix following the incorporation of copper and the formation of the LTQ catalytic cofactor. The propeptide region is cleaved by procollagen C-proteinase at Gly 168 and Asp 169, yielding the propeptide and the 29 kDa active mature enzyme.

allowed for the production of recombinant lysyl oxidase which was purified, in active form, directly from *E. coli* culture [15]. Although this system removed the need for a lengthy refolding procedure and thereby greatly advanced the way by which this enzyme could be purified, it relied on the use of 6 M urea in the purification buffers in order to maintain the enzyme in solution.

The main focus of this work was to develop a purification system whereby the enzyme could be purified directly from *E. coli* but no longer need 6 M urea in the purification buffer. Although the yields and activity of the solubility-tagged enzyme are lower than those our laboratory reported in 2010 [15], the ability to produce an active enzyme without the need for high concentration urea is a significant advancement in the study of mammalian lysyl oxidase. These chimeric enzymes can be used to evaluate mechanical function in engineered ligaments [16] or to generate hydrogels for *in situ*

bone regeneration which requires the cross-linking of lysine residues [17] among many other uses. Herein, we present the study of three different solubility tags, Nus-A, Thioredoxin (Trx), and Glutathione-S-Transferase (GST), and report their effectiveness in the isolation of soluble (nonurea) and active lysyl oxidase. The data indicates that the Nus-A solubility tag is the most effective, yielding 1.5 mg of soluble and active LOX (0.11 U/mg).

2. Materials and Methods

All PCR materials were obtained from New England Biolabs (NEB) with the exception of the primers which were purchased from Midland Certified Reagent Company and the polymerase which was purchased from Agilent. Unless otherwise stated, all materials were purchased from VWR

TABLE 1: Listing of primers used for isolating the LOX gene. Each primer introduced a different restriction site at the $5'$-end in order to insert the resulting PCR product into the corresponding vector in the correct orientation. The reverse primer (MLOXR) was the same for all PCR reactions.

Primer name	Primer sequence	Restriction site
Trx tag (pLOX06 forward primer)	$5'$-AAGGGG**TGGCCA**GGACGACCCTTACAACCCCTAC-$3'$	MscI at $5'$-end
LOX tag (pLOX09 forward primer)	$5'$-AAGGGG**CCCGGG**ACGACCCTTACAACCCC-$3'$	SmaI at $5'$-end
MLOXGSTF (pLOX14 forward primer)	$5'$-AAGGGG**AAGCTT**ATGGACGACCCTTACAACCCC-$3'$	HindIII at $5'$-end
MLOXR (reverse primer for all constructs)	$5'$-AAGGGG**CTCGAG**ATACGGTGAAATTGTGCAGCC-$3'$	XhoI at $3'$-end

International, LLC, or Research Products International Corporation.

2.1. Generation of pLOX06 (Trx), pLOX09 (Nus-A), and pLOX14 (GST) Constructs.

The LOX gene was isolated from plasmid pLOX02 by PCR using the primer sets shown in Table 1. These primer sets introduced the corresponding restriction sites which are also listed in Table 1.

The pET vector system was used to introduce the solubility tags. PCR products and the corresponding vectors were digested with the appropriate restriction enzymes and ligated using NEB's quick ligase kit. pET32b was used to introduce the Trx solubility tag. The insertion points were at the MscI ($5'$-) and XhoI ($3'$-) restriction sites. pET42b was used to introduce the GST solubility tag. The insertion points were at the HindIII ($5'$-) and XhoI ($3'$-) restriction sites. pET43.1b was used to introduce the Nus-A solubility tag. The insertion points were at the SmaI ($5'$-) and XhoI ($3'$-) restriction sites. The ligation reactions were then transformed into NEB 10β competent E. coli cells and the resulting plasmids were isolated using a QIAGEN plasmid isolation kit in sufficient amounts for sequencing and transformation into an expression cell line.

2.2. Expression and Purification of Tagged Lysyl Oxidase.

Three different sets of NEB SHuffle competent E. coli cells each independently containing the pLOX06, pLOX09, and pLOX14 plasmids were plated on LB/ampicillin plates (pLOX06 and pLOX09) or LB/kanamycin (pLOX14) and were incubated at 37°C overnight. Ampicillin concentration was set to 100 μg/mL and kanamycin concentration was set to 50 μg/mL. A single colony was grown overnight in 50 mL of LB media, at 37°C, containing the appropriate concentration of antibiotic. Four Erlenmeyer flasks, each containing 500 mL of Terrific Broth and the appropriate concentration of antibiotic, were inoculated with 10 mL of overnight culture. These cultures were grown, at 37°C, to an O.D.$_{600}$ of 0.8 and induced with 2 mM IPTG. Following induction, the heat was turned off and the cells were further incubated at room temperature overnight. Cells were harvested by centrifugation at 4300 ×g for 15 min in a Beckman Coulter Allegra X-14R centrifuge using a SX4750 swinging bucket rotor. Following centrifugation, cells were frozen at −80°C and lysed by sonication. The lysis/wash buffer was made up of 50 mM Tris-Cl, pH 7.6, 200 mM NaCl, and 20 mM imidazole. The cell lysate was then centrifuged at 4300 ×g. Following

centrifugation, the supernatant was passed through a Ni-NTA gravity column in a 4°C cold room. The column was then washed with 200 mL of lysis/wash buffer to remove any residual cellular proteins. The tagged LOX was eluted from the column with elution buffer made up of 50 mM Tris-Cl, pH 7.6, 200 mM NaCl, and 250 mM imidazole. The isolated and purified enzyme was visualized by SDS-PAGE and the correct molecular weight was verified by using molecular weight markers. In order to ensure that the enzyme's active site was loaded with copper, the purified enzyme was dialyzed against 50 mM Tris-Cl, pH 7.6, 200 mM NaCl, and 1 mM CuSO$_4$ overnight, followed by extensive dialysis against 50 mM Tris-Cl, pH 7.6, 200 mM NaCl, and 2 mM EDTA. This is a modified protocol based on previous studies addressing copper dialysis of lysyl oxidase [18, 19]. EDTA dialysis was carried out for at least three days with buffers being exchanged at a minimum twice per day. Finally, the enzyme was dialyzed against 50 mM Tris-Cl, pH 7.6, and 200 mM NaCl to remove the EDTA.

2.3. Characterization of LOX Activity.

Properly folded LOX samples were prepared in 50 mM Tris-Cl, pH 7.6, and 200 mM NaCl and the following amounts of enzyme were used in activity assays: Trx tagged LOX 76 μg, Nus-A tagged LOX 84 μg, and GST tagged LOX 140 μg. Coupled fluorescence activity assays for H$_2$O$_2$ were carried out at 37°C with 1,5-diaminopentane as the substrate [20]. Substrate concentration was 10 mM per assay. Inhibition of enzyme activity was attempted by adding β-APN during the assay. Assays were run in quadruplicate and specific activity is reported as a mean of all assays. Specific activity was determined by standardization using H$_2$O$_2$, itself standardized by titration with permanganate, and is reported in μmol H$_2$O$_2$ produced per minute per mg of enzyme under the conditions of the assay. Thus this specific activity is not directly comparable to activities reported using different assays or different assay conditions.

2.4. Copper Quantification.

A bicinchoninic acid assay was used to determine copper content [15]. 250 μL of a 300 mg/mL trichloroacetic acid solution was added to 750 μL of a 2x dilution of LOX enzyme. The resulting cloudy solution was centrifuged at 13,400 rpm for 5 min to remove precipitated enzyme. 500 μL of the supernatant was combined with 100 μL of a 0.34 mg/mL ascorbate solution and 400 μL of HEPES/BCA solution (1.8 g NaOH, 7.8 g HEPES, 3 mL of a 6% BCA reagent A solution (Pierce), and 45 mL ddH$_2$O).

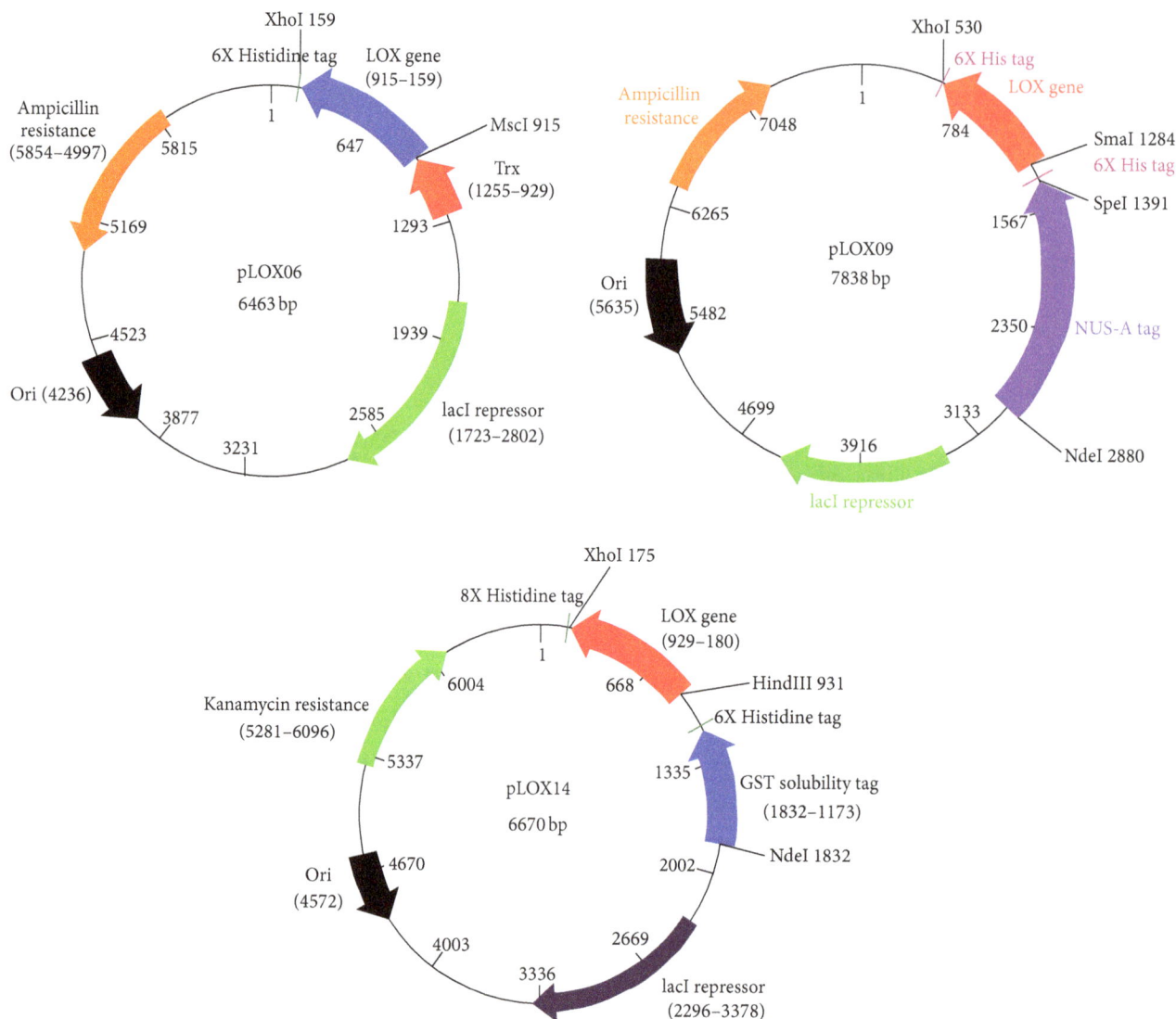

FIGURE 2: Plasmid maps for pLOX06 (Trx), pLOX09 (Nus-A), and pLOX14 (GST). The maps indicate the direction of transcription, the location of the solubility tags, and the insertion points for the wild-type mature LOX gene.

The absorbance of the resulting solution was measured at 362 nm on a Shimadzu UVmini-1240 spectrophotometer. A standard curve was generated by replacing the enzyme with increasing concentrations of copper from 0 to 58 μM in the assay.

2.5. Phenylhydrazine Inhibition. Excess phenylhydrazine was added to a 1 mL solution of LOX and the solution was incubated at room temperature in the dark (microcentrifuge tube was covered with aluminum foil and stored in a drawer) overnight. The following day the solution was scanned from 800 nm to 200 nm on a Shimadzu UV-2401PC spectrophotometer.

3. Results

The DNA cassette corresponding to the mature form of lysyl oxidase, starting at Asp 169, was successfully isolated from plasmid pLOX02 using PCR. This cassette was inserted into pET32b, pET42b, and pET43.1b to generate plasmids pLOX06, pLOX14, and pLOX09, respectively. The plasmid maps for each of these constructs are shown in Figure 2.

Expression was under the control of the T7 promoter in all vectors. Upon induction with 2 mM IPTG at 37°C, LOX was successfully expressed with yields of 1.5 mg for the Nus-A tagged enzyme (0.75 mg/L of media), 7.84 mg for the Trx tagged enzyme (3.93 mg/L media), and 9.33 mg for the GST tagged enzyme (4.67 mg/L of media). Purity was verified by SDS-PAGE and is shown in Figure 3. The calculated molecular weight for each construct was determined using the online bioinformatics program, ProtParam. For the pLOX09 construct (Nus-A) the calculated MW was 89023 Da, for the pLOX06 construct (Trx) the calculated MW was 42,359 Da, and for the pLOX14 construct (GST) the calculated molecular weight was 64380 Da.

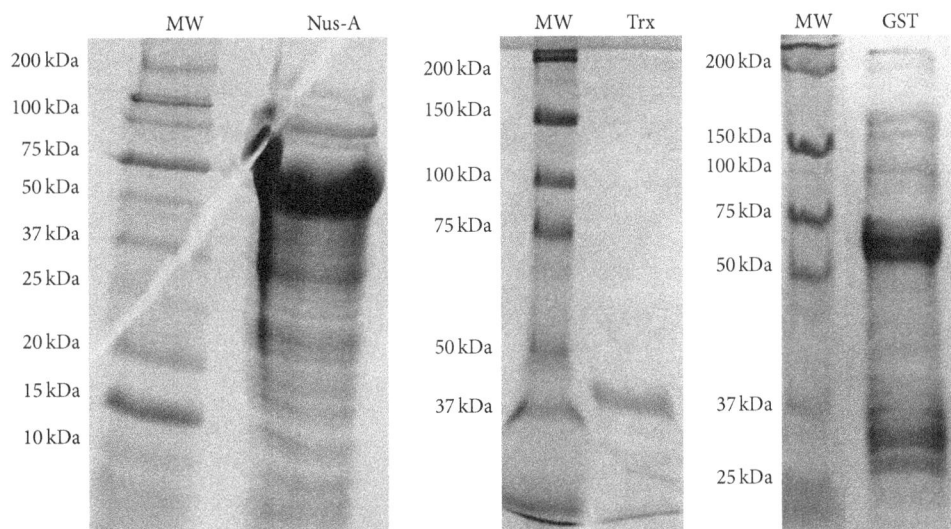

FIGURE 3: SDS-PAGE gel slices showing overexpressed, solubility-tagged lysyl oxidase. The molecular weight of each solubility-tagged enzyme was determined using the ladder on the left.

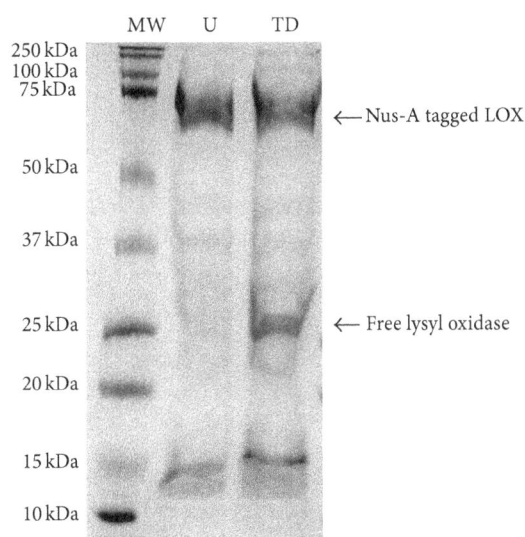

FIGURE 4: Thrombin digest of Nus-A tagged lysyl oxidase. Lane 1: molecular weight marker (MW); Lane 2: undigested tagged enzyme (U); Lane 3: thrombin digested Nus-A tagged LOX (TD).

TABLE 2: Yield, activity, and copper incorporation of each tagged form of lysyl oxidase. The line labeled LOX is data from a previously published paper [15] related to untagged and urea solubilized LOX. This enzyme was not subjected to copper dialysis and hence the lower copper incorporation.

	Protein yield (mg)	Specific activity (U/mg)	Copper incorporation
LOX	10.0	0.31	19%
LOX Nus-A tag	1.5	0.11	68%
LOX Trx tag	7.8	0.03	74%
LOX GST tag	9.3	None Detected	270%

One activity unit is defined as 1μmol of H_2O_2 produced per minute per mg of enzyme at $37°C$.

In order to ascertain the effect of the enzyme on the solubility of LOX, a thrombin cleavage site was engineered between the Nus-A tag and the LOX enzyme. The enzyme was cleaved overnight with thrombin and imaged on 12% SDS-PAGE shown in Figure 4.

Following digestion, the purification of the free LOX enzyme was attempted; however, once the solubility tag was removed, the enzyme crashed out of solution since no urea was used in the buffer.

Following purification, the tagged LOX was assayed for enzymatic activity using a peroxide-coupled assay and 1,5-diaminopentane as the substrate. One unit of amine oxidase activity is defined as the activity resulting in oxidation of

1μmol of 1,5-diaminopentane per minute at $37°C$. The conversion of 1,5-diaminopentane to the corresponding aldehyde was monitored by measuring the increase in fluorescence (Table 2).

This increase is due to the appearance of the Amplex Red product, resorufin. The activity for LOX was found to be 0.11 U/mg for the Nus-A tagged enzyme and 0.032 U/mg for the Trx tagged enzyme, and no detectable activity was observed for the GST tagged enzyme. The activity of the Nus-A tagged enzyme is consistent with the previously reported activity of 0.097 U/mg by Jung et al. [13], although only one-third as active as the previously reported value by our laboratory of the untagged but urea-dependent enzyme [15]. The activity of the Trx enzyme is much lower than that observed by Jung et al. or our laboratory, yet we consider it "active" in that this value is still three times higher than those observed in the initial experiment that identified LTQ as the catalytic cofactor [11]. For the GST tagged enzyme no activity was detected.

Normally, rapid and complete inhibition of *native* LOX is achieved by low levels (0.5 mM) of β-APN [21]. For the two *recombinant* constructs showing activity, inhibition with β-APN was achieved with high concentrations (>2.5 mM) and only after long incubation periods. This behavior has been observed for recombinant LOX in previous experiments by our laboratory as well as others [15].

This enzyme is dependent on Cu(II) for the formation of LTQ and, potentially for activity, the amount of copper present in the enzyme was determined and is presented herein as a percent incorporation. The percentage values represent moles of copper present following dialysis against EDTA per mole of purified LOX enzyme. Copper concentration was determined to be 1.70 μM corresponding to 68.4% copper incorporation for the Nus-A tagged enzyme and 3.14 μM corresponding to 74% copper incorporation for the Trx tagged enzyme. Although the GST tagged lysyl oxidase was dialyzed extensively against EDTA, it appears that this form of the enzyme binds adventitious copper much tighter as copper incorporation was determined to be 200%. In order to ascertain that the solubility tags were not incorporating copper, the empty pET32b, pET42b, and pET43.1b were used to overexpress the Trx, GST, and Nus-A tags, respectively. These tags were treated in the same fashion as the tagged enzyme and assayed in triplicate for copper content. The Trx tag showed 0.70% copper incorporation, the Nus-A tagged had no detectable copper incorporation, and the GST tagged showed 3% copper incorporation.

The presence of the LTQ cofactor was verified by monitoring enzyme inhibition by derivatization with phenylhydrazine. LOX samples at a concentration of 0.18 mg/mL for the Trx tagged enzyme and 0.64 mg/mL for the Nus-A tagged enzyme were incubated with excess phenylhydrazine at room temperature overnight in the dark. Scans were then taken the following day and the appearance of the phenylhydrazone adduct was detected and is shown in Figure 5. Using the established extinction coefficient of 15.4 mM^{-1} cm^{-1} [10], 92% of the LTQ present in the copper-loaded LOX enzyme was titratable with phenylhydrazine.

4. Discussion

The isolation of active lysyl oxidase in large quantities from an overexpression system has been elusive. Attempts have been made to isolate soluble enzyme from Chinese hamster ovary cells [22] which yielded very little enzyme, as well as using *E. coli* [13, 15] which required long refolding procedures and/or the use of 6 M urea in order to solubilize the enzyme.

This work focused on elucidating a method to obtain milligram quantity yields of purified *recombinant* LOX without using urea, in any fashion, in the purification buffers. The data presented here shows that this can be successfully accomplished using an *E. coli* system and nutrient rich media if a solubility tag is tethered to LOX. In particular, the data indicates that the most suitable solubility tag is Nus-A because it yields milligram quantities of the enzyme that is soluble without needing urea, has high activity, and incorporates copper, and the LTQ cofactor is present. Given that both GST and Trx tagged LOX yielded higher amounts of LOX, it appears that the smaller tags interfere less with the overexpression levels; however, the presence of smaller tags

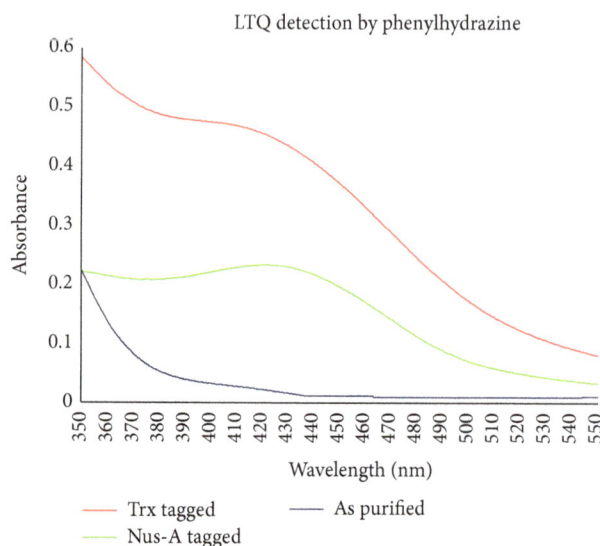

FIGURE 5: Phenylhydrazone adduct that is formed when LTQ reacts with phenylhydrazine. Red curve shows the adduct for the Trx tagged enzyme while the green curve shows the adduct for the Nus-A enzyme. The blue trace is that of LOX as purified prior to reacting with phenylhydrazine.

appears to inhibit the enzyme to a large degree. In the case of the GST tagged lysyl oxidase, it is postulated that the GST tag actually produces a misfold in the enzyme that causes it not only to be inactive but also to bind adventitious copper with high affinity. The copper incorporation of the chimeric enzymes does not appear to be linked to the solubility tags as control experiments show that the tags, in and of themselves, do not incorporate copper.

The presence of the LTQ cofactor, verified and quantified by the formation of a phenylhydrazone adduct, indicates that the Trx and Nus-A solubility tags do not interfere with the formation of LTQ. This is important because the primary function of the tag is to keep the enzyme soluble; hence, it will not be cleaved when carrying out studies on LOX. If the tag is cleaved, the added solubility is removed and the enzyme will crash out of solution.

Future studies on the use of solubility tags with LOX include a comparison of the tagged enzyme with the lysyl oxidase-like isoforms of the enzyme which, unlike LOX, are much more soluble and which retain their N-terminal peptide region.

Lastly, the ability to harvest active LOX in large quantities, now fully soluble in aqueous buffer, puts us one step closer to the ultimate goal with this enzyme: the elucidation of its crystal structure. The tags used in the study were selected specifically because they have been used in previous work to crystallize insoluble proteins without the need to cleave them off [23]. To date the only amine oxidase crystal structures have been of copper-containing amine oxidase enzymes of the TPQ-containing EC1.4.3.6 class, not the EC1.4.3.13 class of LTQ-containing lysyl oxidases. This endeavor has eluded us for far too long and our laboratory is currently working on screening these tagged enzymes in an effort to identify conditions suitable for growing crystals.

Abbreviations

LOX: Lysyl oxidase
LTQ: Lysyl tyrosyl quinone
IPTG: Isopropyl-β-D-thiogalactopyranoside
Trx: Thioredoxin
GST: Glutathione-S-Transferase.

References

[1] H. M. Kagan and P. C. Trackman, "Properties and function of lysyl oxidase," *American Journal of Respiratory Cell and Molecular Biology*, vol. 5, no. 3, pp. 206–210, 1991.

[2] H. A. Lucero and H. M. Kagan, "Lysyl oxidase: an oxidative enzyme and effector of cell function," *Cellular and Molecular Life Sciences*, vol. 63, no. 19-20, pp. 2304–2316, 2006.

[3] H. M. Kagan and W. Li, "Lysyl oxidase: properties, specificity, and biological roles inside and outside of the cell," *Journal of Cellular Biochemistry*, vol. 88, no. 4, pp. 660–672, 2003.

[4] K. Csiszar, "Lysyl oxidases: a novel multifunctional amine oxidase family," *Progress in Nucleic Acid Research and Molecular Biology*, vol. 70, pp. 1–32, 2001.

[5] W. Li, G. Liu, I.-N. Chou, and H. M. Kagan, "Hydrogen peroxide-mediated, lysyl oxidase-dependent chemotaxis of vascular smooth muscle cells," *Journal of Cellular Biochemistry*, vol. 78, no. 4, pp. 550–557, 2000.

[6] P. C. Trackman, D. Bedell-Hogan, J. Tang, and H. M. Kagan, "Post-translational glycosylation and proteolytic processing of a lysyl oxidase precursor," *The Journal of Biological Chemistry*, vol. 267, no. 12, pp. 8666–8671, 1992.

[7] M. V. Panchenko, W. G. Stetler-Stevenson, O. V. Trubetskoy, S. N. Gacheru, and H. M. Kagan, "Metalloproteinase activity secreted by fibrogenic cells in the processing of prolysyl oxidase. Potential role of procollagen C-proteinase," *The Journal of Biological Chemistry*, vol. 271, no. 12, pp. 7113–7119, 1996.

[8] M. I. Uzel, I. C. Scott, H. Babakhanlou-Chase et al., "Multiple bone morphogenetic protein 1-related mammalian metalloproteinases process pro-lysyl oxidase at the correct physiological site and control lysyl oxidase activation in mouse embryo fibroblast cultures," *The Journal of Biological Chemistry*, vol. 276, no. 25, pp. 22537–22543, 2001.

[9] S. N. Gacheru, P. C. Trackman, M. A. Shah et al., "Structural and catalytic properties of copper in lysyl oxidase," *The Journal of Biological Chemistry*, vol. 265, no. 31, pp. 19022–19027, 1990.

[10] J. A. Bollinger, D. E. Brown, and D. M. Dooley, "The formation of lysine tyrosylquinone (LTQ) is a self-processing reaction. Expression and characterization of a *Drosophila lysyl* oxidase," *Biochemistry*, vol. 44, no. 35, pp. 11708–11714, 2005.

[11] S. X. Wang, M. Mure, K. F. Medzihradszky et al., "A crosslinked cofactor in lysyl oxidase: redox function for amino acid side chains," *Science*, vol. 273, no. 5278, pp. 1078–1084, 1996.

[12] M. Mure, S. A. Mills, and J. P. Klinman, "Catalytic mechanism of the topa quinone containing copper amine oxidases," *Biochemistry*, vol. 41, no. 30, pp. 9269–9278, 2002.

[13] S. T. Jung, M. S. Kim, J. Y. Seo, H. C. Kim, and Y. Kim, "Purification of enzymatically active human lysyl oxidase and lysyl oxidase-like protein from *Escherichia coli* inclusion bodies," *Protein Expression and Purification*, vol. 31, no. 2, pp. 240–246, 2003.

[14] M. Ouzzine, A. Boyd, and D. J. Hulmes, "Expression of active, human lysyl oxidase in *Escherichia coli*," *FEBS Letters*, vol. 399, no. 3, pp. 215–219, 1996.

[15] S. E. Herwald, F. T. Greenaway, and K. M. Lopez, "Purification of high yields of catalytically active lysyl oxidase directly from *Escherichia coli* cell culture," *Protein Expression and Purification*, vol. 74, no. 1, pp. 116–121, 2010.

[16] C. A. Lee, A. Lee-Barthel, L. Marquino, N. Sandoval, G. R. Marcotte, and K. Baar, "Estrogen inhibits lysyl oxidase and decreases mechanical function in engineered ligaments," *Journal of Applied Physiology*, vol. 118, no. 10, pp. 1250–1257, 2015.

[17] A. Sánchez-Ferrero, Á. Mata, M. A. Mateos-Timoneda et al., "Development of tailored and self-mineralizing citric acid-crosslinked hydrogels for in situ bone regeneration," *Biomaterials*, vol. 68, pp. 42–53, 2015.

[18] M. S. Kim, S.-S. Kim, S. T. Jung et al., "Expression and purification of enzymatically active forms of the human lysyl oxidase-like protein 4," *The Journal of Biological Chemistry*, vol. 278, no. 52, pp. 52071–52074, 2003.

[19] K. M. Lopez and F. T. Greenaway, "Identification of the copper-binding ligands of lysyl oxidase," *Journal of Neural Transmission*, vol. 118, no. 7, pp. 1101–1109, 2011.

[20] A. H. Palamakumbura and P. C. Trackman, "A fluorometric assay for detection of lysyl oxidase enzyme activity in biological samples," *Analytical Biochemistry*, vol. 300, no. 2, pp. 245–251, 2002.

[21] S. S. Tang, P. C. Trackman, and H. M. Kagan, "Reaction of aortic lysyl oxidase with β-aminopropionitrile," *The Journal of Biological Chemistry*, vol. 258, no. 7, pp. 4331–4338, 1983.

[22] H. M. Kagan, V. B. Reddy, M. V. Panchenko et al., "Expression of lysyl oxidase from cDNA constructs in mammalian cells: the propeptide region is not essential to the folding and secretion of the functional enzyme," *Journal of Cellular Biochemistry*, vol. 59, no. 3, pp. 329–338, 1995.

[23] D. R. Smyth, M. K. Mrozkiewicz, W. J. McGrath, P. Listwan, and B. Kobe, "Crystal structures of fusion proteins with large-affinity tags," *Protein Science*, vol. 12, no. 7, pp. 1313–1322, 2003.

9

Salivary Myeloperoxidase, Assessed by 3,3′-Diaminobenzidine Colorimetry, can Differentiate Periodontal Patients from Nonperiodontal Subjects

Supaporn Klangprapan,[1] Ponlatham Chaiyarit,[2,3] Doosadee Hormdee,[3,4] Amonrujee Kampichai,[4,5] Tueanjit Khampitak,[1] Jureerut Daduang,[6,7] Ratree Tavichakorntrakool,[7,8] Bhinyo Panijpan,[9] and Patcharee Boonsiri[1]

[1]Department of Biochemistry, Faculty of Medicine, Khon Kaen University, Khon Kaen 40002, Thailand
[2]Department of Oral Diagnosis, Faculty of Dentistry, Khon Kaen University, Khon Kaen 40002, Thailand
[3]Research Group of Chronic Inflammatory Oral Diseases and Systemic Diseases Associated with Oral Health, Khon Kaen University, Khon Kaen 40002, Thailand
[4]Department of Periodontology, Faculty of Dentistry, Khon Kaen University, Khon Kaen 40002, Thailand
[5]Dental Department, Fang Hospital, Fang District, Chiangmai 50110, Thailand
[6]Department of Clinical Chemistry, Faculty of Associated Medical Sciences, Khon Kaen University, Khon Kaen 40002, Thailand
[7]Centre for Research and Development of Medical Diagnostic Laboratories, Khon Kaen University, Khon Kaen 40002, Thailand
[8]Department of Clinical Microbiology, Faculty of Associated Medical Sciences, Khon Kaen University, Khon Kaen 40002, Thailand
[9]Faculty of Science, Mahidol University, Bangkok 10400, Thailand

Correspondence should be addressed to Patcharee Boonsiri; patcha_b@kku.ac.th

Academic Editor: Hartmut Kuhn

Periodontal diseases, which result from inflammation of tooth supporting tissues, are highly prevalent worldwide. Myeloperoxidase (MPO), from certain white blood cells in saliva, is a biomarker for inflammation. We report our study on the salivary MPO activity and its association with severity of periodontal diseases among Thai patients. Periodontally healthy subjects ($n = 11$) and gingivitis ($n = 32$) and periodontitis patients ($n = 19$) were enrolled. Assessments of clinically periodontal parameters were reported as percentages for gingival bleeding index (GI) and bleeding on probing (BOP), whereas pocket depth (PD) and clinical attachment loss (CAL) were measured in millimeters and then made to index scores. Salivary MPO activity was measured by colorimetry using 3,3′-diaminobenzidine as substrate. The results showed that salivary MPO activity in periodontitis patients was significantly higher than in healthy subjects ($p = 0.003$) and higher than in gingivitis patients ($p = 0.059$). No difference was found between gingivitis and healthy groups ($p = 0.181$). Significant correlations were observed ($p < 0.01$) between salivary MPO activity and GI ($r = 0.632$, $p < 0.001$), BOP ($r = 0.599$, $p < 0.001$), PD ($r = 0.179$, $p = 0.164$), and CAL ($r = 0.357$, $p = 0.004$) index scores. Sensitivity (94.12%), specificity (54.55%), and positive (90.57%) and negative (66.67%) predictive values indicate that salivary MPO activity has potential use as a screening marker for oral health of the Thai community.

1. Introduction

Periodontal diseases are bacterial infected diseases, which are highly prevalent worldwide [1, 2]. Periodontal status in the oral cavity is classified into healthy and periodontal diseases. The periodontal diseases include gingivitis and periodontitis

[1]. Gingivitis affects only the gingival tissues without alveolar bone destruction, whereas periodontitis leads to alveolar bone destruction [1]. In Thailand, the seventh national oral health survey conducted in the year 2012 reported that children aged 12 and 15 had gingivitis (50.30 and 53.60%, resp.) and adults (age 35–44) and the elderly (age 60–74)

had periodontal diseases for 85.90% and 88.50%, respectively [3].

Periodontal diseases are associated with the systemic diseases such as diabetes mellitus [4], ischemic heart disease, and acute coronary syndromes [5, 6]. Therefore, early detection of periodontal diseases is required for successful treatment and to reduce disease severity and complications.

Several studies reported the correlation of periodontal diseases and some biomarkers including IFN-γ, IL-10, IL-17, IL-1beta, lactoferrin, and myeloperoxidase (MPO) [7, 8]. MPO is released from azurophilic granules of polymorphonuclear cells or neutrophils to catalyze the formation of bactericidal compounds such as hypochlorous acid (HOCl) [9–11]. Increased MPO activity in gingival crevicular fluid (GCF) from patients with periodontal diseases has been reported [12–14]. The objective of this study was to determine MPO activity in saliva and its association with severity of periodontal diseases among Thai patients.

2. Materials and Methods

2.1. Chemicals and Reagents. 3,3'-Diaminobenzidine tetrahydrochloride (DAB) was obtained from Sigma, USA. Hydrogen peroxide (H_2O_2) was purchased from Merck, Germany. The other chemicals used were all of analytical grade.

2.2. Study Population and Selection Criteria. Sixty-two systemically healthy Thai individuals (24 males and 38 females), aged 19–66, who checked up their oral health at Faculty of Dentistry, Khon Kaen University, Thailand, during October-November 2013, were included in the present study. According to clinical classification of periodontal diseases by Armitage [1], assessment of periodontal parameters including gingival bleeding index (GI) and bleeding on probing (BOP) [15] and pocket depth (PD) and clinical attachment loss (CAL) [16] was performed by one examiner. A probe UNC-15 (Hu-Friedy, Chicago, IL) was employed during monitoring. Subjects were divided into 3 groups: periodontally healthy individuals as a control group ($n = 11$); gingivitis patients ($n = 32$); and periodontitis patients ($n = 19$). Inclusion criteria for periodontally healthy individuals were as follows: any individual who had at least 10 remaining teeth, GI $\leq$ 20%, BOP $\leq$ 20% of sites, no PD formation, and no CAL. For gingivitis patients, the inclusion criteria were any individual who had at least 10 remaining teeth, GI > 20%, BOP > 20% of sites, PD $\leq$ 4 mm, and CAL $\leq$ 1 mm. Chronic periodontitis patients were included according to the following criteria: any individual who had at least 10 remaining teeth, GI > 20%, BOP > 20% of sites, PD > 4 mm, and CAL > 1 mm. Upon clinical examination, assessment of PD was inferred to the scores as follows: 1: less than 4 mm; 2: 4 to 6 mm; 3: more than 6 mm [17]. Assessment of CAL was inferred to the scores as follows: 0: no clinical attachment loss; 1: less than 2 mm; 2: 2 to 4 mm; 3: more than 4 mm [18]. The PD index score in each individual was established as follows: each PD score was multiplied by number of sites demonstrating PD score. The sum of multiplied PD scores was divided by the totally measured PD sites. The CAL index score in each

individual was established as follows: each CAL score was multiplied by number of sites demonstrating CAL score. The sum of multiplied CAL scores was divided by the totally measured CAL sites. Those with history of any systemic disease, smoking, current pregnancy or lactation, periodontal therapy or use of antibiotics, or mouth rinse in the previous 3 months were excluded. This study was approved by the human ethics committee, Khon Kaen University (HE551372).

2.3. Saliva Collection. Unstimulated whole saliva samples were collected in the morning (09.00–11.00 am) at the dental clinic, Faculty of Dentistry, Khon Kaen University. No food was allowed for the subjects at least 90 minutes before collection of saliva. Each individual was instructed to rinse the mouth thoroughly with water, followed by expectorating whole saliva into a 50 mL centrifuge tube. A final saliva volume of 3 to 5 mL was obtained from each subject and the sample kept in an icebox. Measurement of salivary pH and MPO activity was performed immediately.

2.4. Colorimetric Assay of Salivary MPO Activity. Salivary MPO activity was measured by using DAB as substrate according to a modified method of Herzog and Fahimi [19]. Each saliva sample (100 μL) was pipetted into 1 mL of the 0.5 mM DAB solution (0.9 g DAB in 50 mL of 0.1 M potassium dihydrogenphosphate pH 4.5). Fifty μL of 6 mM H_2O_2 was added to initiate the reaction. After incubation at room temperature for 20 minutes, 20 μL of 0.1 mM sodium azide was added to stop the reaction. Absorbances were measured at 465 nm (Genesys 20 Thermo Scientific, USA).

2.5. Statistical Analysis. Statistical analyses were performed using SPSS program (version 11.5). The Kolmogorov-Smirnov test and the Shapiro-Wilk test were used to assess the distribution of the investigated data. Comparisons of salivary MPO activity among the three groups (healthy, gingivitis, and periodontitis) were analyzed using one-way ANOVA. Pearson's correlation coefficient was calculated to determine the correlation between salivary MPO activity and severity of periodontal diseases. Two-tailed $p < 0.05$ was considered statistically significant. The sensitivity, specificity, positive predictive value, and negative predictive value were analyzed using STATA program (version 10.1). The cut-off point was obtained according to Youden's index [20].

3. Results

3.1. Characteristics of Study Subjects. Demographic characteristics, salivary pH, and periodontal clinical parameters of the study subjects are provided in Table 1. Kolmogorov-Smirnov test and the Shapiro-Wilk test showed that age, CAL index scores, and PD index scores were not normally distributed. The healthy and gingivitis subjects were significantly younger than periodontitis patients. All periodontal clinical parameters (GI, BOP, PD, and CAL) in periodontal disease patients were significantly higher than those in the control group.

TABLE 1: Demographic characteristics, salivary pH, and periodontal clinical parameters of the study subjects.

Characteristics	Healthy ($n = 11$)	Gingivitis ($n = 32$)	Periodontitis ($n = 19$)
Demographic characteristics			
Age in years[*]	26.72 ± 8.59	27.43 ± 10.36	49.16 ± 13.93[b,c]
Female (%)	81.81	50.00	68.42
Periodontal parameters[*]			
Salivary pH	6.92 ± 0.45	6.95 ± 0.33	6.97 ± 0.36
GI (%)	11.09 ± 4.34	41.06 ± 18.36[a]	56.44 ± 29.58[b,c]
BOP (%)	12.69 ± 3.11	33.04 ± 15.96[a]	52.33 ± 28.13[b,c]
CAL index scores	0.00 ± 0.00	0.91 ± 0.25[a]	1.44 ± 0.42[b,c]
PD index scores	0.97 ± 0.11	1.06 ± 0.01[d]	1.18 ± 0.17[b,c]

GI: gingival index, BOP: bleeding on probing, CAL: clinical attachment loss, and PD: probing depth.
PD index score in each individual = [(1 × number of sites with PD score 1) + (2 × number of sites with PD score 2) + (3 × number of sites with PD score 3)]/total measured PD sites.
CAL index score in each individual = [(1 × number of sites with CAL score 1) + (2 × number of sites with CAL score 2) + (3 × number of sites with CAL score 3)]/total measured CAL sites.
[*]Mean ± SD.
Statistical analysis: One-way ANOVA.
[a]Significant at $p < 0.01$ between healthy and gingivitis groups.
[b]Significant at $p < 0.01$ between healthy and periodontitis groups.
[c]Significant at $p < 0.01$ between gingivitis and periodontitis groups.
[d]Significant at $p < 0.05$ between healthy and gingivitis groups.

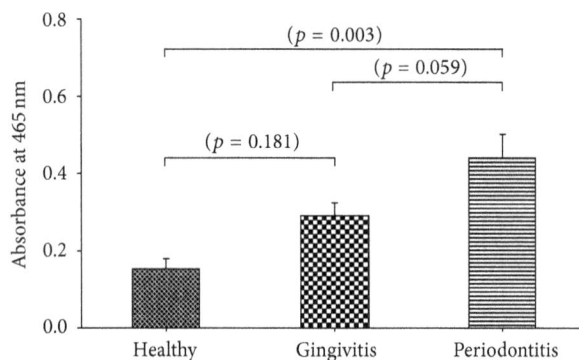

FIGURE 1: Comparison of salivary MPO activity among periodontally healthy individuals ($n = 11$), gingivitis patients ($n = 32$), and periodontitis patients ($n = 19$).

3.2. Measurement of Salivary MPO Activity. The color from the reaction between salivary MPO and DAB was mainly classified as follows: near colorless in periodontally healthy individuals; moderately brown in gingivitis patients; and intensely brown in periodontitis patients. The color change was detected by naked eye and the color intensity was detected by spectrophotometry. Comparison of salivary MPO activity of the 3 studied groups, by using one-way ANOVA analysis, gave the p value of 0.002. In Figure 1, pairwise comparisons by using Scheffe adjustment reveal that salivary MPO activity in periodontitis patients was significantly higher than in healthy subjects ($p = 0.003$). Difference in salivary MPO activity between gingivitis patients and control healthy subjects was observed but not significant ($p = 0.181$). When comparing MPO activity in gingivitis and periodontitis groups, the p value was 0.059 and

the mean difference between these 2 groups was 0.148 (95% CI = [−0.004]–[0.302]) which was considered as clinical significant value. Significant correlations were observed ($p < 0.01$) between salivary MPO activity and severity of periodontal diseases as measured by the clinically periodontal parameters: GI ($r = 0.632$, $p < 0.001$), BOP ($r = 0.599$, $p < 0.001$), PD index scores ($r = 0.179$, $p = 0.164$), and CAL index scores ($r = 0.357$, $p = 0.004$) (Figure 2). Table 2 shows the efficiency of this assay in terms of sensitivity, specificity, and predictive values. The power of differentiation between healthy subjects and periodontal disease patients was estimated at the cut-off value of 0.106 (absorbance at 465 nm) by the receiver operating characteristic curve (ROC) method. The area under ROC curve (AUC) was 0.743 (95% CI = 0.586–0.901). The MPO performance of healthy and gingivitis subjects versus periodontitis patients was also analyzed at the cut-off value of 0.365 (absorbance at 465 nm) (AUC = 0.749, 95% CI = 0.627–0.872). Multiple logistic regression was conducted to investigate the significance of MPO activity in relation to periodontal status by adjusting demographic and clinical status. For healthy versus periodontal patients, one unit increase in MPO activity resulted in an increase in odds of periodontal diseases with the estimate of 1257.68 (95% CI = 0.96 to >999, $p = 0.051$). For healthy and gingivitis subjects versus periodontitis patients, an increase in odds of periodontal diseases with the estimate of 36.23 (95% CI = 1.20–1097.66, $p = 0.039$) was obtained with each unit increase in MPO activity.

4. Discussion

MPO is often used as an inflammatory marker for early detection of several diseases including urinary tract infection [21],

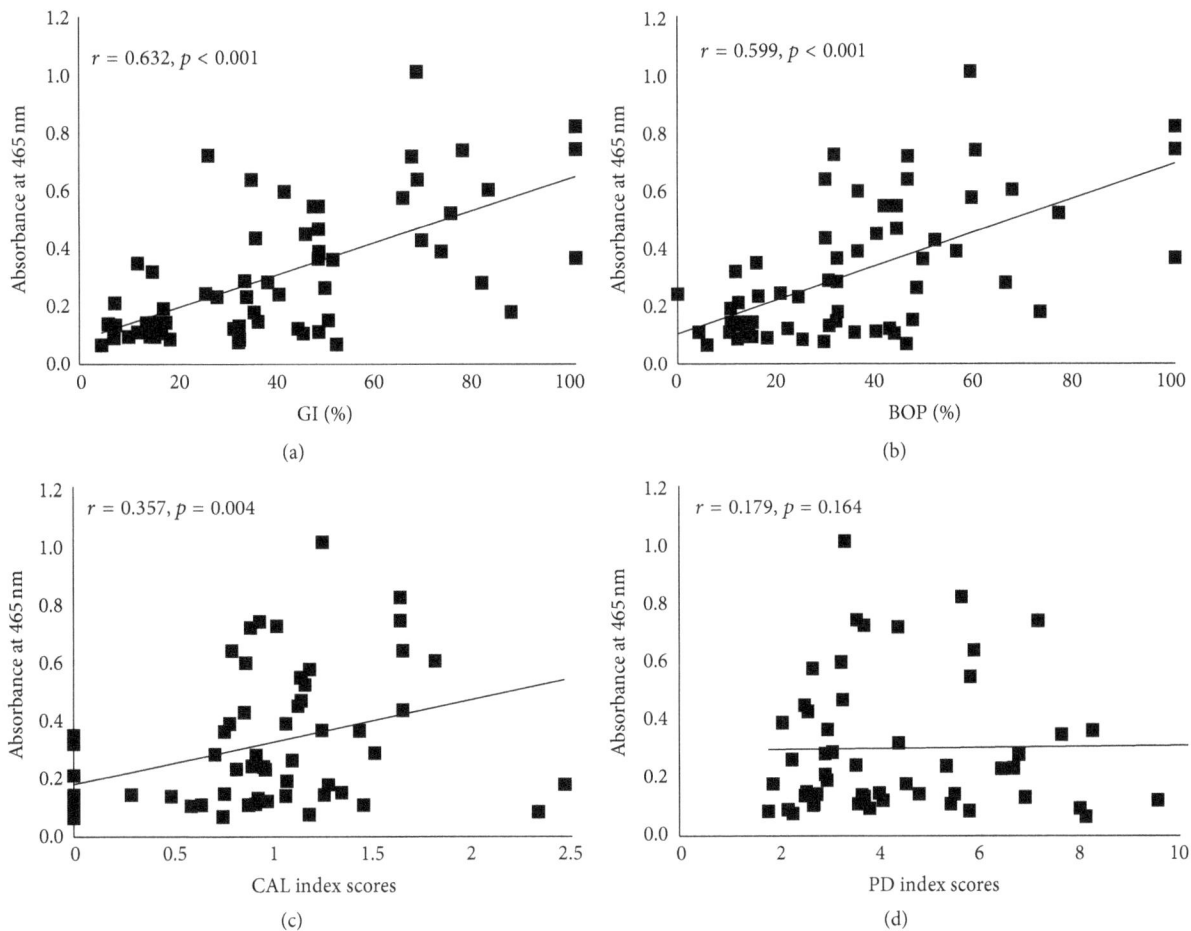

FIGURE 2: Correlations between salivary MPO activity and periodontal clinical parameters including (a) gingival bleeding index (GI), (b) bleeding on probing (BOP), (c) clinical attachment loss (CAL) index scores, and (d) pocket depth (PD) index scores. PD index score in each individual = [(1 × number of sites with PD score 1) + (2 × number of sites with PD score 2) + (3 × number of sites with PD score 3)]/total measured PD sites. CAL index score in each individual = [(1 × number of sites with CAL score 1) + (2 × number of sites with CAL score 2) + (3 × number of sites with CAL score 3)]/total measured CAL sites.

TABLE 2: Diagnostic efficacy of MPO assay in differentiation between healthy subjects and periodontal disease patients.

Comparative groups	ROC curve parameters		Sensitivity (%) (95% CI)	Specificity (%) (95% CI)	Predictive value (%) (95% CI)	
	Cut-off value	AUC (95% CI)			Positive	Negative
Healthy versus gingivitis and periodontitis	0.106	0.743 (0.586–0.901)	94.12 (83.76–98.77)	54.55 (23.38–83.25)	90.57 (79.34–96.87)	66.67 (29.93–92.51)
Healthy and gingivitis versus periodontitis	0.365	0.749 (0.627–0.872)	68.42 (43.45–87.42)	81.40 (66.60–91.61)	61.90 (38.44–81.89)	85.37 (70.83–94.43)

CI: confidence interval.
The cut-off value was calculated using the receiver operating characteristic curve (ROC) method.
Sensitivity, specificity, and positive and negative predictive values were analyzed by using area under ROC (AUC).
ROC plot displayed true positive sensitivity and false positive.

ischemic heart disease and acute coronary heart syndrome [5], and cardiovascular risk in prepubertal obese children [22] and in patients with type 2 diabetes [23]. Periodontal diseases, caused from bacterial infection followed by inflammation, may lead to tooth loss. Therefore, screening of periodontal diseases is important for early treatment. In the present study, the highest MPO activity was observed in saliva from periodontitis patients, followed by gingivitis patients and finally periodontally healthy individuals (Figure 1). MPO was detected in the saliva and GCF. Our findings are consistent with previous studies reporting increased MPO activity in periodontal disease patients [24, 25]. Utilization of saliva samples in the present study was more suitable and practical than using GCF samples in terms of a rapid sample collecting.

Our results demonstrated significantly positive correlations between salivary MPO activity and clinical parameters of periodontal diseases (Figure 2). Salivary MPO activity correlated well with GI and BOP. These findings suggested that MPO activity may be a good candidate marker for detecting the occurrence of inflammation in the periodontal tissues. However, salivary MPO activity showed fair correlation with CAL and PD index scores. Our observations implied that MPO activity might not be a strong candidate marker for detecting alveolar bone loss in the periodontal tissues.

In an attempt to investigate biomarkers for periodontal diseases, the other enzymes which associated with cell injury, including creatine kinase (CK), lactate dehydrogenase (LDH), aspartate aminotransferase (AST), alanine aminotransferase (ALT), gamma glutamyl transferase (GGT), alkaline phosphatase (ALP), and acidic phosphatase (ACP), in saliva from patients with periodontal diseases ($n = 30$) and control group ($n = 20$) were studied by Todorovic et al. [26]. There was a positive correlation between the activity of these salivary enzymes and the gingival index [26]. Analysis of these enzyme activities requires an automatic analyzer, whereas MPO activity assay in the present study can be measured by simple colorimetric method. Salivary MPO activity also showed positive correlation with GI and BOP as shown in Figure 2.

Significant difference in salivary MPO activity was found between periodontitis patients and periodontally healthy individuals ($p = 0.003$) (Figure 1). The present assay did not demonstrate significant differences in salivary MPO activity between gingivitis patients and periodontally healthy individuals ($p = 0.181$) but MPO activity in gingivitis patients tended to be higher than in periodontally healthy individuals. Salivary MPO activity in gingivitis group was lower than in periodontitis group, but not statistically significant ($p = 0.059$); the mean difference (0.148) between gingivitis and periodontitis groups was considered as clinically significant value (95% CI = [−0.004]–[0.302]). Regarding diagnostic value, ROC analysis for distinguishing healthy subjects from subjects with gingivitis and periodontitis yielded an AUC of 0.743 with 94.12% sensitivity and 90.57% positive predictive value (Table 2). However, the specificity and negative predictive value were lower (54.55% and 66.67%, resp.). Hence, these data possibly reflect a considerable number of false positives. In addition, an AUC of 0.749 with 68.42% sensitivity and 81.40% specificity was obtained in differentiating of periodontitis patients from healthy and gingivitis subjects. The observation of reasonable negative predictive value (85.37%) suggests that the assay of salivary MPO activity may serve as a potential screening biomarker, rather than a diagnostic marker for differentiating periodontal patients from nonperiodontal subjects. It should be stressed that periodontal tissues in the patients with periodontitis are gradually destroyed. Even though the statistical significance was not reached, high MPO activity may indicate the risk of periodontal diseases (OR = 1257.68, 95% CI = 0.96 to >999, $p = 0.051$). By contrast, elevated MPO activity is able to predict periodontitis from healthy and gingivitis subjects (OR = 36.23, 95% CI = 1.20–1097.66, $p = 0.039$). Thus salivary MPO activity may be used as a risk indicator for periodontal diseases based on a simple screening method at least to raise awareness of deteriorating oral health, especially in the remote areas. Individuals with positive results from this screening test should be referred to dentists for a complete clinical examination of their oral health status. On this basis, the higher the sensitivity and specificity, the better the diagnostic test. Armitage et al. [27] stated that "there are, however, no preset upper and lower limits of sensitivity and specificity values that determine if a diagnostic test is clinically useful." According to the results of the present study, to improve both sensitivity and specificity of the test, further investigation with a larger sample size is warranted. Taken together, these findings suggest that MPO activity reflects a response to inflammation in periodontal diseases and may be used as an early warning examination tool rather than a definitive diagnostic tool.

5. Conclusions

The present study demonstrated a significant increase in level of salivary MPO activity in periodontal patients using colorimetric method. An association between MPO activity and clinically periodontal parameters was observed. Our finding suggested that MPO may serve as a candidate marker for screening the oral health status of Thai people in rural areas.

Competing Interests

The authors declare that there is no conflict of interests regarding the publication of this paper.

Acknowledgments

This work was supported by the Higher Education Research Promotion and National Research University Project of Thailand, Office of the Higher Education Commission (SHeP-GMS), and Faculty of Medicine, Khon Kaen University (Grant no. I56305). The authors thank Research Group of Chronic Inflammatory Oral Diseases and Systemic Diseases Associated with Oral Health, Khon Kaen University, for saliva specimen collecting. The authors appreciate the help of Professors Paiboon Sithithaworn and Chatanun Eamudomkarn for their help in statistical analysis. They would like to acknowledge Professor David Blair for editing the paper via the Faculty of Medicine Publication Clinic, Khon Kaen University, Thailand.

References

[1] G. C. Armitage, "Periodontal diagnoses and classification of periodontal diseases," *Periodontology 2000*, vol. 34, pp. 9–21, 2004.

[2] P. E. Petersen, "The burden of oral disease: challenges to improving oral health in the 21st century," *Bulletin of the World Health Organization*, vol. 83, no. 1, p. 3, 2005.

[3] Dental Health Division, "The 7th National oral health survey of Thailand report," Department of Health, Ministry of Public Health, 2012, (Thai), http://dental.anamai.moph.go.th/elderly/academic/full99.pdf.

[4] T. Drumond-Santana, F. O. Costa, E. G. Zenóbio, R. V. Soares, and T. D. Santana, "Impact of periodontal disease on quality of life for dentate diabetics," *Cadernos de Saude Publica*, vol. 23, no. 3, pp. 637–644, 2007.

[5] V. Loria, I. Dato, F. Graziani, and L. M. Biasucci, "Myeloperoxidase: a new biomarker of inflammation in ischemic heart disease and acute coronary syndromes," *Mediators of Inflammation*, vol. 2008, Article ID 135625, 4 pages, 2008.

[6] F. A. Scannapieco and R. J. Genco, "Association of periodontal infections with atherosclerotic and pulmonary diseases," *Journal of Periodontal Research*, vol. 34, no. 7, pp. 340–345, 1999.

[7] Q.-Y. Fu, L. Zhang, L. Duan, S.-Y. Qian, and H.-X. Pang, "Correlation of chronic periodontitis in tropical area and IFN-γ, IL-10, IL-17 levels," *Asian Pacific Journal of Tropical Medicine*, vol. 6, no. 6, pp. 489–492, 2013.

[8] P.-F. Wei, K.-Y. Ho, Y.-P. Ho, Y.-M. Wu, Y.-H. Yang, and C.-C. Tsai, "The investigation of glutathione peroxidase, lactoferrin, myeloperoxidase and interleukin-1β in gingival crevicular fluid: implications for oxidative stress in human periodontal diseases," *Journal of Periodontal Research*, vol. 39, no. 5, pp. 287–293, 2004.

[9] G. N. Güncü, T. F. Tözüm, M. B. Güncü et al., "Myeloperoxidase as a measure of polymorphonuclear leukocyte response in inflammatory status around immediately and delayed loaded dental implants: a randomized controlled clinical trial," *Clinical Implant Dentistry and Related Research*, vol. 10, no. 1, pp. 30–39, 2008.

[10] M. B. Hampton, A. J. Kettle, and C. C. Winterbourn, "Inside the neutrophil phagosome: oxidants, myeloperoxidase, and bacterial killing," *Blood*, vol. 92, no. 9, pp. 3007–3017, 1998.

[11] S. J. Klebanoff, "Myeloperoxidase: friend and foe," *Journal of Leukocyte Biology*, vol. 77, no. 5, pp. 598–625, 2005.

[12] L. Karhuvaara, J. Tenovuo, and G. Sievers, "Crevicular fluid myeloperoxidase-an indicator of acute gingival inflammation," *Proceedings of the Finnish Dental Society*, vol. 86, no. 1, pp. 3–8, 1990.

[13] B. Rai, J. R. Rajnish, S. Kharb, S. C. Anand, and K. Laller, "Gingival crevicular fluid myeloperoxidase in non-smoker and smoker periodontitis: a pilot study," *Advances in Medical and Dental Sciences*, vol. 2, pp. 13–15, 2008.

[14] A. M. Marcaccini, P. A. F. Amato, F. V. Leão, R. F. Gerlach, and J. T. L. Ferreira, "Myeloperoxidase activity is increased in gingival crevicular fluid and whole saliva after fixed orthodontic appliance activation," *American Journal of Orthodontics and Dentofacial Orthopedics*, vol. 138, no. 5, pp. 613–616, 2010.

[15] J. Ainamo and I. Bay, "Problems and proposals for recording gingivitis and plaque," *International Dental Journal*, vol. 25, no. 4, pp. 229–235, 1975.

[16] G. C. Armitage, "Clinical evaluation of periodontal diseases," *Periodontology 2000*, vol. 7, pp. 39–53, 1995.

[17] M. V. Vettore, G. D. A. Lamarca, A. T. T. Leão, A. Sheiham, and M. D. C. Leal, "Partial recording protocols for periodontal disease assessment in epidemiological surveys," *Cadernos de Saude Publica*, vol. 23, no. 1, pp. 33–42, 2007.

[18] G. C. Armitage, "Development of a classification system for periodontal diseases and conditions," *Annals of Periodontology*, vol. 4, no. 1, pp. 1–6, 1999.

[19] V. Herzog and H. D. Fahimi, "A new sensitive colorimetric assay for peroxidase using 3,3′-diaminobenzidine as hydrogen donor," *Analytical Biochemistry*, vol. 55, no. 2, pp. 554–562, 1973.

[20] W. J. Youden, "Index for rating diagnostic tests," *Cancer*, vol. 3, no. 1, pp. 32–35, 1950.

[21] P. Ciragil, E. B. Kurutas, and M. Miraloglu, "New markers: urine xanthine oxidase and myeloperoxidase in the early detection of urinary tract infection," *Disease Markers*, vol. 2014, Article ID 269362, 5 pages, 2014.

[22] J. Olza, C. M. Aguilera, M. Gil-Campos et al., "Myeloperoxidase is an early biomarker of inflammation and cardiovascular risk in prepubertal obese children," *Diabetes Care*, vol. 35, no. 11, pp. 2373–2376, 2012.

[23] P. Song, J. Xu, Y. Song, S. Jiang, H. Yuan, and X. Zhang, "Association of plasma myeloperoxidase level with risk of coronary artery disease in patients with type 2 diabetes," *Disease Markers*, vol. 2015, Article ID 761939, 5 pages, 2015.

[24] C. F. Cao and Q. T. Smith, "Crevicular fluid myeloperoxidase at healthy, gingivitis and periodontitis sites," *Journal of Clinical Periodontology*, vol. 16, no. 1, pp. 17–20, 1989.

[25] K. Suomalainen, L. Saxén, P. Vilja, and J. Tenovuo, "Peroxidases, lactoferrin and lysozyme in peripheral blood neutrophils, gingival crevicular fluid and whole saliva of patients with localized juvenile periodontitis," *Oral Diseases*, vol. 2, no. 2, pp. 129–134, 1996.

[26] T. Todorovic, I. Dozic, M. Vicente-Barrero et al., "Salivary enzymes and periodontal disease," *Medicina Oral, Patología Oral y Cirugía Bucal*, vol. 11, no. 2, pp. E115–E119, 2006.

[27] G. C. Armitage, T. D. Rees, T. Blieden et al., "Diagnosis of periodontal diseases," *Journal of Periodontology*, vol. 74, no. 8, pp. 1237–1247, 2003.

10

Characterization of a Hyperthermostable Alkaline Lipase from *Bacillus sonorensis* 4R

Hemlata Bhosale, Uzma Shaheen, and Tukaram Kadam

DST-FIST Sponsored School of Life Sciences, Swami Ramanand Teerth Marathwada University, Nanded 431606, India

Correspondence should be addressed to Hemlata Bhosale; bhoslehemlata@gmail.com

Academic Editor: Jean-Marie Dupret

Hyperthermostable alkaline lipase from *Bacillus sonorensis* 4R was purified and characterized. The enzyme production was carried out at 80°C and 9.0 pH in glucose-tween inorganic salt broth under static conditions for 96 h. Lipase was purified by anion exchange chromatography by 12.15 fold with a yield of 1.98%. The molecular weight of lipase was found to be 21.87 KDa by SDS-PAGE. The enzyme activity was optimal at 80°C with $t_{1/2}$ of 150 min and at 90°C, 100°C, 110°C, and 120°C; the respective values were 121.59 min, 90.01 min, 70.01 min, and 50 min. The enzyme was highly activated by Mg and $t_{1/2}$ values at 80°C were increased from 150 min to 180 min when magnesium and mannitol were added in combination. The activation energy calculated from Arrhenius plot was 31.102 KJ/mol. At 80–120°C, values of ΔH and ΔG were in the range of 28.16–27.83 KJ/mol and 102.79 KJ/mol to 111.66 KJ/mol, respectively. Lipase activity was highest at 9.0 pH and stable for 2 hours at this pH at 80°C. Pretreatment of lipase with $MgSO_4$ and $CaSO_4$ stimulated enzyme activity by 249.94% and 30.2%, respectively. The enzyme activity was greatly reduced by $CoCl_2$, $CdCl_2$, $HgCl_2$, $CuCl_2$, $Pb(NO_3)_2$, PMSF, orlistat, oleic acid, iodine, EDTA, and urea.

1. Introduction

Hyperthermophiles are the group of organisms growing at temperatures between 80 and 110°C. This group is represented by bacterial and archeal species found in all types of terrestrial and marine hot environments. The hyperthermophilic enzymes or thermozymes derived from these organisms exhibit extreme thermostability and highest activity at temperatures above 70°C [1], some being highly active at and above 110°C [2]. Hence, such enzymes are used as model systems for enzyme based research including enzyme research, molecular basis of thermostability, and deciding the upper temperature limit for enzyme function. Thermozymes are extremely stable and active at high temperatures and offer many biotechnological advantages over mesophilic enzymes such as easier purification by heat treatment, higher resistance to chemical denaturants, and reduced risk of microbial contamination [1]. They also offer high reaction rates and process yields by lowering viscosity, causing increased diffusion rates and substrate availability and maintaining favorable equilibrium with endothermal reactions [3]. Thermozymes

isolated from hyperthermophiles growing at the temperature range of 80–110°C are expected to be more thermostable than their mesophilic correspondents as these organisms are in full harmony with the existing thermal conditions and expected to secrete the enzymes that are completely stable at these temperatures to support their physiological processes [4].

Lipases are the most important group of industrial biocatalysts that can be applied both as hydrolases and as synthetases and proved their enormous potential in various biotechnological applications. The unique characters of lipases such as high stability in organic solvents, their broad substrate specificity, and high enantioselectivity greatly increased their demand in industrial market. The current market scenario of hydrolytic enzymes positioned lipases at the top third rank after proteases and amylases and their annual market is targeted to reach about 590.5 million dollars by 2020 [5]. However, most of the industrial processes operate at relatively high temperature and alkaline pH conditions. Hence, the thermoalkalostability of lipases is one of the desired characteristics to endure harsh processing conditions used during industrial applications. Thermostable lipases are the need of food, cosmetics, detergent, and pharmaceutical industries [6, 7].

Most of the research on lipases is concentrated on isolating highly thermostable lipases from different thermophilic microbial sources. While thermostable lipases are known to be produced by *Bacillus* sp. which have unique protein sequence and inherent biochemical properties, lipases from *Burkholderia ambifaria* YCJ01 [8], *Aneurinibacillus thermoaerophilus strain HZ* [9], and *Pseudomonas* sp. [10] are also reported. However, the studies related to thermoalkalostable lipases from hyperthermoalkalophilic organisms are scanty. In the present study we are reporting for the first time the isolation and identification of highly thermostable lipase producing hyperthermophilic strain of *Bacillus sonorensis* 4R. Purification of enzyme and its thermodynamic and biochemical properties are also reported.

2. Materials and Methods

2.1. Materials. All media ingredients, diethylaminoethylcellulose (DEAE-cellulose), phenylmethylsulfonyl fluoride (PMSF), and bovine serum albumin (BSA), were purchased from HiMedia. All other chemicals used were of analytical grade.

2.2. The Lipolytic Organism and Enzyme Production. The lipolytic strain of *Bacillus sonorensis* 4R used in the present study was isolated from Thar Desert ecosystem of Jaisalmer, Rajasthan, India (lat. 27′00N and 71°00E). The strain was grown on alkaline tributyrin inorganic salt agar (g/L: K_2HPO_4, 1; $MgSO_4$, 1; NaCl, 1; ammonium sulphate, 2; $CaCO_3$, 2; $FeSO_4$, 0.001; $MnCl_2$, 0.001; $ZnCl_2$, 0.001; and tributyrin, 10 mL) at pH 9.0, 80°C for 7 days. Lipase production was carried out by growing active culture of 4R (5%) in 2 L glucose-tween inorganic salt broth (g/L: K_2HPO_4, 1; $MgSO_4$, 1; NaCl, 1; ammonium sulphate, 2; $CaCO_3$, 2; $FeSO_4$, 0.001; $MnCl_2$, 0.001; $ZnCl_2$, 0.001; glucose 10, $CaSO_4$, 100 mM; and tween-80, 10 mL) adjusted to pH 9.0. The flasks were incubated at 80°C for 4 days under static conditions. At the end of incubation, the culture broth was centrifuged at 10,000 rpm for 30 min at 4°C to obtain cell-free supernatant. The supernatant was used as crude source of *Bacillus sonorensis* lipase (BSL) and analyzed for lipase activity and protein content.

2.3. Identification of Lipolytic Organism. Bacterial genomic DNA was isolated using geneO-spin Microbial DNA Isolation Kit (geneOmbio Technologies, Pune, India). This DNA was used as template for PCR analysis using the primers 27F: 5′-AGAGTTTGATCMTGGCTCAG-3′ and 1492R: 5′-TACCTTGTTACGACTT-3′. The amplification conditions were 95°C for 10 min, 57°C for 1 min, 72°C for 90 sec, and final amplification at 72°C for 10 min. The PCR products were purified by using a geneO-spin PCR Product Purification Kit (geneOmbio Technologies, Pune, India) and were directly sequenced using an ABI PRISM BigDye Terminator V3.1 Kit (Applied Biosystems, USA). The sequences were analyzed using Sequencing Analysis 5.2 software.

BLAST analysis was performed at BlastN site at NCBI server (http://www.ncbi.nlm.nih.gov/BLAST) and evolutionary relationship of 4R was deduced by constructing phylogenetic tree.

2.4. Lipase Assay and Protein Determination. The assay was performed by using the modified method described by Selvam et al. [11] based on olive oil hydrolysis. To the reaction mixture containing 1 mL of tris-HCl buffer (pH 9.0), 2.5 mL of deionized water, and 3 mL of olive oil emulsion (10% gum arabic emulsified with 5% olive oil), 1 mL of crude enzyme for test and 1 mL of deionized water for blank were added in separate tubes. The reaction mixture was mixed thoroughly by swirling and incubated at 80°C for 30 min. After incubation, enzyme substrate reaction was terminated by addition of 3 mL of 95% ethanol and mixed by swirling. The amount of fatty acids liberated due to lipase activity was estimated by titrating the contents of assay mixture against 0.05 M NaOH using thymolphthalein as a pH indicator. The end point observed was from colorless to light blue. One unit of lipase was defined as the amount of enzyme required to release 1 μmole of fatty acid under assay conditions. Protein content of all fractions was determined by Bradford assay [12] by using BSA as a standard protein.

2.5. BSL Purification. The crude lipase was purified using a two-step procedure including ammonium sulphate precipitation followed by dialysis and DEAE-cellulose ion exchange chromatography. The cell-free supernatant was pretreated at 80°C for 30 min to eliminate the appearance of additional proteins and brought to 80% saturation by adding finely powdered ammonium sulphate. The flask was kept overnight at 4°C and the precipitate was collected by centrifugation at 10,000 rpm for 20 min at 4°C. The precipitate was dissolved in phosphate buffer (0.1 M, pH 9.0) and dialyzed overnight against the same buffer.

The desalted enzyme obtained from dialysis step was loaded on chromatography column (1.5 × 15 cm) packed with DEAE-cellulose and preequilibrated with 0.1 M phosphate buffer (pH 9.0). The enzyme was eluted with linear gradient of NaCl (0.1–0.5 M) in phosphate buffer. The flow rate of column was adjusted to 0.5 mL/min and protein concentration (280 nm, UV Vis Shimadzu) and lipase activity of eluted fractions were determined as mentioned before after desalting.

2.6. Determination of Molecular Mass of BSL. The molecular mass of purified BSL was determined by sodium dodecyl sulphate polyacrylamide gel electrophoresis (SDS-PAGE) technique using HiPer SDS-PAGE Kit (HiMedia) according to the manufacturer's instructions. A broad range of unstained protein standards (insulin [3.5 kda], aprotinin [6.5 kda], lysozyme [14.3 kda], soya bean trypsin inhibitor [20.1 kda], carbonic anhydrase [29.0 kda], ovalbumin [43.0 kda], BSA [66.0 kda], phosphorylase [97.4 kda], and myosin [205.0 kda]) was used as molecular mass makers. The gel was stained with 0.025% Coomassie Brilliant Blue R-250 staining solution provided in the kit and destained overnight by adding 7% acetic acid solution. The molecular mass of purified BSL was determined from a plot between log MW and relative migration values (R_f) of standard protein markers. The activity of purified fraction obtained after electrophoresis was confirmed by zymogram analysis. The gel was prepared by supplementing 1% tributyrin; the sample was loaded and subjected to electrophoresis. The gel was stained and destained as mentioned before and the location of band on gel was observed for presence of clear zone due to tributyrin hydrolysis.

2.7. Effect of Temperature on BSL Activity and Stability.

The optimum temperature for BSL activity was determined over the temperature range of 80–120°C (80, 90, 100, 110, and 120°C) by preincubating aliquots of purified lipase in phosphate buffer (100 mM, pH 9.0) at respective temperatures for 30 min. After incubation, the fractions were cooled on ice and assayed for BSL activity. To determine the effect of temperature on enzyme stability, different aliquots of purified enzyme were preincubated separately at 80–120°C for 3 h in phosphate buffer (100 mM, pH 9.0) and residual activity was measured at intervals of 30 min.

2.8. Effect of Divalent Cations and Polyols on Thermostability of BSL.

The effect of varying concentrations of $CaSO_4$ and $MgSO_4$ and different polyols including glycerol (3C), ethylene glycol (5C), inositol (5C), sorbitol (5C), and mannitol (6C) on thermal stability of BSL was studied by preincubating various enzyme fractions in presence of respective compounds at 80°C for 3 h. The aliquots were withdrawn after every 30 min, ice-cooled, and used for residual activity determination. The activity was compared with initial lipase activity observed before incubation in presence of Ca, Mg, and polyols. The polyol showing improved thermostability was selected over the range of 20–100 mM for further study at 80°C. Similarly, cumulative effect of selected polyol (60 mM) and $MgSO_4$ (80 mM) on thermostability of BSL was determined.

2.9. Thermodynamic Parameters.

The thermodynamic parameters related to BSL activity at elevated temperatures (80–120°C) were determined in terms of half-life ($t_{1/2}$), denaturation constant (K_d), enthalpy of denaturation (ΔH), free energy of denaturation (ΔG), and entropy of denaturation (ΔS). The inactivation rate constants were calculated from a plot of residual activity versus time and used for estimating half-lives. The activation energy of thermal inactivation (E_a) was determined from the Arrhenius plot between $\ln K_d$ and $1/T$ (K) as described before [13]. The values of ΔH, ΔG, and ΔS for inactivation were calculated according to the following equations, respectively, as described by Gummadi [14]:

$$\Delta H = E_a - RT \tag{1}$$

$$\Delta G = -RT \ln \left(\frac{K_d \cdot h}{K_b \cdot T} \right) \tag{2}$$

$$\Delta S = (\Delta H - \Delta G) T, \tag{3}$$

where $R = 8.314 \, \text{JK}^{-1} \, \text{mol}^{-1}$ is the universal gas constant, T is absolute temperature, h is Plank's constant, and K_b is Boltzmann's constant.

2.10. Effect of pH on BSL Activity and Stability.

To determine the effect of pH on BSL activity, the aliquots of enzyme were preincubated in buffers of different pH values (sodium phosphate, 0.1 M, pH 7.5–8; tris-HCl, 0.1 M, 8.5–9.0; carbonate-bicarbonate, 0.1 M, pH 9.5–10.5; sodium phosphate-NaOH, 0.1 M, pH 11-12) for 30 min at 80°C. After incubation the fractions were ice cooled and enzyme activity was determined under assay conditions. To determine the effects of pH on stability, aliquots of BSL were preincubated with buffer of pH 9.0 for 180 min at 80°C and the residual activity was determined at intervals of 20 min.

2.11. Effect of Metal Ions on BSL Activity.

The effects of various metal ions, namely, Ca^{++}, Mg^{++}, Cu^{++}, Pb^{++}, Co^{++}, Cd^{++}, and Hg^{++}, as $CaSO_4$, $MgSO_4$, $CuCl_2$, $Pb(NO_3)_2$, $CoCl_2$, and $HgCl_2$ on BSL activity were studied. The aliquots of BSL (10 μL) were preincubated in presence of different metal ion concentrations (25–150 mM) at 80°C for 30 min and subjected to lipase assay. The effect of metal ions on BSL activity was determined by comparing the enzyme activities in absence of these compounds.

2.12. Effect of Chemical Modulators on BSL Activity.

The effect of different chemical modulators on BSL activity was tested by preincubation of properly diluted enzyme at 80°C for 30 min in presence of selected chemical modulators. The chemical modulators (ethylene diamine tetra-acetic acid (EDTA), urea, PMSF, iodine, orlistat, and oleic acid) were set at 5 mM and after preincubation, BSL activity was determined under assay conditions. The effect of chemical modulators on BSL activity was determined by comparing the enzyme activities in absence of these compounds.

3. Results and Discussion

3.1. Growth and Lipase Production by Bacillus sonorensis 4R.

The lipase producing *Bacillus sonorensis* 4R, isolated from soils of Thar Desert area in Jaisalmer, Rajasthan, India, was detected using tributyrin agar plates. The isolate showed good tributyrin hydrolysis efficiency on plates (28 mm) (Figure 1) as well as in broth (51.33 U/mL) after 4 days of incubation at 80°C. The hyperthermoalkalophilic bacteria optimally grow within temperature range of 80–110°C and are found in all terrestrial and marine hot environments. The selected soil sample of Thar Desert after enrichment in inorganic salt medium supplemented with 1% tributyrin at 80°C and pH 9.0 successfully isolated potential lipase producing thermoalkalophilic strain of 4R.

The isolate was identified as *Bacillus sonorensis* on the basis of its morphological characteristics and 16S rRNA sequencing. The 4R strain has ability to grow at temperature between 80 and 100°C and 8.0–11.0 pH with optimum growth at 80°C and pH 9.0. It appeared as a facultative anaerobe,

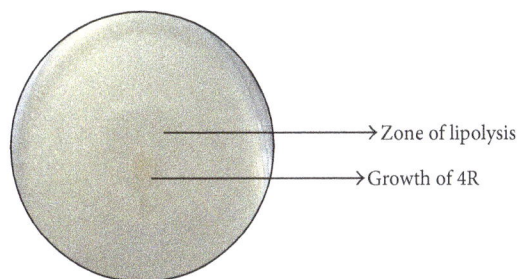

FIGURE 1: Lipase producing bacterial isolate *Bacillus sonorensis* 4R: colonies on tributyrin agar showing zone of lipolysis after 4 days of incubation at 80°C and pH 9.0.

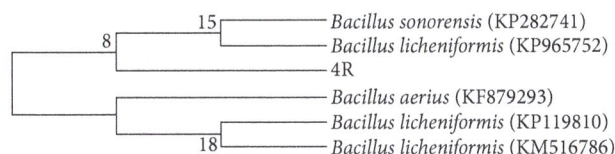

FIGURE 2: Neighbor joining tree based on 16S gene sequencing showing phylogenetic relationship between *Bacillus sonorensis* 4R and related members of the genus *Bacillus*.

Gram-positive long rod, nonmotile, catalase positive bacterium and had the capacity to reduce nitrate and produced acid from glucose, arabinose, xylose, and mannitol. The 1286-base-pair sequence obtained by 16S rRNA sequencing has been deposited to NCBI gene bank database with accession number *KT 368092*. *Bacillus sonorensis* 4R shared the highest homology of 100% with *Bacillus sonorensis* strain ZJY-537. From the phylogenetic analysis it was confirmed that 4R was closely associated with *Bacillus sonorensis* which is a close lineage of *Bacillus licheniformis* and other members of genus *Bacillus* (Figure 2). Hence, the strain 4R was identified as *Bacillus sonorensis*. So far, *Bacillus sonorensis* was reported to be isolated from Sonoran Desert [15] and Kalbadevi estuary, Mumbai [16]. Abundance of thermophilic *Bacillus* species including *Bacillus licheniformis*, *Bacillus aerius*, *Bacillus sonorensis*, *Bacillus subtilis*, and *Bacillus amyloliquefaciens* in hot springs, salt marshes, and desert soil of Morocco has been observed by Aanniz et al. [17]. The study also highlighted the growth of isolated *Bacillus* species at temperature range between 30 and 80°C. All species of *Bacillus sonorensis* showed good growth from 30 to 55°C whereas none of the isolates grew above 70°C. Different studies have reported the dominance of strains of *Bacillus* species in various geothermal habitats including Japanese desert [18] and Atacama Desert soil [19] with special reference of common occurrence of *Bacillus sonorensis* in deserts of Morocco [15]. However, the hyperthermophilic growth at 80°C and above temperatures and lipase production capacity of *Bacillus sonorensis* were not reported earlier. Only few reports on sonorensin, a food preservative [20], and lipopeptide antibiotic production [21] by strain of *Bacillus sonorensis* are available. In this context, we are reporting for the first time the isolation of hyperthermophilic lipase producing *Bacillus sonorensis* from Thar Desert of Rajasthan.

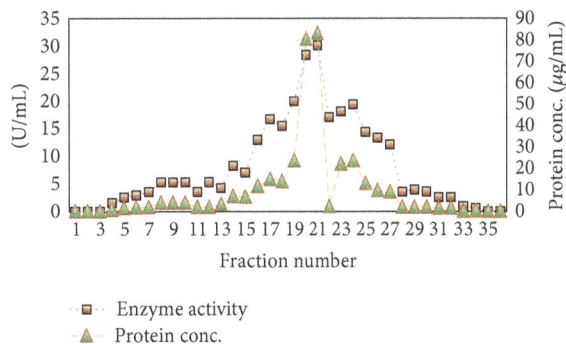

FIGURE 3: Elution profile of BSL for purification on DEAE-cellulose column.

3.2. Purification of BSL. The lipase produced by *Bacillus sonorensis* was purified by using a sequential procedure including salt precipitation, desalting by dialysis, and chromatography on DEAE-cellulose column. The results of the lipase purification profile are summarized in Table 1. The enzyme was finally purified 12.15-fold over crude extract with 1.98% recovery.

Chromatography of lipase on DEAE-cellulose ion exchange column resulted in one prominent peak at the 21st fraction (Figure 3). The active fractions were pooled and the homogeneity of purified enzyme was confirmed by the presence of a single band corresponding to an apparent molecular mass of 21.87 KDa on SDS-PAGE gel (Figure 4(a)). Lipase activity in the purified band was checked by observing presence of lipolysis zone in gels supplemented with 1% tributyrin (Figure 4(b)).

3.3. Effect of Temperature on BSL Activity. The effect of different temperatures on activity of purified lipase is shown in Figure 5(a). *Bacillus sonorensis* produced lipase was more active in temperature range of 80–120°C with more than 50% of its original activity remaining above 90°C up to 120°C after 30 min exposure (Figure 5(b)). The optimum temperature recorded for the lipase activity of TM12350, a recombinant lipase from a hyperthermophilic bacterium *Thermotoga maritima*, was 70°C [22] with maximum activity retained for 60 min at 70°C while maintaining more than its 50% activity within 8 h. At higher temperature the confirmation of enzyme is disrupted which results in reduced affinity sites for substrate [23]. Hence, in the present study when the temperature was increased from 80 to 120°C, a gradual decrease in catalytic activity of BSL was observed. However, the degree by which the activity was decreased was not convincing as BSL retained more than 50% of its original activity at 120°C. The lipase exhibited significant stability at 80°C with a half-life ($t_{1/2}$) of 150 min whereas the values of $t_{1/2}$ reduced to 121.59 min, 90.01 min, 70.01 min, and 50 min, respectively, at 90°C and above temperatures (100°C, 110°C, and 120°C) and at pH 9.0. These characteristics indicated that BSL is a highly thermostable lipase retaining about 50% activity at and above 100°C. This study for the first time showed the highly thermostable nature of lipase produced among *Bacillus* family and probably among all

TABLE 1: Purification summary of hyperthermostable lipase from *B. sonorensis* 4R.

	Protein content (mg/mL)	Total activity	Specific activity (U/mg)	Purification fold	Yield (%)
Crude	0.290	153990	177	1	100
Ammonium sulphate precipitation	0.252	13275	351	1.98	8.62
Dialysis	0.143	5901.5	825	4.67	3.83
DEAE-cellulose column	0.355	3055.96	2152.08	12.15	1.98

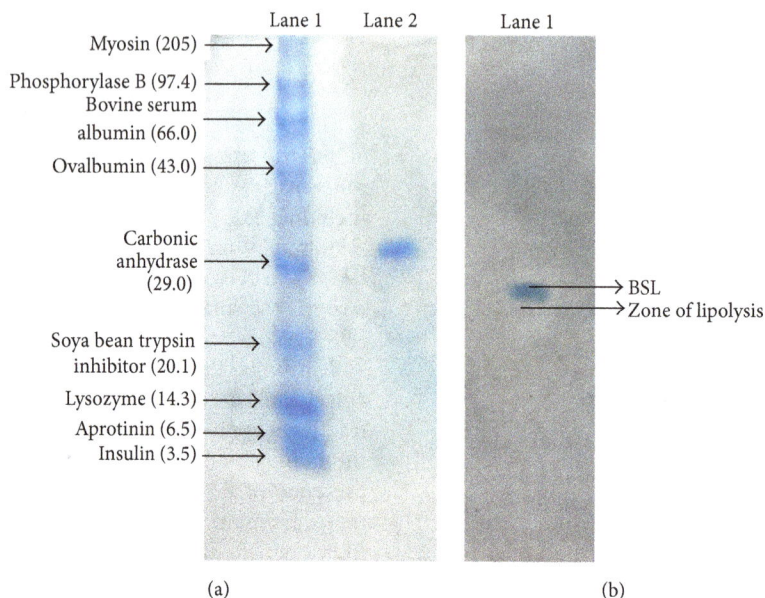

FIGURE 4: SDS-PAGE of hyperthermostable lipase from *Bacillus sonorensis* 4R. (a) Lane 1: standard protein molecular mass markers, Lane 2: purified BSL. (b) Activity characterization of BSL by zymogram analysis.

reported lipases. For lipase produced by *Bacillus licheniformis* MTCC6824, [24] reported $t_{1/2}$ values of 82 min, 75 min, and 48 min at 45°C, 50°C, and 55°C whereas Shariff et al. [25] showed thermoactive nature of L_2 lipase at a temperature range of 55–80°C with temperature optima at 70°C and $t_{1/2}$ of 2 h at 60°C. However, the reported thermostability at alkaline pH (9.0) in the present study was higher where enzyme was retaining its 50% activity at and above 100°C. Lipases at high temperature and alkaline pH are of immense importance in food industry and pharmaceuticals due to their process conditions operating at high temperature (45–50°C) and pH (8.0). The thermostability exhibited by BSL was greater than other thermostable lipases, such as lipase from *Bacillus* species SP42 with $t_{1/2}$ of 45 min at 70°C [26], esterase from *Thermoanaerobacter* sp. with $t_{1/2}$ of 90 min at 70°C [27], and lipase from the hyperthermophilic *Aneurinibacillus thermoaerophilus-HZ* with half-life of 80 min at 70°C. The enzymes activated at and above 40°C are said to undergone thermal activation [10]. In the present study the BSL was only activated at high temperatures (80°C) and the activity was very poor at 40°C indicating thermal activation of BSL (Figure 5(c)). The characteristic of an enzyme to show thermal activation depends on the hydrophobic amino acid

content of the protein and lipases are known to be rich in hydrophobic amino acids [28, 29]. Hence, it is expected that the thermal activation of BSL in the present study might be contributed by its hydrophobic amino acids content.

3.4. Effect of Divalent Cations and Polyols on BSL Activity. The catalytic activity of BSL was greatly increased over control at 80°C in presence of Ca^{2+} and Mg^{2+} at all concentrations (20–100 mM). $CaSO_4$ when used at 60 mM and 80 mM concentrations caused 249.08% and 199% respective enhancement in BSL activity whereas, with increase in incubation time, the activity was gradually decreased and reported absent after 3 h incubation at all concentrations (Figure 6(a)). BSL activity was also found to increase significantly in presence of $MgSO_4$ (80 mM) after 20–100 min exposure. The highest increase in activity was observed after 60 min incubation (423.6%) and thereafter, the activity was slowly reduced (Figure 6(b)).

It has been reported earlier that the molecular size and number of hydroxyl groups per molecule of polyol play an important role in mediating the protection against thermal inactivation [30]. In this study mannitol appeared as the best thermoprotectant at 50 mM concentration as observed in terms of approximately 20% enhancement in residual activity

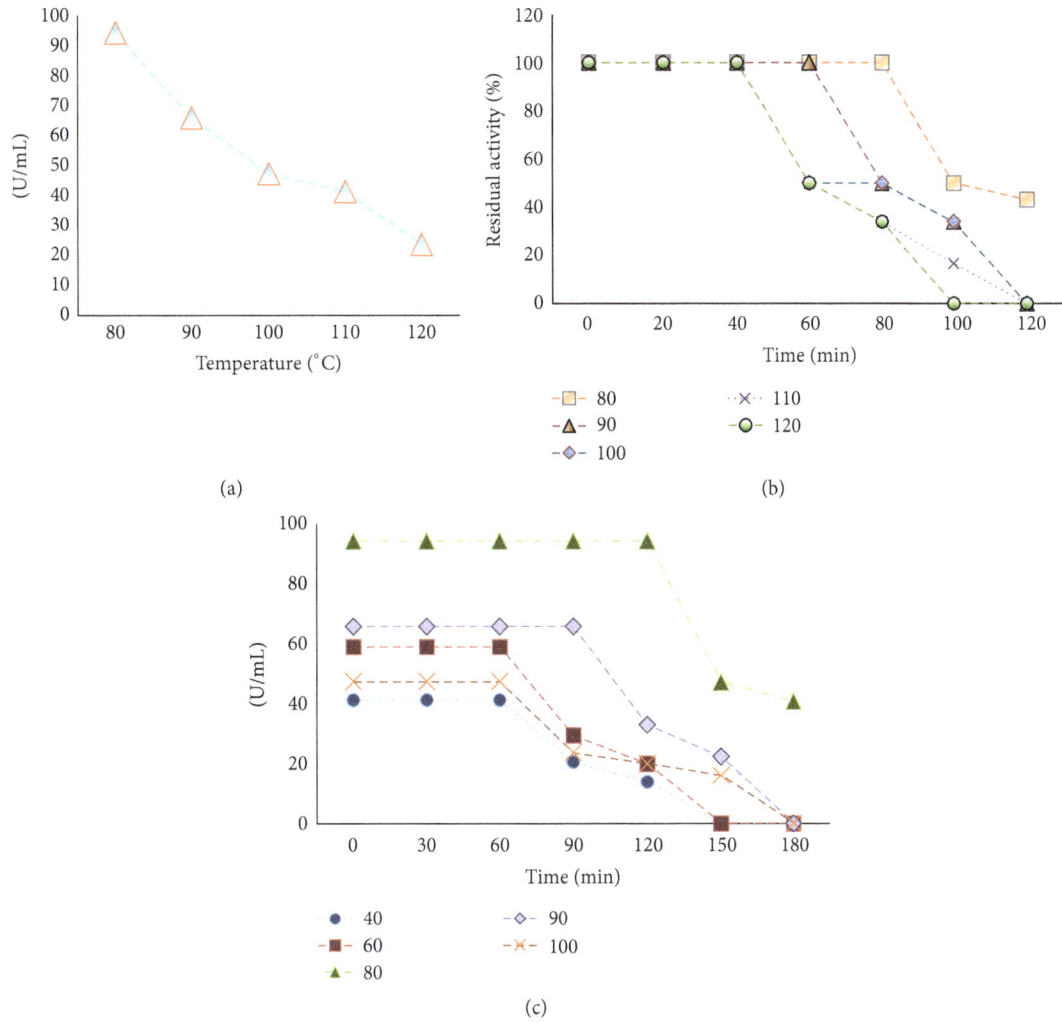

(a)

(b)

(c)

FIGURE 5: (a) Effect of temperature on BSL activity. (b) Thermostability of BSL at temperatures from 80 to 120°C. (c) Thermal activation of BSL at elevated temperatures.

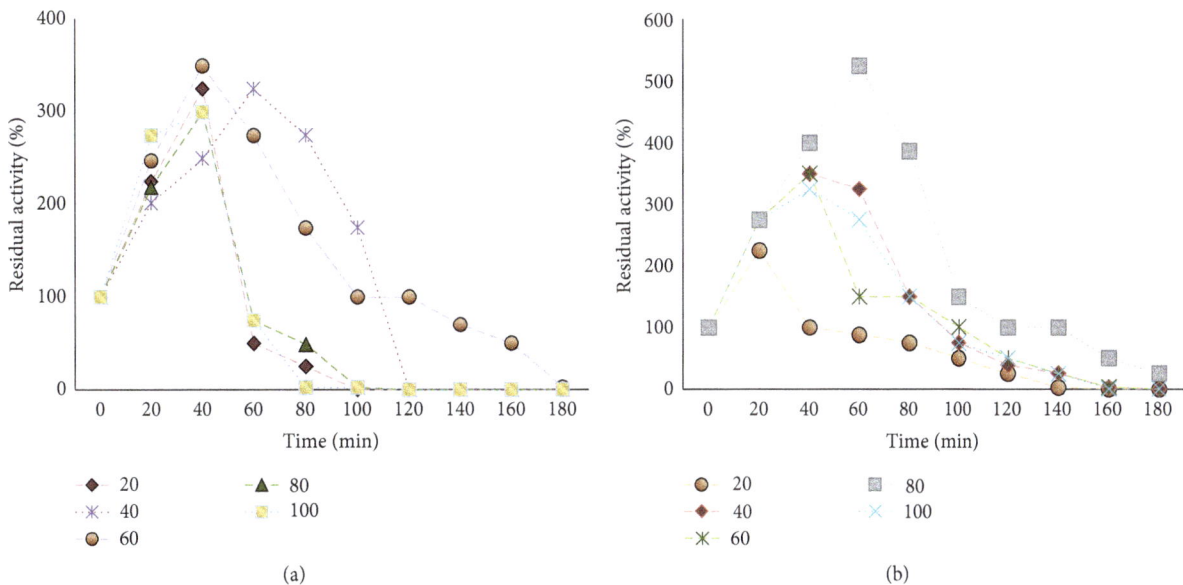

(a)

(b)

FIGURE 6: (a) Effect of CaSO$_4$ (20–100 mM) on BSL thermostability. (b) Effect of MgSO$_4$ (20–100 mM) on BSL thermostability.

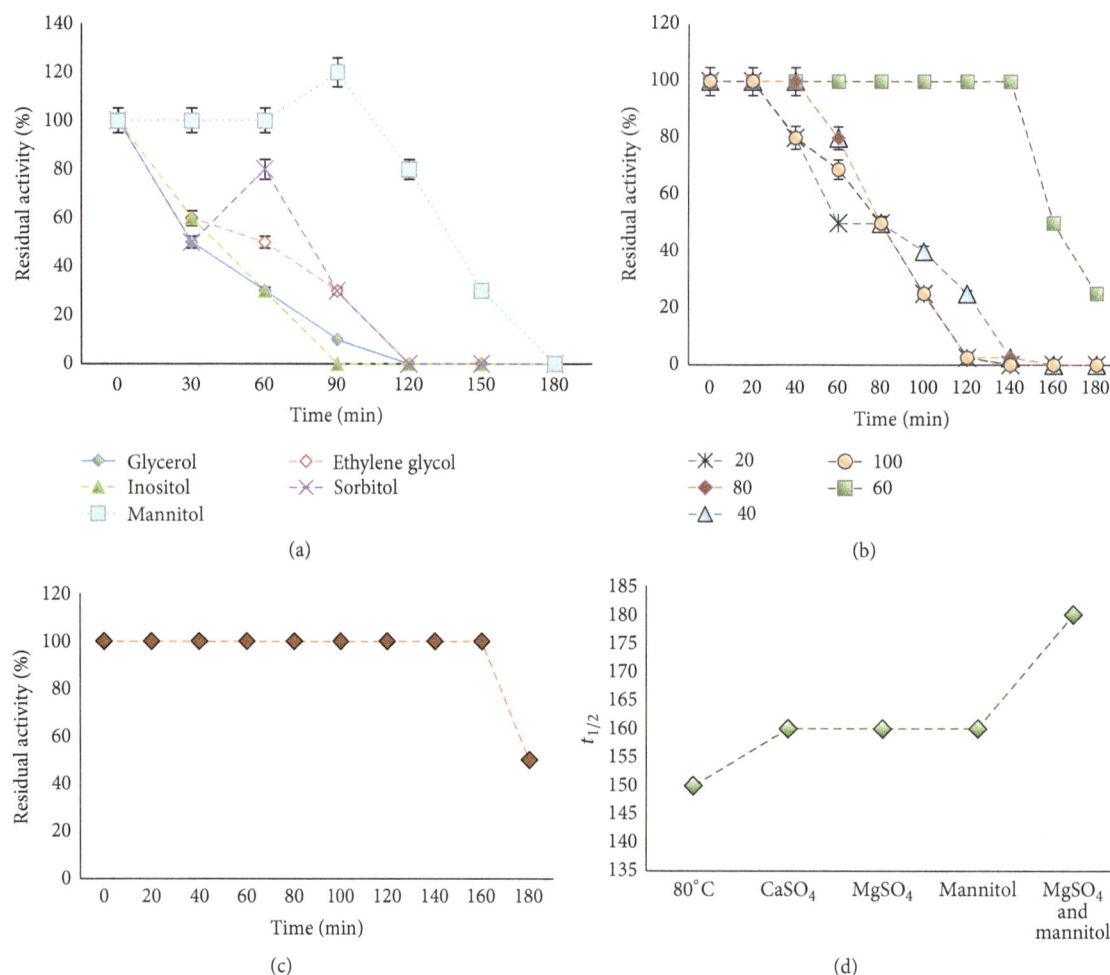

FIGURE 7: (a) Effect of different polyols (50 mM) on BSL activity. (b) Effect of mannitol (20–100 mM) on BSL thermostability. (c) Cumulative effect of $MgSO_4$ (80 mM) and mannitol (60 mM) on BSL thermostability. (d) $t_{1/2}$ values of BSL at 80°C, $CaSO_4$ (60 mM), $MgSO_4$ (80 mM), mannitol (60 mM), and $MgSO_4$ (80 mM) and mannitol (60 mM).

after 90 min exposure while retaining 100% activity when incubated for 60 min (Figure 7(a)). However, with further rise in incubation time, the activity was gradually reduced. The effect of different mannitol concentrations on thermostability of BSL was also evaluated. Increasing mannitol concentration up to 60 mM improved the thermostability of BSL with 49.99% of the original activity remaining after 140 min at 80°C. At higher concentration of mannitol (100 mM) reduced thermostability was observed where 100% of residual activity was retained only for 60 min (Figure 7(b)). Addition of polyols can prevent conformational changes of the enzyme by promoting formation of numerous hydrogen bond or salt bridges between amino acid residues, making the enzyme molecule more rigid and, hence, more resistant to the thermal unfolding [31, 32]. However, the selection of the suitable additive depends on the nature of enzyme and it varies from one enzyme to another. Addition of polyols improves thermostability of lipase from *Bacillus licheniformis* MTCC6824 [24], xylanases from *Trichoderma reesei* QM9414 [33], and xylanase from *A. pullulans* CBS135684 [34]. The effect of

sorbitol on thermostability of lipase has been identified in *Bacillus licheniformis* MTCC6824 [24]. The cumulative effect of $MgSO_4$ (80 mM) and mannitol (60 mM) on thermostability is shown in Figure 7(c). BSL incubated with a combination of $MgSO_4$ and mannitol induced a synergistic effect observed in terms of 100% residual activity of BSL remained after 160 min, as compared to 49.99% when incubated with Mg^{++} or mannitol alone. After 3 h, Mg^{++} and mannitol combination was found to retain approximately 50% of original activity as compared to approximately 25% in presence of Mg^{++} or mannitol alone. The $t_{1/2}$ values at 80°C were increased from 150 min to 180 min when magnesium and mannitol were added in combination (Figure 7(d)). The loss of enzyme activity at elevated temperature ranges is related to changes in enzyme conformation [30, 35]. Improved thermostability of BSL due to Mg^{++} and mannitol, a higher polyhydric alcohol, might be due to hindered denaturation of catalytic site of enzyme caused by hydration resulting in charge rearrangement and ion complexation. Ion complexation of metal ions such as Ca^{++} is a process with favorable entropy factor that

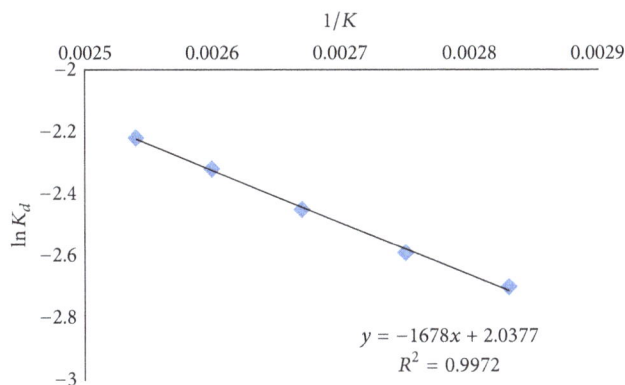

FIGURE 8: Arrhenius plot for determination of the activation energy of BSL.

helps in stabilization of enzymes at high temperatures. The role of Ca^{++} ions in maintenance of stable and active enzyme structure is well stated [24]. However, the Mg and mannitol dependent improvement of hyperthermostable lipases is not reported earlier.

3.5. Thermodynamic Characteristics. The effects of varying temperatures on kinetic and thermodynamic characteristics of BSL are summarized in Table 2. The parameters including half-life period, denaturation constant, entropy, enthalpy, and free energy change at different temperatures were determined from Arrhenius plot as shown in Figure 8. The activation energy calculated from Arrhenius plot was 31.102 KJ/mol at 80°C; the $t_{1/2}$ value was 150 min and it appeared to reduce to 121.59 min at 90°C. Further, increase in temperatures reduced half-lives of BSL to 90.01 min, 70.01 min, and 50 min appearing at 100°C, 110°C, and 120°C, respectively. The values of denaturation constant were increased with increase in temperature from 80 to 120°C. The values of ΔH were not changed significantly from 80 to 120°C and recorded in the range of 28.16–27.83 KJ/mol. ΔG was increased from 102.79 KJ/mol at 80°C to 111.66 KJ/mol at 120°C. It was reported previously that thermodynamically stable proteins exhibit high ΔG [36]. The appearing high values of free energy change in denaturation in the present study confirmed the thermodynamic stability of BSL and its better resistance against thermal unfolding conferred at elevated temperatures. This was further supported by observed entropy values. The change in entropy was not noticeable from 80 to 120°C demonstrating nominal changes in enzyme architecture during thermal unfolding.

3.6. Effect of pH on BSL Activity and Stability. BSL exhibited good activity over the pH range 7.5–11 (Figure 9(a)). The maximum activity was observed at pH 9.0 (118.20 U/mL) followed by pH 9.5 (94.4 U/mL) and pH 10 (70.8 U/mL). A rapid decline in the enzyme activity was observed beyond pH 10.0 whereas the activity was lowest at pH 12 (10 U/mL). At pH 9.0 BSL retained its 100% activity for 120 min (Figure 9(b)) followed by a sequential decrease in original enzyme activity from 140 min (49.99%), 160 min (41.68%), and 180 min (0%)

exposure. Similarly, Jeagar et al. [25] reported maximum lipolytic activity of L_2 lipase towards olive oil as a substrate retaining 50% of its original activity at pH 10.0.

3.7. Effect of Metal Ions on BSL Activity. The effects of varying concentrations of metal ions (25–150 mM) including $CaSO_4$, $MgSO_4$, $CuCl_2$, $Pb(NO_3)_2$, $CoCl_2$, $CdCl_2$, and $HgCl_2$ on BSL activity are shown in Figure 10. A concentration dependent enhancement in lipase activity was found in presence of $MgSO_4$. The residual activity was gradually increased from 100% to 349.94% with increase in $MgSO_4$ concentration from 50 mM to 150 mM, respectively. Addition of $CaSO_4$ at concentration range of 50–150 mM either retained 100% residual activity or caused a marginal enhancement in BSL activity. The activity was reduced drastically in presence of low concentrations of $CuCl_2$ (25–75 mM) and $Pb(NO_3)_2$ (25–50 mM) whereas, at higher concentrations (100–150 mM), BSL was completely inhibited. The inhibition of BSL by Cu^{++}, a transition metal ion, might be due to the changes brought by metal ion in solubility and behavior of ionized fatty acid at the interfaces affecting the catalytic properties of enzyme [37]. At all concentrations of $CoCl_2$, $CdCl_2$, and $HgCl_2$, BSL lost its 100% original activity indicating highly potent inhibitory nature of these metals.

3.8. Effect of Chemical Modulators on BSL Activity. Table 3 shows the effect of different chemical modulators H_2O_2, PMSF, EDTA, bile salts, orlistat, and oleic acid at 5 mM conc. on BSL activity. Among the different compounds tested, BSL activity was greatly reduced by 83.33% in presence of PMSF indicating it as a member of serine family. The inhibition occurring in presence of PMSF might be due to modification of essential serine residue, inducing a direct or indirect change in enzyme confirmation [38, 39]. A considerable inhibition in BSL activity was recorded in presence of EDTA (50.01%) and urea (50.01%). Inhibition in presence of EDTA indicated that the BSL is metalloenzyme [40]. Oleic acid is an end product of olive oil hydrolysis mediated by lipase action. BSL activity was reduced by 80% when enzyme was preincubated in presence of oleic acid. The drastic reduction of BSL activity in presence of oleic acid might be due to the end product inhibition, a regulatory process in which the metabolite formed in downstream reactions inhibits the activities of upstream enzyme [41].

4. Conclusion

In the present study, a highly thermostable alkaline lipase from a desert isolate *Bacillus sonorensis* 4R was isolated and characterized. On the basis of results, BSL can be used as a potential candidate in various industrial and biotechnological sectors with special mention as additives in detergents and food industries, environmental bioremediations, and molecular biology. However, further works relating to improvement in enzyme yield and other kinetic aspects of enzyme activity are required to understand the catalytic properties of this enzyme. Statistical approach based optimization of BSL production and structural elucidation of lipase is in progress.

TABLE 2: Kinetic and thermodynamic parameters for thermal denaturation of BSL.

Temperature (°C)	$t_{1/2}$	K_d	ΔH {enthalpy (KJ/mol)}	ΔG {free energy (KJ/mol)}	ΔS {entropy (J/mol/K)}
80	150	$46 * 10^{-4}$	28.16	102.79	−0.211
90	121.59	$57 * 10^{-4}$	28.08	105.12	−0.212
100	90	$77 * 10^{-4}$	28.00	110.28	−0.220
110	70	$99 * 10^{-4}$	27.91	109.32	−0.212
120	50	$138 * 10^{-4}$	27.83	111.66	−0.211

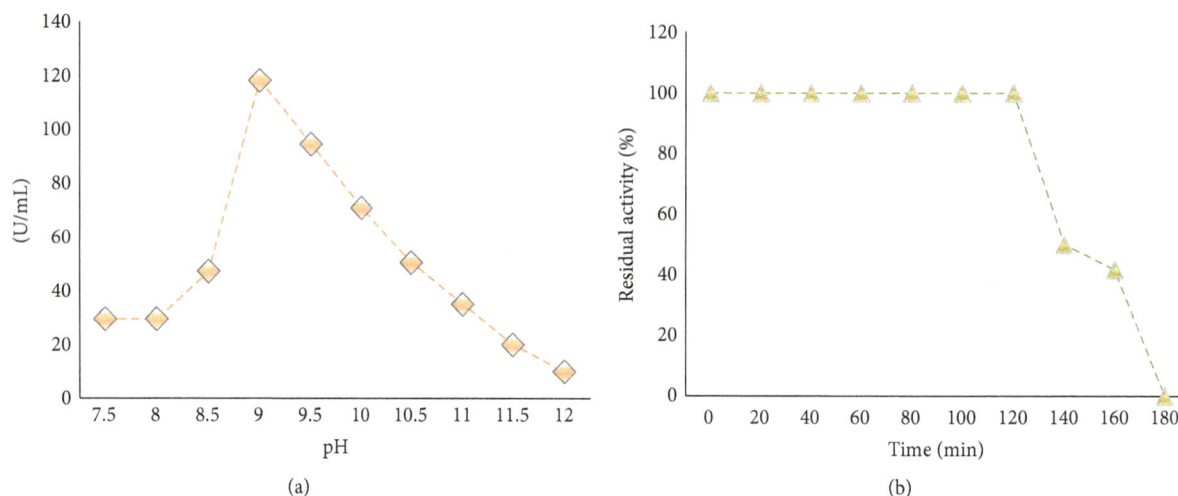

(a)

(b)

FIGURE 9: (a) Effect of pH on BSL activity at 80°C. BSL activity at pH 9.0 was set as 100%. (b) Stability of BSL at pH 9.0. BSL activity without preincubation was set as 100%.

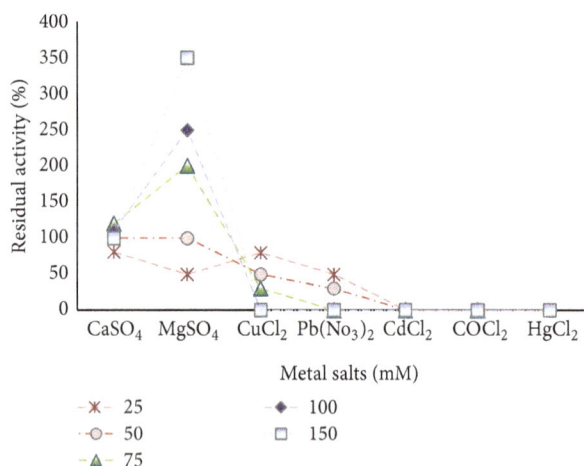

FIGURE 10: Effect of metal ions on BSL activity. BSL activity without amendment of metal ions was set as 100%.

TABLE 3: Effect of chemical modulators on BSL activity. Residual activity of BSL was determined by comparing activities before incubation of BSL in presence of modulators.

Chemical modulators	Residual activity (%)
EDTA	49.99
Urea	49.99
PMSF	16.67
Iodine	20
Orlistat	20
Oleic acid	20

(M.S., India), for providing support and necessary facilities to complete this research work.

Acknowledgment

The authors wish to acknowledge the School of Life Sciences, Swami Ramanand Teerth Marathwada University, Nanded

References

[1] C. Vieille and G. J. Zeikus, "Hyperthermophilic enzymes: sources, uses, and molecular mechanisms for thermostability," *Microbiology and Molecular Biology Reviews*, vol. 65, no. 1, pp. 1–43, 2001.

[2] C. Vieille, D. S. Burdette, and J. G. Zeikus, "Thermozymes," *Biotechnology Annual Review*, vol. 2, pp. 1–83, 1996.

[3] G. D. Haki and S. K. Rakshit, "Developments in industrially important thermostable enzymes: a review," *Bioresource Technology*, vol. 89, no. 1, pp. 17–34, 2003.

[4] D. A. Cowan and R. Fernandez-Lafuente, "Enhancing the functional properties of thermophilic enzymes by chemical modification and immobilization," *Enzyme and Microbial Technology*, vol. 49, no. 4, pp. 326–346, 2011.

[5] Dublin Business Wire and Research and Markets, "Lipase market by source, application and by geography—global forecast to brochure research and markets 2020," The World's Largest Market Research Store, 2015, http://www.researchandmarkets.com/reports/3388717.

[6] P. K. Ghosh, R. K. Saxena, R. Gupta, R. P. Yadav, and S. Davidson, "Microbial lipases: production and applications," *Science Progress*, vol. 79, pp. 119–157, 1996.

[7] R. K. Saxena, P. K. Ghosh, R. Gupta, W. S. Davidson, S. Bradoo, and R. Gulati, "Microbial lipases: potential biocatalysts for the future industry," *Current Science*, vol. 77, no. 1, pp. 101–115, 1999.

[8] C. Yao, Y. Cao, S. Wu, S. Li, and B. He, "An organic solvent and thermally stable lipase from *Burkholderia ambifaria* YCJ01: purification, characteristics and application for chiral resolution of mandelic acid," *Journal of Molecular Catalysis B: Enzymatic*, vol. 85-86, pp. 105–110, 2013.

[9] M. Masomian, R. N. Z. R. A. Rahman, A. B. Salleh, and M. Basri, "A new thermostable and organic solvent-tolerant lipase from *Aneurinibacillus thermoaerophilus* strain HZ," *Process Biochemistry*, vol. 48, no. 1, pp. 169–175, 2013.

[10] P. Rathi, S. Bradoo, R. K. Saxena, and R. Gupta, "A hyperthermostable, alkaline lipase from *Pseudomonas* sp. with the property of thermal activation," *Biotechnology Letters*, vol. 22, no. 6, pp. 495–498, 2000.

[11] K. Selvam, B. Vishnupriya, and S. C. Bose, "Screening and quantification of marine actinomycetes producing industrial enzymes amylase, cellulose, lipase from south coast of India," *International Journal of Pharmaceutical and Biological Archive*, vol. 2, pp. 1481–1485, 2010.

[12] M. M. Bradford, "A rapid and sensitive method for the quantitation of microgram quantities of protein utilizing the principle of protein-dye binding," *Analytical Biochemistry*, vol. 72, no. 1-2, pp. 248–254, 1976.

[13] S. A. Arrhenius, "Über die dissociationswärme und den einfluss der temperatur auf den dissociationsgrad der elektrolyte," *Zeitschrift für Physikalische Chemie*, vol. 4, pp. 96–116, 1889.

[14] S. N. Gummadi, "What is the role of thermodynamics on protein stability?" *Biotechnology and Bioprocess Engineering*, vol. 8, no. 1, pp. 9–18, 2003.

[15] M. M. Palmisano, L. K. Nakamura, K. E. Duncan, C. A. Istock, and F. M. Cohan, "*Bacillus sonorensis* sp. nov., a close relative of *Bacillus licheniformis*, isolated from soil in the Sonoran Desert, Arizona," *International Journal of Systematic and Evolutionary Microbiology*, vol. 51, no. 5, pp. 1671–1679, 2001.

[16] M. Nerurkar, M. Joshi, and R. Adivarekar, "Bioscouring of cotton using lipase from marine bacteria *Bacillus sonorensis*," *Applied Biochemistry and Biotechnology*, vol. 175, no. 1, pp. 253–265, 2015.

[17] T. Aanniz, M. Ouadghiri, M. Melloul et al., "Thermophilic bacteria in Moroccan hot springs, salt marshes and desert soils," *Brazilian Journal of Microbiology*, vol. 46, no. 2, pp. 443–453, 2015.

[18] N.-P. Hua, F. Kobayashi, Y. Iwasaka, G.-Y. Shi, and T. Naganuma, "Detailed identification of desert-originated bacteria carried by Asian dust storms to Japan," *Aerobiologia*, vol. 23, no. 4, pp. 291–298, 2007.

[19] E. D. Lester, M. Satomi, and A. Ponce, "Microflora of extreme arid Atacama Desert soils," *Soil Biology and Biochemistry*, vol. 39, no. 2, pp. 704–708, 2007.

[20] L. Chopra, G. Singh, K. K. Jena, H. Verma, and D. K. Sahoo, "Bioprocess development for the production of sonorensin by *Bacillus sonorensis* MT93 and its application as a food preservative," *Bioresource Technology*, vol. 175, pp. 358–366, 2015.

[21] U. Pandya and M. Saraf, "Isolation and identification of allelochemicals produced by *B. sonorensis* for suppression of charcoal rot of *Arachis hypogaea* L.," *Journal of Basic Microbiology*, vol. 55, no. 5, pp. 635–644, 2015.

[22] R. Tian, H. Chen, Z. Ni et al., "Expression and characterization of a novel thermo-alkalistable lipase from hyperthermophilic bacterium *Thermotoga maritima*," *Applied Biochemistry and Biotechnology*, vol. 176, no. 5, pp. 1482–1497, 2015.

[23] T. Wei, S. Feng, D. Mao, X. Yu, C. Du, and X. Wang, "Characterization of a new thermophilic and acid tolerant esterase from *Thermotoga maritima* capable of hydrolytic resolution of racemic ketoprofen ethyl ester," *Journal of Molecular Catalysis B: Enzymatic*, vol. 85-86, pp. 23–30, 2013.

[24] K. Chakraborty and R. P. Raj, "An extra-cellular alkaline metallolipase from *Bacillus licheniformis* MTCC 6824: purification and biochemical characterization," *Food Chemistry*, vol. 109, no. 4, pp. 727–736, 2008.

[25] F. M. Shariff, R. N. Z. R. A. Rahman, M. Basri, and A. B. Salleh, "A newly isolated thermostable lipase from *Bacillus* sp.," *International Journal of Molecular Sciences*, vol. 12, no. 5, pp. 2917–2934, 2011.

[26] T. H. T. A. Hamid, M. A. Eltaweel, R. N. Z. R. A. Rahman, M. Basri, and A. B. Salleh, "Characterization and solvent stable features of Strep-tagged purified recombinant lipase from thermostable and solvent tolerant *Bacillus* sp. strain 42," *Annals of Microbiology*, vol. 59, no. 1, pp. 111–118, 2009.

[27] J. Zhang, J. F. Liu, J. Zhou, Y. Y. Ren, X. Y. Dai, and H. Xiang, "Thermostable esterase from *Thermoanaerobacter tengcongensis*: high-level expression, purification and characterization," *Biotechnology Letters*, vol. 25, no. 17, pp. 1463–1467, 2003.

[28] H. Klump, J. Di-Ruggiero, M. Kessel, J. B. Park, M. W. W. Adams, and F. T. Robb, "Glutamate dehydrogenase from the hyperthermophile *Pyrococcus furiosus*," *The Journal of Biological Chemistry*, vol. 267, pp. 22681–22685, 1992.

[29] K.-E. Jaeger, S. Ransac, B. W. Dijkstra, C. Colson, M. van Heuvel, and O. Misset, "Bacterial lipases," *FEMS Microbiology Reviews*, vol. 15, no. 1, pp. 29–63, 1994.

[30] L. Cui, G. Du, D. Zhang, and J. Chen, "Thermal stability and conformational changes of transglutaminase from a newly isolated *Streptomyces hygroscopicus*," *Bioresource Technology*, vol. 99, no. 9, pp. 3794–3800, 2008.

[31] S. P. George, A. Ahmad, and M. B. Rao, "A novel thermostable xylanase from *Thermomonospora* sp.: influence of additives on thermostability," *Bioresource Technology*, vol. 78, no. 3, pp. 221–224, 2001.

[32] S. A. Costa, T. Tzanov, A. F. Carneiro, A. Paar, G. M. Gübitz, and A. Cavaco-Paulo, "Studies of stabilization of native catalase using additives," *Enzyme and Microbial Technology*, vol. 30, no. 3, pp. 387–391, 2002.

[33] A. Cobos and P. Estrada, "Effect of polyhydroxylic cosolvents on the thermostability and activity of xylanase from *Trichoderma reesei* QM 9414," *Enzyme and Microbial Technology*, vol. 33, no. 6, pp. 810–818, 2003.

[34] W. Bankeeree, P. Lotrakul, S. Prasongsuk et al., "Effect of polyols on thermostability of xylanase from a tropical isolate of *Aureobasidium pullulans* and its application in prebleaching of rice straw pulp," *SpringerPlus*, vol. 3, article 37, 2014.

[35] D. Fu, C. Li, J. Lu, A. U. Rahman, and T. Tan, "Relationship between thermal inactivation and conformational change of *Yarrowia lipolytica* lipase and the effect of additives on enzyme stability," *Journal of Molecular Catalysis B: Enzymatic*, vol. 66, no. 1-2, pp. 136–141, 2010.

[36] H. N. Bhatti and F. Amin, "Kinetic and hydrolytic characterization of newly isolated alkaline lipase from ganoderma lucidum using canola oil cake as substrate," *Journal of the Chemical Society of Pakistan*, vol. 35, no. 3, pp. 585–592, 2013.

[37] H. Dong, S. Gao, S.-P. Han, and S.-G. Cao, "Purification and characterization of a *Pseudomonas* sp. lipase and its properties in non-aqueous media," *Biotechnology and Applied Biochemistry*, vol. 30, no. 3, pp. 251–256, 1999.

[38] J. Ma, Z. Zhang, B. Wang et al., "Overexpression and characterization of a lipase from *Bacillus subtilis*," *Protein Expression and Purification*, vol. 45, no. 1, pp. 22–29, 2006.

[39] A. Sugihara, T. Tani, and Y. Tominaga, "Purification and characterization of a novel thermostable lipase from *Bacillus* sp.," *Journal of Biochemistry*, vol. 109, no. 2, pp. 211–216, 1991.

[40] G. T. James, "Inactivation of the protease inhibitor phenylmethylsulfonyl fluoride in buffers," *Analytical Biochemistry*, vol. 86, no. 2, pp. 574–579, 1978.

[41] R. M. Denton and C. L. Pogson, *Metabolic Regulation*, Chapmon and Hall, 1976.

11

Immunomodulatory Effects of Chitotriosidase Enzyme

Mohamed A. Elmonem,[1,2] Lambertus P. van den Heuvel,[1,3] and Elena N. Levtchenko[1]

[1]*Department of Pediatric Nephrology & Growth and Regeneration, University Hospitals Leuven, KU Leuven, UZ Herestraat 49, P.O. Box 817, 3000 Leuven, Belgium*
[2]*Department of Clinical and Chemical Pathology, Inherited Metabolic Disease Laboratory, Center of Social and Preventive Medicine, Faculty of Medicine, Cairo University, 2 Ali Pasha Ibrahim Street, Room 409, Monira, P.O. Box 11628, Cairo, Egypt*
[3]*Department of Pediatric Nephrology, Radboud University Medical Center, Post 804, Postbus 9101, 6500 HB Nijmegen, Netherlands*

Correspondence should be addressed to Mohamed A. Elmonem; mohamed.abdelmonem@kasralainy.edu.eg

Academic Editor: Qi-Zhuang Ye

Chitotriosidase enzyme (EC: 3.2.1.14) is the major active chitinase in the human body. It is produced mainly by activated macrophages, in which its expression is regulated by multiple intrinsic and extrinsic signals. Chitotriosidase was confirmed as essential element in the innate immunity against chitin containing organisms such as fungi and protozoa; however, its immunomodulatory effects extend far beyond innate immunity. In the current review, we will try to explore the expanding spectrum of immunological roles played by chitotriosidase enzyme in human health and disease and will discuss its up-to-date clinical value.

1. Introduction

Chitotriosidase enzyme (CHIT1, EC: 3.2.1.14), belonging to the family of 18 glycosyl hydrolases, was the first active chitinase to be discovered in human plasma [1]. Its natural substrate chitin is the second most abundant polysaccharide in nature after cellulose. Chitin is the linear polymer of N-acetylglucosamine and the main component of the cell walls of fungi and protozoa, egg shells of helminthes, and the exoskeletons of arthropods and insects; however, it is completely absent in mammals [2]. Several protein members of the same family were later detected in human plasma and tissues including the enzymatically active acidic mammalian chitinase (AMCase) [3] and chi-lectins, those having a chitin binding domain with no catalytic activity such as chitinase-3 like-1 protein (CHI3L1 or YKL-40), chitinase-3 like-2 protein (CHI3L2 or YKL-39), and oviductal glycoprotein-1 (OVGP1) [4].

In man, chitotriosidase is mainly expressed by different lineages of activated blood and tissue macrophages [5–10] and to a lesser extent by polymorphonuclear leucocytes [11]. The absence of its substrate chitin in the human body

and the exclusive production by immunologically active cells immediately elicited the investigation of chitotriosidase involvement in the innate immunity against chitin coated pathogens [12]. Chitotriosidase was confirmed as an essential factor for the defense against many such organisms as *Plasmodium falciparum* [13], *Wuchereria bancrofti* [14], *Candida albicans* [15], *Madurella mycetomatis* [16], and *Cryptococcus neoformans* [17].

The role of chitotriosidase enzyme inside macrophages is not limited to its chitinolytic activity against the engulfed chitin containing organisms, or even to innate immunity. It has been implicated in the activation and polarization cascades of macrophages, as well as the indirect activation of other immune cells such as T helper cells and eosinophils [17–19]. Recent studies are interested in its immunomodulatory effects through the processes of chitin recognition, antigen presentation, induction of cell mediated immunity and synergistic effects with proteases, and other enzymes to kill different types of pathogens and cancer cells [10, 17–20]. On the other hand, chitotriosidase has been implicated in the pathogenesis of many human diseases through the improper induction of inflammation and faulty tissue remodeling such

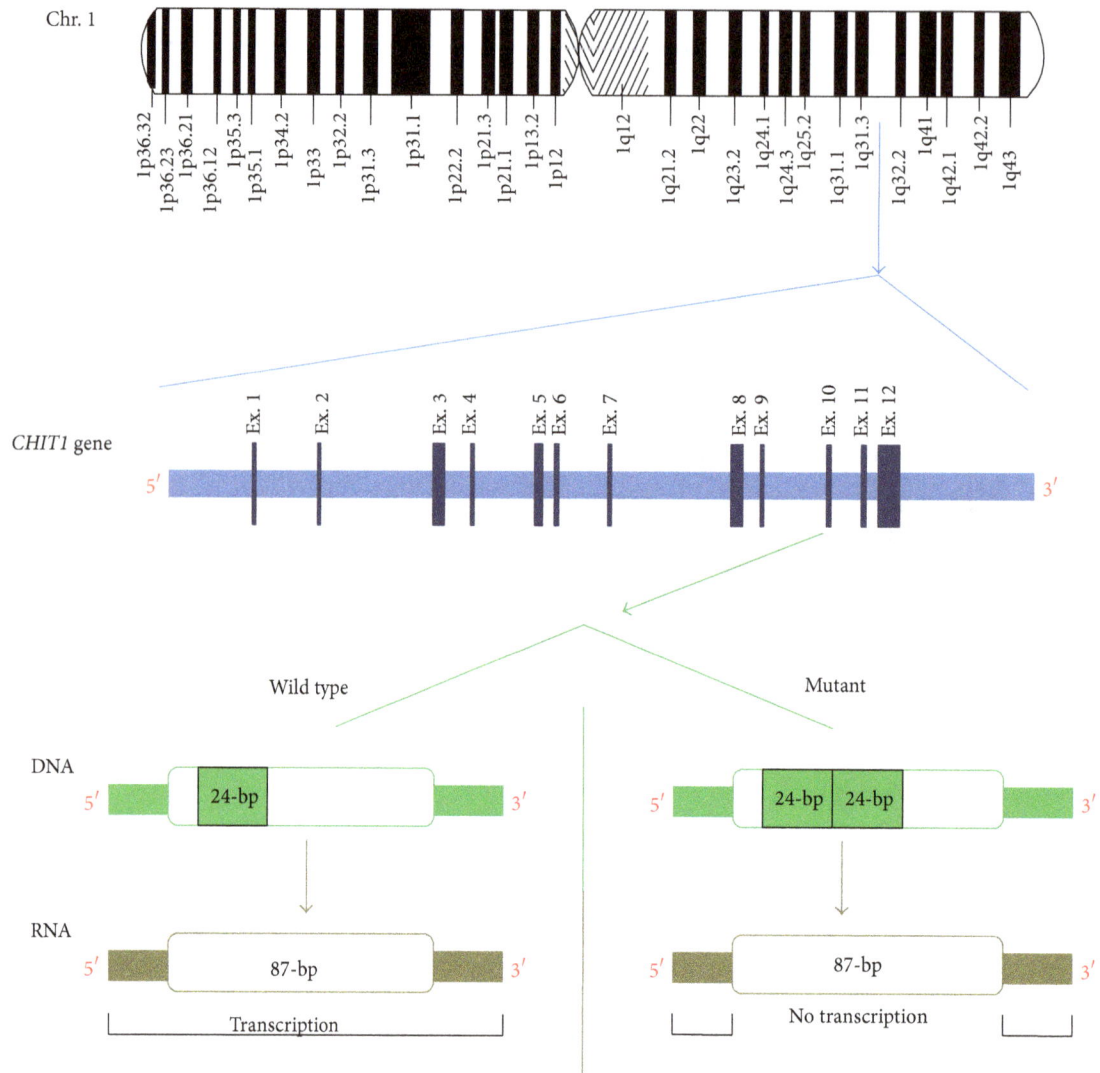

FIGURE 1: Schematic representation of human chromosome 1, *CHIT1* gene on chromosome 1q32 locus, and the common 24-bp duplication mutation at exon 10 of the *CHIT1* gene.

as bronchial asthma, chronic obstructive pulmonary disease (COPD), nonalcoholic fatty liver disease, and neurodegenerative disorders like Alzheimer's disease and amyotrophic lateral sclerosis [19, 21–24].

In the current review, we will provide a summary of basic information about the enzyme and we will discuss its immunomodulatory effects in humans over both innate and acquired immunity together with its current and possible future clinical applications.

2. Genetics

All the genes encoding active chitinases and chitinase-like proteins are clustered in two loci on the human chromosome 1 (1p13 and 1q32). This genetic clustering displays a high degree of conservation among different mammals indicating an evolutionary relationship through common ancestral gene duplication events [4].

The human chitotriosidase gene *(CHIT1)*, extending over 20 kilobases at locus 1q32, consists of 12 exons translating a 466-amino acid protein [5, 25]. A common 24-bp duplication mutation in exon 10 of the *CHIT1* gene (Figure 1), leading to alternate splicing and in-frame deletion of 87 nucleotides (29 amino acids), is responsible for almost all detected enzyme deficiencies in different populations when homozygously mutated [25, 26]. Although the mutated 24-bp duplication allele is fairly common in Caucasian individuals (4–6% homozygous and 30–50% heterozygous), it is extremely rare in Sub-Saharan African individuals (0–2% only heterozygous), suggesting the evolutionary advantage of keeping the wild type enzyme in areas with high degrees of endemic parasitic and fungal disease loads [26]. Interestingly, the 24-bp duplication mutation is much more common in the Far East in populations of Japanese, Chinese, and Korean ancestries as almost 30% of them are homozygous, while 50% are heterozygous for the mutation [27, 28].

Other relatively common functional mutations have been also reported in the *CHIT1* gene like G102S, G354R, and A442V; however, when homozygously mutated, they are only associated with mild to moderate decrease in enzyme activity [27].

3. Chemistry and Modes of Action

Chitotriosidase enzyme has two major forms (a 50-kilodalton form dominant in blood stream and a 39-kilodalton form dominant in tissues), both having equal chitinolytic activities. The 50-kilodalton protein (466 amino acids) is initially produced by macrophages containing the 39-kilodalton N-terminus having the catalytic domain and the C-terminus having the chitin-binding domain connected together by a short hinge region. The 39-kilodalton protein (387 amino acids) is either cleaved post-translationally in the lysosome of macrophages or less commonly formed through differential RNA processing [25].

The enzyme was initially thought to be an exochitinase because it can hydrolyze chitotriose residues and was termed chitotriosidase based on this observation. However, recent structural and binding modes studies revealed the enzyme to be more of an endochitinase rather than an exochitinase [29, 30]. Furthermore, the strong binding affinity of chitotriosidase to its substrate is also responsible for the relatively high transglycosylation activity of the enzyme even in the absence of excess substrate concentrations, making chitotriosidase a complete independent chitinolytic machinery, and this is in accordance with its anticipated physiological role as a potent immunological defense weapon against microorganisms containing chitin [31, 32].

4. Stimulatory Signals

Although chitotriosidase enzyme is relatively recently linked to human pathology, many aspects of its intracellular and extracellular mechanistic actions, effector and affector molecules, and diseases influenced by the increase or decrease of its expression have been investigated. After its initial association with innate immunity against chitin coated pathogens it was rapidly identified as one of the major protein products of activated macrophages and hence an important nonspecific marker of macrophage activation [12]. Chitotriosidase activity was several hundred-fold elevated in the plasma of patients with the inflammatory based lysosomal storage disorder Gaucher's disease [1], in which macrophages play an essential role in the clearance of the disease sphingolipid storage material. The lipid laden macrophage becomes the disease histopathological pathognomonic cell in the bone marrow and tissues or what is known as the Gaucher cell. Furthermore, chitotriosidase is significantly elevated in some infections caused by bacterial and viral pathogens lacking its natural substrate chitin [33, 34].

Several intracellular pathways have been proposed to explain the stimulatory molecules and the activation cascades of chitotriosidase enzyme inside human macrophages. Figure 2 provides a simple schematic representation of a macrophage with different proposed stimulatory signals for chitotriosidase expression as well as some of the main intra- and intercellular activities of the enzyme.

Chitin naturally is a potent activator of chitotriosidase expression, either phagocytized by the macrophage in the cell walls of different fungi and protozoa or directly introduced to macrophages through an unidentified receptor [35, 36]. Chitin and its small hydrolyzed particles when cocultured with macrophages can stimulate the production of TNF-α and IFN-γ [37] which can both increase the expression and activity of chitotriosidase enzyme inside macrophages [38]. Lipopolysaccharide (LPS) which is an important component of the bacterial cell wall was also confirmed as a potent stimulant of chitotriosidase transcription mainly through the NF-κB signaling pathway [38, 39], which is also responsible for exerting the effects of TNF-α; however, IFN-γ most probably produces its effect through stimulating the Jak-Stat signaling pathway [40]. Another possible mechanism of chitotriosidase activation by bacteria is through the bacterial peptidoglycan product muramyl dipeptide (MDP), activating the NOD2 signaling pathway which is also implicated in the expression of chitotriosidase enzyme inside macrophages [41, 42].

Prolactin hormone which is structurally related to many human cytokines and is involved in the regulation of monocyte/macrophage functions was also shown to increase macrophage chitotriosidase production [43]. Through studying signal pathway inhibitors, prolactin was shown to stimulate chitotriosidase expression through multiple signaling pathways including the mitogen activated protein kinase (MAPK), PI3 kinase (PI3K/Akt), and the protein tyrosine kinase (PTK) pathways [44].

The inflammasome system is also expected to play some role in the activation of chitotriosidase expression as both ingested bacterial MDP and cystine crystals can stimulate macrophages through the NLRP3 inflammasome system [45, 46], leading eventually to the production of IL-1β, which either can stimulate the NF-κB signaling pathway directly or indirectly induces the production of TNF-α [47], thus stimulating the expression of chitotriosidase.

Another efficient way to activate macrophages and induce chitotriosidase expression is through the paracrine effect of natural killer cells (NK cells), which when exposed to cells infected with viral, bacterial, or fungal pathogens or even neoplastic cells can produce large amounts of INF-γ and TNF-α [36, 48] in the vicinity of macrophages, both increasing the expression and release of chitotriosidase.

5. Immunological Effects and Clinical Perspectives

Chitotriosidase enzyme expression increases exponentially during the normal monocyte to macrophage maturation process showing a peak of expression between the 5th day and the 7th day of culture [18] and is recently detected to be expressed in both macrophage polarization forms (M1 and M2). M1 macrophages, or classically activated macrophages, are mainly directed to promote inflammation, kill the invading pathogens. and stimulate tissue fibrosis following injury. On the other hand, M2 macrophages, or alternatively

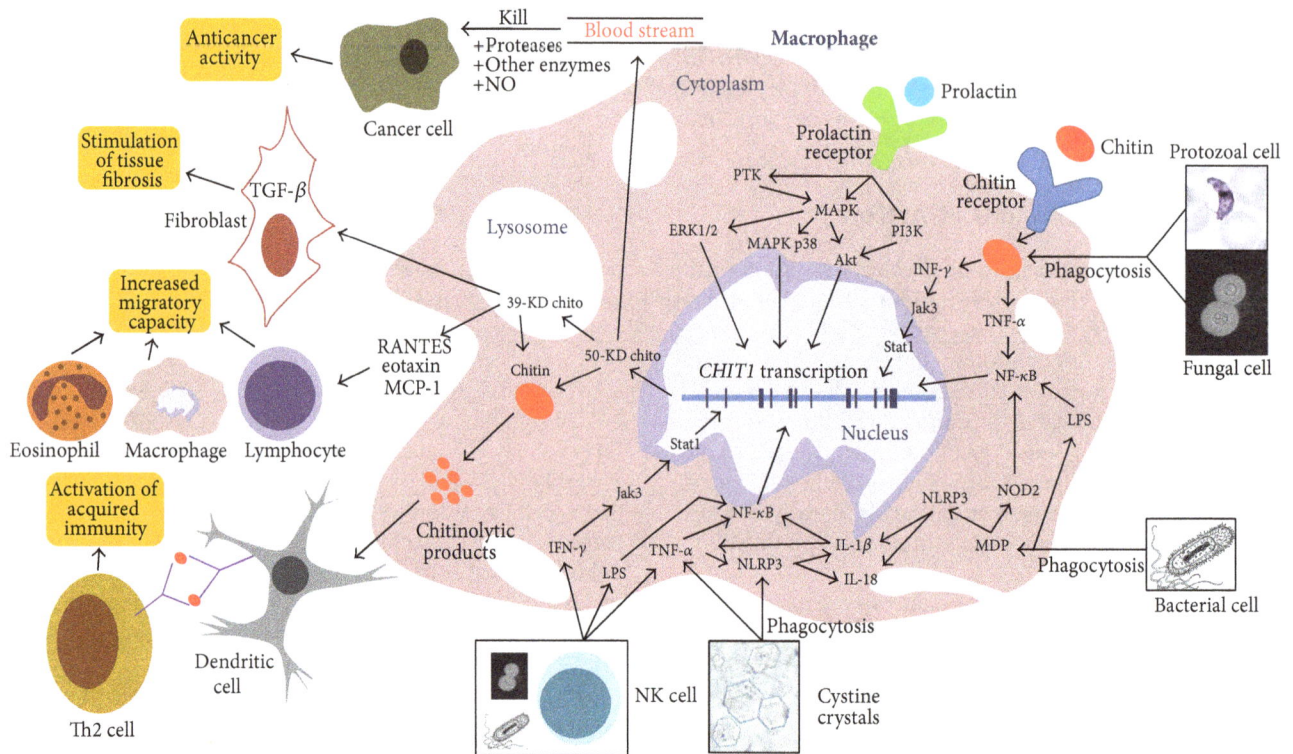

FIGURE 2: Schematic representation of a human macrophage showing different stimuli leading to the increased expression and release of chitotriosidase enzyme. An elaborate description of the processes implicated in increased chitotriosidase expression, as well as its immunological effects, is provided in the text. 39-KD chito: chitotriosidase (39 kilodalton protein); 50KD-chito: chitotriosidase (50 kilodalton protein); ERK1/2: extracellular signal regulated kinases 1/2; IL-1β: interleukin-1β; IL-18: interleukin-18; INF-γ: interferon-gamma; Jak: Janus kinase; LPS: lipopolysaccharide; MAPK: mitogen activated protein kinase; MCP-1: monocyte chemotactic protein-1; MDP: muramyl dipeptide; NF-κB: nuclear factor-kB; NK cell: natural killer cell; NLRP3: NOD-like receptor family, pyrin domain containing 3; NO: nitric oxide; NOD2: nucleotide-binding oligomerization domain-containing protein 2; PI3K: PI3 kinase; PTK: protein tyrosine kinase; RANTES: regulated on activation, normal T cell expressed and secreted; Stat; signal transducers and activators of transcription; TGF-β: transforming growth factor-beta; Th2 cell: T helper type 2 cell; TNF-α: tumor necrosis factor-alpha.

activated macrophages, provide regulatory signals to protect the host from an exaggerated inflammatory response and promote tissue remodeling and healing [49]. The fact that chitotriosidase is expressed almost equally in both forms denotes its regulatory roles over processes far beyond the hydrolysis of chitin in pathogens. Further studies are still needed to clarify the role of chitotriosidase enzyme in the alternatively activated M2 macrophages.

Dendritic cells are the most important antigen presenting cells in the human body and one of the members of the monocyte/macrophage lineage. Although AMCase is widely expressed in different tissues, especially in the epithelial cells of the gastrointestinal tract and lungs [3], chitotriosidase was the only implicated active chitinase in the process of chitin recognition and antigen presentation [17]. Recent evidence suggests that the induction of human T helper 2 cells (Th2) in response to pulmonary cryptococcal infection is totally dependent on chitin cleavage by chitotriosidase and that CD11b+ conventional dendritic cells act as antigen presenting cells for the specifically fragmented chitin products [17]. Furthermore, chitotriosidase mRNA and protein concentrations were significantly elevated in mature dendritic cells without

chitin sensitization as compared to immature dendritic cells [10], implying that chitotriosidase might be also playing a role in the process of antigen presentation regardless of the presence of chitin.

Recently, activated macrophages and chitotriosidase elevations were implicated in the pathogenesis of nephropathic cystinosis, another lysosomal storage disorder characterized by cystine crystal accumulation inside macrophages in different body organs. Cystine crystals in vitro when incubated with human monocyte derived macrophages were able to activate macrophages in a concentration dependent manner evidenced by the increased concentrations of TNF-α and the concomitant activities of chitotriosidase enzyme in culture supernatant and in cell homogenate. Furthermore, plasma chitotriosidase activities in cystinotic patients correlated positively with leucocytes cystine concentrations, making it a potential target for the disease therapeutic monitoring [50]. Cystinosis was the first crystal based disease with the confirmed involvement of chitotriosidase enzyme in its pathogenesis, making it an interesting target to investigate in other more common crystal related disorders such as gout and hyperoxaluria.

TABLE 1: Human diseases associated with elevated chitotriosidase enzyme.

Disease group	Disease	Proposed clinical value (sample type)	References
Lysosomal storage diseases	Gaucher	Screening, therapeutic monitoring (P/S)	[1, 62]
	Niemann-Pick A/B and C	Screening, therapeutic monitoring (P/S)	[65]
	Cystinosis	Therapeutic monitoring (P/S)	[50]
	Fabry	Therapeutic monitoring (P/S)	[66]
	Krabbe	Screening, marker of severity (P/S)	[67]
	Wolman	Therapeutic monitoring (P/S)	[68]
	Farber	Screening (P/S)	[69]
	GM1	Screening (P/S)	[70]
	Sialidosis type II	Screening (P/S)	[71]
Infectious diseases	Systemic fungal infections: *Candida albicans, Madurella mycetomatis, and Cryptococcus neoformans*	Prognosis, therapeutic monitoring (P/S)	[15–17]
	Malaria	Prognosis (P/S)	[13]
	Filariasis	Screening (P/S)	[14]
	Tuberculosis	Prognosis, therapeutic monitoring (P/S)	[72]
	Brucellosis	Therapeutic monitoring (P/S)	[73]
	Leprosy	Prognosis, therapeutic monitoring (P/S)	[74]
	Crimean-Congo hemorrhagic fever	Prognosis (P/S)	[34]
Respiratory diseases	Asthma	Marker of severity (P/S)	[19, 21]
	COPD	Marker of severity (P/S, BAL)	[21, 51]
	Interstitial lung disease	Screening, marker of severity (BAL)	[52, 75]
Endocrinological diseases	Diabetes	Marker of endothelial damage (P/S)	[76]
		Marker of nephropathy progression (P/S)	[53]
Cardiovascular diseases	Atherosclerosis	Marker of severity (P/S)	[77, 78]
	Stroke	Prognosis (P/S)	[79]
	Coronary artery disease	Prognosis (P/S)	[80]
	Erectile dysfunction	Marker of severity (P/S)	[81]
Neurological diseases	Amyotrophic lateral sclerosis	Screening, marker of severity (P/S, CSF)	[24, 82]
	Alzheimer's disease	Prognosis, marker of severity (CSF)	[23, 83]
	Cerebral adrenoleukodystrophy	Prognosis (P/S, CSF)	[84]
	Neuromyelitis optica	Screening (CSF)	[50]
	Multiple sclerosis	Screening, prognosis (CSF)	[50]
Gynecological and obstetrical diseases	PCOS	Prognosis (P/S)	[85]
	Endometriosis	Marker of severity (P/S)	[86]
	Preeclampsia	Marker of fetal compromise (UC)	[87]
Miscellaneous	NAFLD	Marker of severity (P/S)	[22]
	FMF	Screening, marker of severity (P/S)	[88]
	β-Thalassemia	Marker of severity, therapeutic monitoring (P/S)	[89]
	Sarcoidosis	Marker of severity, therapeutic monitoring (P/S)	[63]
	Acute appendicitis	Screening (P/S)	[90]
	Juvenile idiopathic arthritis	Screening, marker of severity (SV)	[91]
	Prostate cancer	Prognosis (P/S)	[92]

BAL: bronchoalveolar lavage; COPD: chronic obstructive pulmonary disease; CSF: cerebrospinal fluid; FMF: familial Mediterranean fever; GM1: gangliosidosis M1; NAFLD: nonalcoholic fatty liver disease; PCOS: polycystic ovarian syndrome; P/S: plasma or serum; SV: synovial fluid; UC: umbilical cord blood.

Chitotriosidase also mediates many inflammatory processes through the direct stimulation of different inflammatory mediators such as IL-8, MMP9 (collagenase type IV), MCP-1 (CCL2), RANTES (CCL5), and eotaxin (CCL11), thus increasing the migratory capacity of many immunological cells including T lymphocytes, macrophages, and eosinophils [51, 52]. Levels of chitotriosidase activities strongly correlated with the concentrations of IL-1β and TNF-α in the bronchoalveolar lavage (BAL) of COPD patients supporting the hypothesis of a mutual regulation cascade in the production of these inflammatory mediators [52]. Furthermore, chitotriosidase was involved in the induction of fibrosis in the murine model of interstitial lung disease as the bleomycin-induced pulmonary fibrosis was significantly reduced in $Chit1^{-/-}$ mice and significantly enhanced in lungs from $Chit1$ overexpressing transgenic mice. This effect is explained by the activation of fibroblasts through enhancing TGF-β and increasing the expression of TGF-β receptors 1 and 2 leading to the activation of the Smad and MAPK/ERK signaling pathways [53]. Similar effects have been clinically observed with other human diseases characterized by faulty tissue remodeling and abnormal healing such as nonalcoholic fatty liver disease, bronchial asthma, and diabetic nephropathy in which chitotriosidase plasma activities strongly correlated with disease progression and/or the degree of tissue fibrosis [21, 22, 54]. Targeting the chitotriosidase activation cascade through the administration of specific antibodies or pan-chitinase inhibitors significantly ameliorates inflammation and fibrosis in several animal models of autoimmune diseases, most likely by suppressing the chitotriosidase dependent release of different cytokines and chemokines [55]; however, the safety of this approach in humans is not yet determined as it might increase the susceptibility to fungal and protozoal infections.

Chitinases and chi-lectins could play a detrimental role in human cancer development, especially CHI3L1 (YKL-40) which has been associated with increased tumor angiogenesis and bad prognosis in many human neoplasms such as breast, lung, and cervical cancers [56–58]. On the other hand, chitinases are also believed to have some anticancer cell activities [59]. Macrophages were always considered as a primary defense line against neoplastic cells, but the exact mechanisms beyond this action were not very clear [60]. Speculations were made about a combined effect of released NO and H_2O_2; however, there was no much evidence to support this hypothesis [20]. Recently, bacterial and human chitinases were both confirmed as having strong synergistic effects with protease enzymes produced by macrophages to dissolve mucin [61]. This mucolytic activity selectively attacked the altered mucin in the cancer cell wall of animal models and not the healthy cell mucin [59]. It is too early to speculate about the therapeutic applications of this observation as many explanatory mechanistic studies are needed to determine the specificity and the exact molecular and chemical targets of this process.

Chitotriosidase is currently an established or a candidate screening marker, severity marker, and/or therapeutic monitor for over 40 different diseases, inherited and acquired. Table 1 provides a summary of human diseases clinically associated with macrophage activation and chitotriosidase enzyme production. Being a nonspecific marker of macrophage activation and deficient in about 6% of normal population clearly limits chitotriosidase usability as a screening marker for many diseases; however, in established diagnosed nondeficient patients, chitotriosidase is an excellent marker to monitor compliance and response to treatment [62, 63], especially when the treatment targets inflammatory pathways. And in some diseases its marked elevations made it also a quite beneficial screening marker as in Gaucher's disease even when other lysosomal storage disorders are suspected [64].

6. Conclusions

Being a sensitive biomarker of macrophage activation, chitotriosidase activity is currently more and more commonly used in clinical practice to evaluate the status and response to treatment of inflammatory based diseases in which macrophages play a significant role. The interesting mixture of harmful and beneficial immunological effects made human chitinases and chitotriosidase enzyme an intriguing point of research. If proven safe in humans, the development of targeted therapies either to suppress chitotriosidase activity in autoimmune and inflammatory disorders, or to specifically enhance its targeted activity to kill cancer cells or to potentiate immunity against certain infections will not be far in the future.

References

[1] C. E. M. Hollak, S. van Weely, M. H. J. van Oers, and J. M. F. G. Aerts, "Marked elevation of plasma chitotriosidase activity. A novel hallmark of Gaucher disease," *The Journal of Clinical Investigation*, vol. 93, no. 3, pp. 1288–1292, 1994.

[2] R. N. Tharanathan and F. S. Kittur, "Chitin—the undisputed biomolecule of great potential," *Critical Reviews in Food Science and Nutrition*, vol. 43, no. 1, pp. 61–87, 2003.

[3] R. G. Boot, E. F. C. Blommaart, E. Swart et al., "Identification of a novel acidic mammalian chitinase distinct from chitotriosidase," *The Journal of Biological Chemistry*, vol. 276, no. 9, pp. 6770–6778, 2001.

[4] A. P. Bussink, D. Speijer, J. M. F. G. Aerts, and R. G. Boot, "Evolution of mammalian chitinase(-like) members of family 18 glycosyl hydrolases," *Genetics*, vol. 177, no. 2, pp. 959–970, 2007.

[5] R. G. Boot, G. H. Renkema, A. Strijland, A. J. Van Zonneveld, and J. M. F. G. Aerts, "Cloning of a cDNA encoding chitotriosidase, a human chitinase produced by macrophages," *The Journal of Biological Chemistry*, vol. 270, no. 44, pp. 26252–26256, 1995.

[6] M. A. Seibold, S. Donnelly, M. Solon et al., "Chitotriosidase is the primary active chitinase in the human lung and is modulated by genotype and smoking habit," *Journal of Allergy and Clinical Immunology*, vol. 122, no. 5, pp. 944–950, 2008.

[7] L. Malaguarnera, M. Di Rosa, A. M. Zambito, N. dell'Ombra, F. Nicoletti, and M. Malaguarnera, "Chitotriosidase gene expression in Kupffer cells from patients with non-alcoholic fatty liver disease," *Gut*, vol. 55, no. 9, pp. 1313–1320, 2006.

[8] R. G. Boot, T. A. E. van Achterberg, B. E. van Aken et al., "Strong induction of members of the chitinase family of proteins in atherosclerosis: chitotriosidase and human cartilage gp-39 expressed in lesion macrophages," *Arteriosclerosis, Thrombosis, and Vascular Biology*, vol. 19, no. 3, pp. 687–694, 1999.

[9] M. Di Rosa, D. Tibullo, M. Vecchio et al., "Determination of chitinases family during osteoclastogenesis," *Bone*, vol. 61, pp. 55–63, 2014.

[10] M. Di Rosa, D. Tibullo, D. Cambria et al., "Chitotriosidase expression during monocyte-derived dendritic cells differentiation and maturation," *Inflammation*, vol. 38, no. 6, pp. 2082–2091, 2015.

[11] L. Bouzas, J. C. Guinarte, and J. C. Tutor, "Chitotriosidase activity in plasma and mononuclear and polymorphonuclear leukocyte populations," *Journal of Clinical Laboratory Analysis*, vol. 17, no. 6, pp. 271–275, 2003.

[12] M. van Eijk, C. P. A. A. van Roomen, G. H. Renkema et al., "Characterization of human phagocyte-derived chitotriosidase, a component of innate immunity," *International Immunology*, vol. 17, no. 11, pp. 1505–1512, 2005.

[13] R. Barone, J. Simporé, L. Malaguarnera, S. Pignatelli, and S. Musumeci, "Plasma chitotriosidase activity in acute *Plasmodium falciparum* malaria," *Clinica Chimica Acta*, vol. 331, no. 1-2, pp. 79–85, 2003.

[14] E. H. Choi, P. A. Zimmerman, C. B. Foster et al., "Genetic polymorphisms in molecules of innate immunity and susceptibility to infection with Wuchereria bancrofti in South India," *Genes and Immunity*, vol. 2, no. 5, pp. 248–253, 2001.

[15] M. Vandevenne, V. Campisi, A. Freichels et al., "Comparative functional analysis of the human macrophage chitotriosidase," *Protein Science*, vol. 20, no. 8, pp. 1451–1463, 2011.

[16] P. E. Verwer, C. C. Notenboom, K. Eadie et al., "A polymorphism in the chitotriosidase gene associated with risk of mycetoma due to *Madurella mycetomatis* mycetoma: a retrospective study," *PLOS Neglected Tropical Diseases*, vol. 9, no. 9, Article ID e0004061, 2015.

[17] D. L. Wiesner, C. A. Specht, C. K. Lee et al., "Chitin recognition via chitotriosidase promotes pathologic type-2 helper T cell responses to cryptococcal infection," *PLoS Pathogens*, vol. 11, no. 3, Article ID e1004701, 2015.

[18] M. Di Rosa, G. Malaguarnera, C. De Gregorio, F. Drago, and L. Malaguarnera, "Evaluation of CHI3L-1 and CHIT-1 expression in differentiated and polarized macrophages," *Inflammation*, vol. 36, no. 2, pp. 482–492, 2013.

[19] K. W. Kim, J. Park, J. H. Lee et al., "Association of genetic variation in chitotriosidase with atopy in Korean children," *Annals of Allergy, Asthma and Immunology*, vol. 110, no. 6, pp. 444–449, 2013.

[20] X.-Q. Pan, "The mechanism of the anticancer function of M1 macrophages and their use in the clinic," *Chinese Journal of Cancer*, vol. 31, no. 12, pp. 557–563, 2012.

[21] A. J. James, L. E. Reinius, M. Verhoek et al., "The chitinase proteins YKL-40 and chitotriosidase are increased in both asthma and COPD," *American Journal of Respiratory and Critical Care Medicine*, 2015.

[22] M. Di Rosa, K. Mangano, C. De Gregorio, F. Nicoletti, and L. Malaguarnera, "Association of chitotriosidase genotype with the development of non-alcoholic fatty liver disease," *Hepatology Research*, vol. 43, no. 3, pp. 267–275, 2013.

[23] C. Rosén, C. H. Andersson, U. Andreasson et al., "Increased levels of chitotriosidase and YKL-40 in cerebrospinal fluid from patients with Alzheimer's disease," *Dementia and Geriatric Cognitive Disorders Extra*, vol. 4, no. 2, pp. 297–304, 2014.

[24] A. M. Varghese, A. Sharma, P. Mishra et al., "Chitotriosidase—a putative biomarker for sporadic amyotrophic lateral sclerosis," *Clinical Proteomics*, vol. 10, no. 1, article 19, 2013.

[25] R. G. Boot, G. H. Renkema, M. Verhoek et al., "The human chitotriosidase gene. Nature of inherited enzyme deficiency," *The Journal of Biological Chemistry*, vol. 273, no. 40, pp. 25680–25685, 1998.

[26] L. Malaguarnera, J. Simporè, D. A. Prodi et al., "A 24-bp duplication in exon 10 of human chitotriosidase gene from the sub-Saharan to the Mediterranean area: role of parasitic diseases and environmental conditions," *Genes and Immunity*, vol. 4, no. 8, pp. 570–574, 2003.

[27] P. Lee, J. Waalen, K. Crain, A. Smargon, and E. Beutler, "Human chitotriosidase polymorphisms G354R and A442V associated with reduced enzyme activity," *Blood Cells, Molecules, and Diseases*, vol. 39, no. 3, pp. 353–360, 2007.

[28] K. H. Woo, B. H. Lee, S. H. Heo et al., "Allele frequency of a 24 bp duplication in exon 10 of the CHIT1 gene in the general Korean population and in Korean patients with Gaucher disease," *Journal of Human Genetics*, vol. 59, no. 5, pp. 276–279, 2014.

[29] F. Fusetti, H. von Moeller, D. Houston et al., "Structure of human chitotriosidase. Implications for specific inhibitor design and function of mammalian chitinase-like lectins," *Journal of Biological Chemistry*, vol. 277, no. 28, pp. 25537–25544, 2002.

[30] K. B. Eide, A. R. Lindbom, V. G. H. Eijsink, A. L. Norberg, and M. Sørlie, "Analysis of productive binding modes in the human chitotriosidase," *FEBS Letters*, vol. 587, no. 21, pp. 3508–3513, 2013.

[31] L. W. Stockinger, K. B. Eide, A. I. Dybvik et al., "The effect of the carbohydrate binding module on substrate degradation by the human chitotriosidase," *Biochimica et Biophysica Acta*, vol. 1854, no. 10, pp. 1494–1501, 2015.

[32] F. Fadel, Y. Zhao, R. Cachau et al., "New insights into the enzymatic mechanism of human chitotriosidase (CHIT1) catalytic domain by atomic resolution X-ray diffraction and hybrid QM/MM," *Acta Crystallographica Section D: Biological Crystallography*, vol. 71, no. 7, pp. 1455–1470, 2015.

[33] I. Labadaridis, E. Dimitriou, M. Theodorakis, G. Kafalidis, A. Velegraki, and H. Michelakakis, "Chitotriosidase in neonates with fungal and bacterial infections," *Archives of Disease in Childhood: Fetal and Neonatal Edition*, vol. 90, no. 6, pp. F531–F532, 2005.

[34] Y. G. Kurt, T. Cayci, P. Onguru et al., "Serum chitotriosidase enzyme activity in patients with Crimean-Congo hemorrhagic fever," *Clinical Chemistry and Laboratory Medicine*, vol. 47, no. 12, pp. 1543–1547, 2009.

[35] Y. Shibata, L. A. Foster, W. J. Metzger, and Q. N. Myrvik, "Alveolar macrophage priming by intravenous administration of chitin particles, polymers of N-acetyl-D-glucosamine, in mice," *Infection and Immunity*, vol. 65, no. 5, pp. 1734–1741, 1997.

[36] C. L. Bueter, C. A. Specht, and S. M. Levitz, "Innate sensing of chitin and chitosan," *PLoS Pathogens*, vol. 9, no. 1, Article ID e1003080, 2013.

[37] Y. Shibata, W. J. Metzger, and Q. N. Myrvik, "Chitin particle-induced cell-mediated immunity is inhibited by soluble mannan: mannose receptor-mediated phagocytosis initiates IL-12 production," *The Journal of Immunology*, vol. 159, no. 5, pp. 2462–2467, 1997.

[38] L. Malaguarnera, M. Musumeci, M. Di Rosa, A. Scuto, and S. Musumeci, "Interferon-gamma, tumor necrosis factor-α, and lipopolysaccharide promote chitotriosidase gene expression in human macrophages," *Journal of Clinical Laboratory Analysis*, vol. 19, no. 3, pp. 128–132, 2005.

[39] O. Sharif, V. N. Bolshakov, S. Raines, P. Newham, and N. D. Perkins, "Transcriptional profiling of the LPS induced NF-κB response in macrophages," *BMC Immunology*, vol. 8, article 1, 2007.

[40] K. Schroder, P. J. Hertzog, T. Ravasi, and D. A. Hume, "Interferon-γ: an overview of signals, mechanisms and functions," *Journal of Leukocyte Biology*, vol. 75, no. 2, pp. 163–189, 2004.

[41] J. P. Boyle, R. Parkhouse, and T. P. Monie, "Insights into the molecular basis of the NOD2 signalling pathway," *Open Biology*, vol. 4, no. 12, Article ID 140178, 2014.

[42] M. van Eijk, S. S. Scheij, C. P. A. A. van Roomen, D. Speijer, R. G. Boot, and J. M. F. G. Aerts, "TLR- and NOD2-dependent regulation of human phagocyte-specific chitotriosidase," *FEBS Letters*, vol. 581, no. 28, pp. 5389–5395, 2007.

[43] L. Malaguarnera, M. Musumeci, F. Licata, M. Di Rosa, A. Messina, and S. Musumeci, "Prolactin induces chitotriosidase gene expression in human monocyte-derived macrophages," *Immunology Letters*, vol. 94, no. 1-2, pp. 57–63, 2004.

[44] M. Di Rosa, A. M. Zambito, A. R. Marsullo, G. Li Volti, and L. Malaguarnera, "Prolactin induces chitotriosidase expression in human macrophages through PTK, PI3-K, and MAPK pathways," *Journal of Cellular Biochemistry*, vol. 107, no. 5, pp. 881–889, 2009.

[45] F. Martinon, L. Agostini, E. Meylan, and J. Tschopp, "Identification of bacterial muramyl dipeptide as activator of the NALP3/cryopyrin inflammasome," *Current Biology*, vol. 14, no. 21, pp. 1929–1934, 2004.

[46] G. Prencipe, I. Caiello, S. Cherqui et al., "Inflammasome activation by cystine crystals: implications for the pathogenesis of cystinosis," *Journal of the American Society of Nephrology*, vol. 25, no. 6, pp. 1163–1169, 2014.

[47] K. Lieb, C. Kaltschmidt, B. Kaltschmidt et al., "Interleukin-1β uses common and distinct signaling pathways for induction of the interleukin-6 and tumor necrosis factor α genes in the human astrocytoma cell line U373," *Journal of Neurochemistry*, vol. 66, no. 4, pp. 1496–1503, 1996.

[48] D. Artis and H. Spits, "The biology of innate lymphoid cells," *Nature*, vol. 517, no. 7534, pp. 293–301, 2015.

[49] M. Di Rosa, G. Malaguarnera, C. De Gregorio, F. D'Amico, M. C. Mazzarino, and L. Malaguarnera, "Modulation of chitotriosidase during macrophage differentiation," *Cell Biochemistry and Biophysics*, vol. 66, no. 2, pp. 239–247, 2013.

[50] M. A. Elmonem, S. H. Makar, L. van den Heuvel et al., "Clinical utility of chitotriosidase enzyme activity in nephropathic cystinosis," *Orphanet Journal of Rare Diseases*, vol. 9, no. 1, article 155, 2014.

[51] J. Correale and M. Fiol, "Chitinase effects on immune cell response in neuromyelitis optica and multiple sclerosis," *Multiple Sclerosis*, vol. 17, no. 5, pp. 521–531, 2011.

[52] S. Létuvé, A. Kozhich, A. Humbles et al., "Lung chitinolytic activity and chitotriosidase are elevated in chronic obstructive pulmonary disease and contribute to lung inflammation," *The American Journal of Pathology*, vol. 176, no. 2, pp. 638–649, 2010.

[53] C. G. Lee, E. L. Herzog, F. Ahangari et al., "Chitinase 1 is a biomarker for and therapeutic target in scleroderma-associated interstitial lung disease that augments TGF-β1 signaling," *Journal of Immunology*, vol. 189, no. 5, pp. 2635–2644, 2012.

[54] M. A. Elmonem, H. S. Amin, R. A. El-Essawy et al., "Association of chitotriosidase enzyme activity and genotype with the risk of nephropathy in type 2 diabetes," *Clinical Biochemistry*, 2015.

[55] T. E. Sutherland, R. M. Maizels, and J. E. Allen, "Chitinases and chitinase-like proteins: potential therapeutic targets for the treatment of T-helper type 2 allergies," *Clinical and Experimental Allergy*, vol. 39, no. 7, pp. 943–955, 2009.

[56] R. Shao, Q. J. Cao, R. B. Arenas, C. Bigelow, B. Bentley, and W. Yan, "Breast cancer expression of YKL-40 correlates with tumour grade, poor differentiation, and other cancer markers," *British Journal of Cancer*, vol. 105, no. 8, pp. 1203–1209, 2011.

[57] X. W. Wang, C. L. Cai, J. M. Xu, H. Jin, and Z. Y. Xu, "Increased expression of chitinase 3-like 1 is a prognosis marker for non-small cell lung cancer correlated with tumor angiogenesis," *Tumor Biology*, vol. 36, no. 2, pp. 901–907, 2015.

[58] N. Ngernyuang, R. A. Francescone, P. Jearanaikoon et al., "Chitinase 3 like 1 is associated with tumor angiogenesis in cervical cancer," *International Journal of Biochemistry and Cell Biology*, vol. 51, no. 1, pp. 45–52, 2014.

[59] X. Q. Pan, C. C. Shih, and J. Harday, "Chitinase induces lysis of MCF-7 cells in culture and of human breast cancer xenograft B11-2 in SCID mice," *Anticancer Research*, vol. 25, no. 5, pp. 3167–3172, 2005.

[60] B. Bonnotte, N. Larmonier, N. Favre et al., "Identification of tumor-infiltrating macrophages as the killers of tumor cells after immunization in a rat model system," *Journal of Immunology*, vol. 167, no. 9, pp. 5077–5083, 2001.

[61] N. N. Sanders, V. G. H. Eijsink, P. S. van den Pangaart et al., "Mucolytic activity of bacterial and human chitinases," *Biochimica et Biophysica Acta—General Subjects*, vol. 1770, no. 5, pp. 839–846, 2007.

[62] E. Shemesh, L. Deroma, B. Bembi et al., "Enzyme replacement and substrate reduction therapy for Gaucher disease," *Cochrane Database of Systematic Reviews*, no. 3, Article ID CD010324, 2015.

[63] M. Harlander, B. Salobir, M. Zupančič, M. Dolenšek, T. Bavčar Vodovnik, and M. Terčelj, "Serial chitotriosidase measurements in sarcoidosis—two to five year follow-up study," *Respiratory Medicine*, vol. 108, no. 5, pp. 775–782, 2014.

[64] M. A. Elmonem, D. I. Ramadan, M. S. M. Issac, L. A. Selim, and S. M. Elkateb, "Blood spot versus plasma chitotriosidase: a systematic clinical comparison," *Clinical Biochemistry*, vol. 47, no. 1-2, pp. 38–43, 2014.

[65] M. Ries, E. Schaefer, T. Lührs et al., "Critical assessment of chitotriosidase analysis in the rational laboratory diagnosis of children with Gaucher disease and Niemann-Pick disease type A/B and C," *Journal of Inherited Metabolic Disease*, vol. 29, no. 5, pp. 647–652, 2006.

[66] A. C. Vedder, J. Cox-Brinkman, C. E. M. Hollak et al., "Plasma chitotriosidase in male Fabry patients: a marker for monitoring lipid-laden macrophages and their correction by enzyme replacement therapy," *Molecular Genetics and Metabolism*, vol. 89, no. 3, pp. 239–244, 2006.

[67] E. Dimitriou, M. Cozar, I. Mavridou, D. Grinberg, L. Vilageliu, and H. Michelakakis, "The spectrum of Krabbe disease in

Greece: biochemical and molecular findings," *JIMD Reports*, 2015.

[68] Ž. Reiner, O. Guardamagna, D. Nair et al., "Lysosomal acid lipase deficiency—an under-recognized cause of dyslipidaemia and liver dysfunction," *Atherosclerosis*, vol. 235, no. 1, pp. 21–30, 2014.

[69] M. Muranjan, S. Agarwal, K. Lahiri, and M. Bashyam, "Novel biochemical abnormalities and genotype in Farber disease," *Indian Pediatrics*, vol. 49, no. 4, pp. 320–322, 2012.

[70] A. Wajner, K. Michelin, M. G. Burin et al., "Comparison between the biochemical properties of plasma chitotriosidase from normal individuals and from patients with Gaucher disease, GM1-gangliosidosis, Krabbe disease and heterozygotes for Gaucher disease," *Clinical Biochemistry*, vol. 40, no. 5-6, pp. 365–369, 2007.

[71] A. Caciotti, M. Di Rocco, M. Filocamo et al., "Type II sialidosis: review of the clinical spectrum and identification of a new splicing defect with chitotriosidase assessment in two patients," *Journal of Neurology*, vol. 256, no. 11, pp. 1911–1915, 2009.

[72] C. Tasci, S. Tapan, S. Ozkaya et al., "Efficacy of serum chitotriosidase activity in early treatment of patients with active tuberculosis and a negative sputum smear," *Therapeutics and Clinical Risk Management*, vol. 8, pp. 369–372, 2012.

[73] O. Coskun, S. Oter, H. Yaman, S. Kilic, I. Kurt, and C. P. Eyigun, "Evaluating the validity of serum neopterin and chitotriosidase levels in follow-up brucellosis patients," *Internal Medicine*, vol. 49, no. 12, pp. 1111–1118, 2010.

[74] A. Iyer, M. van Eijk, E. Silva et al., "Increased chitotriosidase activity in serum of leprosy patients: association with bacillary leprosy," *Clinical Immunology*, vol. 131, no. 3, pp. 501–509, 2009.

[75] E. Bargagli, M. Margollicci, A. Luddi et al., "Chitotriosidase activity in patients with interstitial lung diseases," *Respiratory Medicine*, vol. 101, no. 10, pp. 2176–2181, 2007.

[76] A. Sonmez, C. Haymana, S. Tapan et al., "Chitotriosidase activity predicts endothelial dysfunction in type-2 diabetes mellitus," *Endocrine*, vol. 37, no. 3, pp. 455–459, 2010.

[77] M. Artieda, A. Cenarro, A. Gañán et al., "Serum chitotriosidase activity is increased in subjects with atherosclerosis disease," *Arteriosclerosis, Thrombosis, and Vascular Biology*, vol. 23, no. 9, pp. 1645–1652, 2003.

[78] T. Kologlu, S. K. Ucar, E. Levent, Y. D. Akcay, M. Coker, and E. Y. Sozmen, "Chitotriosidase as a possible marker of clinically evidenced atherosclerosis in dyslipidemic children," *Journal of Pediatric Endocrinology and Metabolism*, vol. 27, no. 7-8, pp. 701–708, 2014.

[79] A. Bustamante, C. Dominguez, V. Rodriguez-Sureda et al., "Prognostic value of plasma chitotriosidase activity in acute stroke patients," *International Journal of Stroke*, vol. 9, no. 7, pp. 910–916, 2014.

[80] B. S. Yildiz, B. Barutcuoglu, Y. I. Alihanoglu et al., "Serum chitotriosidase activity in acute coronary syndrome," *Journal of Atherosclerosis and Thrombosis*, vol. 20, no. 2, pp. 134–141, 2013.

[81] M. R. Safarinejad and S. Safarinejad, "Plasma chitotriosidase activity and arteriogenic erectile dysfunction: association with the presence, severity, and duration," *The Journal of Sexual Medicine*, 2010.

[82] V. Pagliardini, S. Pagliardini, L. Corrado et al., "Chitotriosidase and lysosomal enzymes as potential biomarkers of disease progression in amyotrophic lateral sclerosis: a survey clinic-based study," *Journal of the Neurological Sciences*, vol. 348, no. 1-2, pp. 245–250, 2015.

[83] B. Olsson, C. Malmeström, H. Basun et al., "Extreme stability of chitotriosidase in cerebrospinal fluid makes it a suitable marker for microglial activation in clinical trials," *Journal of Alzheimer's Disease*, vol. 32, no. 2, pp. 273–276, 2012.

[84] P. J. Orchard, T. Lund, W. Miller et al., "Chitotriosidase as a biomarker of cerebral adrenoleukodystrophy," *Journal of Neuroinflammation*, vol. 8, article 144, 2011.

[85] A. Aydogdu, I. Tasci, S. Tapan et al., "Women with polycystic ovary syndrome have increased plasma chitotriosidase activity: a pathophysiological link between inflammation and impaired insulin sensitivity?" *Experimental and Clinical Endocrinology and Diabetes*, vol. 120, no. 5, pp. 261–265, 2012.

[86] I. Alanbay, H. Coksuer, C. M. Ercan et al., "Chitotriosidase levels in patients with severe endometriosis," *Gynecological Endocrinology*, vol. 28, no. 3, pp. 220–223, 2012.

[87] Ü. Aksoy, H. Aksoy, G. Açmaz, M. Babayiğit, and Ö. Kandemir, "Umbilical artery serum chitotriosidase concentration in pregnancies complicated by preeclampsia and relationship between chitotriosidase levels and fetal blood flow velocity," *Hypertension in Pregnancy*, vol. 32, no. 4, pp. 401–409, 2013.

[88] A. Taylan, O. Gurler, B. Toprak et al., "S1000A12, chitotriosidase, and resolvin D1 as potential biomarkers of familial mediterranean fever," *Journal of Korean Medical Science*, vol. 30, no. 9, pp. 1241–1245, 2015.

[89] R. Barone, G. Bertrand, J. Simporè, M. Malaguarnera, and S. Musumeci, "Plasma chitotriosidase activity in β-thalassemia major: a comparative study between Sicilian and Sardinian patients," *Clinica Chimica Acta*, vol. 306, no. 1-2, pp. 91–96, 2001.

[90] A. Acar, M. Keskek, F. K. Işman, M. Kucur, and M. Tez, "Serum chitotriosidase activity in acute appendicitis: preliminary results," *American Journal of Emergency Medicine*, vol. 30, no. 5, pp. 775–777, 2012.

[91] J. K. H. Brunner, S. Scholl-Bürgi, D. Hössinger, P. Wondrak, M. Prelog, and L.-B. Zimmerhackl, "Chitotriosidase activity in juvenile idiopathic arthritis," *Rheumatology International*, vol. 28, no. 9, pp. 949–950, 2008.

[92] M. Kucur, F. K. Isman, C. Balci et al., "Serum YKL-40 levels and chitotriosidase activity as potential biomarkers in primary prostate cancer and benign prostatic hyperplasia," *Urologic Oncology: Seminars and Original Investigations*, vol. 26, no. 1, pp. 47–52, 2008.

12

Characterization of Pectinase from *Bacillus subtilis* Strain Btk 27 and its Potential Application in Removal of Mucilage from Coffee Beans

Oliyad Jeilu Oumer[1] and Dawit Abate[2]

[1]*Department of Biology, Ambo University, P.O. Box 19, Ambo, Ethiopia*
[2]*College of Natural Science, Addis Ababa University, P.O. Box 1176, Addis Ababa, Ethiopia*

Correspondence should be addressed to Oliyad Jeilu Oumer; oliyad.jeilu@ambou.edu.et

Academic Editor: Toshihisa Ohshima

The demand for enzymes in the global market is projected to rise at a fast pace in recent years. There has been a great increase in industrial applications of pectinase owing to their significant biotechnological uses. For applying enzymes at industrial scale primary it is important to know the features of the enzyme. Thus, this study was undertaken with aims of characterizing the pectinase enzyme from *Bacillus subtilis strain Btk27* and proving its potential application in demucilisation of coffee. In this study, the maximum pectinase activity was achieved at pH 7.5 and 50°C. Also, the enzyme activity was found stimulated with Mg2+ and Ca2+ metal ions. Moreover, it was stable on EDTA, Trixton-100, Tween 80, and Tween 20. Since *Bacillus subtilis* strain Btk27 was stable in most surfactants and inhibitors it could be applicable in various industries whenever pectin degradation is needed. The enzyme Km and Vmax values were identified as 1.879 mg/ml and 149.6 U, respectively. The potential application of the enzyme for coffee processing was studied, and it is found that complete removal of mucilage from coffee beans within 24 hours of treatment indicates the potential application in coffee processing.

1. Introduction

Biotechnological answers for environmental sustainability are modern solutions that help in the growth of the nation and are a boon for the welfare of human beings for the present and forthcoming generations. Biotechnology operations for enzyme production are no longer academic; it is a potentially useful alternative proposition for the future [1]. In this regard, pectinolytic enzymes can be applied in various industrial sectors wherever the degradation of pectin is required for a particular process. Several microorganisms have been used to produce different types of pectinolytic enzymes [2]. Microbial pectinases account for 25% of the global food and industrial enzyme sales [3, 4] and their market is increasing day by day. These are used extensively for fruit juice clarification, juice extraction, manufacture of pectin free starch, refinement of vegetable fibers, degumming of natural fibers, and wastewater treatment and as an analytical tool in the assessment of plant

products [5, 6]. Pectinase treatment accelerates tea fermentation and also destroys the foam forming property of instant tea powders by destroying pectins. They are also used in coffee fermentation to remove mucilaginous coat from coffee beans [7, 8].

Parenthetically, Ethiopia is the original home of *Coffea arabica* L. and, thus, possesses the largest diversity in coffee genetic resources. Coffee is critical to the Ethiopian economy, since over 25% of the Ethiopia population depends on coffee for its livelihood. As per the past few years data, coffee production accounted on average for about 5% of Gross Domestic Product (GDP). Though Ethiopian exports continue to be dominated by basic commodities, share of coffee in total exports has shrunk from 53% to 31% during 2000–2012 [9].

Regardless of the importance of the crop, poor postharvest processing techniques largely contribute to the decline in coffee quality. The traditional processing practices employed by producers have imparted a negative impact on Ethiopian coffee quality. So far, few research attempts have been made to optimize with regard to fermentation for wet processing of coffee. Conventional coffee processing uses water to remove mucilage from coffee beans by natural fermentation. Quite often the mucilage breakdown is not complete even after 36–72 hour of fermentation. If the coffee beans are fermented for long hours, stinker beans (over fermented beans) develop. Most quality defects of coffee are attributed to incomplete mucilage removal and uncontrolled fermentation [10].

Previously, we screened microorganisms for the pectinase activity and identified *Bacillus subtilis strain* Btk 27 as potent pectinase producer. And we extensively studied the parameters for maximal pectinase production. The main aims of this study are to characterize the pectinase from *Bacillus subtilis strain* Btk 27 and testing the potential application in removal of mucilage from coffee beans.

2. Material and Methods

2.1. Inoculum Preparation.
Fresh culture of *Bacillus subtilis strain* Btk 27 was inoculated into sterilized YEP medium with pH of 7.0 ± 0.5. The inoculated flask was incubated at $30°C$ on a rotary shaker at 120 rpm. Culture was grown in 50 ml media in 250 ml Erlenmeyer flasks.

2.2. Production of Pectinase.
In 250 ml conical flask, 5.0 g of wheat bran was moistened by 75% of distilled water and autoclaved at $121°$ for 15 minute. The flasks were inoculated with 2.0 ml of overnight-grown seed culture of *Bacillus subtilis strain Btk 27*, mixed well to evenly distribute the inoculum, and incubated at $30°C$ for 48 h.

2.3. Extraction of Pectinase from Solid Substrate.
Extraction of pectinase from ssf was done according to the method of Xiros et al. (2008) [11]. After 48 h of incubation 50 ml of distilled water was added into the solid substrate and the flasks are shaken for 1 h at 120 rpm on orbital shaker thoroughly and slurry is formed. Then, the flasks were kept at $4°C$ for 30 min under static conditions to facilitate the enzyme extraction. The slurry was centrifuged at $10,000 g$ for 10 min at $4°C$, and the clear supernatant was collected to assay the pectinase activity. The pectinase activity was determined in the supernatant as U/g of solid substrate used. The pectinase enzyme assay was based on the determination of reducing sugars produced as a result of enzymatic hydrolysis of pectin by dinitrosalicylic acid reagent (DNS) method (Miller, 1959). The enzyme unit was defined as the amount of enzyme that catalyzes μmol of galacturonic acid per minute (μmol min^{-1}) under the assay conditions. Relative activity was calculated as the percentage of enzyme activity of the sample with respect to the sample for which maximum activity was obtained.

Relative Activity

$$= \text{Activity of sample (U)} \qquad (1)$$
$$\times \frac{100}{\text{Maximum enzyme activity (u)}}.$$

2.4. Effect of Substrate Specificity on Pectinase Activity.
The effect of substrate specificity on pectinase enzyme activity was determined by incubating 100 μL of suitably diluted enzyme with 900 μL of different substrates like Apple pectin, Citrus pectin, Xylan, and Galactose. These substrates were prepared in 0.1 M of phosphate buffer (pH 7.5) with 0.5% w/v concentration. The reaction mixture was incubated at $50°C$ for 10 minute and the enzyme activity assayed.

2.5. Effect of pH on Pectinase Activity.
The effect of pH on pectinase activity was determined by incubating 900 μL of substrate at different pHs with 100 μL of suitably diluted enzyme at $50°C$ for 10 min and followed by assaying the enzyme activity. Substrate (0.5% w/v Citrus Pectin) was prepared at different pH values (pH 4.5–9.5) using different buffers (0.1 M) such as sodium acetate buffer, pH 4.5–6.0, phosphate buffer, pH 6.0–7.9, Tris-HCl buffer, pH 7.5–9.0, and glycine NaOH buffer, pH 8.5–10.0.

2.6. Effect of Temperature on Pectinase Activity.
The effect of temperature on pectinase enzyme was evaluated by incubating the reaction mixture (900 μL of substrate at different pHs with 100 μL of suitably diluted enzyme) at different temperatures in the range of 30–80$°C$ for 10 min with 5$°C$ interval and the enzyme activity was assayed.

2.7. Effect of Surfactants and Inhibitors on Pectinase Activity.
The effect of surfactants and inhibitors including mercaptoethanol, EDTA (1 mM), SDS (1%, w/v), Tween (20 and 80; 0.1%, v/v), and Triton X-100 (0.1%, v/v) on pectinase enzyme activity was studied by directly incorporating them into the enzyme substrate system. And then, the reaction mixture was incubated at $50°C$ for 10 min and the enzyme activity was assayed.

2.8. Effect of Metal Ions on Pectinase Activity.
The effect of metal ions on pectinase activity was studied by directly incorporating them into the enzyme substrate system at a final concentration of 5 mM. Metal ions which were examined for their effect are Ca^{2+}, Mg^{2+}, Co^{2+}, Cu^{2+}, Fe^{3+}, and Mn^{2+}. The reaction mixture was incubated at $50°C$ for 10 min and the enzyme activity was assayed.

2.9. Thermostability of the Enzyme.
The effect of enzyme stability under optimized temperature and optimized pH was studied by incubating the reaction mixture at various time intervals ranging 30, 60, 90, 120, 150, and 180 min.

2.10. Michaelis-Menten Constant (Km) and V max Values.
The K_m and V_{max} values were determined by measuring the reaction velocity at different concentrations of the substrate (Citrus Pectin). First stock solution of Citrus Pectin which was of 10 mg/ml concentration was prepared with appropriate buffer (phosphate, pH 7.5). Then the stock solution was diluted by appropriate volume of buffer to make the final mg/ml Citrus Pectin concentrations listed in Table 1. The appropriate mg/ml Citrus Pectin (900 μl) was incubated with

TABLE 1: mg/ml concentrations of Citrus Pectin to determine the Km and Vmax values.

Volume of stock solution (μl)	Volume of buffer (μl)	mg/ml of Citrus Pectin
180	720	2
360	540	4
540	360	6
720	180	8
900	0	10

TABLE 2: Effect of substrate specificity on pectinase activity.

Substrate	Enzyme activity (U/g)*
Apple Pectin	441.53 ± 13.3^{a}
Citrus Pectin	1272.4 ± 25.5^{b}
Xylan	697.23 ± 11.73^{c}
Galactose	0.0 ± 0.0^{d}

(i) *Values are mean ± S.D. of 3 replicates; (a) values followed by different superscripts are significantly different at $P < 0.05$; (b) values followed by same superscripts are not significantly different at ($P < 0.05$).

100 μl of suitably diluted enzyme at 50°C for 10 minute and the pectinase enzyme activity was assayed.

The relationship between substrate (mg/ml of Citrus Pectin) and velocity (pectinase enzyme activity) was plotted using GraphPad Prism 5 software. The Km and Vmax values were calculated using nonlinear regression.

2.11. Removal of Mucilage from Coffee Beans Using Pectinase. Fresh coffee beans were harvested and pulped manually. The pulped beans were soaked with the enzyme mixture under static conditions until the mucilage was removed. Complete demucilisation was observed by hand feel as per traditional method; finally the demucilised coffee beans were washed and sun dried. To compare the enzymatic demucilisation with natural fermentation, the pulped coffee beans were soaked with water without enzyme addition.

3. Results

3.1. Effect of Substrate Specificity. The effect of substrate specificity on the activity of pectinase enzyme was determined by incubating the pectinase enzyme with different substrates (Table 2). The highest activity was observed when Citrus pectin was used as substrate. The effect of Citrus Pectin was significantly higher than the other tasted substrates.

3.2. Effect of pH. The effect of pH on pectinase activity was studied by incubating reaction mixture (Citrus Pectin and pectinase) at different pH values (pH 4.5–9.5). It was observed that the pectinase enzyme from *Bacillus subtilis* strain Btk 27 had highest activity at pH of 7.5 (Figure 1).

3.3. Effect of Temperature. The effect of temperature on pectinase enzyme was evaluated by incubating the reaction mixture at different temperatures in the range of 30–80°C.

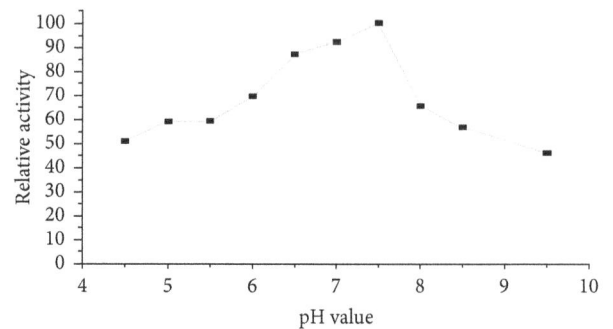

FIGURE 1: Effect of pH on activity of pectinase.

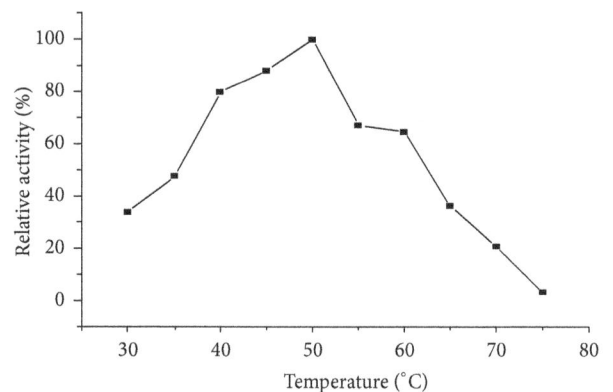

FIGURE 2: Effect of temperature on activity of pectinase.

TABLE 3: Effect of inhibitors and surfactants on pectinase activity.

Surfactant and Inhibitor	Enzyme activity (U/g)*
Control	1272.4 ± 25.5^{a}
EDTA	2103.3 ± 11.5^{b}
Mercaptoethanol	0.0 ± 0.0^{c}
SDS	697.6 ± 5.1^{d}
Trixton-100	1715.9 ± 8.5^{ab}
Tween 20	1954.4 ± 7.8^{b}
Tween 80	1277.5 ± 11.1^{a}

(i) *Values are mean ± S.D. of 3 replicates; (a) values followed by different superscripts are significantly different at $P < 0.05$; (b) values followed by same superscripts are not significantly different at ($P < 0.05$).

The maximum pectinase activity observed was at 50°C (Figure 2).

3.4. Effect of Inhibitors and Surfactants on Pectinase Activity. The effect of surfactants and inhibitors on pectinase activity was studied by directly incorporating them into the enzyme substrate system. Among the tasted surfactants and inhibitors, EDTA, Trixton-100, Tween 80, and Tween 20 enhanced the pectinase activity with relative activity of 165.3, 134.9, 100.4, and 153.6 (%), respectively. It was observed that the presence of Mercaptoethanol and SDS in the enzyme substrate system decreased pectinase activity significantly (Table 3).

TABLE 4: Effects of metal ions on pectinase activity.

Metal Ion	Enzyme activity (U/g)*
CaCl2	1684.6 ± 20.0^a
CoCl2	1618.5 ± 9.3^a
FeCl2	1528.2 ± 15.1^a
MgCl2	1739.3 ± 31.8^a
MnCl2	944.0 ± 38.7^a
Control	1272.4 ± 25.5^a

(i) *Values are mean ± S.D. of 3 replicates; (a) values followed by different superscripts are significantly different at $P < 0.05$; (b) values followed by same superscripts are not significantly different at ($P < 0.05$).

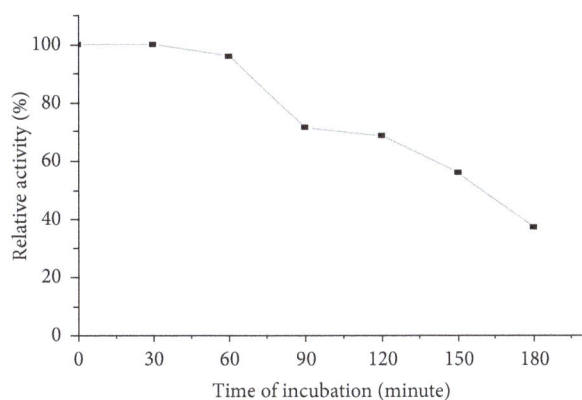

FIGURE 4: Michaelis-Menten Kinetics.

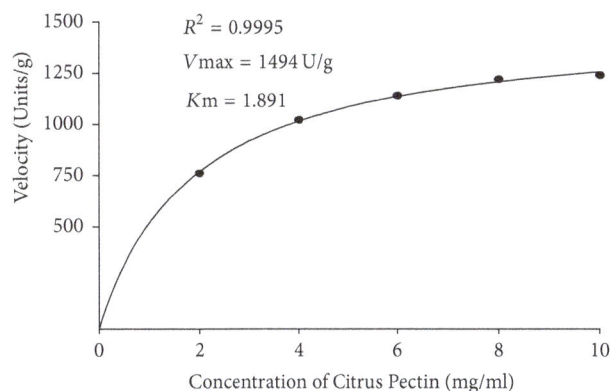

FIGURE 3: Enzyme stability.

FIGURE 5: Removal of mucilage from coffee beans using natural fermentation (right) and using pectinase enzyme (left).

3.5. Effect of Metal Ions. The effect of metal ions on pectinase activity was studied by directly incorporating them into the enzyme substrate system at a final concentration of 5 mM. The highest relative activities observed were 136.7% and 132.4% in the presence of Mg2+ and Ca2+ metal ions, respectively. The lowest activity observed was with the presence of Mn2+ metal ion (Table 4). However, the effect of these tested metal ions on pectinase activity was not significant.

3.6. Thermostability of the Enzyme. The stability of pectinase enzyme under optimized temperature and pH was studied by incubating the reaction mixture at various time intervals (Figure 3). It was observed that the enzyme was stable with 100% relative activity until 60 minutes of incubation. However, beyond 60 minutes of incubation, the enzyme activity declined.

3.7. Michaelis-Menten Constant (Km) and Vmax Values. The Km and Vmax values of the enzyme were determined by measuring the reaction velocity at different concentrations of the substrate (Citrus Pectin). The relation between reaction velocity and the substrate concentration was analyzed with nonregression analysis. The regression coefficient (R^2) was equal to 0.999 which describes the concentrations of Citrus Pectin and velocity (enzyme activity) readings were positively correlated (Figure 4). From the nonregression analysis, Km and Vmax values were identified as 1.891 mg/ml and 1494 U/g, respectively.

3.8. Potential Application of Pectinase on Demucilisation of Coffee Beans. Fresh coffee beans were harvested and pulped manually. Half of the pulped beans were soaked into water that contained the crude pectinase whereas the other half were subjected to natural fermentation. Complete demucilisation was observed within 24 hours of incubation on pectinase treated coffee beans (Figure 5). However, in case of natural fermentation demucilisation was not completed even within 36 hour of fermentation.

4. Discussions

The maximum pectinase activity was observed when Citrus Pectin was used as substrate. Similarly, Celestino et al., (2006) [12] reported that novel pectinase enzyme from *Acrophialophora nainiana* showed the highest substrate activity on Citrus Pectin. Thus, it can be inferred that pectinase have high affinity for Citrus Pectin compared to others which are used in this study.

The optimum pH for pectinase activity was recorded at pH 7.5. Reports have shown pectinase activity to be highest around alkaline pH [13, 14]. Similar study on *B. stearothermophilus* showed optimum pectinase activities at pH 7.5 [15]. Moreover, pectinase from *Bacillus* sp. DT7 was maximally stable under alkaline conditions of pH 7.5–8.5 [16]. Therefore, this pectinase will have potential applications whenever alkaline pectin degradation is needed such as in coffee processing, paper and pulp industry, and Pectic waste water treatment.

The maximum pectinase activity was observed at 50°C; with further increase of temperature, the pectinase activity was decreased. This may be a result of thermal denaturation of the enzyme possibly due to disruption of noncovalent linkages, including hydrophobic interactions [17]. Likewise, Phutela et al. (2005) [18] reported an optimum temperature of 60°C for thermophiles *A. fumigates* pectinase. Alana et al., (1990) also reported that *Penicillium italicum* pectinase activity increase up to 50°C. The result might indicate that pectinase from *Bacillus subtilis* strain Btk27 is thermophilic enzyme.

Surfactant agent stability of the enzyme is one of the important parameters enabling enzymes to be used in different types of industries. In this study, the pectinase activity was stimulated on EDTA, Trixton-100, Tween-20, and Tween-80, whereas SDS significantly decreased pectinase activity. Moreover, Mercaptoethanol completely inhibited pectinase activity. Li et al., (2012) [19] reported that Tween-80 and Tween-20 stimulated the polygalacturonase activity. Zu-ming et al. (2008) [20] stated also surfactants such as Tween-80 and Tween-20 had stimulatory effects on pectinase activity. On the contrary, Amid et al. (2014) [21] reported that SDS, Trixton-100, and Tween-20 significantly reduced, Mercaptoethanol significantly increased, and EDTA had no significant effect on thermoalkaline pectinase. According to Zohdi and Amid, (2013) [22] most of the surfactants which interact with proteins cause distinct electrostatic and hydrophobic regions and alter the secondary or tertiary structure of enzymes. The stimulatory effect of some surfactants may be probably that the surfactants might improve the turnover number of pectinase by increasing the contact frequency between the active site of the enzyme and the substrate by lowering the surface tension of the aqueous medium [23]. Since Bacillus *subtilis* strain Btk27 was stable in most surfactants and inhibitors it could be applicable in various industries whenever pectin degradation is needed.

Among the metal ions, Mg^{2+}, Zn^{2+}, Co^{2+}, and Fe^{2+} increased pectinase activity whereas Mn^{2+} decreased the pectinase activity; however their effect was not significant. Metal ions like Ca^{2+} and Mg^{2+} might play a vital role in maintaining the active confirmation of alkaline endo polygalacturonase to stimulate the activity [24]. Alana et al., (1990) [25] reported that Ca^{2+}, Mg^{2+}, Zn^{2+}, and Mn^{2+} did not affect pectin lyase activity of *P. italicum* at 5 mM. This discrepancy in the divalent metal ion preference suggested that the enzymes might have differential flexibility in the active site. Beg and Gupta, (2003) [26] reported that metal ions such as Mg^{2+} and Ca^{2+} might play a vital role in maintaining the active confirmations of the alkaline pectinase to stimulate the activity.

Pectinase from *Bacillus subtilis* strain Btk27 was stable with 100% relative activity until 60 minutes of incubation. However, above 60 minutes of incubation the enzyme activity declined. Çelik et al. (2010) [27] reported that purified enzyme was stable and retained its full activity until 1 hour incubation period but the activity was reduced to 20% after 1 hour incubation. Gummadi and Panda (2003) [28] stated that the stability of pectinases is affected by both physical parameters (pH and temperature) and chemical parameters

(inhibitors or activators). The thermal inactivation of enzymes is always due to denaturation of enzyme [29].

In enzymatic reaction, the kinetic parameter is also important, which describes enzyme efficiency. In this study, Vmax and Km values were 149.6 U and 1.88 mg/ml, respectively. Saad et al. (2007) [30] reported a Km of 1.88 mg/mL and Vmax of 0.045 mole/mL/min for *Mucor rouxii*. Celestino et al., (2006) [12] also reported that *Acrophialophora nainiana* had a Km value of 4.22 mg/ml. Moreover, Laha et al. (2014) [31] reported that *P. chrysogenum* had Km and Vmax values of 1.0 mg/mL and 78 U, respectively. Pectinase from Bacillus *subtilis* strain Btk27 relatively has the highest affinity for substrate due to its lowest Km; it also has the highest utility of pectin substrate as a result of its highest Vmax. As a result of this high binding of pectinase from *Bacillus subtilis* strain Btk27 with pectin substrate, small quantity of the enzyme will digest a considerably high amount of substrate. This may therefore reduce the cost for the enzyme in industrial use.

Pectinase are used in coffee processing to remove the mucilaginous coat from the coffee beans [32]. However, there is no reported application of pectinase in Ethiopia for processing coffee to date. In this study, pectinase was applied in small scale coffee processing, and complete removal of mucilage from coffee beans within 24 hours of incubation was observed. Murthy and Naidu (2011) [33] reported complete demusilisation of Robusta coffee within 36 hour of incubation. The enzyme treatment significantly reduces the fermentation time and holds up coffee quality loss due to traditional coffee processing.

5. Conclusion

The pectinase from *Bacillus subtilis* strain Btk27 was alkaline, thermophilic, and stable with many of tasted surfactants. In addition, It was observed that the pectinase from *Bacillus subtilis* strain Btk27 has huge promising potential in removal of mucilage from coffee beans.

Authors' Contributions

This research is undertaken by the corresponding author, Oliyad Jeilu Oumer, under supervision of Dr. Dawit Abate.

Acknowledgments

The authors are delighted to acknowledge Addis Ababa University and Ambo University for their cooperation during this study.

References

[1] J. B. van Beilen and Z. Li, "Enzyme technology: an overview," *Current Opinion in Biotechnology*, vol. 13, no. 4, pp. 338–344, 2002.

[2] R. S. Jayani, S. K. Shukla, and R. Gupta, "Screening of bacterial strains for polygalacturonase activity: its production by bacillus sphaericus (MTCC 7542)," *Enzyme Research*, vol. 2010, Article ID 306785, 5 pages, 2010.

[3] R. S. Jayani, S. Saxena, and R. Gupta, "Microbial pectinolytic enzymes: a review," *Process Biochemistry*, vol. 40, no. 9, pp. 2931–2944, 2005.

[4] H. A. Murad and H. H. Azzaz, "Microbial pectinases and ruminant nutrition," *Research Journal of Microbiology*, vol. 6, no. 3, pp. 246–269, 2011.

[5] I. Alkorta, C. Garbisu, M. J. Llama, and J. L. Serra, "Industrial applications of pectic enzymes: a review," *Process Biochemistry*, vol. 33, no. 1, pp. 21–28, 1998.

[6] S. A. Singh, M. Ramakrishna, and A. G. Appu Rao, "Optimisation of downstream processing parameters for the recovery of pectinase from the fermented bran of Aspergillus carbonarius," *Process Biochemistry*, vol. 35, no. 3-4, pp. 411–417, 1999.

[7] C. Sieiro, B. García-Fraga, J. López-Seijas, A. F. Da Silva, and T. G. Villa, "Microbial Pectic Enzymes in the Food and Wine In dustry, Food Ind. Process. - Methods Equip," 2012, http://www.intechopen.com/books/food-industrial-processes-methods-and-equipment/microbial-pectic-enzymes-in-the-food-and-wine-industry.

[8] G. Hoondal, R. Tiwari, R. Tewari, N. Dahiya, and Q. Beg, "Microbial alkaline pectinases and their industrial applications: a review," *Applied Microbiology and Biotechnology*, vol. 59, no. 4-5, pp. 409–418, 2002.

[9] M. Atingi-Ego and M. Miyazaki, "the Federal Democratic Republic of Ethiopia," *International Monetary Fund*, vol. 14, pp. 1–76, 2014.

[10] S. Avallone, B. Guyot, J.-M. Brillouet, E. Olguin, and J.-P. Guiraud, "Microbiological and biochemical study of coffee fermentation," *Current Microbiology*, vol. 42, no. 4, pp. 252–256, 2001.

[11] C. Xiros, E. Topakas, P. Katapodis, and P. Christakopoulos, "Hydrolysis and fermentation of brewer's spent grain by Neurospora crassa," *Bioresource Technology*, vol. 99, no. 13, pp. 5427–5435, 2008.

[12] S. M. C. Celestino, S. Maria de Freitas, F. Javier Medrano, M. Valle de Sousa, and E. X. F. Filho, "Purification and characterization of a novel pectinase from Acrophialophora nainiana with emphasis on its physicochemical properties," *Journal of Biotechnology*, vol. 123, no. 1, pp. 33–42, 2006.

[13] E. Namasivayam, D. John Ravindar, K. Mariappan, A. jiji, M. Kumar, and R. L. Jayaraj, "Production of extracellular pectinase by bacillus cereus isolated from market solid waste," *Journal of Bioanalysis and Biomedicine*, vol. 3, no. 3, pp. 70–75, 2011.

[14] A. Kumar and R. Sharma, "Production of alkaline pectinase by bacteria (Cocci sps.) isolated from decomposing fruit materials," *Enzyme*, vol. 4, pp. 1–5, 2012.

[15] N. Torimiro, "Full length research paper a comparative study of pectinolytic enzyme production by bacillus species," *African Journal of Biotechnology*, vol. 12, no. 46, pp. 6498–6503, 2013.

[16] D. R. Kashyap, S. Chandra, A. Kaul, and R. Tewari, "Production , puri ® cation and characterization of pectinase from a Bacillus sp . DT7," 2000.

[17] F. Amin, H. N. Bhatti, I. Ahmad Bhatti, and M. Asgher, "Utilization of wheat bran for enhanced production of exo-polygalacturonase by penicillium notatum using response surface methodology," *Pakistan Journal of Agricultural Sciences*, vol. 50, no. 3, pp. 469–477, 2013.

[18] U. Phutela, V. Dhuna, S. Sandhu, and B. S. Chadha, "Pectinase and polygalacturonase production by a thermophilic Aspergillus fumigatus isolated from decomposting orange peels," *Brazilian Journal of Microbiology*, vol. 36, no. 1, pp. 63–69, 2005.

[19] S. Li, X. Yang, S. Yang, M. Zhu, and X. Wang, "Technology prospecting on enzymes: application, marketing and engineering," *Computational and Structural Biotechnology Journal*, vol. 2, no. 3, p. e201209017, 2012.

[20] L. I. Zu-ming, J. I. N. Bo, and Z. Hong-xun, "Purification and characterization of three alkaline endopolygalacturonases from a newly isolated bacillus gibsonii," *The Chinese Journal of Process Engineering*, vol. 8, pp. 4–9, 2008.

[21] M. Amid, Y. Manap, and K. Zohdi, "Purifcation and characterisation of thermo-alkaline pectinase enzyme from hylocereus polyrhizus," *European Food Research and Technology*, vol. 239, no. 1, pp. 21–29, 2014.

[22] N. K. Zohdi and M. Amid, "Optimization of extraction of novel pectinase enzyme discovered in red pitaya (Hylocereus polyrhizus) peel," *Molecules*, vol. 18, no. 11, pp. 14366–14380, 2013.

[23] Q. K. Beg, B. Bhushan, M. Kapoor, and G. S. Hoondal, "Production and characterization of thermostable xylanase and pectinase from Streptomyces sp. QG-11-3," *Journal of Industrial Microbiology and Biotechnology*, vol. 24, no. 6, pp. 396–402, 2000.

[24] Y. Li, N. I. Haddad, S. Yang, and B. Mu, "Variants of Lipopeptides Produced by Bacillus licheniformis HSN221 in Different Medium Components Evaluated by a Rapid Method ESI-MS," *International Journal of Peptide Research and Therapeutics*, vol. 14, no. 3, pp. 229–235, 2008.

[25] A. Alana, I. Alkorta, J. B. Dominguez, M. J. Llama, and J. L. Serra, "Pectin lyase activity in a Penicillium italicum strain," *Applied and Environmental Microbiology*, vol. 56, no. 12, pp. 3755–3759, 1990.

[26] Q. K. Beg and R. Gupta, "Purification and characterization of an oxidation-stable, thiol-dependent serine alkaline protease from Bacillus mojavensis," *Enzyme and Microbial Technology*, vol. 32, no. 2, pp. 294–304, 2003.

[27] H. Nadaroglu, Taskın E., A. Adıgüzel, M. Güllüce, and N. Demir, "Production of a novel pectin lyase from Bacillus pumilus (P9), purification and characterization and fruit juice application," *Romanian Biotechnological Letters*, vol. 15, no. 2, pp. 5167–5176, 2010, https://www.rombio.eu/rbl2vol15/15%20-Nazan%20Demir.pdf.

[28] S. N. Gummadi and T. Panda, "Purification and biochemical properties of microbial pectinases—a review," *Process Biochemistry*, vol. 38, no. 7, pp. 987–996, 2003.

[29] M. V. V. de Andrade, A. B. Delatorre, S. A. Ladeira, and M. L. L. Martins, "Production and partial characterization of alkaline polygalacturonase secreted by thermophilic Bacillus sp. SMIA-2 under submerged culture using pectin and corn steep liquor," *Ciencia e Tecnologia de Alimentos*, vol. 31, no. 1, pp. 204–208, 2011.

[30] N. Saad, M. Briand, C. Gardarin, Y. Briand, and P. Michaud, "Production, purification and characterization of an endopolygalacturonase from Mucor rouxii NRRL 1894," *Enzyme and Microbial Technology*, vol. 41, no. 6-7, pp. 800–805, 2007.

[31] S. Laha, D. Sarkar, and S. Chaki, "Optimization of production and molecular characterization of pectinase enzyme produced from penicillium chrysogenum," *Scholars Academic Journal of Biosciences*, vol. 2, no. 5, pp. 326–335, 2014.

[32] D. R. Kashyap, P. K. Vohra, S. Chopra, and R. Tewari, "Applications of pectinases in the commercial sector: a review," *Bioresource Technology*, vol. 77, no. 3, pp. 215–227, 2001.

[33] P. S. Murthy and M. M. Naidu, "Improvement of robusta coffee fermentation with microbial enzymes," *European Journal of Applied Sciences*, vol. 3, no. 4, pp. 130–139, 2011.

13

Recycle of Immobilized Endocellulases in Different Conditions for Cellulose Hydrolysis

D. F. Silva,[1] **A. F. A. Carvalho,**[1] **T. Y. Shinya,**[1] **G. S. Mazali,**[1]
R. D. Herculano,[2] **and P. Oliva-Neto**[1]

[1]*Biological Science Department, Universidade Estadual Paulista (UNESP), Avenida Dom Antônio, 2100 Bairro,*
 Parque Universitário, 19806-900 Assis, SP, Brazil
[2]*Bioprocess & Biotechnology Department, Universidade Estadual Paulista (UNESP), Rod. Araraquara-Jaú Km 1 Bairro,*
 Machados, 14800-901 Araraquara, SP, Brazil

Correspondence should be addressed to D. F. Silva; douglasfsilva@gmail.com

Academic Editor: Raffaele Porta

The immobilization of cellulases could be an economical alternative for cost reduction of enzyme application. The derivatives obtained in the immobilization derivatives were evaluated in recycles of paper filter hydrolysis. The immobilization process showed that the enzyme recycles were influenced by the shape (drop or sheet) and type of the mixture. The enzyme was recycled 28 times for sheets E' and 13 times for drops B'. The derivative E' showed the highest stability in the recycle obtaining 0.05 FPU/g, RA of 10%, and FPU Yield of 1.64 times, higher than FPU spent or Net FPU Yield of 5.3 times, saving more active enzymes. The derivative B showed stability in recycles reaching 0.15 FPU/g of derivative, yield of Recovered Activity (RA) of 25%, and FPU Yield of 1.57 times, higher than FPU spent on immobilization or Net PFU Yield of 2.81 times. The latex increased stability and resistance of the drops but did not improve the FPU/gram of derivative.

1. Introduction

The possible use of renewable fuels has aroused an increasing interest all over the world [1]. This fact has happened due to the positive impacts of replacing fossil fuels with renewable energy. Biofuels are renewable, available, and ecologically friendly [1]. As already reported by Vásquez et al. [2], new studies are being done to develop biotechnological processes that allow the use of lignocellulosic biomass waste, like corn and rice straw and bagasse from sugar cane and pulp industry waste, among others abundantly produced in the world. These residues will be used for the production of biofuels such as second-generation bioethanol. The production of sugarcane in 2016/17 will be increased by 2.9% in relation to the previous season. In absolute numbers, production of 684.77 million tons of sugar cane is estimated, compared to 665.59 million tons in 2015/16 [3]. Considering also that the sugar and alcohol industries generate 135 kg of dried bagasse

per ton of crushed cane [4], the total bagasse generated in this season was about 80.5 billion tons.

In order to carry out the hydrolysis of these residues, it is necessary to use cellulases, which are usually produced by filamentous fungi [5]. This hydrolysis is made by reducing the biomass into mainly glucose and xylose, which in turn can be fermented by facultative microorganisms such as *Saccharomyces cerevisiae*, and then converted into bioethanol [2, 5].

The enzymatic conversion of cellulose to glucose is difficult due to the physical nature of the substrate, which is composed mainly of insoluble crystalline fibers (microfibrils) in which the hydrogen bonds hold the molecules together. These fibers are embedded in a matrix of hemicellulose and lignin [6] decreasing the accessibility of cellulolytic enzymes (Beguin, 1990). The cellulases produced by fungi have three main components: endoglucanases that hydrolyze internal β1,4 D-glycosidic bonds; the cellobiohydrolases (exocellulase) which produce cellobiose from nonreducing ends from cellulose; the β-glucosidases (cellobiases) which convert cellobiose to glucose. For effective hydrolysis of cellulose a consortium of these enzymes in a synergistic action are required (Lynd et al., 2002) [7].

However, the high cost of these enzymes has limited the economic viability of their use in industrial bioprocesses [8]. Therefore, to make them economically feasible, the reuse of immobilized enzyme may be an alternative [9, 10].

Alginate is a natural polysaccharide widely used as support in immobilization by microencapsulation technologies and composed of alternating chains of α-L-guluronic acid and β-D-mannuronic acid residues [11, 12]. Alginate supports are usually made by cross-linking the carboxyl group of the α-L-guluronic acid with a cationic gelling solution such as calcium chloride or barium chloride [13], mixed or not with the solution containing the biocatalyst, depending on the derivative of immobilization [14].

Chitosan is a cationic biopolymer obtained by deacetylation of chitin. This polymer has two functional groups, amino and hydroxyl residues, being used as sites of reaction and coordination. This polysaccharide comprises a linear sequence of sugar monomers β-(1-4) 2-acetamide-2-deoxy-D-glucose (N-acetylglucosamine) bases [10, 15, 16]. Natural rubber latex (NRL, cis-1,4-polyisoprene) extracted from *Hevea brasiliensis* has been widely used as a raw material in the manufacturing of gloves, condoms, balloons, and other medical and dental devices. However, recently several new biomedical applications have been proposed using a different manufacturing process [17, 18]. Therefore, due to the porosity of the membrane, ease in handling, low cost, and the possibility of numerous modifications, the latex can be an excellent support for enzyme immobilization.

Hybrid chitosan-alginate is reported in the literature as a support for enzyme immobilization [19]. Chitosan is a polycationic polymer and alginate is a polyanionic polymer, so ionic interactions between them allow rigid gels to form [20]. Busto et al. [21] immobilized microbial endo-β-glucanase in alginate beads retaining 75% of its original activity. Wu et al. [22] immobilized cellulase in nanofibrous of PVA membranes by electrospinning retaining 36% of initial activity after six cycles of reuse. Chang et al. [23] used mesoporous silica nanoparticles (MSNs) in cellulase immobilization for conversion of cellulose into glucose. Cellulase chemically linked to MSNs exhibited a large pore size which was responsible for effective cellulose-to-glucose conversion exceeding 80% yield and excellent stability. Romo-Sánchez et al. [24] working with cellulases and xylanases immobilization performed 19 cycles maintaining 64% of the enzymatic activity. Zhang et al. [25] immobilized cellulases on modified silica gel and obtained significant activity over multiple reuses, with 82% and 31% of activity after 7 and 15 recycles, respectively. Song et al. [26] using super paramagnetic nanoparticles immobilized cellulases and reported 85% and 43% of the initial immobilized enzyme activities after being recycled 3 and 10 times, respectively.

This work studies the immobilization and recycling of cellulases, aiming at future applications in the production of second-generation ethanol.

2. Materials and Methods

2.1. Reagents, Supports, and Enzyme Used in Immobilization Process. The supports used in immobilization process of cellulases were chitosan ($C_{12}H_{24}N_2O_9$) high molecular weight (Aldrich® code 419419-50 G); sodium alginate (Labsynth® code A1089.01.AF); zeolite (mx/n [(AlO$_2$) × (SiO$_2$) y]·wH$_2$O) (Sigma); cationic and anionic exchanger resin (Amberlite® MB-20, Dow Chemical Comp., US) and polystyrene (Styrofoam®); natural latex (extracted in the farm of ESALQ-USP, Piracicaba, SP); 25% (m/m) glutaraldehyde (Nuclear, Brazil); calcium chloride (Vetec) and acetate buffer (Impex, pH 5.6). As substrate for determination of hydrolytic activity, Filter Paper Whatman no. 1 was used in the technique of FPase. The enzyme cellulase (EC3.2.1.4, 1,4-β-endoglucanase, ROHAMENT®CL, AB Enzymes, Darmstadt, Germany) from *Trichoderma reesei* was used for immobilization.

2.2. Maintenance of Latex. The latex was extracted at BDF Rubber Latex Co. Ltd (producer and distributor of concentrated rubber latex), Guarantã, Brazil. The latex solution extracted from *Hevea brasiliensis* consisted of a mixture of different clones. After extraction, ammonia was used to keep the latex liquid, and this material was centrifuged at 8000 rpm. The centrifugation was important since this process decreased some proteins contained in natural latex that cause allergic reactions [17].

2.3. The Immobilization Process

2.3.1. 1,4-β-Endoglucanase Immobilization on Activated Carbon, Zeolite, Ion Exchange Resin, and Polystyrene. Preliminary tests with some supports (activated carbon, zeolite, ion exchange resin, and polystyrene) were conducted for cellulase immobilization to determine which would be more efficient in the process of enzyme immobilization in sheets (derivatives E, E$'$, J, and J$'$, Table 1). Polystyrene was previously treated with the following procedure: autoclavation for 15 min at 120°C, and subsequently maintenance in 1:2 (w·v^{-1}) 50% (v·v^{-1}) ethanol solution for 30 min at 28°C. Then this support was washed with deionized water (1000 mL), modified derivative by Hou et al. [27]. 4 mL of endoglucanase (0.43 FPU/mL) was used in the immobilization process. This amount was 1% (w·w^{-1}) of enzyme (calculated as protein in the enzyme solution) based on powder support. The mixture was maintained for 24 hours under bland agitation at 25°C in 250 mL flask in shaker. After this step, the solid was separated from the liquid phase by filtering, and in both samples the protein concentration (Bradford), cellulolytic activity (FPU, Filter Paper Unit), and the yield of immobilization were determined. The best derivative was selected to be used in some derivatives of immobilization listed in Table 1. After the immobilization, the derivatives were packed at 5°C in acetate buffer (pH 5.6).

2.3.2. Production of Drop and Sheet Derivatives. The production of derivatives follows the modified methodology of Albarghouthi et al. [28] and Tanriseven and Doğan [29].

TABLE 1: The derivatives of cellulase immobilization.

Type of supports	Shape	Names of derivatives
Calcium alginate +1% enzyme	Drop	A
Calcium alginate + 1% enzyme	Drop	A′ (glutaraldehyde[3])
Calcium alginate + chitosan + 1% enzyme	Drop	B
Calcium alginate + chitosan + 1% enzyme	Drop	B′ (glutaraldehyde)
Calcium alginate + 1% enzyme	Sheet	C
Calcium alginate + 1% enzyme	Sheet	C′ (glutaraldehyde)
Calcium alginate + chitosan + 1% enzyme	Sheet	D
Calcium alginate + chitosan + 1% enzyme	Sheet	D′ (glutaraldehyde)
Calcium alginate + adsorbent[1] + 1% enzyme	Sheet	E
Calcium alginate + adsorbent[1] + 1% enzyme	Sheet	E′ (glutaraldehyde)
Calcium alginate (EAAP[2])	Drop	F
Calcium alginate (EAAP)	Drop	F′ (glutaraldehyde)
Calcium alginate + chitosan (EAAP)	Drop	G
Calcium alginate + chitosan (EAAP)	Drop	G′ (glutaraldehyde)
Calcium alginate (EAAP)	Sheet	H
Calcium alginate (EAAP)	Sheet	H′ (glutaraldehyde)
Calcium alginate + chitosan (EAAP)	Sheet	I
Calcium alginate + chitosan (EAAP)	Sheet	I′ (glutaraldehyde)
Calcium alginate + adsorbent (EAAP)	Sheet	J
Calcium alginate + adsorbent (EAAP)	Sheet	J′ (glutaraldehyde)
Calcium alginate + 3% latex + chitosan-calcium + 1% enzyme	Drop	K
Calcium alginate + 5% latex + chitosan-calcium + 1% enzyme	Drop	L
Calcium alginate + 10% latex + chitosan-calcium + 1% enzyme	Drop	M

[1]The adsorbent was cation and anion exchanger resin (Amberlite, MB-20); [2]EAAP: enzyme adsorbed after drop production; [3]treatment with 0.5% glutaraldehyde for 1 hour.

Drop. Derivatives A, A′, F, and F′ were prepared by dripping the 3% $(w \cdot v^{-1})$ sodium alginate in 0.15 M $CaCl_2$ $(1 : 2 - v \cdot v^{-1})$. Derivatives B, B′, G, and G′ were prepared by dripping in 3% $(w \cdot v^{-1})$ sodium alginate mixed with chitosan-acetic acid (1% $(w \cdot v^{-1})$ chitosan in 1% $(w \cdot v^{-1})$ acetic acid) in 0.15 M $CaCl_2$ $(1 : 2 - v \cdot v^{-1})$. The derivatives were kept for 1 hour under gentle agitation at 25°C with or without activation with glutaraldehyde (Table 1).

Sheet. These derivatives were obtained by three procedures: (a) 3% $(w \cdot v^{-1})$ sodium alginate (derivatives C, C′, G, and G′); (b) 3% $(w \cdot v^{-1})$ sodium alginate mixed with chitosan-acetic acid (1% $(w \cdot v^{-1})$ chitosan in 1% $(w \cdot v^{-1})$ acetic acid) (derivatives D, D′, I, and I′); (c) 3% $(w \cdot v^{-1})$ sodium alginate mixed with chitosan-acetic acid (1% $(w \cdot v^{-1})$ chitosan in 1% $(w \cdot v^{-1})$ acetic acid) and 2% $(w \cdot v^{-1})$ of the *adsorbent* (E, E′, J, and J′). Solutions containing the supports were homogenized under constant mechanical stirring. The sheets were transferred to a polypropylene Petri dish (100 mm in diameter) in a layer of 2.5 mm thickness and dried in oven at 30°C for a period between 24 hours, reaching a thickness of 0.5 mm. Then 30 mL of 0.15 M $CaCl_2$ was added on dried derivative which remained under mild agitation for 12 hours at 25°C. The derivatives were kept for 1 hour under gentle agitation at 25°C with or without activation with glutaraldehyde (Table 1).

2.4. Immobilization of 1,4-β-Endoglucanase by Entrapment in Alginate, Chitosan, and Latex. The 1,4-β-endoglucanase was immobilized by two types of procedure in solid (drop or sheet) support: (a) 1% $(v \cdot w^{-1})$ cellulase solution (calculated as protein based on mass of support) was mixed with the solution of sodium alginate and after 0.15 M $CaCl_2$ solution was used for precipitation and formation of drops or sheets (A, A′, B, B′, C, C′, D, D′, E, and E′); (b) mixing sodium alginate (combined or not with other supports) with 0.15 M $CaCl_2$ solution for the precipitation and formation of drops or sheets. After that, the drops or sheets were mixed with 1% $(w \cdot w^{-1})$ cellulase solution, calculated as protein based on the mass of support (F, F′, G, G′, H, H′, I, I′, J, and J′). Subsequently, part of derivatives was activated by the immersion in 0.5% glutaraldehyde water solution for 1 hour at 25°C [10]. After that, the derivatives were washed with deionized water (1000 mL) and packed in 5°C acetate buffer (pH 5.6) for use in tests of enzymatic activity. Cellulolytic activity (FPase, Filter Paper method) in derivatives was determined in one or more recycles (reuses).

The blend of alginate/chitosan/latex for endoglucanase immobilization was made in drops (3, 5, 7, and 10% $m \cdot m^{-1}$ latex) (Table 1, letters K, L, M, and N), with 3% $(w \cdot v^{-1})$ alginate solution and 1% $(w \cdot v^{-1})$ enzyme. The solution of chitosan-calcium was composed of 1% $(w \cdot v^{-1})$ chitosan, 1% $(v \cdot v^{-1})$ acetic acid, and 0.15 M $CaCl_2$, and the derivatives

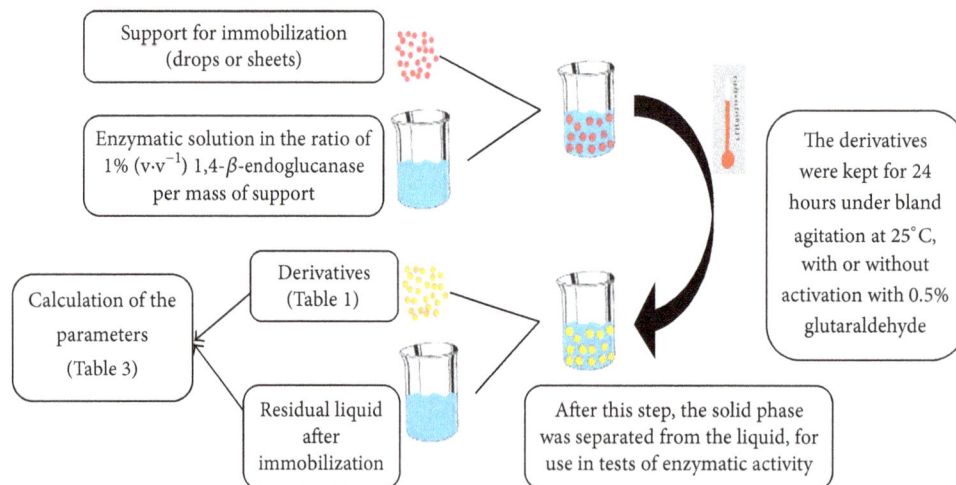

FIGURE 1: Flowchart of the enzyme immobilization process.

were kept for 24 hours under bland agitation at 25°C in 250 mL flask in shaker. After this process, the derivatives were washed with deionized water (1000 mL) and packed in 5°C in acetate buffer (pH 5.6) for use in tests of the cellulolytic activity. The enzymatic stability was determined for successive recycles.

2.5. Enzyme Immobilization in Drops and Sheets. The 1,4-β-endoglucanase was also immobilized in two types of solid supports: drops in oval shape (Table 1, letters A, A$'$; B, B$'$; F, F$'$; G, G$'$) and sheets (Table 1, letters C, C$'$; D, D$'$; E, E$'$; H, H$'$; I, I$'$). The drops A, A$'$ and F, F$'$ were prepared by dripping 3% (w·v^{-1}) sodium alginate on 0.15 M CaCl$_2$ solution (1:2, v·v^{-1}). Derivatives B, B$'$ and G, G$'$ were prepared by dripping 3% (w·v^{-1}) sodium alginate mixed with 1% (w·v^{-1}) chitosan in 1% (w·v^{-1}) acetic acid with addition of 0.15 M CaCl$_2$ (1:2 - v·v^{-1}). In both derivatives the drops were kept for 24 hours under bland agitation at 25°C, with or without activation with 0.5% glutaraldehyde as previously described (Table 1). The drops presented a diameter of 5-6 mm.

The sheets were obtained by three procedures: (a) solution of sodium alginate pure (derivative C, C$'$; G, G$'$), (b) mixture of the solution of alginate and chitosan (1:1) (derivatives D, D$'$; I, I$'$), and (c) sodium alginate solution with addition of 2% (w·v^{-1}) of the best support obtained in the preliminary tests with activated carbon, zeolite, ion exchange resin, or polystyrene, derivatives E, E$'$; J, J$'$. Solutions containing the supports were homogenized under constant mechanical stirring. After that, these sheets were transferred to a polypropylene Petri dish (100 mm in diameter) in a layer of 2.5 mm thickness and dried in oven at 25°C for 24 hours, reaching a thickness of 0.5 mm. Then, 30 mL of 0.15 M CaCl$_2$ was added on dried sheet which remained under bland agitation for 24 hours at 25°C, with or without activation with 0.5% glutaraldehyde as previously described (Table 1). The immobilization process is summarized in the flowchart (Figure 1).

2.6. Enzyme Immobilization in Drops with Latex. The drops (K, L, M, and N) were prepared by dropping the 3% (w·v^{-1}) sodium alginate solution, latex (3, 5, and 10%, v·v^{-1}), and 1% (w·v^{-1}) enzyme in solution chitosan-calcium (1:2, v·v^{-1}); the drops were kept for 24 hours under bland agitation at 25°C in 250 mL flask in shaker (Table 1). After this process, the derivatives were washed with deionized water (1000 mL) and packed in 5°C in acetate buffer (pH 5.6) for use in testing of the cellulolytic activity.

2.7. The Procedures of Derivatives with Latex in the enzyme Immobilization Process. The derivatives were prepared with latex in different concentrations as follows: (a) 5% latex (v·v^{-1}) as a stabilizing agent, different chitosan concentrations (0.5 and 1% w·v^{-1} in 1% of acetic acid, v·v^{-1}), and sodium alginate (w·v^{-1}) (0 and 3%). 1% of enzyme solution (w·v^{-1}) was added in 3% sodium alginate (w·v^{-1}) and 5% latex (v·v^{-1}); 3% sodium alginate (w·v^{-1}) + 0.5% chitosan (w·v^{-1}, 1% of acetic acid, v·v^{-1}) + 5% latex (v·v^{-1}); 3% sodium alginate (w·v^{-1}) + 1% chitosan (w·v^{-1}, 1% of acetic acid, v·v^{-1}) + 5% latex (v·v^{-1}); and 1% chitosan (w·v^{-1} in 1% of acetic acid, v·v^{-1}) + 5% latex (v·v^{-1}) (Table 1). In these derivatives, the drops were kept for 24 hours under bland agitation at 25°C in 250 mL flask in shaker. After this process, the derivatives were washed with deionized water (1000 mL) and packed at 5°C in acetate buffer (pH 5.6) for use in the tests of the cellulolytic activity.

2.8. Temperature. The temperatures of enzyme immobilization process (5, 10, 15, 25, 35, and 45°C) were performed with the best derivative in 24 hours in water bath (Tecnal, Piracicaba, SP, Brazil). Subsequently, the derivatives were washed with deionized water (1000 mL) and packed at 5°C in acetate buffer (pH 5.6) for use in the tests of the cellulolytic activity.

2.9. Analytical Procedure

2.9.1. Determination of Total Cellulose Activity (Filter Paper Activity, FPase).
The enzymatic activity of free or immobilized endoglucanase was measured using the technique of FPase [30]. One unit of Filter Paper (FPU) activity was defined as the amount of enzyme releasing 1 μmol of reducing sugar from Filter Paper per mL per min under the conditions described [31].

2.9.2. Determination of Protein and Sugar Concentration.
Protein was determined by Bradford method [32]. The determination of reducing sugars (glucose) liberated by hydrolysis of cellulose was carried out by 3,5-dinitrosalicylic acid (DNS) method under alkaline conditions [33].

2.10. Fourier Transform Infrared Spectroscopy (FTIR).
Fourier transform infrared spectroscopy (FTIR) was obtained to show the functional groups of drops (calcium alginate, latex, chitosan-calcium, and enzyme). The samples were measured directly by Attenuated Total Reflection (ATR) method, which is an efficient method for obtaining infrared information for the sample surface. The samples were characterized using a TENSOR 27 (Bruker, Germany) (500–6000 cm^{-1}) with a resolution of 4 cm^{-1}. Each reagent was analyzed separately and then together. As each group absorbs infrared radiation at a characteristic frequency, the inference of the presence of each group was possible comparing the radiation intensity versus frequency graph. Consequently, with this procedure, the determination of chemical interaction of materials was possible. The software Origin Pro 8® was used to make the statistical analysis of the data.

2.11. Compression Tests.
The deformation and strength measures obtained by universal testing machine were analyzed to determine the stiffness of materials. Young's modulus or modulus of elasticity is obtained by the equation of the graph of stress and strain within the elastic limit of the reversible deformation; the function of this equation is called BiDoseResp ($y = A1 + (A2 - A1)[p/1 + 10^{(LOGx01-x)h1} + 1 - p/1 + 10^{(LOGx02-x)h2}]$) [34, 35], which is the slope of the line [36]. Then this technique can measure the property of linear elastic solid materials. It measures the force (per unit area) that is needed to stretch (or compress) a material sample. A constant Young's modulus applies only to linear elastic materials. The material whose Young's modulus is very high can be approximated as rigid [36, 37]. The compression tests were carried out in a Universal Testing EMIC DL 2000 fitted with 10 kgf load cell at a speed. The cross-head speed employed was 10 mm/min. At least a triplicate of the samples was tested, and the average and standard deviation were reported. Prior to the tests, the samples were conditioned at 25°C. The mechanical compression test was conducted to examine the resistance to degradation of the drops. In this step, tests were performed with three different compositions: 3% sodium alginate (w·v^{-1}) + 1% chitosan (w·v^{-1}, 1% of acetic acid, v·v^{-1}); 3% sodium alginate (w·v^{-1}) + 1% chitosan (w·v^{-1} in 1% of acetic acid, v·v^{-1}) + 5% latex (v·v^{-1}); 3% sodium alginate (w·v^{-1}) + 1% chitosan (w·v^{-1} in 1% of acetic acid, v·v^{-1}) + 10% latex (v·v^{-1}).

2.12. Calculation of the Parameters of Immobilization.
The calculation of the parameters of immobilization and immobilized protein was calculated according to Silva et al. [10].

(i) Immobilized Protein Yield (IPY). It was calculated as percentage of immobilized protein based on difference of supplied protein (P_0) and the protein remaining in residual liquid after immobilization (P_f), divided by P_0 according to

$$\text{IPY (\%)} = \left[\frac{\left(P_0 - P_f \right)}{P_0} \right] \times 100. \tag{1}$$

(ii) Enzyme Immobilization Yield (IY). It was calculated as percentage of immobilized enzyme based on the difference of supplied enzyme (U_0) and the one remaining in the liquid after the immobilization (U_f) (both expressed in FPU) divided by U_0 according to

$$\text{IY (\%)} = \left[\frac{\left(U_0 - U_f \right)}{U_0} \right] \times 100. \tag{2}$$

(iii) Recovered Activity (RA). It was calculated as percentage of immobilized enzyme (expressed in FPU) in the support (U_{support}) divided by the difference between U_0 and U_f, according to

$$\text{RA (\%)} = \left[\frac{\left(U_{\text{support}} \right)}{\left(U_0 - U_f \right)} \right] \times 100. \tag{3}$$

The RA can be considered more precise immobilization yield than IY since the first is based on the enzyme activity measured in the derivatives (immobilized enzymes). There is a loss of activity in the immobilized enzyme, or not all enzymes remain active in the derivative.

(iv) Lost Activity (LA). Since U_f can be recovered in another immobilization, LA represents the percentage of the lost enzymes in the total immobilization process, due to the difference of U_0 and the sum of U_{support} and U_f, according to

$$\text{LA (\%)} = \left[\frac{\left(U_0 - \left(U_{\text{support}} + U_f \right) \right)}{U_0} \right] \times 100. \tag{4}$$

(v) FPU Yield (%). The increase in the FPU activity: this parameter was calculated by the sum of FPU in all recycled derivatives in enzymatic hydrolysis of Filter Paper (ΣFPU), divided by U_0, according to

$$\text{FPU Yield} = \left[\frac{\left(\Sigma\text{FPU} \right)}{\left(U_0 \right)} \right], \tag{5}$$

expressed in number of times.

(vi) Net FPU Yield. It is the net yield in FPU from ΣFPU of the all recycled derivatives in enzymatic hydrolysis of Filter Paper in relation to the amount of supplied enzyme without U_f. This parameter was calculated according to

$$\text{Net FPU Yield} = \frac{\Sigma\text{FPU}}{\left(U_0 - U_f\right)}, \qquad (6)$$

expressed in number of times.

2.13. Statistical Treatment. Tests for enzymatic activity were conducted in triplicate, and data was submitted for analysis of variance (ANOVA), and the means were compared by Tukey test, using the program GraphPad Instat, Version 3.05 (Rutgers University). Curves and graphs were made in OriginLab software, Version 9.1, and Excel 2010. The treatments were analyzed statistically and considered significant at $p < 0.05$.

3. Results

3.1. Characterization of Commercial 1,4-β-Endoglucanase. The commercial 1,4-β-endoglucanase (ROHAMENT, CL) was analyzed by the total cellulase (Filter Paper activity, FPase) and total protein concentration (Bradford). In this way, the volumetric activity (FPU/mL) and specific activity (FPU/g protein) were determined in 3, 5, 12, and 20% $(\text{v}\cdot\text{v}^{-1})$ cellulase solution in acetate buffer at 40, 50, and 60°C (Table 2). The highest values of cellulase activity were obtained at 50°C with 20% $(\text{v}\cdot\text{v}^{-1})$ enzyme solution (0.67 ± 0.00 FPU/mL), although with a lower specific activity (72.94 ± 0.09 FPU/g protein) when compared to 3% $(\text{v}\cdot\text{v}^{-1})$ enzyme solution (0.34 ± 0.00 FPU/mL and 245.95 ± 0.15 FPU/g protein). Despite these higher results at 50°C, there were no statistical differences between 50 and 60°C ($p > 0.05$), but a significant difference at 40°C ($p < 0.05$). There was a decrease of more than 30% in activity in 20% $(\text{v}\cdot\text{v}^{-1})$ enzyme solution when comparing 40°C and 50°C (0.45 ± 0.02 and 0.67 ± 0.05, resp.). Therefore, the increase of the activity did not follow the same proportion of the increase of enzyme concentration, probably due to the limitation of the substrate concentration for the excess of enzyme.

3.2. Immobilization of 1,4-β-Endoglucanase on Activated Carbon, Zeolite, Ion Exchange Resin, and Polystyrene. Preliminary tests of the cellulase immobilization on activated carbon, zeolite, ion exchange resin, and polystyrene supports were performed for the selection of a promising solid support (Table 3). The highest and significant ($p < 0.05$) enzyme activity was obtained with the use of ion exchange resin (0.32 ± 0.02 FPU/g support) representing almost double. Therefore, this support was used in the derivatives of the sheets E, E$'$; J, J$'$ according to Table 1.

Despite zeolite and polystyrene showing a better yield of enzyme immobilization (IY) (83.3% and 68.8%, resp.), their Recovered Activity (RA) was, respectively, only 13.4% and 20%, reflecting the higher Lost Activity (LA) in FPU per gram of support. The best adsorption of endoglucanase was on the ion exchange resin. Even with this last support showing

that IY of 33.34% reached higher RA (81.6%) due to lower LA (6.14%), reflecting positively in the immobilized activity. The activated carbon showed no immobilization yields (Table 3).

3.3. Immobilization of 1,4-β-Endoglucanase by Pure and Mixed Alginate. These derivatives were made by adding the enzyme during the preparation of the derivatives before entrapment with CaCl$_2$. Table 3 shows the results with the drops (A, A$'$, B, B$'$, K, L, and M) and sheets (C, C$'$, D, D$'$, E, and E$'$).

The drops of the derivatives B and B$'$ (Table 3) showed, respectively, the highest FPase activity (0.15 and 0.17 FPU/g of support) and IY (56.1 and 55.4%), although with a great loss of activity (LA, 42.1% and 39.6%, resp.). The content of the immobilized protein (IP) and Immobilized Protein Yield (IPY) were almost the same (IP, 12 mg protein/g support and IPY, 86%) in all derivatives (A, A$'$, B, and B$'$).

The sheet D showed higher active enzyme (0.12 FPU/g of support) than the sheets C, C$'$, D$'$, E, and E$'$ and higher RA (54.1%), lower LA (10.3%), and higher IY (22.4%) when compared to the other derivatives. The content of the immobilized protein (IP) and protein yield (IPY) in derivative E$'$ were better (IP 11.2% and IPY 93.8%) when compared to the others (Table 3).

In the derivatives of drops immobilization with the addition of latex on supports of alginate and chitosan, IPY was gradually increasing to 85, 98, and 99%, respectively, according to the increase of latex concentration (3, 5, and 10% latex) in immobilization derivatives (Table 3). This increase was also followed by the increase of cellulose activity from 0.018 to 0.084 FPU/g support. However, in the drops of the derivative B with only alginate and chitosan (without latex) the immobilized active enzyme was higher (0.153 FPU/g) but with a lesser RA (25%) and higher LA (42.1%) (Table 3). The latex makes the adsorption of the enzyme on support difficult. This hypothesis should be considered since in these immobilizations the total protein immobilized in the support (IP) was very low when compared with the other derivatives without latex. For example, in the drops of the derivatives B and B$'$ with only alginate and chitosan (without latex), the IP were 12 mg/g of support for each derivative, while only 1.9–2.1 mg/g of support was obtained for the derivatives with latex (derivatives K, L, and M) (Table 3). One possibility of explaining this problem was the presence of lysozyme/chitinase activity from *Hevea brasiliensis* latex [38] which could change the chemical structure of alginate and chitosan on some level.

3.4. Enzyme Immobilization by Immersion of Solid Supports in 1,4-β-Endoglucanase Solution. The results of the immobilization process by immersion of solid supports (drops or sheets) in enzyme solution are shown in Table 4 for the drops (F, F$'$, G, and G$'$) and sheets (H, H$'$, I, I$'$; and J, J$'$, sheets).

The drops of the derivatives F and F$'$ (Table 3) showed higher FPase activity (0.13 and 0.093 FPU/g of support) and although in both derivatives similar results were obtained, their IY showed a great difference. The derivative F$'$ showed a better IY (51.5%) than F (15.4%). However, the Recovered Activity (RA) of F$'$ was lower than the

TABLE 2: Cellulase activity of 1,4-β-endoglucanase (ROHAMENT, CL) in different concentrations and temperatures in pH 5.6.

Enzyme concen. (%) (v·v⁻¹)	Protein concen. (g·mL⁻¹·10⁻⁴)	40°C		50°C²		60°C²	
		FPU/mL	FPU/g	FPU/mL	FPU/g	FPU/mL	FPU/g
3	13.83 ± 1.15^a	$0.12 \pm 0.01^{a,a'}$	$90.75 \pm 1.1^{a,a'}$	$0.34 \pm 0.05^{a,b'}$	$245.95 \pm 10.4^{a,b'}$	$0.30 \pm 0.04^{a,b'}$	$236.95 \pm 12.2^{a,b'}$
5	37.43 ± 3.01^b	$0.16 \pm 0.00^{b,a'}$	$44.93 \pm 5.05^{b,a'}$	$0.43 \pm 0.03^{b,b'}$	$117.28 \pm 9.1^{b,b'}$	$0.41 \pm 0.02^{b,b'}$	$108.28 \pm 10.1^{b,b'}$
12	77.89 ± 09.05^c	$0.37 \pm 0.03^{c,a'}$	$48.76 \pm 3.10^{b,a'}$	$0.61 \pm 0.05^{c,b'}$	$78.50 \pm 7.0^{c,b'}$	$0.59 \pm 0.03^{c,b'}$	$69.50 \pm 09.0^{c,b'}$
20	92.71 ± 12.03^d	$0.45 \pm 0.02^{d,a'}$	$49.21 \pm 3.09^{b,a'}$	$0.67 \pm 0.05^{c,b'}$	$72.94 \pm 5.0^{c,b'}$	$0.65 \pm 0.05^{c,b'}$	$71.0 \pm 5.0^{c,b'}$

Different letters indicate that they are statistically different ($p < 0.05$), without lines indicate comparison between concentrations, and with lines between temperatures.

TABLE 3: Parameters of cellulase immobilized in different derivatives.

Derivatives	IP[1]	IPY[2]	FPU/g[3]	IY[5] (%)	RA[6] (%)	LA[7] (%)
Control-free enzyme (5%)			0.438^4	100	100	—
Drops						
A (calcium alginate)	11.9	85.6	0.141 ± 0.01^a	13.4	96.0	0.53
A′ (calcium alginate + glut.[8])	12.0	86.1	0.077 ± 0.02^b	13.5	52.2	6.47
B (calcium alginate + chitosan)	12.0	86.3	0.153 ± 0.06^a	56.1	25.0	42.1
B′ (calcium alginate + chitosan + glut.)	12.0	86.5	0.172 ± 0.01^a	55.4	28.5	39.6
F (calcium alginate)	5.0	48.8	0.134 ± 0.03^a	15.4	76.0	3.7
F′ (calcium alginate + glut.)	5.3	52.0	0.093 ± 0.02^a	51.5	16.0	43.3
G (calcium alginate + chitosan)	4.9	48.6	0.080 ± 0.00^b	13.5	51.4	6.6
G′ (calcium alginate + chitosan + glut.)	5.3	52.4	0.086 ± 0.01^b	13.4	55.7	6.0
K (calcium alginate + chitosan + 3% latex)	1.9	84.9	0.018 ± 0.00^c	49.37	44.8	27.2
L (calcium alginate + chitosan + 5% latex)	2.0	98.4	0.066 ± 0.01^b	42.6	62.2	16.0
M (calcium alginate + chitosan + 10% latex)	2.1	99.5	0.084 ± 0.03^b	45.8	73.0	12.3
Sheets						
C (calcium alginate)	9.3	94.7	0.063 ± 0.02^m	36.3	12.6	31.8
C′ (calcium alginate + glut.)	9.6	97.5	0.050 ± 0.00^m	53.5	6.78	49.9
D (calcium alginate + chitosan)	5.9	33.1	0.120 ± 0.03^n	22.4	54.1	10.3
D′ (calcium alginate + chitosan + glut.)	5.9	33.1	0.015 ± 0.01^o	17.8	8.77	16.2
E (calcium alginate + resin)	10.6	88.8	0.085 ± 0.01^m	36.2	14.7	30.9
E′ (calcium alginate + resin + glut.)	11.2	93.8	0.049 ± 0.01^m	30.8	10.0	27.7
H (calcium alginate)	2.7	27.9	0.105 ± 0.02^n	28.5	15.0	24.2
H′ (calcium alginate + glut.)	3.2	32.4	0.093 ± 0.00^m	35.3	10.7	31.5
I (calcium alginate + chitosan)	0.1	1.1	0.047 ± 0.00^m	55.4	3.5	53.5
I′ (calcium alginate + chitosan + glut.)	4.5	45.8	0.089 ± 0.03^m	46.1	7.9	42.4
J (calcium alginate + resin)	1.4	15.0	0.075 ± 0.00^m	23.1	13.2	20.0
J′ (calcium alginate + resin + glut.)	1.6	16.0	0.091 ± 0.00^m	24.6	15.2	20.8
Other derivatives						
Activated carbon	0.0	0.0	0.00 ± 0.00	0.0	0.0	100.0
Zeolite	0.46	4.6	0.13 ± 0.01^b	83.3	13.4	72.1
Ion exchange resin	0.97	9.7	0.32 ± 0.02^c	33.3	81.7	6.1
Polystyrene	2.26	22.7	0.16 ± 0.05^b	68.8	20.0	55.0
3% calcium alginate + 5% latex	1.7	33.5	0.0025 ± 0.00^a	35.0	100.0	—
3% calcium alginate + 0.5% chitosan + 5% latex	2.0	71.5	0.026 ± 0.00^b	48.0	49.2	24.9
3% calcium alginate + 1% chitosan + 5% latex	2.0	98.4	0.066 ± 0.00^c	42.6	62.2	16.0
1% chitosan + 5% latex[9]	0.0	0.0	0.0	0.0	0.0	100.0

[1] Amount of immobilized protein (mg protein/gram of support); [2] Immobilized Protein Yield (%); [3] Enzymatic activity (FPU) per gram of support; [4] Enzymatic activity (FPU/mL) of enzyme solution (5%); [5] Immobilization Yield (%); [6] Recovered Activity (%); [7] Lost Activity (%); [8] Treatment with 0.5% glutaraldehyde (%); [9] there was no solubilization of chitosan in latex. Obs. different letters indicate that they are statistically different ($p < 0.05$).

TABLE 4: Parameters of cellulase immobilized (derivative B′) in different temperatures.

Temperature of immobilization (°C)[1]	IP[2]	IPY[3]	FPU/g[4]	IY[5] (%)	RA[6] (%)	LA[7] (%)
5	11.9	85.8	0.087 ± 0.01^a	81.7	9.7	73.7
10	12.1	87.2	0.091 ± 0.03^a	80.4	10.3	72.0
15	11.9	85.6	0.093 ± 0.05^a	53.7	15.8	45.2
25	12.0	86.5	0.172 ± 0.01^b	55.4	28.5	39.6
35	12.0	86.4	0.065 ± 0.10^c	57.0	10.4	51.1
45	12.1	87.0	0.069 ± 0.08^c	57.9	10.9	51.6

[1] Derivative B′ (alginate + chitosan + glut.); [2] amount of immobilized protein (mg protein/gram of support); [3] Immobilized Protein Yield (%); [4] Activity FPU/g of support; [5] Immobilization Yield (%); [6] Recovered Activity (%); [7] Lost Activity (%). Obs. different letters indicate that they are statistically different ($p < 0.05$).

derivative F (RA 16% and 76%, resp.). This last derivative showed a low Lost Activity (LA, 3.7%), and consequently the cellulolytic activity was higher when compared to the derivative treated with glutaraldehyde (F′), in which the Lost Activity (LA) was 43.3%. The derivatives G and G′ showed IPY of 48.6% and 52.4% respectively, although they had lower IY (13.4 and 13.5%) and RA (51.4 and 55.7%) (Table 3).

Among the sheets (Table 3) prepared by enzyme immersion, higher protein immobilization was verified in derivative I′ (IP 4.5 mg/g and IPY 45.8%). However, the derivatives with added enzyme during the preparation of support were superior, since the derivative C′ showed IP of 9.6 mg/g and IPY of 97.5%. In sheets H and H′ higher cellulolytic activity (0.105 and 0.093 FPU/g support) than the others using enzyme immersion was verified. These sheets demonstrated an improvement in IY, but the presence of glutaraldehyde (derivative H′) caused a decrease in the RA (10.74%) if compared with RA in derivative H (15%). The derivatives I and I′ showed, respectively, higher IY (55.4% and 46.1%), but also higher LA (53.5% and 42.4%).

3.5. Effect of Temperature on Immobilization of 1,4-β-Endoglucanase.

In the test of the temperature of immobilization derivative B′ was used since with this derivative was obtained the best results of FPU per gram of derivative (Table 4). The parameter IPY was similar for all temperatures, but the FPU/g was significantly higher (0.172 FPU/g of support) at 25°C than at other temperatures ($p < 0.05$). Similarly, the RA was the highest (28.5%) and the LA was the lowest value (39.6%). Therefore, these results indicated 25°C as more efficient in the immobilization process.

3.6. Immobilization of Endoglucanase in Alginate, Chitosan, and Latex.

The immobilization of endoglucanase in hybrid support with alginate, chitosan, and latex was evaluated in two concentrations of chitosan (0.5 and 1%), with and without sodium alginate (Table 3).

The Immobilized Protein Yield (IPY) showed that the highest the concentration of chitosan, the greater this parameter: 3% sodium alginate ($w \cdot v^{-1}$) + 5% latex ($v \cdot v^{-1}$) (IPY 33.5%); 3% sodium alginate ($w \cdot v^{-1}$) + 0.5% chitosan ($w \cdot v^{-1}$, 1% of acetic acid, $v \cdot v^{-1}$) + 5% latex ($v \cdot v^{-1}$) (IPY 71.5%); 3% sodium alginate ($w \cdot v^{-1}$) + 1% chitosan ($w \cdot v^{-1}$, 1% of acetic acid, $v \cdot v^{-1}$) + 5% latex ($v \cdot v^{-1}$) (IPY 98.4%). For the derivative with only chitosan, no reliable results were observed, probably because an unstable drop was obtained.

The derivative prepared with only alginate showed the highest RA (100%), but when comparing the amount of FPU per gram of support (0.0025 FPU/g of support) this value was 69 times lower than the derivative B′ (0.172 FPU/g of support) (Table 3). The derivatives using 0.5% and 1% chitosan showed better results when compared to the drops without chitosan, respectively, 0.026 and 0.066 FPU/g of support ($p < 0.05$). Therefore, based on these results the synergism between chitosan and alginate requires a better and more effective enzyme immobilization with higher yields, as already shown

at derivative B′. However, if the resistance and flexibility were also considered, alginate, chitosan, and latex are more interesting, according to the following tests.

3.7. Compression Tests.

The addition of latex into samples influenced the mechanical behavior. The latex became stiffer and resistant with a greater plastic deformation (more ductile), leading to higher interfacial interactions, acting as reinforcement or creating cross-linking [39]. The test of mechanical compression in three different compositions of latex in the drops was performed to determine their resistance to degradation: (a) 3% sodium alginate ($w \cdot v^{-1}$) + 1% chitosan ($w \cdot v^{-1}$, 1% of acetic acid, $v \cdot v^{-1}$); (b) 3% sodium alginate ($w \cdot v^{-1}$) + 1% chitosan ($w \cdot v^{-1}$, 1% of acetic acid, $v \cdot v^{-1}$) + 5% latex ($v \cdot v^{-1}$); (c) 3% sodium alginate ($w \cdot v^{-1}$) + 1% chitosan ($w \cdot v^{-1}$, 1% of acetic acid, $v \cdot v^{-1}$) + 10% latex ($v \cdot v^{-1}$). Therefore, the effect of latex concentration in the mechanical strength of these drops was evaluated, taking into account the fact that the more sturdy support for the immobilization of the enzyme suffers less degradation. Figure 2(a) refers to the behavior of support alginate and chitosan in the test of compression, with the x-axis being the characteristic compressive strength of the material and the y-axis being the deformation ($y = 1.4902x + 0.1907$, $R^2 = 0.9426$). In Figure 2(b) ($y = 2.3351x + 0.0322$, $R^2 = 0.9565$) and Figure 2(c) ($y = 1.1876x + 0.2122$, $R^2 = 0.9787$) the compression of the drops was analyzed according to the presence of latex (5 and 10%, $v \cdot v^{-1}$, resp.). The drops prepared with 3% ($w \cdot v^{-1}$) sodium alginate and 1% chitosan (without latex) showed the mean of the elasticity modulus 1.787 ± 1.033 MPa. However, the drops with 5 or 10% of latex ($v \cdot v^{-1}$) showed higher elasticity modulus (mean of 2.310 ± 0.160 Pa and 2.500 ± 1.140 MPa, resp.) than the drops without latex. Therefore, the addition of 5 and 10% latex in the samples influenced the mechanical behavior, increasing this parameter by 1.3 and 1.4 times, respectively.

3.8. FTIR Analysis of Endoglucanase Immobilized in Drops of Alginate, Chitosan, and Latex.

Each functional group of chemical compounds absorbs a characteristic frequency in the infrared region [40]. The infrared spectrogram can be used to characterize the functional groups of an unknown material. Spectra were obtained from the latex, sodium alginate, and chitosan and mixtures of these materials with each other and with the commercial enzyme used in this study. Figure 3 shows the infrared spectra by Fourier transform, using module of "Attenuated Total Reflectance" (ATR) in the region 500–5000 cm^{-1}.

Figure 3 showed significant absorption spectra for the materials, blends, and enzyme; the main bands are 1060 cm^{-1}, 1090 cm^{-1}, between 1070 and 1100 cm^{-1}, between 1400 and 1500 cm^{-1}, 1600 cm^{-1}, 2600 cm^{-1}, 3000 cm^{-1}, 3200 cm^{-1}, and 3400 cm^{-1}. The chitosan spectrum (Figure 3) had two distortions between 1600 cm^{-1} and 3400 cm^{-1} related to a primary amine (NH_2) obtained by deacetylation of chitin, a band near 3000 cm^{-1} (C-H bonds, organic compound) and 1400 cm^{-1} (alkyl groups and carboxylate O-C=O). Other bands were presented in 1060 cm^{-1} (C-O stretch vibrational primary

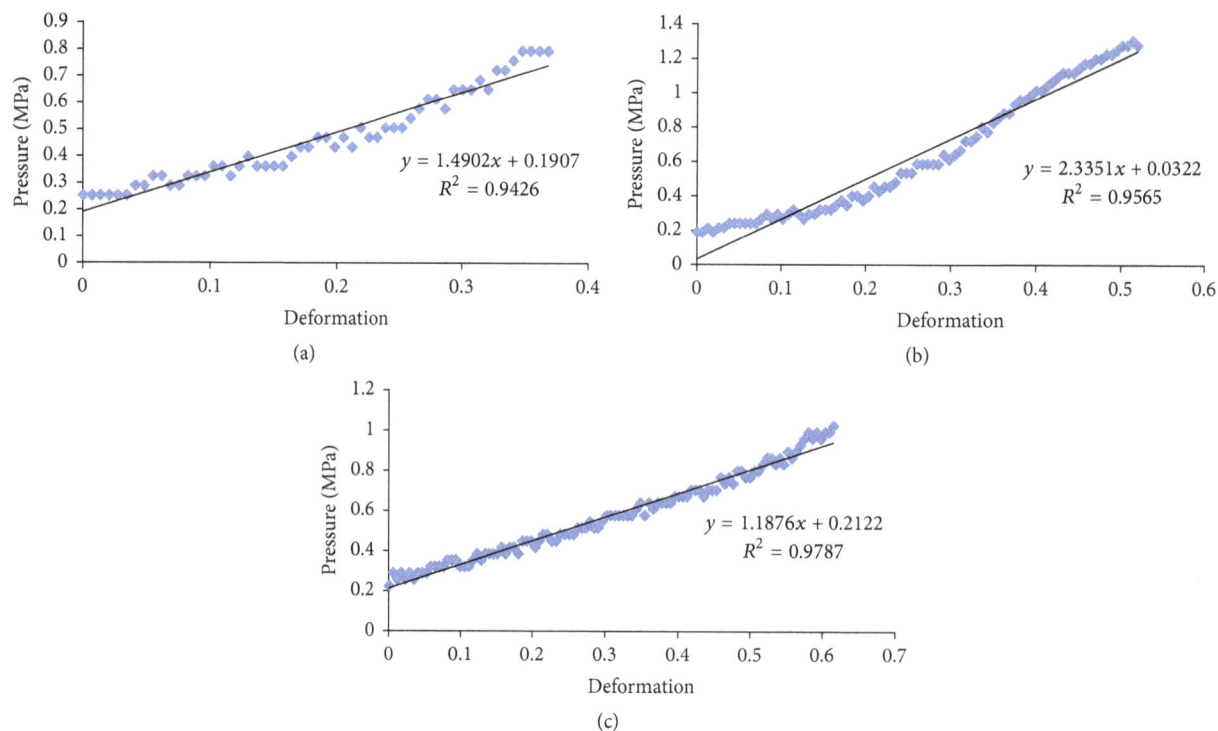

FIGURE 2: Different supports in behavior under compressive force. Supports are composed of (a) 3% sodium alginate (w·v^{-1}) + 1% chitosan (w·v^{-1} - 1% of acetic acid - v·v^{-1}); (b) 3% sodium alginate (w·v^{-1}) + 1% chitosan (w·v^{-1} - 1% of acetic acid - v·v^{-1}) + 5% latex (v·v^{-1}) in behavior under compressive force; (c) 3% sodium alginate (w·v^{-1}) + 1% chitosan (w·v^{-1} - 1% of acetic acid - v·v^{-1}) + 10% latex (v·v^{-1}).

alcohol); 1090 cm^{-1} (vibrational stretch ether group); and 1070–1100 cm^{-1} (aliphatic amines) [41].

The FTIR spectrum of sodium alginate is presented in Figure 3. Initially, the absorption peak observed at 3200–3500 cm^{-1} corresponds to stretching of hydroxyl groups, 2936 cm^{-1} is due to C-H stretching, and 1026 cm^{-1} is due to C-O-C stretching [42].

The interaction of sodium alginate with calcium chloride in mixtures causes a small change in the band near 1600 cm^{-1}. Due to the connection and mixture with alginate the peaks related to the calcium chloride disappeared indicating the connection. The chitosan showed a band in 2000 cm^{-1} and 2600 cm^{-1}, and vibration around 3200–3300 cm^{-1}, and in the region between 1000 and 1700 cm^{-1}. The IR of the blend chitosan/alginate there is bands in 2000 cm^{-1} and 2600 cm^{-1} and vibration between 1000 and 1700 cm^{-1}. However, when the enzyme was mixed with this blend the bands in 2000 cm^{-1} and 2600 cm^{-1} were absent, indicating that there was a chemical interaction between enzyme and this blend (Figure 3).

The latex was employed in this study due to some interesting characteristics such as easy manipulation, low cost, and high mechanical resistance. The FTIR spectra of polymers blends (latex + chitosan, latex + alginate, and latex + chitosan + alginate) were analyzed. The pure latex spectrum (Figure 3(b)) had significant absorption near 2800 cm^{-1}, which equals the bonds C-H present in abundance in the material. Another slight distortion that can be seen is 3200–3500 cm^{-1}, indicating the presence of hydroxyl group.

The band of absorption near 1400 to 1500 cm^{-1} indicated the presence of N-O and links that may be related to the ammonia present in the latex mixture to keep it liquid. In summary, FTIR spectral data confirmed the chemical stability of natural latex in alginate and chitosan blends [17].

However, Figure 3(c) presents FTIR spectra of blend (latex + alginate) before and after enzyme encapsulation. The most pronounced effect is a decrease of band absorbance at 2852–2925 cm^{-1} and 2961 cm^{-1} corresponding to CH$_2$ symmetric and CH$_3$ asymmetric stretching vibrations. It indicates the interaction between molecular chain of enzyme and this blend.

3.9. Recycle of Derivatives in Enzymatic Reaction of the Paper Filter as Substrate. The recycles of the drops (Figure 4) and sheets (Figure 5) that reacted with Filter Paper in water solution are presented. The recycles of all derivatives were quantified by the maximum number of recycles and the sum of FPU/g of support (Table 5). Derivatives A, A', B, and B' achieved greater stability over successive recycles of cellulolytic activity (Figure 4(a)). The drops achieved up to 13 recycles of immobilized enzyme in the derivatives B and B', while in the derivatives F, F', G, and G' the cellulolytic activity was missed in the 5th recycle (Figure 4(a)). In the derivative B better results of its reuses was verified, and the total activity summed over 13 cycles was 1.59 FPU/g of support, while only 1.01 FPU/g of derivative B was obtained in the immobilized enzyme recycle. Therefore, there was an FPU Yield of 1.57 times higher in the cellulolytic activity than the total enzyme supplied for this immobilization method.

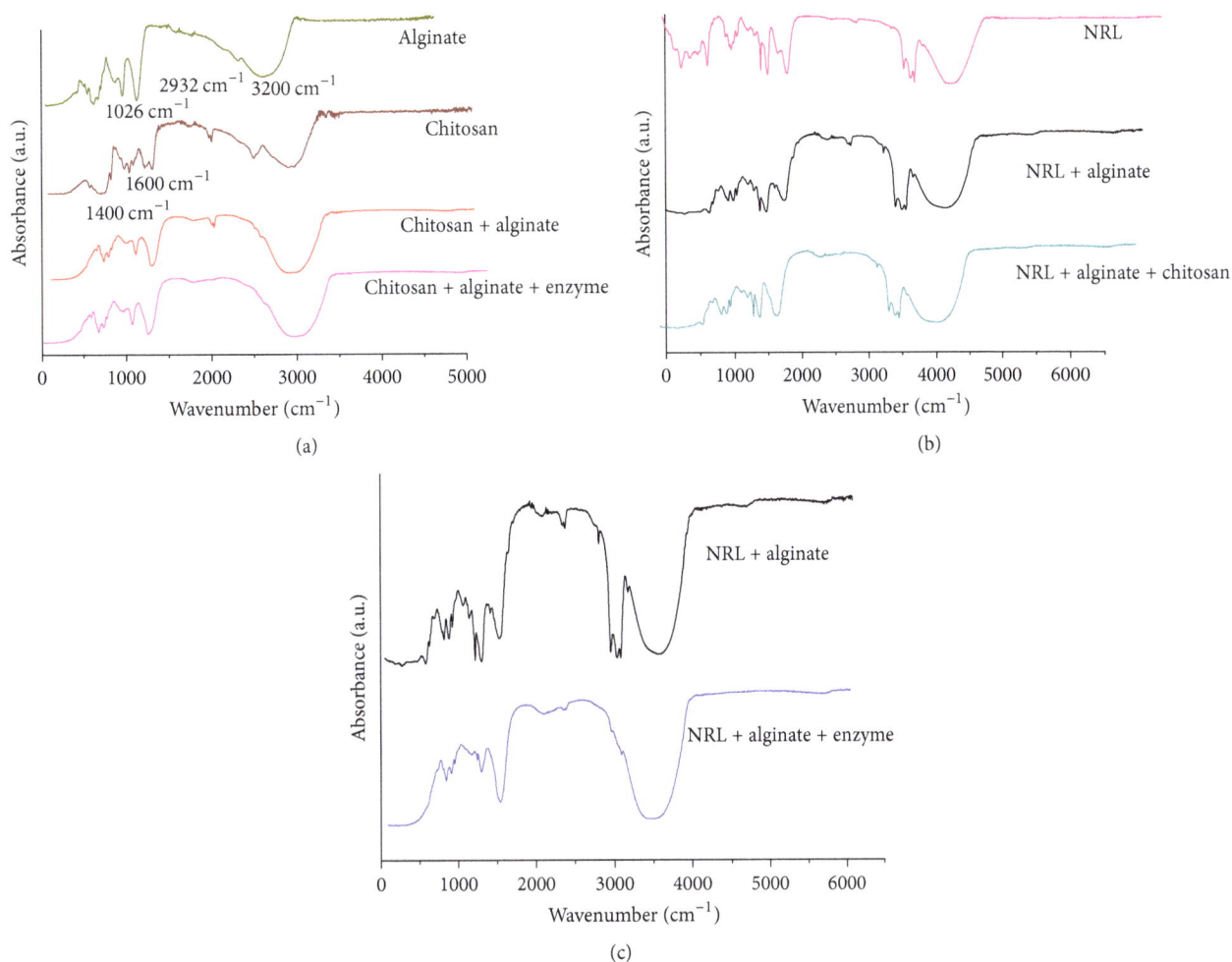

Figure 3: Infrared spectroscopy: (a) alginate, chitosan, blend, and enzyme loaded blend; (b) natural latex: NRL, NRL + alginate, and NRL + alginate + chitosan; (c) NRL + alginate and NRL + alginate + enzyme.

Table 5: Number of recycles of celluloses immobilized in pure and hybrid derivatives on reaction with paper filter as substrate.

Derivat[1]	Glut.[2]	Recycle number				Sum of FPU in all recycles (FPU/g)			
		Enzyme added during derivative preparation		Enzyme immerged on support		Enzyme added during derivative preparation		Enzyme immerged on support	
		Sheet	Drop	Sheet	Drop	Sheet	Drop	Sheet	Drop
Calcium align.	No	27 (C)	6 (A)	4 (H)	4 (F)	1.72	0.36	0.25	0.20
	Yes	27 (C′)	7 (A′)	4 (H)	4 (F′)	1.37	0.17	0.25	0.13
Calcium align. + chitosan	No	25 (D)	13 (B)	1 (I)	4 (G)	0.81	1.59	0.05	0.14
	Yes	25 (D′)	13 (B′)	5 (I′)	4 (G′)	0.74	1.62	0.43	0.12
Calcium align. + resin	No	28 (E)	NP[c]	4 (J)	NP	1.95	NP	0.16	NP
	Yes	28 (E′)	NP	5 (J′)	NP	2.30	NP	0.24	NP
3% latex	No	NP	8 (K)	NP	NP	NP	0.17	NP	NP
5% latex	No	NP	6 (L)	NP	NP	NP	0.42	NP	NP
10% latex	No	NP	4 (M)	NP	NP	NP	0.19	NP	NP

[1] The total of endoglucanase supplied to prepare each derivative was A, A′, F, F′, G, G′ = 1.10 FPU/g support, B/B′, K/L/M − 1.01 FPU/g support; C/C′, E, E′ = 1.4 FPU/g support, D, D′ = 0.98 FPU/g support, G, G′ = 1.1 FPU/g support, H, H′ = 1.22 FPU/g support; I, I′ = 1.81 FPU/g support; J/J′ = 0.868 FPU/g support; [2] Treated with 0.5% glutaraldehyde in 1 hour. [c] Not performed.

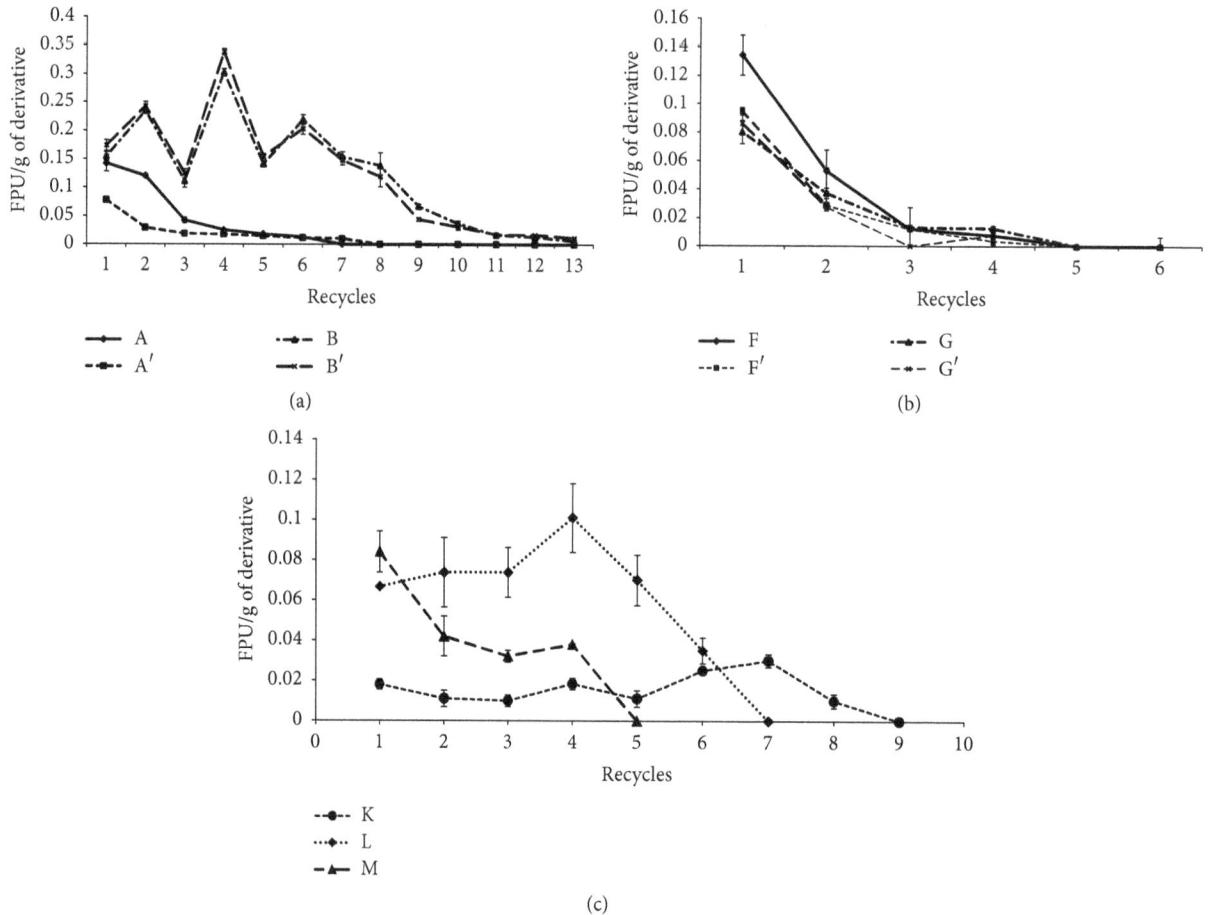

(a)

(b)

(c)

FIGURE 4: Recycle of the cellulases in drops on reactions with paper filter: (a) enzyme immobilized during the drop in production; (b) enzyme immobilized after of the drop was produced; (c) enzyme immobilized during the production of drop including latex (3, 5, and 10%). Derivatives are described in Table 1.

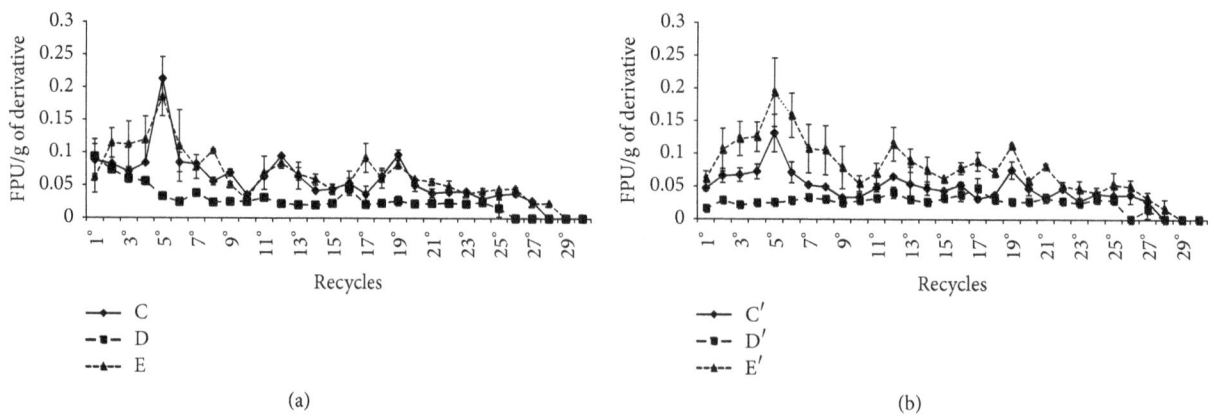

(a)

(b)

FIGURE 5: Recycle of cellulases on sheets in reactions with Filter Paper: (a) derivatives prepared with the enzyme adsorbed during preparation of sheets without glutaraldehyde; (b) derivatives treated with glutaraldehyde. Derivatives are described in Table 1.

However, if the Recovered FPU in liquid after immobilization was considered, the Net FPU Yield was 2.81 times (Table 6).

The stability of cellulolytic sheets was also evaluated by their recycles in reactions with paper filter as substrate and determination of FPase. The sheets (Figures 5(a) and 5(b))

showed greater stability in successive recycles of enzyme reaction. The sheets E and E′ were more efficient than the others since 28 reutilizations of derivatives were performed. The sheets H, H′, I, I′, J, and J′ showed a low stability since their recycles showed no activity after the 6th recycle. The

TABLE 6: General overview of immobilization process in derivatives B and E$'$.

Item	Data	Drop B	Sheet E$'$
1	Number of recycles	13	28
2	Mass of derivative	15 g	3 g
3	Total enzyme supplied (U_0)	15.15 total FPU	4.2 total FPU
4	Supplied Enzyme per gram	1.01 FPU	0.047 FPU
5	Activity in liquid after the immobilization (U_f)	6.65 total FPU	2.90 total FPU
6	Total activity in support ($U_{support}$)	2.12 FPU	0.14 FPU
7	Total Lost Activity	6.38 FPU	1.16 FPU
8	$U_0 - U_f$	8.5 total FPU	1.3 total FPU
9	Sum of FPU in all recycles	23.85 total FPU	6.90 total FPU
10	FPU Yield	1.57 times	1.64 times
11	Net FPU Yield	2.81 times	5.31 times

Item 4 = item 3/item 2; item 7 = item 3 − (item 6 + item 5); item 8 = item 3 − item 5; item 10 = item 9/item 3; item 11 = item 9/8.

sheets treated with glutaraldehyde have proved to be superior. The sheet E showed 1.95 FPU/g of support considering the total FPase activity summed in 28 cycles, but E$'$ reached 2.30 FPU/g of support, while only 1.4 FPU/g of derivatives E or E$'$ were supplied for the enzyme immobilization, or FPU Yield in sheet E was 1.64 times higher than the spent FPU (U_0) (Table 6). In addition, the net balance between the enzyme spent for the immobilization and the sum of enzyme obtained with the reuse in sheet E$'$ (6.9 total FPU) means an increase of 5.3 times. These results demonstrate a great economy in the use of this enzyme by this method of enzyme immobilization.

The drops K with 3% latex (v·v^{-1}) reached a higher stability during successive recycles of cellulolytic activity. The average of total activity was 0.17 FPU/g derivative, obtained by the sum of residual activities in 8 recycles (Figure 4(c)). The drops L with 5% latex (v·v^{-1}) showed absence of cellulolytic activity in the 6th cycle, but the average of total activity was 0.42 FPU/g of support by the sum of residual activities in 5th recycle. An average of 0.19 FPU/g of support was obtained in the drops M with 10% latex (v·v^{-1}) by the sum of 4 recycles of the immobilized enzyme (Figure 4(c)).

4. Discussion

Some aspects of endoglucanase immobilization process were studied, like shape (sheet or drop), the type of polymer used in support, the use of glutaraldehyde as cross-linking agent, and the step where the enzyme is added to produce the derivative. The evaluation of the best immobilization procedures was based on four main aspects: (a) cellulolytic activities effectively incorporated in the derivatives (FPU/g support), or the Recovered Activity (RA) which demonstrates the efficiency of the immobilization process, (b) the interaction of the latex in immobilization and stability of the supports, (c) the enzyme recycles, showing the stability of the chemical interaction of enzyme and solid support, and (d) the increase in the FPase activity in recycled derivatives, and its results in the economy of spending enzyme.

4.1. Immobilization on Activated Carbon, Zeolite, Ion Exchange Resin, and Polystyrene. Despite zeolite and polystyrene showing a better yield of enzyme immobilization,

their Recovered Activity was low, probably due to the type of chemical interaction between the substrate and the enzyme, which was only through an adsorption. This type of immobilization is not selective and the electrostatic interaction can occur anywhere in the enzyme leading to inactivation of the catalytic site. Another consequence is the weak bonds between the carrier and the enzyme, contributing to the enzyme release during the reaction [43].

4.2. Effect of Chitosan and Glutaraldehyde in the Drops of Immobilized Endoglucanase. The presence of chitosan with alginate in drops B and B$'$ was probably responsible for the superior adsorption of the supplied enzyme compared to the derivatives A and A$'$, even with the great losses in residual liquid after immobilization (U_f). The drop B$'$ was superior to B probably due to the presence of glutaraldehyde increasing positive groups (NH_2) in combination with chitosan. In fact, there is a cross-linking reaction and the formation of a Schiff base (imine) by one glutaraldehyde molecule with one amino group, enhancing its aldol condensation with other glutaraldehyde molecules. The final cross-linked structure would be a linear aldol-condensed oligomer of glutaraldehyde, with several Schiff base linkages branching off [44]. The derivatives of the hybrid gels presented higher value of enzyme Immobilization Yield (IY), probably due to the presence of reactive free amine groups present in the chitosan and a greater affinity for proteins (Monteiro, Junior, 1999).

The addition of enzyme during the preparation of the drops probably makes the contact of cellulase and chitosan easier, and this reaction was probably responsible for the increase of IY. The RA (96%) in the drop A was superior than drops B and B$'$ (25% and 28.5%). If the catalytic site of the enzyme reacts with the amino groups of chitosan during the enzyme immobilization, these reactions could lead to the inactivation of the enzyme. This hypothesis is probably responsible for the increase of LA in the derivatives with chitosan, since LA was 42% in derivative B and 39.6% in derivative B$'$, comparing with only 6.47% in derivative A$'$ and only 0.53% in derivative A (Table 3).

4.3. Effect of Glutaraldehyde in the Sheets of Immobilized Endoglucanase. The enzyme Immobilization Yield (IY, 17.8%)

and Recovered Activity (RA, 8.8%) in sheet D′ were lower than sheet D due to the glutaraldehyde (Table 3). In the derivatives C and C′ (alginate), higher values of IY, respectively, 36.3 and 53.5%, were observed. However, in these derivatives, there was a low RA (12.6 and 6.78%) and cellulolytic activity (0.063 and 0.05 FPU/g support) due to higher Lost Activity (LA, 31.8 and 49.9%). Therefore, the glutaraldehyde increased IY but decreased RA. This fact probably occurred due to the denaturant action caused by this agent on the enzyme indicating the sensitivity of endoglucanase to glutaraldehyde [45, 46]. Several enzymes have different behaviors in the presence of glutaraldehyde [45]. Spagna et al. [47] confirmed the degrading action of glutaraldehyde acting as enzyme inhibitor, inducing a total or partial loss of enzyme activity. A reticulation of xylanases and cellulases in drops of chitin-chitosan with glutaraldehyde was evaluated in concentrations from 0.125% to 1.5% for 0.5 hours. The most effective glutaraldehyde concentration was 0.125% and higher values prompted reduced activity [24]. This result confirms the inhibitory effect of glutaraldehyde verified in the present work.

In other derivatives (D′, E′, and I′) of enzyme immobilization in sheets with glutaraldehyde, a low IY was also verified. However, these results contrast with those obtained with the drop A′; therefore the drops are less sensitive to glutaraldehyde than sheets. The immobilization of cellulases in alginate sheets presented in this word is unprecedented, with no comparative studies in the literature.

4.4. Effect of Temperature in Immobilization of Cellulase.

The decrease in the Immobilization Yield (IY) studied in extreme temperatures can be related to the physical characteristics of the reactants which make up the derivative. Thus, alginate viscosity variations are related to the temperature at which they are subjected, where every 5.6°C increase led to a reduction of 12% of viscosity [48]. In the same way the temperature decrease leads to an increase in viscosity of alginate. Therefore, the physical change that occurs in the alginate with the temperature variation could be related to the low efficiency of enzyme immobilization, since the glycosidic bonds were broken [48]. The physical properties of chitosan are also modified in different temperatures (Zanira-Mora et al., 2014), and the temperature changes the binding properties of the substrate with the enzyme component [49].

4.5. Immobilization Using Hydrogels.

Chitosan was always necessary for effective immobilization, probably due to the presence of reactive free amine groups and this characteristic justifies a higher affinity for proteins (Monteiro-Junior and Airoldi, 1999) [10].

The interaction between the blend of chitosan and alginate is important in the formation of a hybrid gel. The biochemical constitution of chitosan was compounded by a deacetylated biopolymer (N-acetyl-D-glucosamine) which leads to a high number of reactive free amine groups. These charges are responsible for higher affinity for proteins [50]. However, the high amount of reactive free amine groups also

gives higher solubility to chitosan [51]. The blend produced by these two compounds shows higher physical stability of the support, allowing greater resistance after enzyme recycles.

Zhang et al. [52] encapsulated α-transglycosidase in supports of chitosan-alginate under ideal conditions. The results were similar to the enzyme in free form, but with greater stability of pH and temperature. In another study, Ramirez et al. [53] immobilized pectinase in alginate-chitosan and had 70% Protein Yield and 60% Recovered Activity, while maintaining catalytic capability around 50% even after 9 cycles of reuse. Saleem et al. [54] immobilized endoglucanase in polyacrylamide gel and achieved 53.4% Immobilization Yield (IY), near the value obtained in this study for immobilization in alginate-chitosan (42.63% and 48.08%). Furthermore, the enzyme activity remained even after five repeated cycles of applications. Silva et al. [10] working with immobilization of papain on chitin treated with PEI (polyethyleneimine) and chitosan cross-linked with TPP (tripolyphosphate) managed 6.07% and 15.7%, respectively, for yields of active immobilized enzyme. These results were lower than those obtained in the present work.

The activity FPU/g of derivatives produced with latex (drops K, L, and M) was approximately 1.93% lower than other values obtained by the immobilization derivatives without latex. In this case, the latex did not help in the increase of the Immobilization Yield (IY), even with the increase of Immobilized Protein Yield (IPY). The free amine groups of chitosan and cellulases may have been inhibited by the latex (Monteiro Jr, 1999), [10].

4.6. The Immersion of the Drops and Sheets in Enzyme Solution.

The presence of chitosan in drops G and G′ was responsible for the major adsorption of the supplied enzyme, although a lower YI and RA were obtained when compared to the derivative F (Table 3). The drop G′ was superior to drop G probably due to the action of glutaraldehyde in the enzyme.

The immersion of sheets in enzyme solution disfavors the protein immobilization on the support. The IP and IPY in these derivatives were lower than the derivatives of the enzyme mixed during the production of sheets. This fact shows the importance of the preparation method of the derivatives. Sheets were less efficient than drops produced by enzyme immersion on support. The drop F (only alginate) showed 0.134 FPU/g support, RA of 76.01%, and only 3.7% LA, while sheet H showed only 0.105 FPU/g support, RA of 15%, and 24.2% LA.

4.7. Sheets × Drops of Immobilized Endoglucanase.

The content of the immobilized protein (IP) in drops was better than in sheets (Table 3), but the sheet E′ showed higher IPY than the others probably due to the presence of ion exchange resin and cross-linking with glutaraldehyde. However, the cellulolytic activities in sheets (sheet H, 0.105 FPU/g, or sheet E, 0.085 FPU/g) were lower than in the best drops (drop B′, 0.172 FPU/g, drop B, 0.153 FPU/g). The comparison of RA and LA between sheet and drop of immobilized endoglucanase revealed a higher RA (76–96%) and lower LA (0.53–3.7%) in the drops (A and F), than the best sheets (D and H, RA = 15–54% and LA 10–24%). However, there is an important difference between these two processes. The process of sheet production had a final drying step (25°C for 24 hours) which did not occur with the process of drops production. This drying may compromise the catalytic site of the enzyme leading to an increase of LA and a decrease of RA. On the other hand, this evaluation is just for one cycle of reaction. The strength of these adsorptions must also be evaluated by the recycle of these derivatives to conclude which process is more advantageous.

4.8. Latex Blend: The Ability to Resist Deformation and Mechanical Compression Test. The high values of elasticity for only sodium alginate + chitosan (1.787 MPa) demonstrated that the use of calcium alginate in the hydrogel matrix formation provides an increase in the stiffness and compressive strength, according to de Moura et al. [36]. However, chitosan is important for deformation. The mechanical behavior of polymer blends with chitosan and collagen was studied by Tonhi and Plepis [55]. They observed a larger deformation suffered in these blends with higher amounts of chitosan. In the present work, the results obtained without latex were lower than 5% latex (2.31 MPa) or 10% latex (2.50 MPa). The presence of the latex increased this parameter compared to derivatives only with alginate and chitosan, demonstrating that the latex blends become more elastic and resistant to deformation. Fresh natural latex is composed of 33% of hydrocarbons as *cis*-1,4-polyisoprene, and this product undergoes processes to achieve up to 60% of rubber in the composition providing a wide elastic characteristic [37]. Simões et al. [56] assert that the natural latex has its own characteristics such as elasticity, plasticity, and wear resistance. Therefore, the addition of latex to immobilization increased their resistance to degradation.

4.9. FTIR Analysis in Different Materials. The bands at $1060 \, \text{cm}^{-1}$ are linked chemical bonds of the primary alcohol at $1090 \, \text{cm}^{-1}$ to vibrational stretch ether group, and at $1070–1100 \, \text{cm}^{-1}$ related to aliphatic amines (Torres et al., 2005). The interaction of sodium alginate with calcium chloride in the mixture caused a small change, which can be analyzed by the band near $1600 \, \text{cm}^{-1}$ [57]. The mixture of chitosan with alginate (Figure 3(a)) is found to change after complexation between both, where the bands represent the symmetrical and asymmetrical connections CO_2^- [58]. Moreover, the bands in the region between 1700 and $1000 \, \text{cm}^{-1}$ characteristic of alginate and absent in chitosan, after this mixture still present, indicate an effective interaction between the polymers and the prevalence of alginate in the end bracket [58].

The mixture of chitosan + alginate (Figure 3(a)) has the bands representing the symmetrical and asymmetrical connections CO_2^- change very slightly after complexation [58]. The spectrum of the mixtures also shows changes relative to the chitosan spectrum, with the bands relative to the primary amino groups (NH_2) at about $3400 \, \text{cm}^{-1}$ (Figure 3(a)). The changes in the amino and carboxyl groups may indicate the bond between the compounds. There are bands in the region between 2600 and $3400 \, \text{cm}^{-1}$ relating to

ammoniums groups in chitosan curve that is absent in the alginate curve and lost in this blend, indicating interactions among these carbohydrates. The band of $2600 \, \text{cm}^{-1}$ is only present in chitosan and the blend. Moreover, the bands in the region between 1700 and $1000 \, \text{cm}^{-1}$ in the alginate curve, still present and absent in complexing with chitosan, indicate an effective interaction between the polymers and the prevalence of alginate in the end bracket [41, 58].

The band at $2800 \, \text{cm}^{-1}$ is present in latex and the blend latex has more alginate. However, when the chitosan was mixed, this band was absent, indicating the interaction between chitosan and latex (Figures 3(a) and 3(b)). The vibration in $2800 \, \text{cm}^{-1}$ is characteristic of chitosan [17]. However, in the curves where there is the presence of cellulolytic enzyme, the disappearance of the bands of some other component was observed (Figure 3), such as ammonium groups of chitosan (around $3300 \, \text{cm}^{-1}$), carbon-hydrogen bonds of latex, and carboxyl groups of alginate. These changes may indicate that interactions actually occur between the enzyme and the support.

4.10. Recycle of Derivatives in Enzymatic Reaction with Filter Paper as Substrate. The derivatives prepared by immersion of supports in endoglucanase solution proved to be more unstable after their recycles on paper filter reaction. The number of recycles in drops or sheets was inferior to the method of inclusion enzyme during the preparation of derivatives, independently of the treatment with glutaraldehyde, type of support (pure or hybrid), or the shape (drop or sheet).

The presence of glutaraldehyde was not relevant either in the immobilization of drops or sheets, since the number of enzyme recycles was the same for all derivatives using the preparation method by immersion of supports in endoglucanase solution. In contrast, the number of recycles was influenced by the shape, since the sheets were recycled up to 28 times (Figure 5), while drops only up to 13 times (Figures 4(a) and 4(b)).

The type of the mixture of the supports was also relevant in the immobilization process, since drops of the blend alginate and chitosan showed higher activity. The drops B showed greater stability in recycles, with a total of 13 reuses, FPU Yield of 1.57 times, and Net PFU Yield of 2.81 times based on used enzyme. However, the sheet E′ showed the highest stability in 28 reuses of enzyme, PFU Yield of 1.64 times, and Net FPU Yield of 5.31 times proving to be the most efficient treatment saving more enzyme by the immobilization process (Figures 3 and 4). These results were better than those found in the literature. In a study using cellulases obtained from *Bacillus subtilis* TD6 immobilized in calcium alginate only 4 reuses of cellulases were obtained [59]. In another study with a *β*-1,3-glucanase from *Trichoderma harzianum* immobilized in calcium alginate up to 7 reuses of these enzymes were achieved without losing the enzymatic activity (El-Katatny, 2008). Zhang et al. [60] working with immobilization of lipase in alginate hydrogel beads maintained a high activity only in five cycles. Saleem et al. [54] showed only 6 times of enzyme reuse with immobilizing endoglucanase in polyacrylamide gel. Romo-Sánchez et al. [24] immobilized cellulases and xylanases in different polymers in two steps. In the first, the improvement of the adsorption of these enzymes in alginate-chitosan and chitin-chitosan was performed. In a second step, the reticulation (adsorption and cross-linking) was performed after improvement of the adsorption conditions for the chitin-chitosan support by adding 0.125% glutaraldehyde to the chitosan-enzyme system for 0.5 hours. In these conditions they obtained enzyme stability up to 19 cycles retaining 64% activity.

The drops of alginate, chitosan, and latex although demonstrating a high Immobilized Protein Yield (IPY) and moderate stability, did not overcome the performance of the derivative B (without latex), with 13 recycles and 1.59 FPU/g in the sum of all recycles, since, in the derivative using 3% latex (drop K) with 8 recycles, only 0.17 FPU/g of support was obtained. The drops of 3% latex (v·v^{-1}) showed greater stability than 5 or 10% latex (v·v^{-1}) (8, 6, and 5 recycles, resp.), but with a lower cellulolytic activity in each cycle, dropping more than 50% in the first recycle. Therefore, the presence of latex in the concentrations evaluated did not favor the immobilization of cellulases and its stability, probably due to the interaction of the enzyme with the support or the action of some carbohydrases presented in the natural latex [38]. The improvement of this technology is possible by the study of purity of cellulases, the ratio of enzyme and support, quality of the supports, pH, time of immobilization, and other functional agents. The cellulosic residues such as cardboard, paper, newspaper, and pretreated lignocellulosic residues could be used to produce fermentable and low cost sugars with wide application in the biotechnology industry, as well as reduction of waste and environmental damage.

The results obtained in the present work showed the immobilization of endocellulase and its recycle are technically possible, preferably using the enzyme added during the step of preparation of derivatives. The number of reuses of immobilized enzyme was relevant and this method must be considered to be applied on an industrial scale.

Authors' Contributions

All authors of this paper contributed sufficiently to the scientific work and therefore share collective responsibility and accountability for the results.

Acknowledgments

The authors wish to thank the CNPq (Conselho Nacional de Desenvolvimento Científico e Tecnológico) and FAPESP (Fundação para a Pesquisa do Estado deSão Paulo) for financial support.

References

[1] P. Oliva-Neto, C. Dorta, A. F. A. Carvalho, V. M. G. Lima, and DF. Silva, "The Brazilian technology of fuel ethanol fermentation—yeast inhibition factors and new perspectives to improve the technology," in *Materials and Processes for Energy: Communicating Current Research and Technological Developments*, A. Méndez-Vilas, Ed., pp. 371–379, Formatex, Badajoz, Spain, 1st edition, 2013.

[2] M. P. Vásquez, J. N. C. da Silva, M. B. de Souza, and N. Pereira, "Enzymatic hydrolysis optimization to ethanol production by simultaneous saccharification and fermentation," *Applied Biochemistry and Biotechnology*, vol. 137–140, no. 1–12, pp. 141–153, 2007.

[3] CONAB, National Company of The Brazilian Government of Supply, 2016, http://www.conab.gov.br/OlalaCMS/uploads/arquivos/16_08_18_12_03_30_boletim_cana_portugues_-_2o_lev_-_16-17.pdf.

[4] M. Brienzo, A. F. Siqueira, and A. M. F. Milagres, "Search for optimum conditions of sugarcane bagasse hemicellulose extraction," *Biochemical Engineering Journal*, vol. 46, no. 2, pp. 199–204, 2009.

[5] S. W. Kang, Y. S. Park, J. S. Lee, S. I. Hong, and S. W. Kim, "Production of cellulases and hemicellulases by *Aspergillus niger* KK2 from lignocellulosic biomass," *Bioresourse Techonology*, vol. 91, no. 2, pp. 153–156, 2004.

[6] A. Pandey, C. R. Soccol, P. Nigam, and V. T. Soccol, "Biotechnological potential of agro-industrial residues. I: sugarcane bagasse," *Bioresource Technology*, vol. 74, no. 1, pp. 69–80, 2000.

[7] M. Dashtban, M. Maki, K. T. Leung, C. Mao, and W. Qin, "Cellulase activities in biomass conversion: measurement methods and comparison," *Critical Reviews in Biotechnology*, vol. 30, no. 4, pp. 302–309, 2010.

[8] S. Datta, L. R. Christena, and Y. R. S. Rajaram, "Enzyme immobilization: an overview on techniques and support materials," *3 Biotech*, vol. 3, no. 1, pp. 1–9, 2013.

[9] X. L. Shen and L. M. Xia, "Production and immobilization of cellobiase from *Aspergillus niger* ZU-07," *Process Biochemistry*, vol. 39, no. 11, pp. 1363–1367, 2004.

[10] D. F. Silva, H. Rosa, A. F. A. Carvalho, and P. Oliva-Neto, "Immobilization of papain on chitin and chitosan and recycling of soluble enzyme for deflocculation of *Saccharomyces cerevisiae* from bioethanol distilleries," *Enzyme Research*, vol. 2015, Article ID 573721, 10 pages, 2015.

[11] G. Fundueanu, C. Nastruzzi, A. Carpov, J. Desbrieres, and M. Rinaudo, "Physico-chemical characterization of Ca-alginate microparticles produced with different methods," *Biomaterials*, vol. 20, no. 15, pp. 1427–1435, 1999.

[12] F. Velten, C. Laue, and J. Schrezenmeir, "The effect of alginate and hyaluronate on the viability and function of immunoisolated neonatal rat islets," *Biomaterials*, vol. 20, no. 22, pp. 2161–2167, 1999.

[13] O. Smidsrod and G. Skaka-Break, "Alginate as immobilization matrix for cell," *Trends of Biotechnology*, vol. 8, pp. 71–79, 1990.

[14] M. Rucka, B. Turkiewicz, M. Tomaszewska, and N. Chlubek, "Hydrolysis of sunflower oil by means of hydrophobic membrane with lipolytic activity," *Biotechnology Letters*, vol. 11, no. 3, pp. 167–172, 1989.

[15] L. Wang, E. Khor, and L.-Y. Lim, "Chitosan-alginate-CaCl$_2$ system for membrane coat application," *Journal of Pharmaceutical Sciences*, vol. 90, no. 8, pp. 1134–1142, 2001.

[16] J. S. Mao, H. F. Liu, Y. J. Yin, and K. De Yao, "The properties of chitosan-gelatin membranes and scaffolds modified with hyaluronic acid by different methods," *Biomaterials*, vol. 24, no. 9, pp. 1621–1629, 2003.

[17] R. D. Herculano, L. C. Tzu, C. P. Silva et al., "Nitric oxide release

using natural rubber latex as matrix," *Materials Research*, vol. 14, no. 3, pp. 355–359, 2011.

[18] C.-F. Lee, M.-L. Hsu, C.-H. Chu, and T.-Y. Wu, "Synthesis and characteristics of poly(methyl methacrylate-*co*-methacrylic acid)/Poly(methacrylic acid-*co*-N-isopropylacrylamide) thermosensitive semi-hollow latex particles and their application to drug carriers," *Journal of Polymer Science Part A: Polymer Chemistry*, vol. 52, no. 23, pp. 3441–3451, 2014.

[19] L. Gómez, H. L. Ramírez, M. L. Villalonga, J. Hernández, and R. Villalonga, "Immobilization of chitosan-modified invertase on alginate-coated chitin support via polyelectrolyte complex formation," *Enzyme and Microbial Technology*, vol. 38, no. 1-2, pp. 22–27, 2006.

[20] M. L. Huguet and E. Dellacherie, "Calcium alginate beads coated with chitosan: effect of the structure of encapsulated materials on their release," *Process Biochemistry*, vol. 31, no. 8, pp. 745–751, 1996.

[21] M. D. Busto, N. Ortega, and M. Perez-Mateos, "Characterization of microbial endo-β-glucanase immobilized in alginate beads," *Acta Biotechnologica*, vol. 18, no. 3, pp. 189–200, 1998.

[22] L. Wu, X. Yuan, and J. Sheng, "Immobilization of cellulase in nanofibrous PVA membranes by electrospinning," *Journal of Membrane Science*, vol. 250, no. 1-2, pp. 167–173, 2005.

[23] R. H.-Y. Chang, J. Jang, and K. C.-W. Wu, "Cellulase immobilized mesoporous silica nanocatalysts for efficient cellulose-to-glucose conversion," *Green Chemistry*, vol. 13, no. 10, pp. 2844–2850, 2011.

[24] S. Romo-Sánchez, C. Camacho, H. L. Ramirez, and M. Arévalo-Villena, "Immobilization of commercial cellulase and xylanase by different methods using two polymeric supports," *Advances in Bioscience and Biotechnology*, vol. 5, no. 6, pp. 517–526, 2014.

[25] D. Zhang, H. E. Hegab, Y. Lvov, L. Dale Snow, and J. Palmer, "Immobilization of cellulase on a silica gel substrate modified using a 3-APTES self-assembled monolayer," *SpringerPlus*, vol. 5, article 48, pp. 1–20, 2016.

[26] Q. Song, Y. Mao, M. Wilkins, F. Segato, and R. Prade, "Cellulase immobilization on superparamagnetic nanoparticles for reuse in cellulosic biomass conversion," *AIMS Bioengineering*, vol. 3, no. 3, pp. 264–276, 2016.

[27] C. Hou, H. Zhu, D. Wu et al., "Immobilized lipase on macroporous polystyrene modified by PAMAM-dendrimer and their enzymatic hydrolysis," *Process Biochemistry*, vol. 49, no. 2, pp. 244–249, 2014.

[28] M. Albarghouthi, D. A. Fara, M. Saleem, T. El-Thaher, K. Matalka, and A. Badwan, "Immobilization of antibodies on alginate-chitosan beads," *International Journal of Pharmaceutics*, vol. 206, no. 1-2, pp. 23–34, 2000.

[29] A. Tanriseven and E. Doğan, "Immobilization of invertase within calcium alginate gel capsules," *Process Biochemistry*, vol. 36, no. 11, pp. 1081–1083, 2001.

[30] M. Mandels, R. E. Andreotti, and C. Roche, "Measurement of scarifying cellulose," *Biotechnology and Bioengineering Symposium*, vol. 6, p. 1471, 1976.

[31] T. K. Ghose, "Measurement of cellulase activities," *Pure and Applied Chemistry*, vol. 59, no. 2, pp. 257–268, 1987.

[32] M. M. Bradford, "Determination of total proteins," *Analytical Biochemistry*, vol. 72, p. 248, 1976.

[33] G. L. Miller, "Use of dinitrosalicylic acid reagent for determination of reducing sugar," *Analytical Chemistry*, vol. 31, no. 3, pp. 426–428, 1959.

[34] W. J. Price, S. A. Leigh, S. M. Hsu, T. E. Patten, and G.-Y. Liu, "Measuring the size dependence of young's modulus using force modulation atomic force microscopy," *Journal of Physical Chemistry A*, vol. 110, no. 4, pp. 1382–1388, 2006.

[35] K. Miyake, N. Satomi, and S. Sasaki, "Elastic modulus of polystyrene film from near surface to bulk measured by nanoindentation using atomic force microscopy," *Applied Physics Letters*, vol. 89, no. 3, Article ID 031925, 2006.

[36] M. F. S. F. de Moura, R. D. S. G. Campilho, and J. P. M. Gonçalves, "Crack equivalent concept applied to the fracture characterization of bonded joints under pure mode I loading," *Composites Science and Technology*, vol. 68, no. 10-11, pp. 2224–2230, 2008.

[37] L. F. Valadares, C. A. P. Leite, and F. Galembeck, "Preparation of natural rubber-montmorillonite nanocomposite in aqueous medium: evidence for polymer-platelet adhesion," *Polymer*, vol. 47, no. 2, pp. 672–678, 2006.

[38] P. A. Jekel, J. B. H. Hartmann, and J. J. Beintema, "The primary structure of hevamine, an enzyme with lysozyme/chitinase activity from *Hevea brasiliensis* latex," *European Journal of Biochemistry*, vol. 200, no. 1, pp. 123–130, 1991.

[39] H. D. Murbach, G. J. Ogawa, F. A. Borges et al., "Ciprofloxacin release using natural rubber latex membranes as carrier," *International Journal of Biomaterials*, vol. 2014, Article ID 157952, 7 pages, 2014.

[40] S. G. Kazarian and K. L. A. Chan, "ATR-FTIR spectroscopic imaging: recent advances and applications to biological systems," *Analyst*, vol. 138, no. 7, pp. 1940–1951, 2013.

[41] L. Vitali, K. C. Justi, M. C. M. Laranjeira, and V. T. Fávere, "Impregnation of chelating agent 3,3-bis-N,N bis-(carboxymethyl)aminomethyl-o-cresolsulfonephthalein in biopolymer chitosan: adsorption equilibrium of Cu(II) in aqueous medium," *Polímeros*, vol. 16, no. 2, pp. 116–122, 2006.

[42] S. Mandal, S. S. Kumar, B. Krishnamoorthy, and S. K. Basu, "Development and evaluation of calcium alginate beads prepared by sequential and simultaneous methods," *Brazilian Journal of Pharmaceutical Sciences*, vol. 46, no. 4, pp. 785–793, 2010.

[43] K. Buchholz, V. Kasche, and U. T. Bornscheuer, *Biocatalysts and Enzyme Technology*, Wiley-Blackwell, Weinheim, Germany, 2nd edition, 2012.

[44] I. Migneault, C. Dartiguenave, M. J. Bertrand, and K. C. Waldron, "Glutaraldehyde: behavior in aqueous solution, reaction with proteins, and application to enzyme crosslinking," *BioTechniques*, vol. 37, no. 5, pp. 790–802, 2004.

[45] G. B. Broun, "Chemically aggregated enzymes," *Methods in Enzymology*, vol. 44, pp. 263–280, 1976.

[46] L. T. Nguyen and K.-L. Yang, "Uniform cross-linked cellulase aggregates prepared in millifluidic reactors," *Journal of Colloid and Interface Science*, vol. 428, pp. 146–151, 2014.

[47] G. Spagna, F. Andreani, E. Salatelli, D. Romagnoli, and P. G. Pifferi, "Immobilization of α-L-arabinofuranosidase on chitin and chitosan," *Process Biochemistry*, vol. 33, no. 1, pp. 57–62, 1998.

[48] M. V. Matos, J. M. M. Pérez, and G. R. Lao, "Estudio de la influencia del campo magnético y la temperatura en la cinética de secado del alginato de sódio," *Tecnologiaquímica*, vol. 20, no. 3, 2000.

[49] Z. Gan, T. Zhang, Y. Liu, and D. Wu, "Temperature-triggered enzyme immobilization and release based on cross-linked gelatin nanoparticles," *PLoS ONE*, vol. 7, no. 10, Article ID e47154, 2012.

[50] O. A. C. Monteiro Jr. and C. Airoldi, "Some studies of crosslinking chitosan-glutaraldehyde interaction in a homogeneous system," *International Journal of Biological Macromolecules*, vol. 26, no. 2-3, pp. 119–128, 1999.

[51] M. W. Anthonsen, K. M. Vårum, and O. Smidsrød, "Solution properties of chitosans: conformation and chain stiffness of chitosans with different degrees of N-acetylation," *Carbohydrate Polymers*, vol. 22, no. 3, pp. 193–201, 1993.

[52] L. Zhang, Y. Jiang, X. Sun, X. Shi, W. Cheng, and Q. Sun, "Immobilized transglucosidase in biomimetic polymer-inorganic hybrid capsules for efficient conversion of maltose to isomaltooligosaccharides," *Biochemical Engineering Journal*, vol. 46, no. 2, pp. 186–192, 2009.

[53] H. L. Ramirez, A. I. Briones, J. Úbeda, and M. Arevalo, "Immobilization of pectinase by adsorption on an alginate-coated chitin support," *Biotecnologia Aplicada*, vol. 30, no. 2, pp. 101–104, 2013.

[54] M. Saleem, M. H. Rashid, A. Jabbar, R. Perveen, A. M. Khalid, and M. I. Rajoka, "Kinetic and thermodynamic properties of an immobilized endoglucanase from *Arachniotus citrinus*," *Process Biochemistry*, vol. 40, no. 2, pp. 849–855, 2005.

[55] E. Tonhi and A. M. G. Plepis, "Obtentionand characterization of polymer blends Collagen-chitosan," *Química Nova*, vol. 25, no. 6, pp. 943–948, 2002.

[56] R. D. Simões, A. E. Job, D. L. Chinaglia et al., "Structural characterization of blends containing both PVDF and natural rubber latex," *Journal of Raman Spectroscopy*, vol. 36, no. 12, pp. 1118–1124, 2005.

[57] G. Lawrie, I. Keen, B. Drew et al., "Interactions between alginate and chitosan biopolymers characterized using FTIR and XPS," *Biomacromolecules*, vol. 8, no. 8, pp. 2533–2541, 2007.

[58] B. Sarmento, D. Ferreira, F. Veiga, and A. Ribeiro, "Characterization of insulin-loaded alginate nanoparticles produced by ionotropic pre-gelation through DSC and FTIR studies," *Carbohydrate Polymers*, vol. 66, no. 1, pp. 1–7, 2006.

[59] D. Andriani, C. Sunwoo, H.-W. Ryu, B. Prasetya, and D.-H. Park, "Immobilization of cellulase from newly isolated strain *Bacillus subtilis* TD6 using calcium alginate as a support material," *Bioprocess and Biosystems Engineering*, vol. 35, no. 1-2, pp. 29–33, 2012.

[60] S. Zhang, W. Shang, X. Yang et al., "Immobilization of lipase with alginate hydrogel beads and the lipase-catalyzed kinetic resolution of α-phenyl ethanol," *Journal of Applied Polymer Science*, vol. 131, no. 8, Article ID 40178, 2014.

14

Insight into the Mechanistic Basis of the Hysteretic-Like Kinetic Behavior of Thioredoxin-Glutathione Reductase (TGR)

Juan L. Rendón ⓘ, Mauricio Miranda-Leyva, Alberto Guevara-Flores, José de Jesús Martínez-González, Irene Patricia del Arenal, Oscar Flores-Herrera, and Juan P. Pardo

Departamento de Bioquímica, Facultad de Medicina, Universidad Nacional Autónoma de México, Apartado Postal 70-159, 04510, D.F. México, Mexico

Correspondence should be addressed to Juan L. Rendón; jrendon@bq.unam.mx

Academic Editor: Paul Engel

A kinetic study of thioredoxin-glutathione reductase (TGR) from *Taenia crassiceps* metacestode (cysticerci) was carried out. The results obtained from both initial velocity and product inhibition experiments suggest the enzyme follows a two-site ping-pong bi bi kinetic mechanism, in which both substrates and products are bound in rapid equilibrium fashion. The substrate GSSG exerts inhibition at moderate or high concentrations, which is concomitant with the observation of hysteretic-like progress curves. The effect of NADPH on the apparent hysteretic behavior of TGR was also studied. At low concentrations of NADPH in the presence of moderate concentrations of GSSG, atypical time progress curves were observed, consisting of an initial burst-like stage, followed by a lag whose amplitude and duration depended on the concentration of both NADPH and GSSG. Based on all the kinetic and structural evidence available on TGR, a mechanism-based model was developed. The model assumes a noncompetitive mode of inhibition by GSSG in which the disulfide behaves as an affinity label-like reagent through its binding and reduction at an alternative site, leading the enzyme into an inactive state. The critical points of the model are the persistence of residual GSSG reductase activity in the inhibited GSSG-enzyme complexes and the regeneration of the active form of the enzyme by GSH. Hence, the hysteretic-like progress curves of GSSG reduction by TGR are the result of a continuous competition between GSH and GSSG for driving the enzyme into active or inactive states, respectively. By using an arbitrary but consistent set of rate constants, the experimental full progress curves were successfully reproduced *in silico*.

1. Introduction

Thioredoxin-glutathione reductase (TGR E.C. 1.8.1.B1) represents an interesting splicing variant of the animal thioredoxin reductase (TR), featured by the presence of a glutaredoxin-like domain appended at the N-terminal end of the TrxR module [1, 2]. As a member of the disulfide reductase family, TGR is a NADPH-dependent homodimeric flavoenzyme with a dithiol/disulfide redox center located at the *si* face of the FAD prosthetic group [2]. As in mammalian TR [3], a selenocysteine residue located at the C-terminal end of the enzyme is also catalytically essential [4]. Unlike typical TR, however, TGR is also able to reduce oxidized glutathione (GSSG) at significant rates and to perform thiol/disulfide exchanges [1, 4], thus making it a multifunctional enzyme. Such additional catalytic capabilities are dependent on the presence of the Grx-like domain. Both the Grx-like domain and the C-terminal redox center of the enzyme are essential in the reduction of GSSG, as derived from site-directed mutagenesis studies [4–7]. Interestingly, the two redox centers are far away from one another as revealed by the three-dimensional structure of TGR from the blood fluke *Schistosoma mansoni* [8, 9], suggesting that TGR acts as a two-site enzyme during GSSG reduction. It has been proposed that, during the catalytic cycle of TGR with GSSG as the substrate, electrons must be shuttled from the reduced selenol/thiol couple toward the redox center of the Grx-like domain [9, 10], involving a large conformational transition of the C-terminal arm of the neighbor subunit.

The presence of the enzyme has been demonstrated in animals [1, 4], human [11], as well as in the parasitic representatives of the flatworms [12–17]. The latter organisms lack the typical glutathione reductase (GR) and TR enzymes, and TGR is the only disulfide reductase involved in the regeneration of the reduced state of both thioredoxin (trx) and glutathione. Hence, it has been considered as a potential drug target for an antihelminthic therapy [18–20]. However, in spite of its unusual tertiary structure and multifunctional nature, no detailed kinetic study of TGR has been carried out. In a previous work the purification and general kinetic properties of TGR from the larval stage (cysticerci) of *Taenia crassiceps* were reported [14]. Particularly noticeable was the existence of an atypical kinetic behavior with GSSG as the substrate, characterized by a lag time in the time courses at moderate or high concentrations of the disulfide. The magnitude of the lag time depended on enzyme concentration and of the presence of disulfide reducing reagents [14]. Such atypical kinetics was considered as a kind of hysteresis [21]. The same kinetic phenomenon was reported in TGR from larval *Echinococcus granulosus* [5] and *Taenia solium* cysticerci [17] and in the enzyme from the adult stage of the flukes *Fasciola hepatica* [15], *Fasciola gigantica* [16], and *Schistosoma mansoni* [7]. Thus, the GSSG-dependent atypical kinetics appear to be a common feature in TGR. Even in the recombinant enzyme from human such atypical kinetic behavior was observed [11]. However, no detailed molecular mechanism for such phenomenon is yet available. It was proposed that the atypical time courses of TGR were dependent on the covalent modification through glutathionylation of two structural cysteine residues of the enzyme [5]. Such model, however, shows serious faults as was demonstrated [15]. On the other hand, the potential role that the substrate NADPH could play in the atypical kinetics of TGR has not been investigated. In all the works reporting the apparent hysteretic behavior of the enzyme, NADPH concentrations of 100 μM or higher have been used [5, 7, 11, 14–17]. In order to elucidate the molecular basis of the hysteretic behavior of the enzyme, in the present work, the kinetic mechanism of wild type TGR from *T. crassiceps* was investigated. Furthermore, the effect of NADPH on the atypical kinetic behavior of TGR was also studied. Based on both the kinetic and the crystallographic evidence on TGR, a comprehensive model consistent with all the experimental observations is put forward and tested through *in silico* simulations. All the experimental observations regarding the kinetic behavior of the enzyme were successfully reproduced by the model.

2. Materials and Methods

2.1. Reagents. $2'5'$ ADP–Sepharose 4B was obtained from Amersham Pharmacia Biotech (Uppsala, Sweden). All others chemicals were obtained from Sigma Chemical Company (St. Louis, MO, USA) and used without further purification. Water purified by reverse osmosis was used in the preparation of solutions.

2.2. Growth of T. crassiceps Cysticerci. The HYG strain of *T. crassiceps* was used as a source of TGR. The cysticerci growth conditions, as well as its extraction, rupture, and preparation of a crude homogenate have been described elsewhere [22].

2.3. Enzyme. TGR from the cytosolic fraction of larval *T. crassiceps* was purified to homogeneity as previously described [14]. Enzyme solutions were stored at -20°C until use.

2.4. Enzyme Assays. The disulfide reductase activity assays of TGR were carried out in an Agilent 8453 uv/visible spectrophotometer (Hewlett Packard) fitted with a thermostated cell holder. All the kinetic experiments were performed at 25°C in 0.1 M Tris/HCl buffer (pH 7.8) containing 1 mM EDTA (buffer A) in a final volume of 1 mL.

The GSSG reductase activity of TGR was determined by following the decrease in absorbance at 340 nm as a consequence of NADPH oxidation [23]. Both NADPH and GSSG were incubated in buffer A during two min in order to obtain the baseline. Then, the reaction was started by adding a small enzyme aliquot. An extinction coefficient at 340 nm of 6220 M^{-1} cm^{-1} for NADPH was used in the calculations of initial velocities. In those experiments dealing with the effect of GSSG on the reduction of oxidized thioredoxin (Trx), human Trx was used as substrate. In this case, the reaction mixture contained 100 μM NADPH, human Trx at the corresponding concentration, and 2 mM GSSG at a final volume of 120 μL in buffer A. The reaction was started by adding a small TGR aliquot.

The concentration of both NADPH and NADP$^+$ was determined spectrophotometrically by reading the absorbance of an aliquot of the corresponding stock solution at either 340 nm (ε = 6220 M^{-1} cm^{-1}) or 259 nm (ε = 18000 M^{-1} cm^{-1}) for NADPH or NADP$^+$, respectively. As regards GSSG, its concentration in the stock solution was calculated through enzyme assays by mixing an aliquot of the disulfide with TGR in the presence of an excess of NADPH. After exhaustion of GSSG, its concentration was determined from the total change in absorbance at 340 nm.

2.5. Protein Determination. The monomer concentration in the stock solutions of TGR was determined from its absorbance at 462 nm using a molar extinction coefficient of 11.3 mM^{-1} cm^{-1} for protein bound FAD [24].

2.6. Steady-State Kinetics. In order to elucidate the kinetic mechanism of TGR, initial velocity data were obtained by varying the concentration of both NADPH and GSSG [25]. For the determination of the kinetic mechanism with GSSG as the substrate, concentrations up to 60 μM of the disulfide were used in order to avoid the strong substrate inhibition observed at moderate or high concentrations of GSSG. This decision is warranted because the initial portion of the saturation curve is very sensitive to discriminate between sequential and ping-pong kinetic mechanisms [25]. A global fitting of data through nonlinear regression to the rate equation for either a ping-pong bi bi (see (1)) or ordered bi bi (see (2)) kinetic mechanisms in the absence of products was then performed.

$$v = \frac{Vm\,[A]\,[B]}{Km_B\,[A] + Km_A\,[B] + [A]\,[B]} \tag{1}$$

$$v = \frac{Vm\,[A]\,[B]}{K_{ia}Km_B + Km_B\,[A] + Km_A\,[B] + [A]\,[B]} \tag{2}$$

where A represents NADPH, B corresponds to either GSSG or DTNB, and Km_A and Km_B are the corresponding Michaelis-Menten constants, while Kia represents the dissociation constant for A. The inhibitory ability of NADP$^+$ on the GSSG reductase activity of TGR was determined by analyzing its effect on the initial velocities under steady-state conditions. NADP$^+$ was incubated in buffer A with NADPH and GSSG under nonhysteretic conditions and the reaction was started by adding a small enzyme aliquot. Data obtained in the presence of the product NADP$^+$ as the inhibitor were fitted to either a competitive (see (3)) or an uncompetitive (see (4)) model of inhibition:

$$v = \frac{V_m [A]}{Km_A \left(1 + [P]/K_{is}\right) + [A] \left(1 + Km_B/[B]\right)} \quad (3)$$

$$v = \frac{V_m [B]}{Km_B + [B] \left\{1 + \left[\left(Km_A/[A]\right)\left(1 + [P]/K_{ii}\right)\right]\right\}} \quad (4)$$

where A and B are defined as above and P stands for NADP$^+$, while K_{is} and K_{ii} represent slope and intercept inhibition constants, respectively. These two latter equations were derived by using the King-Altman method [26] for a ping-pong bi bi kinetic mechanism in which NADP$^+$ acts as a dead-end inhibitor through the formation of a complex with the unmodified form of the enzyme. Finally, to gain insight into the GSSG-dependent substrate inhibition of TGR, initial velocity data obtained over a broad range of both NADPH and GSSG concentrations were fitted to

$$v = \frac{Vm_1 [A][B] + Vm_2 [A][B]^2/Km_{B'}}{Km_B [A] + Km_A [B] + [A][B] + [A][B]^2/Km_{B'} + [A][B]^3/Km_{B'} K_i + Km_A [B]^2/Km_{B'}} \quad (5)$$

where A, B, Km_A, and Km_B are defined as above and Vm_1 represents the catalytic pathway followed by the enzyme at low concentrations of GSSG, while Vm_2 corresponds to the alternative minor catalytic pathway obtained at high concentrations of GSSG. Km_B' is a second Km value for substrate B, while Ki is the inhibitor constant for GSSG acting as inhibitor (see the corresponding model under discussion). An initial estimate for the Ki value was obtained by plotting initial velocity data obtained over a broad range of GSSG concentrations in a semilog fashion [27] and the following equation:

$$K_i = S_1 + S_2 - 4S_m \quad (6)$$

where S_m represent the concentration of substrate at the maximum point of the curve, while S_1 and S_2 correspond to the substrate concentration at the two points where velocity is half that at the maximum.

In those kinetic experiments carried out at high concentrations of GSSG, the magnitude of the apparent lag time observed in the full time progress curves was estimated as described elsewhere [28].

2.7. Statistical Analysis. Fitting of data to the different velocity equations was made by nonlinear regression analysis using Sigma Plot software. No weighting of data was applied. Kinetic parameters are given as mean ± standard deviation.

2.8. Model Discrimination Analysis. When the initial velocity patterns did not allow a clear distinction between alternative kinetic mechanisms, a model discrimination analysis was needed. The following rules were used in the data analysis.

(I) The initial selection of a particular kinetic model was based on visual inspection of the corresponding double-reciprocal plot. When this preliminary analysis did not lead to a clear model discrimination, data were fitted to both alternative velocity equations.

(II) If the two alternative kinetic models had equal or very similar χ^2 values, then the model consistent with additional

kinetic evidence (i.e., inhibitory patterns) was chosen as the most plausible.

2.9. Global Data Fitting Procedure. In order to compare the experimental full progress curves obtained under different concentrations of NADPH, GSSG, and enzyme, with that predicted by the mechanistic model (see discussion), a curve fitting procedure was performed. In this case the fitting procedures were carried out with the Dynafit software [29] version 4. The following conditions were fixed.

(i) In all cases, the best set of rate constants pertaining to the ping-pong reaction cycle were fixed (see Supplementary Materials), representing ten microscopic rate constants (reactions 1 to 6 of the model).

(ii) In the fitting procedure of the full progress curves obtained at moderate or high concentrations of GSSG, two rate constants were allowed to be fitted in order to find the best fit values. One of such constants pertains to a reaction associated with the reversible binding of GSSG to the inhibitory site (reactions 7 and 9 of the model), while the other microscopic rate constant was chosen from any of the two reactions responsible for the atypical full progress curves (reactions 10 and 12 or reactions 11 and 13, see Figure 11). The same conditions were used in those cases where a global fitting of several full progress curves was carried out.

(iii) For long full time courses, the best fitting value for either GSSG or enzyme concentrations, or both, was also searched, allowing a variation of ±10% of the experimental value.

2.10. In Silico Analysis. Simulation of the full time courses of TGR was performed with the Dynafit software [29] version 4. The particular conditions used in the modelling are described in Supplementary Materials.

2.11. Docking Analysis. For molecular docking, the protein structure of *S. mansoni* TGR (PDB code 2X8C) was obtained

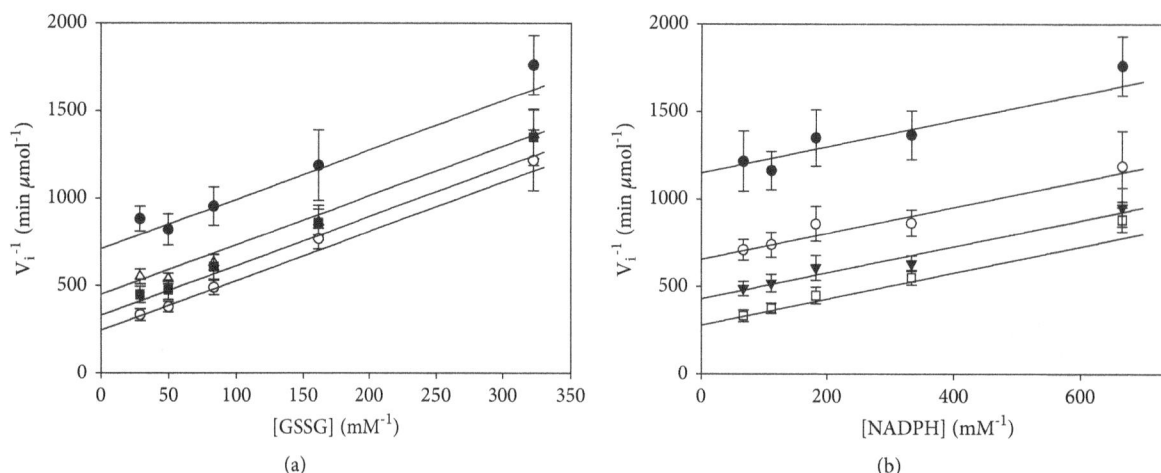

FIGURE 1: **Initial velocity patterns of *T. crassiceps* TGR.** Enzyme assays were carried out as described under Materials and Methods at 25°C and pH 7.8 in the presence of low concentrations of GSSG. (a) GSSG as the variable substrate at the following fixed concentrations of NADPH: (●) 1.5 μM; (△) 3μM; (■) 5.5 μM; (○) 15 μM. (b) NADPH as the variable substrate at the following fixed concentrations of GSSG: (●) 3.1 μM. (○) 6.2 μM; (▼) 12 μM; (□) 36 μM. The final enzyme subunit concentration was 6.1 nM. Continuous lines were obtained from the corresponding double-reciprocal form of equation (1) using the parameters resultant from the global fitting of data (each point represents mean ± standard deviation (n=6).

from the RCSB Protein Data Bank (http://www.rcsb.org). The structure of GSSG was recovered from the *Homo sapiens* glutathione S-transferase μ2 (PDB code 1YKC) using the Hic-up server (http://xray.bmc.uu.se/hicup/). The structures of TGR and GSSG were processed with AutoDockTools (ADT) version 1.5.4 (http://mgltools.scripps.edu) [30]. Essential hydrogen atoms and Kollman united atom charges were added to the protein and then saved in PDBQT file format, for input into Auto Dock Vina version 1.1.2 (http://vina.scripps.edu) [31]. For TGR, the search space was defined as a grid box of size 24 × 20 x 20 Å with a grid spacing of 0.375 Å, centered on the putative binding site.

3. Results

3.1. Steady-State Kinetic Study of TGR. Figure 1 shows the results of the initial velocity experiments presented as double-reciprocal plots. With either GSSG (Figure 1(a)) or NADPH (Figure 1(b)) as the variable substrate a parallel pattern of lines was obtained, suggesting TGR follows a ping-pong bi bi kinetic mechanism. When GSSG was replaced with DTNB, an artificial substrate indicative of TR activity, identical parallel double-reciprocal plots were obtained (data not shown). The global fitting procedure of the initial velocity data to either a ping-pong or an ordered sequential kinetic mechanism resulted in essentially identical kinetic parameters. However, fitting of data to the rate equation for an ordered mechanism required a too low Kia value (4.98 × 10^{-17} μM) with a very high variation coefficient (7.4 × 10^{17}%). To confirm the ping-pong kinetic model suggested by the initial velocity patterns, a product inhibition study with NADP$^+$ was performed. The results are shown in Figure 2. With NADPH as the variable substrate, double-reciprocal plots showed an intersecting pattern (Figure 2(a)); however, the graphical analysis of the data did not allow a clear distinction between the competitive or the mixed-type modes of inhibition. Similarly, the statistical results of the global fitting procedure showed very similar χ^2 values (1.49 versus 1.4 for the competitive or the mixed-type inhibition, respectively). In order to elucidate this point, the effect of NADP$^+$ on the initial velocity patterns, with GSSG as the variable substrate, was also analyzed. The resultant double-reciprocal plots showed a parallel lines pattern, strongly suggesting an uncompetitive inhibition (Figure 2(b)). Clearly, the above inhibition patterns are not the expected ones for a typical ping-pong bi bi kinetic mechanism, in which NADP$^+$ is expected to act as a mixed-type inhibitor against NADPH and as a competitive inhibitor against GSSG, respectively [25, 32]. To clarify these apparently contradictory results an additional experiment, in which the concentrations of both NADPH and GSSG were varied together while maintaining their concentrations at a constant ratio was performed. The results of such experiment revealed linear double-reciprocal plots (Figure 3), consistent with a ping-pong bi bi kinetic mechanism [25]. Hence, the unexpected inhibition patterns obtained with NADP$^+$ are due to the formation of a dead-end complex between NADP$^+$ and the form of the enzyme to which NADPH binds. As a corollary of such proposal, it is expected that the dissociation of NADP$^+$ from the reduced form of the enzyme during the normal catalytic cycle would be an irreversible process. To test this prediction, the effect of NADP$^+$ at a low concentration on the initial velocity pattern, with GSSG as the variable substrate, was analyzed. The results revealed no effect of NADP$^+$ on the slope of the double-reciprocal plots (data not shown), confirming the lack of a reversible connection during the NADP$^+$ dissociation step. Hence, it can be concluded that the inhibition of the product NADP$^+$ on the GSSG reductase activity of TGR is

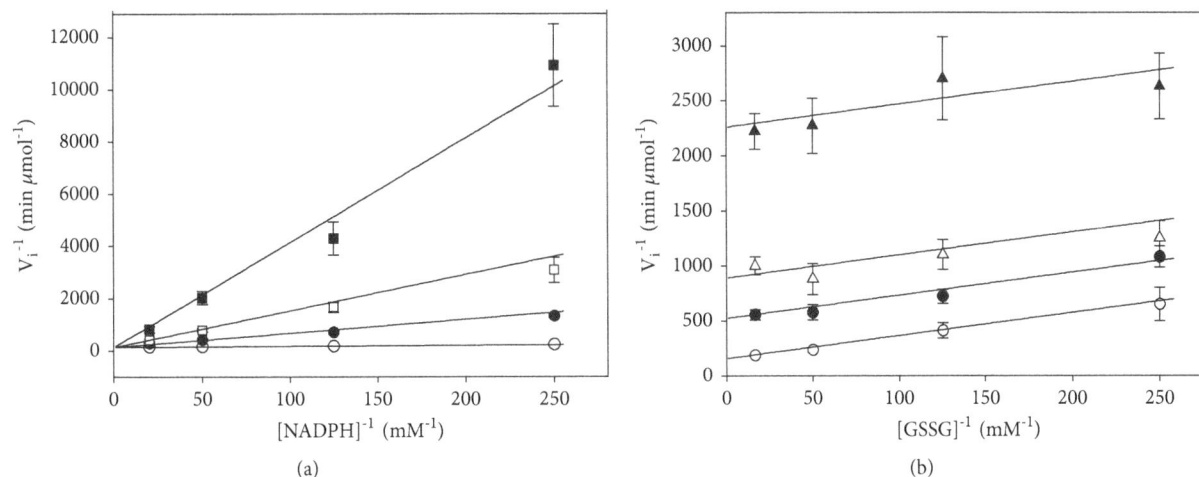

(a)

(b)

FIGURE 2: **Product inhibition patterns of *T. crassiceps* TGR by NADP$^+$.** Enzyme assays and incubation conditions were as described under Materials and Methods. (a) NADPH as the variable substrate at a constant GSSG concentration (70 μM) and the following fixed concentrations of NADP$^+$: (O) 0; (●) 0.5 mM; (□) 1.5 mM; (■) 5 mM. Continuous lines were obtained from the corresponding double-reciprocal form of equation (3) using the parameters resulting from the global fitting of data. (b) GSSG as the variable substrate at a constant NADPH concentration (20 μM) and the following fixed concentrations of NADP$^+$: (O) 0; (●) 0.8 mM; (△) 1.6 mM; (▲) 5 mM. Continuous lines were obtained from the corresponding double-reciprocal form of equation (4) using the parameters resulting from the global fitting of data. In all of these inhibition experiments, the final concentration of enzyme subunit was 6.1 nM. Each point represents mean ± standard deviation (n=6).

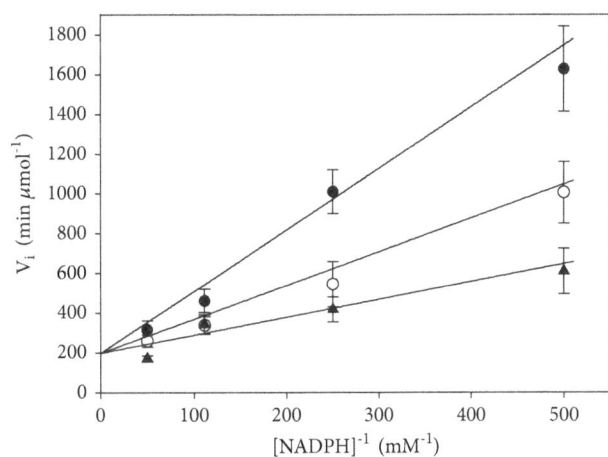

FIGURE 3: **Double-reciprocal plot of the initial velocity data.** Each line was obtained by varying the concentration of both NADPH and GSSG such that [GSSG] = χ [NADPH]. The corresponding χ factor values were as follows: (●) 1; (O) 2.5; (▲) 8. Final concentration of enzyme subunit was 5.3 nM. Lines represent the result of the global adjustment of initial velocity data to the corresponding equation [27]. Each point represents mean ± standard deviation (n=6).

the result of the formation of a dead-end complex with the unmodified enzyme, explaining the competitive inhibition pattern of NADP$^+$ with NADPH as the variable substrate. In Table 1 the kinetic parameters K_m, k_{cat} as well as the specificity constant k_{cat}/K_m for both NADPH and GSSG of *T. crassiceps* TGR are summarized. These values are consistent with those previously reported [14].

Reduced glutathione was a very poor inhibitor product. At 8 mM GSH, barely 10% inhibition was detected. Between

1 and 4 mM GSH, a moderate but consistent activating effect (up to 25%) was observed. Due to the micromolar concentrations of GSSG present in the GSH stock solutions, its effect as an inhibitor product was not further explored.

3.2. GSSG-Dependent Substrate Inhibition of TGR. In the presence of moderate or high concentrations of the substrate GSSG, the disulfide reductase activity of *T. crassiceps* TGR is strongly inhibited. Such inhibition is concomitant with the appearance of the hysteretic-like progress curves [14], suggesting both phenomena could be mechanistically related. Figure 4 shows the saturation curves obtained with either GSSG or NADPH as the variable substrate in a broad range of concentrations. A strong inhibitory effect with GSSG at any constant concentration of NADPH was observed (Figure 4(a)); in contrast, no inhibition with NADPH was detected, even at concentrations as high as 30 times K_m (Figure 4(b)). To warrant the kinetic analysis of the initial velocity data obtained at high concentrations of GSSG in the presence of the atypical time courses, it was necessary to perform *in silico* simulations using the mechanism-based model (see discussion) either in the presence or in the absence of the reactions involved in the generation of the hysteretic-like full time courses. The results of the simulations revealed no significant difference when the initial velocity data obtained under the two above-mentioned conditions were compared (Fig. S1). Therefore, it can be concluded that, in the initial stages of the reaction carried out at a high GSSG concentration, steady-state conditions can be assumed. Thus, the initial velocity data shown in Figure 4 were fitted to (5) to obtain the kinetic parameters Vm_2, Km'_B, and Ki. Results of the global fitting procedure revealed the Vm_2 value is barely 2.5% of Vm_1. As regards Ki, the resultant figure of 331 ± 104

TABLE 1: Summary of kinetic parameters of *T. crassiceps* TGR.

Parameter	Experimental value[a]	Theoretical value[b]
K_m NADPH	$3.5 \pm 0.4\ \mu M$ (n = 6)	$1.8\ \mu M$
K_m GSSG	$14.4 \pm 2.3\ \mu M$ (n = 6)	$6.6\ \mu M$
K_m' GSSG	$100 \pm 22\ \mu M$ (n = 6)	$47\ \mu M$
k_{cat}	$12 \pm 2.4\ s^{-1}$ (n = 6)	$7.4\ s^{-1}$
k_{cat}/K_m NADPH	$3.43 \times 10^6\ s^{-1}\ M^{-1}$	$4.12 \times 10^6\ s^{-1}\ M^{-1}$
k_{cat}/K_m GSSG	$0.83 \times 10^6\ s^{-1}\ M^{-1}$	$1.12 \times 10^6\ s^{-1}\ M^{-1}$
Ki GSSG	$331 \pm 79\ \mu M$ (n = 6)	$512\ \mu M$
Ki NADP$^+$	55.3 ± 3.5 (n = 4)[c]	nd
Ki NADP$^+$	37.6 ± 5.1 (n = 4)[d]	nd

a: determined by fitting the corresponding rate equation to initial velocity data. Data represent mean ± standard deviation of the n replicates (in parenthesis); b: determined from the values of the theoretical rate constants and the definition of the corresponding kinetic parameter as derived from the mechanism-based model; c: an uncompetitive inhibition regarding GSSG; d: competitive inhibition regarding NADPH; nd not determined.

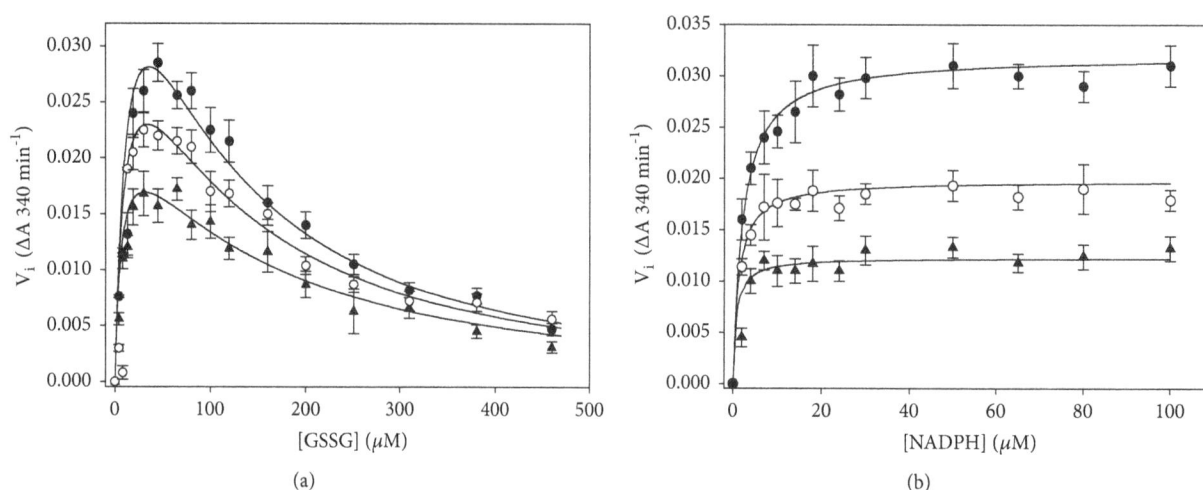

(a)

(b)

FIGURE 4: **Dependence of initial velocities of *T. crassiceps* TGR on either GSSG (a) or NADPH (b) concentrations.** Data were obtained as described under Materials and Methods. (a) GSSG as the variable substrate at the following constant concentrations of NADPH: (▲) 3 μM; (○) 8 μM; (●) 40 μM. (b) NADPH as the variable substrate at the following constant concentrations of GSSG: (▲) 4 μM; (○) 10 μM; (●) 35 μM. In all the enzyme assays, the final concentration of enzyme subunit was 14 nM. In order to avoid overlapping of data, only results obtained at three constant concentrations of the corresponding fixed substrate is shown. Continuous lines were obtained from equation (5) using the parameters resulting from the global adjustment of data. Each point represents mean ± standard deviation (n=6).

μM is consistent with the value of $310 \pm 47\ \mu M$ obtained by (6). A summary of these kinetics parameters is shown in Table 1.

3.3. Effect of NADPH on the Hysteretic-Like Kinetic Behavior of TGR. To analyze the effect of NADPH on the atypical kinetic behavior of *T. crassiceps* TGR, enzyme assays of GSSG reductase activity under a variety of initial concentrations of NADPH, GSSG, and enzyme were carried out. Figure 5 shows representative full progress curves. At low concentrations of both substrates (Figures 5(a) and 5(b), left traces), conventional profiles of NADPH consumption were observed. By increasing the concentration of either NADPH or GSSG, however, complex profiles of the progress curves resulted (Figures 5(a) and 5(b), middle traces), which were particularly noticeable between 100 μM and 300 μM GSSG. In such profiles three kinetic stages of NADPH consumption can be discerned, a first burst-like stage in which the rate

of NADPH consumption was relatively fast, followed by a second stage where a temporary inhibition of the GSSG reductase activity was evident, and a final third stage in which the enzyme activity was regained, leading the reaction into an apparent steady-state condition. The relative amplitude, as well as the duration of the first and second stages of the reaction, was strongly dependent on the concentration of both substrates as well as the enzyme concentration. At 4.7 μM NADPH and 270 μM GSSG (Figure 5(a), middle trace) the temporary inhibition stage was already detectable, and the initial fast stage of NADPH oxidation was clearly observable. By contrast, at 46.5 μM NADPH and 310 μM GSSG (Figure 5(c), right trace), the initial fast consumption of NADPH was barely detectable and the temporary inhibition stage was extended over ten minutes. At any NADPH concentration where the GSSG-dependent substrate inhibition is observed, increasing the initial concentration of GSSG in the reaction mixture resulted in a decrease in both the

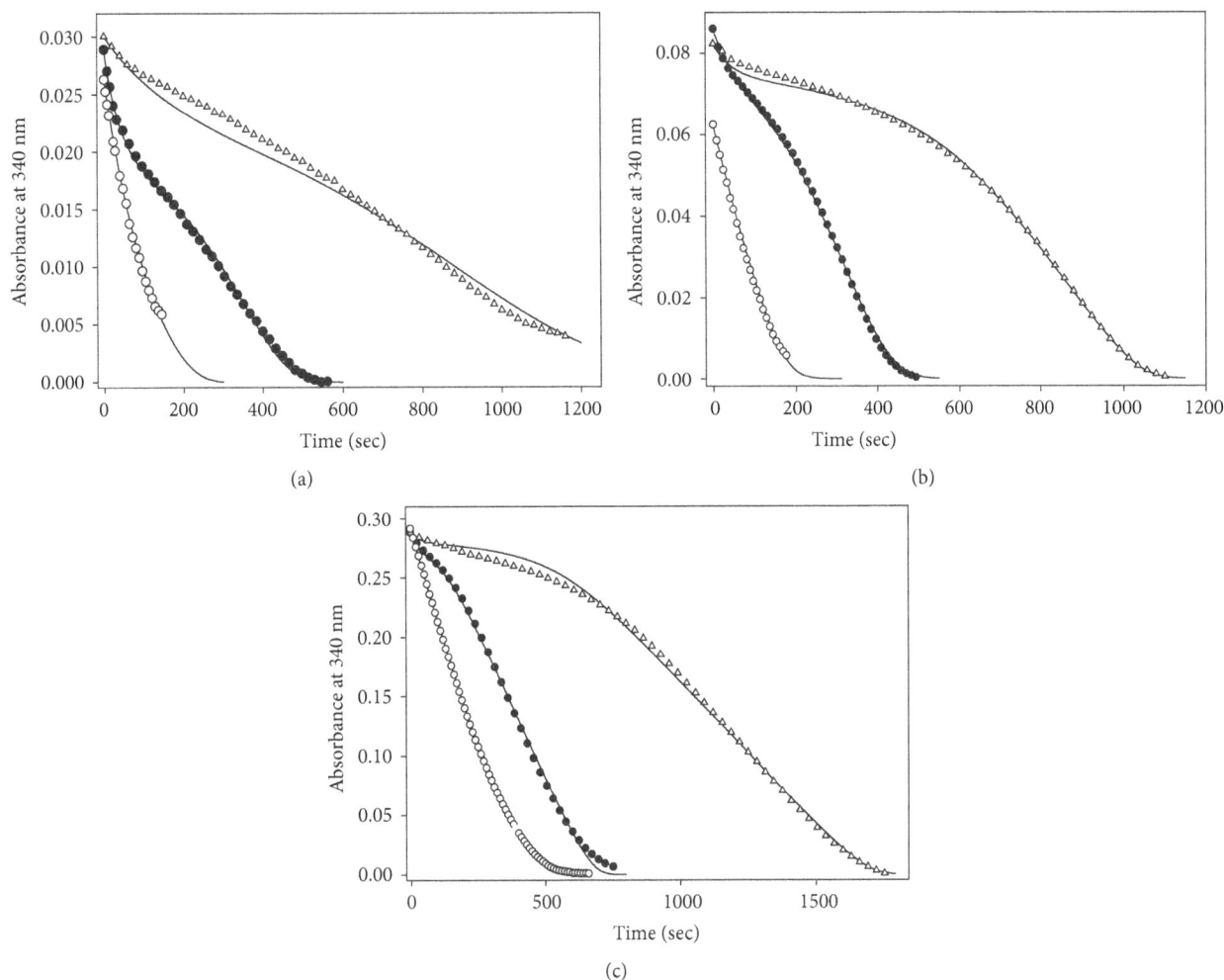

FIGURE 5: **Full time progress curves of *T. crassiceps* TGR as a function of both NADPH and GSSG concentration**. Enzyme assays and incubation conditions were as described under Materials and Methods. The particular concentration of NADPH, GSSG, and enzyme used in the corresponding enzyme assays were as follows: Panel (a): (O) 4.2 μM NADPH, 5 μM GSSG, 14.7 nM TGR; (●) 4.7 μM, 270 μM GSSG, 11.5 nM TGR; (Δ) 4.9 μM NADPH, 500 μM GSSG, 11.5 nM TGR. Panel (b): (O) 10.1 μM NADPH, 20 μM GSSG, 14.7 nM TGR; (●) 13.9 μM NADPH, 200 μM GSSG, 11.5 nM TGR; (Δ) 13.4 μM NADPH, 510 μM GSSG, 11.5 nM TGR. Panel (c): (O) 47 μM NADPH, 120 μM GSSG, 11.5 nM TGR; (●) 46.5 μM NADPH, 200 μM GSSG, 11 nM TGR; (Δ) 46 μM NADPH, 310 μM GSSG, 6.6 nM TGR. Continuous lines represent fitting of experimental data points to the mechanism-based model using the best set of rate constants (see Materials and Methods for details of the conditions used in the fitting procedure).

initial velocity and the relative amplitude of the first stage, concomitant with a significant rise in the duration of the inhibited stage. An identical effect was obtained by increasing the concentration of NADPH at a constant but high enough GSSG concentration. Figure 6 shows the dependence of the apparent lag time on the concentration of GSSG at different NADPH concentrations. Clearly, its magnitude is determined by the concentration of both substrates. The complex full progress curves were observed at any NADPH concentration when the GSSG concentration was high enough.

By increasing the enzyme concentration in the reaction mixture under conditions in which an apparent lag phase was observed resulted in a significant rise in the amplitude of the initial fast stage, concomitant with a decrease in the lag time as previously reported [14]. Attempts to fit the full time course

data to the single exponential equation derived for simple hysteretic kinetics [28] were unsuccessful.

The above results strongly suggest that, under conditions in which complex time courses were observed, continuous changes in the relative abundance of both active and inactive complexes of the enzyme were occurring. It is worth noting that, under any condition in which an atypical full progress curve was observed, the magnitude of the velocity calculated from the maximal slope at the apparent steady-state segment of the curve was significantly lower than that predicted from the rate equation for a ping-pong bi bi kinetic mechanism (Figure 7), both at a low (Figure 7(a)) and at a high NADPH (Figure 7(b)) concentration. Even when the accumulation of NADP$^+$ and the depletion of the substrates were taken into account, the same results were obtained. Thus, in the apparent

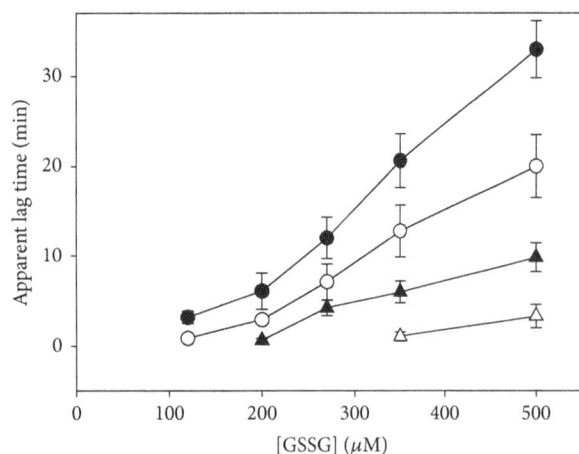

FIGURE 6: **Dependence of the apparent lag time on the concentration of both NADPH and GSSG.** The magnitude of the relaxation time was estimated as described under Materials and Methods from full time courses performed at 25°C and pH 7.8 at the following NADPH concentrations: (Δ) 5.5 μM; ($\blacktriangle$) 9 μM; ($\bigcirc$) 15.5 μM; ($\bullet$) 50 μM. Each point represents mean $\pm$ standard deviation (n=6).

steady-state segment of the progress curves, the inhibition of TGR by GSSG was still present, and the degree of inhibition depended on the initial concentrations of both NADPH and GSSG.

3.4. Effect of GSSG on Trx Reduction. When the alternative biological disulfide Trx was assayed at relatively high concentrations (up to 170 μM), no evidence for substrate inhibition or hysteretic kinetics was observed [14]. Hence, it can be concluded that both the substrate inhibition and the apparent hysteretic kinetic of TGR are due only to GSSG. However, the addition of a high concentration of GSSG to the assay mixture for TR activity resulted in inhibition (Figure 8). With Trx as the alternative substrate, the profile of the full time course obtained in the presence of 2 mM GSSG was similar to those described for the reduction of GSSG under hysteretic conditions, showing a clear temporary inhibition stage (Figure 8). By increasing the concentration of Trx in the reaction mixture, however, the magnitude of the lag time was significantly shortened.

3.5. Docking Analysis with GSSG. From the X-ray crystallographic analysis of *S. mansoni* TGR a potential zone for GSSG binding was revealed [8]. It is located at the *si* face of FAD and is characterized by a high density of positively charged residues (K124, K128, R450, and R454), giving a surface electrostatic potential which is more similar to the corresponding site on GR than that of the homologous TR [8]. Docking analysis with TGR from both *S. mansoni* [10] and *F. gigantica* [16] revealed the feasibility for GSSG binding at this putative alternative site. The results of such studies suggest that GSSG is able to interact with the positively charged residues K124 and R450 through electrostatic interactions. We have confirmed these results with TGR from *T. crassiceps* and the associated binding enthalpy was estimated as – 7.1 kcal

mol^{-1}. This figure is significantly lower than the value of – 24.6 kcal mol^{-1} obtained by isothermic titration calorimetry for the binding of GSSG to the active site of human GR [33], suggesting a weaker interaction of GSSG at the corresponding site on TGR. In this sense, the value of 385 μM for K_i obtained in the present work from the substrate inhibition data is consistent with such result. Figure 9(b) shows the location on the enzyme at which GSSG is potentially bound as inhibitor. Its disulfide bond is located midway between the redox active dithiol of the enzyme and the catalytically essential selenocysteine residue of the neighbor subunit. Assuming that electrostatic interactions are involved in the GSSG binding at the putative inhibitory site, then it can be guessed that the ionic strength of the medium could modify the atypical full progress curves of TGR. Experiments to test such prediction were carried out. Figure 10 shows the effect of NaCl on both the atypical profile of the hysteretic-like progress curves and on the apparent lag time. Clearly, by increasing the ionic strength in the reaction mixture a significant decrease in the magnitude of the apparent lag time, concomitant with a slight increase in the amplitude of the burst-like stage, was obtained. Thus, the involvement of electrostatic interactions in GSSG binding at the inhibitory site is strongly suggested. Although the diminution in the size of the apparent lag time due to NaCl resulted in an increase in the amplitude of the burst-like stage, the initial velocities measured were still far below that expected in the absence of inhibition. This is the result of an inhibitory effect by NaCl on the reductase activity overlapped with its ability to modify the lag time. Such inhibition was confirmed by analyzing the effect of NaCl on the GSSG reductase activity of TGR at a low concentration of GSSG (80 μM). In the range from 50 mM to 600 mM NaCl, an inhibition up to 70% of the initial velocity was observed.

4. Discussion

TGR represents an atypical case in the disulfide reductase family of enzymes. Although the enzyme retains the homodimeric nature which is common to this set of oxidoreductases, it is outstanding in tertiary structure, total number of redox centers, and wide substrate specificity. Thus, in TGR a Grx-like domain has been appended to the N-terminal end of the animal TR module, conferring to the enzyme additional catalytic abilities, notably the reduction of GSSG at significant rates, as well as catalysis of thiol-disulfide exchange reactions, including deglutathionylation of mixed disulfides protein-glutathione [1, 4, 6]. Furthermore, the dithiol/disulfide motif of the Grx-like domain of TGR adds to the FAD prosthetic group and the N- and C-terminal redox centers of the TR module, giving to the enzyme the potential ability to store up to eight reducing equivalents in its maximal reduction state. In addition to the above features, TGR displays unusual full time courses of enzyme activity at moderate or high concentrations of GSSG [5, 14–17], which have been considered as hysteretic behavior [5, 14]. To explain such atypical kinetics of TGR, a model based on covalent modification through glutathionylation of the structural cysteine residues 88 and 354 of the enzyme from *E. granulosus* [5] was proposed.

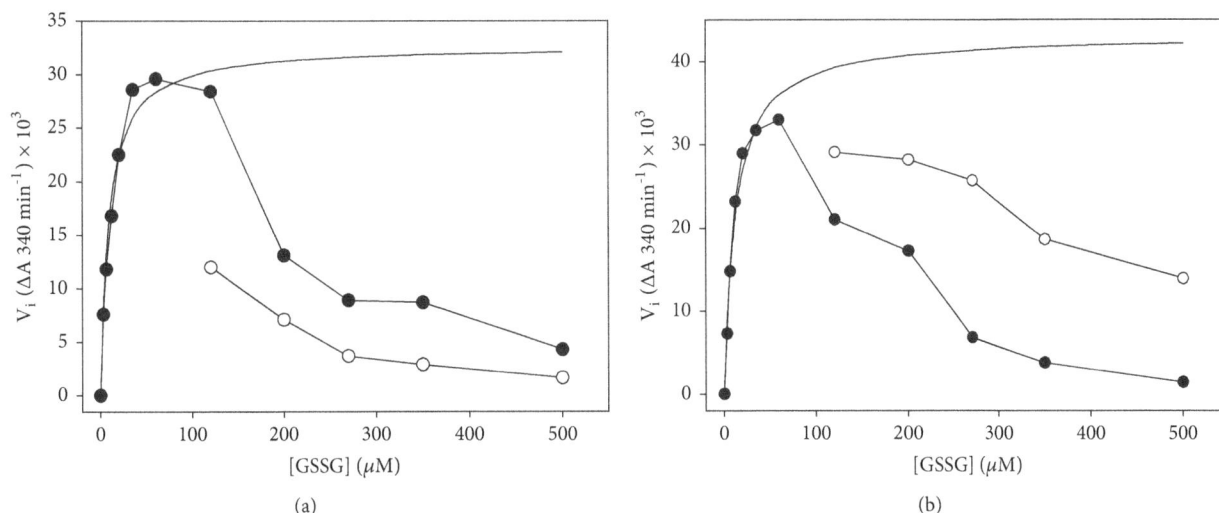

(a)

(b)

FIGURE 7: **Dependence of the initial and the apparent steady-state velocities on both NADPH and GSSG concentration**. Initial (●) and apparent steady-state (○) velocities were obtained from full progress curves by varying the concentration of GSSG at the following initial concentrations of NADPH: (a) 9 μM; (b) 50 μM. Continuous spline lines represent the initial velocities predicted by the rate equation for a ping-pong bi bi kinetic mechanism in the absence of substrate inhibition. Each point represent the average of six experiments.

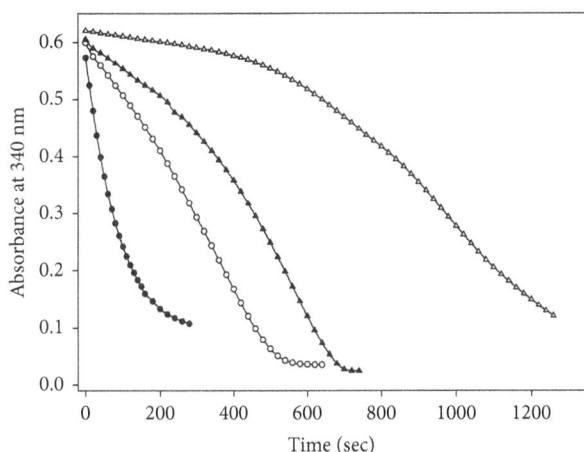

FIGURE 8: **Effect of GSSG on the reduction of human Trx by *T. crassiceps* TGR**. Reaction mixtures were prepared in buffer A by mixing Trx with 100 μM NADPH either in the presence or in the absence of 2 mM GSSG. After two minutes, enzyme was added to start the reaction. The final volume of the reaction mixture was 120 μL. (△) 2 mM GSSG, no trx; (●) 150 μM Trx, no GSSG; (▲) 2 mM GSSG plus 60 μM Trx; (○) 2 mM GSSG plus 150 μM Trx. The final concentration of enzyme was 37 nM.

However, such model is not supported by the experimental observations. In this sense, TGR from the flukes *F. hepatica* [15] and *S. mansoni* [7] also shows atypical full time progress curves of reductase activity at high concentrations of GSSG, even though they lack the two above noted cysteine residues. An alternative hypothesis, based on the dithiol/disulfide redox motif of the Grx-like domain, was put forward to explain the hysteretic behavior of the enzyme [6]. Thus, when the C-terminal cysteine residue of the dithiol/disulfide redox center of the Grx domain (Cys34 of *E. granulosus* TGR)

was replaced with serine, the lag time observed at a high concentration of GSSG was fully abolished. However, by decreasing the protein concentration of the mutant enzyme in the reaction mixture, the hysteretic-like progress curve was regained [6]. In these two hypotheses, the role of NADPH in the hysteretic behavior of TGR was not considered.

The results described in the present work show that TGR from *T. crassiceps* cysticerci follows a ping-pong bi bi kinetic mechanism with the NADPH/GSSG couple as the substrates, in which dissociation of NADP$^+$ from the enzyme during the catalytic cycle is an irreversible event. Interestingly, the inhibition patterns obtained with NADP$^+$ are consistent with a variant of the ping-pong bi bi kinetic mechanism typical for two-site enzymes [34–36]. According to such mechanism, binding of the substrates on the enzyme occurs at separate sites under rapid equilibrium conditions, requiring a mobile carrier in order to transfer the corresponding chemical group between both substrate binding sites. The three-dimensional structure of *S. mansoni* TGR [8, 9] is consistent with such proposal. In this enzyme, the role of the mobile carrier is played by the C-terminal end of the neighbor subunit, where the essential selenocysteine residue is located. Further, in a two-site enzyme, dissociation of the first product (i.e., NADP$^+$) can occur either before or after binding of the second substrate (i.e., GSSG), explaining the atypical inhibition patterns observed with *T. crassiceps* TGR. Thus, TGR can be considered as an additional example of enzymes where the active site is split into halves.

On the other hand, the initial velocity data obtained in a broad range of GSSG concentrations revealed a strong substrate inhibition (Figure 4(a)). The overlap of the latter with the atypical progress curves could, in principle, reject a kinetic analysis of the initial rate data by the steady-state formalism. However, the results of the *in silico* simulations performed either in the presence or in the absence of the

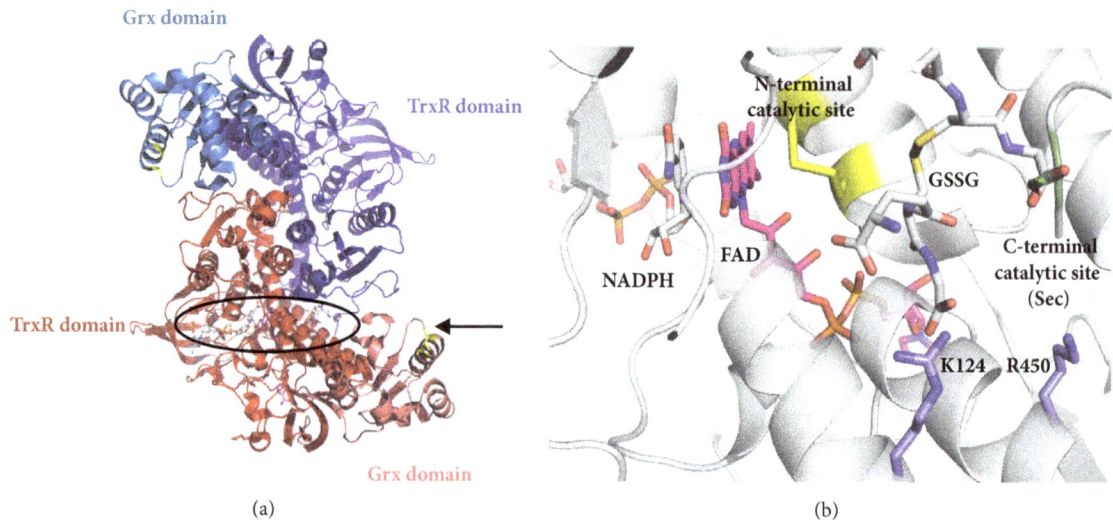

FIGURE 9: (a) Dimeric structure of *S. mansoni* TGR as derived from X-ray crystallography. Monomers of the enzyme are shown in either blue or red and the location of the glutaredoxin domains is indicated. The enzyme was placed such that the region in which both the NADPH and the FAD binding sites, enclosed in an oval, could be viewed. (b) Amplified view showing details of the potential binding site for GSSG as inhibitor as derived from docking studies. The reducing substrate NADPH, as well as the prosthetic group FAD and the redox active disulfide (in yellow), are shown in stick. The catalytically essential selenocysteine residue of the partner subunit is shown in green.

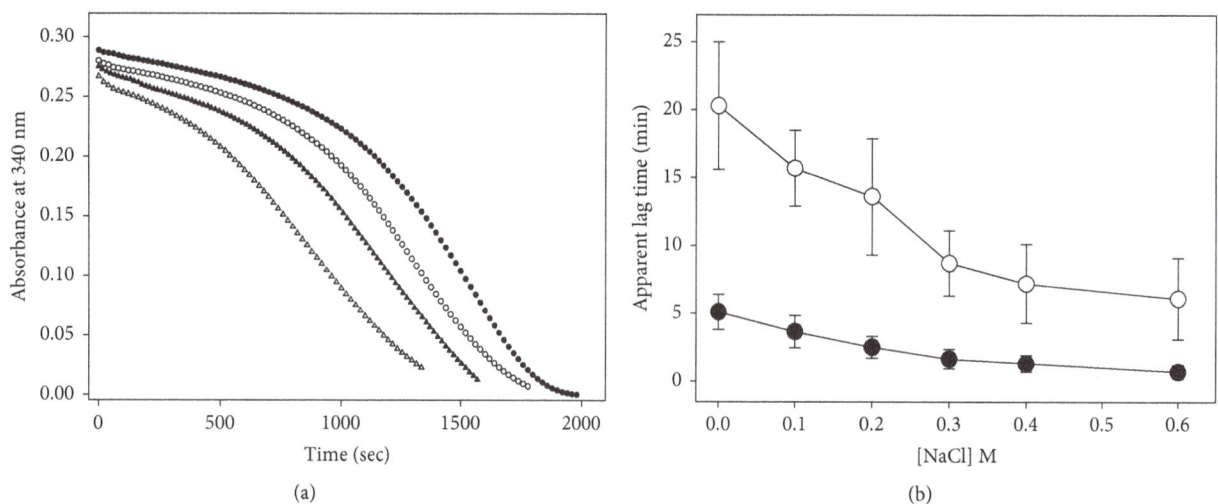

FIGURE 10: **Effect of ionic strength on the apparent lag time of *T. crassiceps* TGR**. Enzyme assays and incubation conditions were as described under Materials and Methods. (a) Representative full progress curves of GSSG reductase activity determined in the presence of 50 μM NADPH and 600 μM GSSG at the following concentrations of NaCl: ($\bullet$) control without NaCl; ($\circ$) 0.2 M NaCl; ($\blacktriangle$) 0.4 M NaCl; ($\triangle$) 0.6 M NaCl. To avoid overlapping of traces, curves were displaced vertically. (b) Dependence of the apparent lag time on NaCl concentration. Enzyme assays were carried out in the presence of either 300 μM GSSG ($\bullet$) or 600 μM GSSG ($\circ$). Each point represents mean $\pm$ standard deviation (n=6).

reactions responsible for the unusual kinetic behavior of TGR (see below) revealed such analysis is warranted. As shown in Fig. S1, differences in the initial velocity data are observed at GSSG concentrations above 50 μM, reaching a maximum (about 8%) at the highest concentrations of GSSG, both at a low and at a high NADPH concentration. Such differences are in the range of the experimental uncertainty obtained in the determination of initial velocities.

As regards the hysteretic-like full progress curves of TGR observed at moderate or high concentrations of GSSG, the

results obtained in the present work revealed the presence of an additional complexity. Thus, the enzyme assays performed at low concentrations of the electron donor NADPH showed an initial burst-like stage, followed by the lag stage. The initial stage of fast NADPH consumption is barely noticeable in the enzyme assays carried out in the presence of high concentrations of both GSSG and NADPH [5, 6, 14, 16, 17]. Due to its minor contribution to the full progress curves of reductase activity under these conditions, such initial stage was not considered in any previous work dealing with the

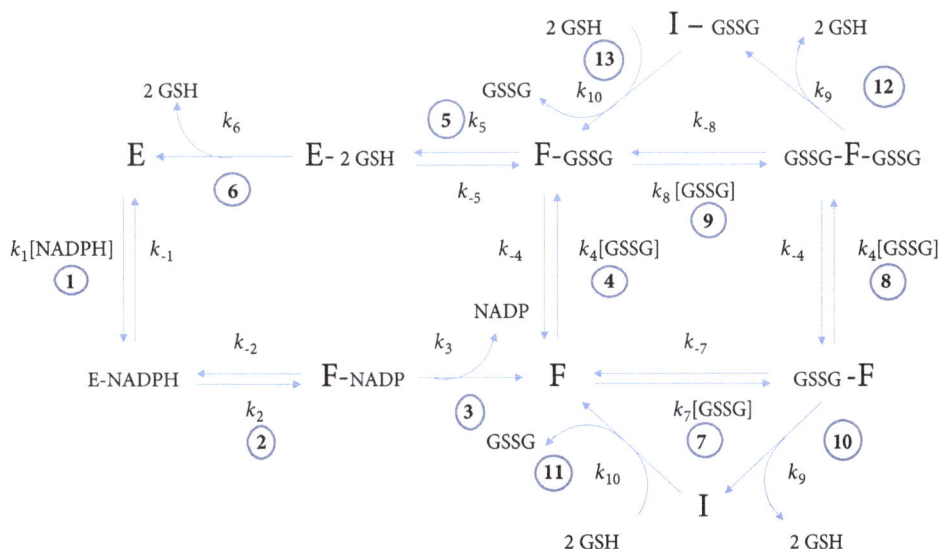

FIGURE 11: **Mechanism-based model for GSSG reductase activity of *T. crassiceps* TGR.** Reactions 1 to 6 pertain to the ping-pong bi bi kinetic mechanism; reactions 7 to 9 correspond to the reversible inhibitory branch in which the substrate GSSG acts as noncompetitive inhibitor. Reactions 10 and 12 represent residual GSSG reductase activity of the inhibited complexes while reactions 11 and 13 correspond to reactivation of the inactive intermediaries I and I-GSSG by GSH. Reactions 1, 4, and 8 of the model correspond to those segments working under rapid equilibrium conditions.

atypical kinetic behavior of TGR. As was demonstrated in the present work, the relative contribution of both the burst-like and the lag-like stages to the time course depends on the concentration of both NADPH and GSSG.

In order to build a comprehensive model to explain the atypical kinetic behavior of TGR, the following observations regarding the normal performance of the enzyme must be stated. These are based on all the kinetic and structural evidence available. Figure 9 shows a full view of dimeric TGR (Figure 9(a)) as well as a detailed view (Figure 9(b)) at the TR module region where both the NADPH binding site and the N-terminal dithiol/disulfide redox active motif are located.

(i) In the GSSG reductase activity of TGR, the Grx-like domain of the enzyme plays an essential role [1, 4, 7]. Hence, its dithiol/disulfide redox motif must be in the reduced state. Thus, in addition to the reduced redox centers typical of the high molecular weight TR, in TGR additional reducing equivalents are needed in order to catalyze GSSG reduction. Therefore, during the catalytic cycle the enzyme must oscillate between states with a reduction degree higher than the two-electron reduced state. Recent work with *S. mansoni* TGR [9] supports the existence of the four-electron (EH_4) reduced species as an intermediary in its normal functioning.

(ii) Natural variants of the enzyme (e.g., mouse testes TGR) in which the C-terminal cysteine residue of the redox motif at the Grx-like domain is absent are fully functional in GSSG reduction [1]. This fact strongly suggests that in the GSSG reduction pathway by TGR, the Grx-like domain is able to catalyze only a thiol-disulfide exchange reaction with either GSSG or protein-glutathione mixed disulfides. In this proposal, the N-terminal cysteine residue of such motif will be involved in the nucleophilic attack on the disulfide

bond of GSSG, leading to the formation of a glutathione-enzyme mixed disulfide [4]. Experiments carried out with TGR mutants of both *E. granulosus* and *S. mansoni*, in which either the N-terminal or the C-terminal cysteine residue of the Grx redox motif was replaced [6, 7], support such proposal. In this sense, it is worth noting that, for typical Grx, only the N-terminal redox active cysteine is required in thiol-disulfide exchange reactions [37, 38].

(iii) As with the related TR, both subunits of TGR are required during the catalytic cycle of the enzyme. Thus, in order to regenerate the reduced state of the nucleophilic cysteine of subunit A, electrons must be shuttled from the selenol-thiol redox center, located at the C-terminal end of subunit B [9]. However, as was revealed by the X-ray crystallographic studies with *S. mansoni* TGR (Figure 9(a)), the two redox centers are distant from one another [8, 9]. Thereby, in order for the electron transfer to take place, a large conformational change of the C-terminal arm of the enzyme is needed. Molecular dynamic simulations suggest such change is feasible [10]. In a recent work with *E. granulosus* TGR it is proposed that the Grx-like domain of the enzyme participates in the conformational change during the aforementioned electron transfer [37].

(iv) The X-ray crystallographic structures of both *S. mansoni* [8] and *E. granulosus* [37] TGR have revealed the existence of a potential second site for GSSG binding, located on the TR module of the enzyme at the FAD binding domain (Figure 9(b)). This region is characterized by a high charge density whose electrostatic potential is similar to the binding site for GSSG on GR [8]. Hence, the putative interaction of GSSG on this site must involve electrostatic interactions. The results dealing with the effect of the ionic strength on the apparent lag time reported in the present work support the latter conclusion, while the docking studies have revealed that such interaction is both structurally and thermodynamically feasible [10, 16].

Regarding the complex kinetic behavior of TGR, the following experimental facts must be considered.

(i) The profiles of the full progress curve of NADPH consumption and the magnitude of the apparent lag time are dependent on the concentration of the substrates NADPH and GSSG. Thus, the amplitude of both the initial burst-like stage and the inhibited segment of the progress curves can be modified by changing the initial concentration of the substrates in the reaction mixture.

(ii) Both the magnitude of the apparent lag time and the profile of the full progress curves of TGR are not those expected for a typical hysteretic enzyme. For the latter, the reported lag times are in the range of seconds or a few minutes and the shape of the curve in the transition zone is due to a single exponential transition [28, 38–41].

(iii) The presence of disulfide reducing reagents (e.g., GSH, DTT, cysteine) at micromolar concentrations in the reaction mixture lead either to a decrease in the magnitude of the apparent lag time or its full abolition [5, 14, 15]. In this sense, in the presence of 50 μM GSH saturation curves with GSSG as the variable substrate was fully hyperbolic. Up to a concentration of 300 μM of the disulfide no inhibition was observed (data not shown). Such evidence strongly suggests that in the atypical kinetic behavior of TGR thiol/disulfide exchange reactions are involved.

(iv) The concentration of the enzyme in the reaction mixture is also critical for the presence or absence of the atypical progress curves as well as for the amplitude of the initial fast stage of the reaction. The higher the enzyme concentration, the lesser the magnitude of the apparent lag time [5, 14].

Based on all the above kinetic and the structural evidence on TGR, a mechanism-based model was built to explain its atypical kinetic properties (Figure 11). The major features of the model are briefly described:

(i) According to the results obtained in the present work, the normal catalytic cycle of the enzyme (reactions 1 to 6 in Figure 11) is based on a two-site ping-pong bi bi kinetic mechanism, in which both NADPH and GSSG bind at different sites under rapid equilibrium conditions. Such sites are located at the TR module and the grx-like domain of the enzyme for NADPH and GSSG, respectively, as derived from the X-ray crystallographic studies of *S. mansoni* [8, 9] and *E. granulosus* [37] TGR.

(ii) Dissociation of the products NADP$^+$ and GSH during the normal catalytic cycle (reactions 3 and 6 in Figure 11) is assumed to be irreversible steps, as revealed by the results of the product inhibition studies in the present work. However, both compounds will be able to act as competitive inhibitors of either NADPH or GSSG, respectively, through binding to their corresponding rapid equilibrium segment.

(iii) GSSG acts as a noncompetitive inhibitor by binding to the alternative site of either the reduced form of the enzyme (F) or the F-GSSG binary complex (reactions 7 and 9 in Figure 11), leading to the formation of the GSSG-F binary and the GSSG-F-GSSG ternary complexes. Binding

of GSSG at the alternative site is the cause for the substrate inhibition shown in Figure 4(a). Although an alternative and simpler uncompetitive inhibition for GSSG also explains the atypical progress curves of enzyme activity of TGR, the noncompetitive inhibition pattern is more consistent with the existence of two sites. Such conclusion is supported by the statistical results of model discrimination. The potential site at which GSSG binds as inhibitor is located on the TR module at the *si* face of FAD (Figure 9(b)), near the conventional disulfide/dithiol redox center typical of the disulfide reductase family of enzymes. The affinity of the enzyme for GSSG at this alternative binding site is significantly lower ($Ki = 331 \pm 79$ μM) compared with the affinity of the grx-like domain ($Km = 14.4 \pm 2.3$ μM). GSSG bound at the inhibitory site will block the electron flow from FAD to the C-terminal redox center, thus inhibiting its reduction.

(iv) It is proposed that both the GSSG-F binary and the GSSG-F-GSSG ternary complexes of the enzyme have the ability to reduce GSSG at the low-affinity alternative site (reactions 10 and 12 in Figure 11), albeit at a low rate. In such reaction the conventional N-terminal dithiol/disulfide redox center must be involved. Experimental evidence supporting this proposal is available. Thus, mutants of *S. mansoni* TGR lack either the essential selenocysteine residue or the redox active cysteines of the Grx domain [7], and also *E. granulosus* mutants in which the N-terminal cysteine residue of the Grx domain has been replaced with serine [6] are still able to catalyze the reduction of GSSG, although at a very low rate. The reported turnover numbers were 0.19 ± 0.03 s^{-1} and 0.6 ± 0.03 s^{-1} for the *S. mansoni* and *E. granulosus* enzymes, respectively. Evidence for residual GSSG reductase activity in *T. crassiceps* TGR was obtained by incubating an enzyme aliquot in the presence of a 4:1 molar excess of the irreversible inhibitor auranofin [14]. Under these conditions, a very low but measurable reductase activity was detected (turnover number 0.08 s^{-1}). Hence, in the absence of either the redox active motif of the Grx-like domain or the selenocysteine residue, TGR will be still able to reduce GSSG in a GR-like fashion. The low activity of GSSG reductase at the alternative site can be explained as a result of a nonoptimal distance of the disulfide bond of GSSG after binding from the nucleophilic cysteine of the redox active motif (Figure 9(b)). In this sense, the strong dependence of the redox reactions on the distance between the electron donor and the acceptor centers is well characterized [42].

(v) As a result of the residual disulfide reductase activity of both the GSSG-F binary complex and the GSSG-F-GSSG ternary complexes, the enzyme population will be directed into the inactive states I and I-GSSG (reactions 10 and 12 in Figure 11) in which its redox centers shall be in the fully oxidized state. This proposal is in agreement with the observation that there was no glutathionylation of any catalytically essential cysteine residue when *E. granulosus* TGR was incubated with GSSG at a high concentration (1 mM) either in the presence or absence of NADPH [5]. The formation of these two fully inactive complexes explains the sudden decrease in slope in the progress curves of the reaction observed under a variety of concentrations of both NADPH and GSSG (Figure 5). By omitting reactions 10

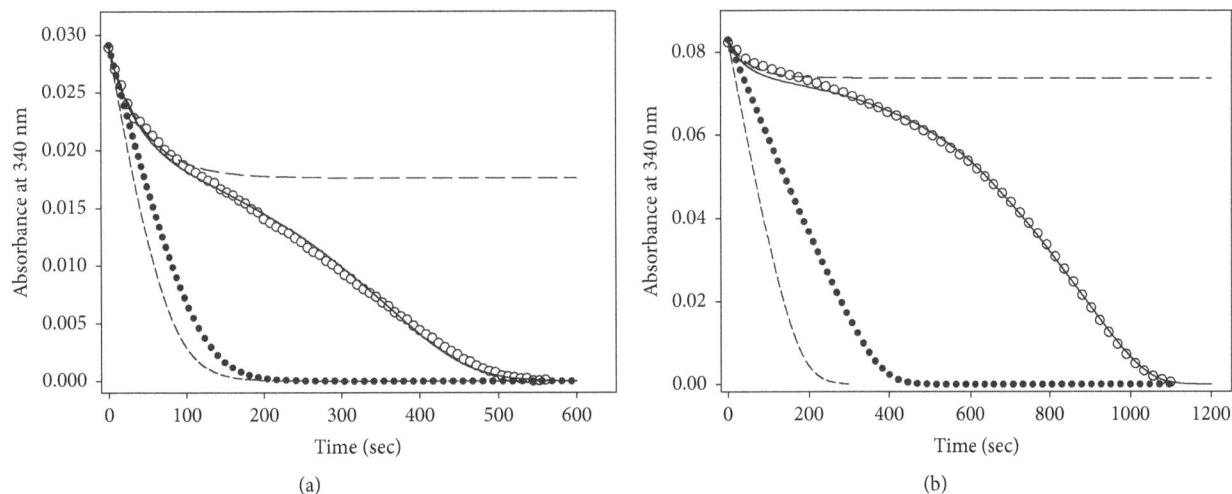

(a)

(b)

FIGURE 12: **Predictive value of the mechanism-based model.** Experimental data points (○) were compared with the full progress curves predicted by either the whole model (continuous line), the model without reactions 10-11 and 12-13 (●), or the model without reactions 11 and 13 (long dash line). The short dash line represents the profile predicted by the model in the absence of any substrate inhibition. (a) Experimental full progress curve determined at 4.7 μM NADPH and 270 μM GSSG. (b) Experimental full progress determined at 13.4 μM NADPH and 510 μM GSSG. In both cases an enzyme concentration of 11.5 nM was used. For both panels (a) and (b) the predicted full progress curves were obtained by the set of theoretical rate constants shown in Table S1 except k_9 (= 0.072 s^{-1}) for panel (a) and k_{-8} (= 2.43 s^{-1}) and k_9 (= 0.03 s^{-1}) for panel (b).

and 12 (and hence reactions 11 and 13) from the model, the resultant full progress curves obtained by *in silico* simulations revealed conventional profiles of NADPH consumption in which GSSG acts as an reversible inhibitor but without any atypical behavior (Figure 12 and Fig. S2).

(vi) The residual GSSG reductase activity of inhibited TGR will result in a slow but continuous increase in the concentration of the product GSH. The latter will be able to reactivate both I and I-GSSG complexes by reducing its redox active centers (reactions 11 and 13 in Figure 11) through thiol/disulfide exchange reactions, leading to a gradual increase in the concentration of the catalytically competent F and F-GSSG complexes and hence reversing the GSSG-dependent inhibition of TGR. The observation that the addition of thiol compounds such as GSH, cysteine, or DTT to the reaction mixture can abolish the hysteretic-like progress curves [5, 14, 15] strongly suggests disulfide bonds are involved in the inactivation of the enzyme by GSSG. Thus, the atypical time progress curves of GSSG reductase activity of TGR observed in the presence of moderate or high concentrations of the substrate GSSG will be the result of a continuous competition between the latter and GSH for driving the enzyme into inactive or active complexes. Such competition explains the large size that the apparent lag time can reach, extending over 1 h at high concentrations of both NADPH and GSSG. As a corollary, it can be concluded that the slope observed at the reactivating segment does not correspond to an authentic steady-state velocity. This point must be stressed, because, in some works dealing with the atypical kinetic behavior of TGR [11, 16, 37], the maximal slope in a full time progress curve of GSSG reductase activity has been mistakenly noted as a true steady-state velocity.

By deleting from the model the reactivation of the enzyme by disulfide reducing compounds (reactions 11 and 13 of Figure 11), no hysteretic-like profile of the full progress curves is obtained, as revealed by the *in silico* simulations (Figure 12 and Fig. S3). Instead, the shape of the traces are the expected ones for an enzyme catalyzed reaction in the presence of an irreversible inhibitor [43, 44].

(vii) The effect of NADPH on the hysteretic-like full progress curves of TGR is explained as a result of providing the enzyme species (i.e., the reduced form F) to which GSSG can bind, both as substrate and as inhibitor. Hence, GSSG will behave as an uncompetitive inhibitor regarding NADPH. The unusual profile of the full progress curves observed at low concentrations of NADPH (Figure 5) but at high GSSG concentrations in which a significant burst stage is present, will be the result of the slow binding of GSSG as inhibitor to that fraction of the enzyme population in the reduced state (i.e., F and F-GSSG). It is worth noting that the best values obtained from the fitting procedure for the rate constants k_7 and k_8 (Table S1) suggest GSSG behave as a slow-binding inhibitor [45, 46]. Thus, in the first stages of the reaction in the presence of moderate concentrations of GSSG, the kinetic competence between the catalytic and the inhibitory reaction cycles will result in the presence of a burst-like stage.

(viii) The effect of a high concentration of GSSG on the reduction of the alternative disulfide substrate Trx by TGR further supports the proposal that the binding of GSSG occurs at the putative site located in the neighborhood of the FAD prosthetic group (Figure 9(b)). The ability of Trx to reverse the GSSG-dependent inhibition will be the result of both a competition with GSSG for the catalytically competent

F form of the enzyme as well as the chemical reduction of GSSG by reduced Trx, leading to a gradual increase in GSH concentration.

The present model also can furnish an explanation of the results obtained with the *Eg*TGR C34S mutant. As reported by Bonilla et al. [6], the full time course of GR activity of such mutant in the presence of a high concentration of GSSG (1 mM) did not shows any trace of hysteretic-like behavior. By decreasing the enzyme concentration, however, the atypical profile of GSSG reductase activity was regained. Such experimental observation is explained as a result of the replacement of cysteine 34 by serine producing a more catalytically efficient enzyme (i.e., a higher turnover number). Thus, in this variant of TGR the normal catalytic cycle will be able to compete successfully with the inhibitory cycle, resulting in no observation of atypical progress curves, even at GSSG concentrations at which the inhibitory stage persists at times as long as 30 min with the wild type enzyme.

To test the predictive value of the model, simulations *in silico* of both initial velocity patterns and full progress curves were carried out. The conditions used to obtain a selected set of the specific rate constants are described in Supplementary Materials. As shown in the supplementary figures, all the experimental observations on the atypical kinetic behavior of TGR are reproduced by the model. These include the GSSG-dependent substrate inhibition (Figure S1), as well as the effect of the concentration of both GSSG and NADPH (Figures S4), GSH (Figure S6), and enzyme concentration (Figure S5), on the time progress curves. Simulations of the full progress curves using particular experimental conditions reproduce with a high consistency the experimental traces obtained under the same conditions of GSSG, NADPH and enzyme concentrations, as shown in Figures 5 and 12. Even for long time courses obtained at high concentrations of both NADPH and GSSG (Figure 5(c)), the profiles of the progress curve are consistently reproduced.

Thus, it can be concluded that the original designation of hysteretic behavior refer to the atypical full time progress curves of TGR was not correctly applied. Although apparently complex, the phenomenon is simply due to a continuous competition between a substrate (GSSG) and a product (GSH) for driving the enzyme into inactive or active states, resulting in atypical hysteretic-like progress curves. We have named this kind of kinetic behavior as pseudohysteresis. This kind of atypical progress curve of enzyme activity adds to other bizarre kinetic phenomena such as the true hysteretic behavior [21, 28] damping oscillatory hysteresis [47], and instability of the reaction product [48].

A comment concerning a recently published paper dealing with recombinant human TGR [11]. In such work, the effect of high concentrations of GSSG on enzyme activity was tested. As shown in the Figure 6 of the paper, hysteretic-like progress curves were obtained. These data further support the view that the pseudohysteretic phenomenon is common to all TGRs, independently of the presence of one or two cysteine residues at the redox active motif of the Grx-like domain.

Finally, although the substrate inhibition of TGR by moderate or high concentrations of GSSG may, in principle, appear contradictory with the function of the enzyme, it must be taken into account that the degree of inhibition is dependent on the concentration of both GSH and GSSG. As described above, 50 μM GSH is high enough to avoid the GSSG-dependent inhibition of the enzyme. From the total concentration of glutathione of 1.2 mM and a GSH/GSSG ratio of 131 determined in *T. crassiceps* cysticerci under basal conditions [49] it appears that the atypical kinetic behavior of the enzyme has no physiological significance. However, taking into account the fact that under in vivo conditions the parasite is under attack by the defense system of the host, the substrate inhibition of TGR by GSSG could play a potential physiological role in the parasite survival. Thus, a sudden oxidative challenge could lead to an increase in the concentration of GSSG with a concomitant decrease in the GSH concentration, producing a temporary and partial inhibition of the enzyme activity and thus allowing the protection of essential sulfhydryl groups of proteins from oxidation through its conjugation with glutathione. In this sense, it is worth noting that the deglutathionylation activity of *E. granulosus* TGR is also inhibited by high concentrations of GSSG [6]. It is also possible that a chemical species such as the superoxide anion could participate as inhibitor of TGR, contributing to the maintenance of a relatively high, albeit temporary, concentration of GSSG.

5. Conclusions

(i) Thioredoxin-glutathione reductase from *T. crassiceps* follows a two-site ping-pong bi bi kinetic mechanism with GSSG as the substrate. The latter exerts a strong but temporary substrate inhibition, resulting in hysteretic-like progress curves.

(ii) NADPH plays a critical role in the atypical kinetic behavior of the enzyme, by supplying the reduced form of the enzyme to which GSSG binds.

(iii) A mechanism-based model explaining all the kinetic observations of TGR was developed. A key point of the model is the presence of a low-affinity second binding site for GSSG, with the ability to reduce GSSG at a low rate.

(iv) From the *in silico* simulations of the model, it becomes clear that the hysteretic-like progress curves of the enzyme are the result of a continuous competition between GSH and GSSG for driving the enzyme into active or inactive pathways. Hence, the atypical full progress curves of GSSG reductase activity of TGR must not be considered as a kind of hysteretic behavior.

(v) As a corollary of the model, it must be stressed that the maximal slope observed in the atypical full progress curves does not represent the steady-state of enzyme activity.

Abbreviations

GSSG: Disulfide form of glutathione
DTNB: 5,5′-Dithiobis(2-nitrobenzoic acid)
trx: Thioredoxin
grx: Glutaredoxin
GR: Glutathione reductase
TrxR: Thioredoxin reductase
TGR: Thioredoxin-glutathione reductase.

Acknowledgments

This research was supported by Dirección General de Asuntos del Personal Académico (DGAPA, UNAM) through Research Grants IN 219414 and IN 218816.

Supplementary Materials

Figure S1. In silico simulation of initial velocities of T. crassiceps TGR with GSSG as the disulfide substrate at two different NADPH concentrations. Simulations were based on the model of Figure 11 by using the rate constants shown in Table S1. NADPH concentrations of 5 μM (circles) and 50 μM (triangles) were used. A value of 11.5 nM for TGR concentration was used in the simulation. Open symbols represent data obtained by omitting reactions 10 to 13 from the model. *Figure S2. In silico simulation showing the effect of omitting reactions 10 to 13 from the model on the full progress curves of T. crassiceps TGR.* Simulations were based on the model of Figure 11 by using the rate constants shown in Table S1. GSSG concentrations used were as follows: (●) 120 μM; (○) 300 μM; (▲) 500 μM; (△) 800 μM. NADPH and enzyme concentrations were 50 μM and 11 nM, respectively. *Figure S3. In silico simulation showing the effect of omitting reactions 11 and 13 from the model on the full progress curves of T. crassiceps TGR.* Simulations were based on the model of Figure 11 by using the rate constants shown in Table S1. GSSG concentrations used were as follows: (●) 140 μM; (○) 220 μM; (▲) 400 μM; (△) 550 μM. NADPH and enzyme concentrations were 15 μM and 11 nM, respectively. *Figure S4. In silico simulation showing the effect of varying both NADPH and GSSG concentrations on the profile of the full progress curves of T. crassiceps TGR.* Simulations were carried out at the following concentrations of NADPH: (a) 5 μM; (b) 15 μM; (c) 50 μM. In all cases, the following concentrations of GSSG were used: (●) 60 μM; (○) 120 μM; (▲) 200 μM; (△) 300 μM. An enzyme concentration of 11.5 nM was used. *Figure S5. In silico simulation showing the effect of varying enzyme concentration on the profile of the full progress curves by T. crassiceps TGR.* Simulations were based on the model of Figure 11 using the rate constants shown in Table S1. (a) 7 μM NADPH; (b) 50 μM NADPH. In both cases the following enzyme concentrations were used: (△) 8.5 nM; (▲) 11.5 nM; (○) 20 nM; (●) 60 nM. A value of 350 μM for GSSG concentration was used. *Figure S6. In silico simulation showing the effect of varying GSH concentration on the profile of the full progress curves by T. crassiceps TGR.* Simulations

were based on the model of Figure 11 using the rate constants shown in Table S1. The following GSH concentrations were used: (△) none; (▲) 2 μM; (○) 10 μM; (●) 90 μM. Values of 500 μM and 11.5 nM for the concentration of GSSG and enzyme, respectively, were used. *Table S1. Best fitting theoretical rate constants used in the simulation of the model. (Supplementary Materials)*

References

[1] Q.-A. Sun, L. Kirnarsky, S. Sherman, and V. N. Gladyshev, "Selenoprotein oxidoreductase with specificity for thioredoxin and glutathione systems," *Proceedings of the National Acadamy of Sciences of the United States of America*, vol. 98, no. 7, pp. 3673–3678, 2001.

[2] S. Prast-Nielsen, H.-H. Huang, and D. L. Williams, "Thioredoxin glutathione reductase: Its role in redox biology and potential as a target for drugs against neglected diseases," *Biochimica et Biophysica Acta (BBA) - General Subjects*, vol. 1810, no. 12, pp. 1262–1271, 2011.

[3] L. Zhong and A. Holmgren, "Essential role of selenium in the catalytic activities of mammalian thioredoxin reductase revealed by characterization of recombinant enzymes with selenocysteine mutations," *The Journal of Biological Chemistry*, vol. 275, no. 24, pp. 18121–18128, 2000.

[4] Q.-A. Sun, D. Su, S. V. Novoselov, B. A. Carlson, D. L. Hatfield, and V. N. Gladyshev, "Reaction mechanism and regulation of mammalian thioredoxin/glutathione reductase," *Biochemistry*, vol. 44, no. 44, pp. 14528–14537, 2005.

[5] M. Bonilla, A. Denicola, S. V. Novoselov et al., "Platyhelminth mitochondrial and cytosolic redox homeostasis is controlled by a single thioredoxin glutathione reductase and dependent on selenium and glutathione," *The Journal of Biological Chemistry*, vol. 283, no. 26, pp. 17898–17907, 2008.

[6] M. Bonilla, A. Denicola, S. M. Marino, V. N. Gladyshev, and G. Salinas, "Linked thioredoxin-glutathione systems in platyhelminth parasites: Alternative pathways for glutathione reduction and deglutathionylation," *The Journal of Biological Chemistry*, vol. 286, no. 7, pp. 4959–4967, 2011.

[7] H.-H. Huang, L. Day, C. L. Cass, D. P. Ballou, C. H. Williams, and D. L. Williams, "Investigations of the catalytic mechanism of thioredoxin glutathione reductase from Schistosoma mansoni," *Biochemistry*, vol. 50, no. 26, pp. 5870–5882, 2011.

[8] F. Angelucci, A. E. Miele, G. Boumis, D. Dimastrogiovanni, M. Brunori, and A. Bellelli, "Glutathione reductase and thioredoxin reductase at the crossroad: the structure of Schistosoma mansoni thioredoxin glutathione reductase," *Proteins: Structure, Function, and Genetics*, vol. 72, no. 3, pp. 936–945, 2008.

[9] F. Angelucci, D. Dimastrogiovanni, G. Boumis et al., "Mapping the catalytic cycle of Schistosoma mansoni thioredoxin glutathione reductase by X-ray crystallography," *The Journal of Biological Chemistry*, vol. 285, no. 42, pp. 32557–32567, 2010.

[10] M. Sharma, S. Khanna, G. Bulusu et al., "Comparative modeling of thioredoxin glutathione reductase," *Journal of Molecular Graphics and Modelling*, vol. 27, pp. 665–675, 2009.

[11] C. Brandstaedter, K. Fritz-Wolf, S. Weder et al., "Kinetic characterization of wild-type and mutant human thioredoxin glutathione reductase defines its reaction and regulatory mechanisms," *FEBS Journal*, vol. 285, no. 3, pp. 542–558, 2018.

[12] H. M. Alger and D. L. Williams, "The disulfide redox system of Schistosoma mansoni and the importance of a multifunctional enzyme, thioredoxin glutathione reductase," *Molecular and Biochemical Parasitology*, vol. 121, no. 1, pp. 129–139, 2002.

[13] A. Agorio, C. Chalar, S. Cardozo, and G. Salinas, "Alternative mRNAs arising from trans-splicing code for mitochondrial and cytosolic variants of *Echinococcus granulosus* thioredoxin glutathione reductase," *The Journal of Biological Chemistry*, vol. 278, no. 15, pp. 12920–12928, 2003.

[14] J. L. Rendón, I. P. del Arenal, A. Guevara-Flores, A. Uribe, A. Plancarte, and G. Mendoza-Hernández, "Purification, characterization and kinetic properties of the multifunctional thioredoxin-glutathione reductase from *Taenia crassiceps* metacestode (cysticerci)," *Molecular and Biochemical Parasitology*, vol. 133, no. 1, pp. 61–69, 2004.

[15] A. Guevara-Flores, J. P. Pardo, and J. L. Rendón, "Hysteresis in thioredoxin-glutathione reductase (TGR) from the adult stage of the liver fluke *Fasciola hepatica*," *Parasitology International*, vol. 60, no. 2, pp. 156–160, 2011.

[16] A. Gupta, M. Kesherwani, D. Velmurugan, and T. Tripathi, "*Fasciola gigantica* thioredoxin glutathione reductase: Biochemical properties and structural modeling," *International Journal of Biological Macromolecules*, vol. 89, pp. 152–160, 2016.

[17] A. Plancarte and G. Nava, "Purification and kinetic analysis of cytosolic and mitochondrial thioredoxin glutathione reductase extracted from *Taenia solium* cysticerci," *Experimental Parasitology emphasizes*, vol. 149, pp. 65–73, 2015.

[18] A. N. Kuntz, E. Davioud-Charvet, A. A. Sayed et al., "Thioredoxin glutathione reductase from *Schistosoma mansoni*: An essential parasite enzyme and a key drug target," *PLoS Medicine*, vol. 4, no. 6, pp. 1071–1085, 2007.

[19] J. J. Martínez-González, A. Guevara-Flores, G. Álvarez, J. L. Rendón-Gómez, and I. P. Del Arenal, "In vitro killing action of auranofin on *Taenia crassiceps* metacestode (cysticerci) and inactivation of thioredoxin-glutathione reductase (TGR)," *Parasitology Research*, vol. 107, no. 1, pp. 227–231, 2010.

[20] L. Song, J. Li, S. Xie et al., "Thioredoxin glutathione reductase as a novel drug target: evidence from *Schistosoma japonicum*," *PLoS ONE*, vol. 7, no. 2, Article ID e31456, 2012.

[21] C. Frieden, "Kinetic aspects of regulation of metabolic processes. The hysteretic enzyme concept.," *The Journal of Biological Chemistry*, vol. 245, no. 21, pp. 5788–5799, 1970.

[22] C. Larralde, E. Sciutto, J. Grun et al., "Biological determinants of host-parasite relationship in mouse cysticercosis caused by *Taenia crassiceps*:Influence of sex, major histocompatibility complex and vaccination," in *Cell Function and Disease*, L. E. Cañedo, L. E. Todd, L. Packer, and J. Jas, Eds., pp. 325–332, Plenum Publishing Corporation, New York, NY, USA, 1989.

[23] I. Carlberg and B. Mannervik, "Glutathione reductase," *Methods in Enzymology*, vol. 113, pp. 484–490, 1985.

[24] E. S. J. Arnér, L. Zhong, and A. Holmgren, "Preparation and assay of mammalian thioredoxin and thioredoxin reductase," *Methods in Enzymology*, vol. 300, pp. 226–239, 1998.

[25] I. H. Segel, *Enzyme kinetics*, Wiley, New York, NY, USA, 1975.

[26] E. L. King and C. Altman, "A schematic method of deriving the rate laws for enzyme-catalyzed reactions," *The Journal of Physical Chemistry C*, vol. 60, no. 10, pp. 1375–1378, 1956.

[27] W. Wallace Cleland, "Substrate Inhibition," *Methods in Enzymology*, vol. 63, no. C, pp. 500–513, 1979.

[28] K. E. Neet and G. Robert Ainslie, "Hysteretic Enzymes," *Methods in Enzymology*, vol. 64, no. C, pp. 192–226, 1980.

[29] P. Kuzmič, "DynaFit-A Software Package for Enzymology," *Methods In Enzymology*, vol. 467, no. C, pp. 247–280, 2009.

[30] G. M. Morris, R. Huey, W. Lindstrom et al., "AutoDock4 and AutoDockTools4: Automated docking with selective receptor flexibility," *Journal of Computational Chemistry*, vol. 30, pp. 2785–2791, 2009.

[31] O. Trott and A. J. Olson, "AutoDock Vina: improving the speed and accuracy of docking with a new scoring function, efficient optimization and multithreading," *Journal of Computational Chemistry*, vol. 31, no. 2, pp. 455–461, 2010.

[32] F. B. Rudolph, "Inhibition and abortive complex formation," in *Methods in Enzymology*, D. L. Purich, Ed., vol. 63, pp. 411–436, Academic Press, New York, NY, USA, 1979.

[33] W. Janes and G. E. Schulz, "Role of the Charged Groups of Glutathione Disulfide in the Catalysis of Glutathione Reductase: Crystallographic and Kinetic Studies with Synthetic Analogues," *Biochemistry*, vol. 29, no. 16, pp. 4022–4030, 1990.

[34] P. F. Cook and W. W. Cleland, *Enzyme Kinetics and Mechanism*, Garland Science Publishing, London, UK, 2007.

[35] D. B. Northrop, "Transcarboxylase. VI. Kinetic analysis of the reaction mechanism.," *The Journal of Biological Chemistry*, vol. 244, no. 21, pp. 5808–5819, 1969.

[36] D. S. Sem and C. B. Kasper, "Kinetic Mechanism for the Model Reaction of NADPH-Cytochrome P450 Oxidoreductase with Cytochrome c," *Biochemistry*, vol. 33, no. 40, pp. 12012–12021, 1994.

[37] G. Salinas, W. Gao, Y. Wang et al., "The enzymatic and structural basis for inhibition of *Echinococcus granulosus* Thioredoxin Glutathione Reductase by gold(I)," *Antioxidants & Redox Signaling*, vol. 27, no. 18, pp. 1491–1504, 2017.

[38] G. Kim and D. J. Graves, "On the Hysteretic Response of Rabbit Skeletal Muscle Phosphorylase Kinase," *Biochemistry*, vol. 12, no. 11, pp. 2090–2095, 1973.

[39] R. Iyengar, "Hysteretic activation of adenylyl cyclases. II. Mg ion regulation of the activation of the regulatory component as analyzed by reconstitution," *The Journal of Biological Chemistry*, vol. 256, no. 21, pp. 11042–11050, 1981.

[40] B. Schobert and J. K. Lanyi, "Hysteretic behavior of an ATPase from the archaebacterium, *Halobacterium saccharovorum*," *The Journal of Biological Chemistry*, vol. 264, no. 22, pp. 12805–12812, 1989.

[41] S. Lebreton, B. Gontero, L. Avilan, and J. Ricard, "Memory and imprinting effects in multienzyme complexes II. Kinetics of the bienzyme complex from *Chlamydomonas reinhardtii* and hysteretic activation of chloroplast oxidized phosphoribulokinase," *European Journal of Biochemistry*, vol. 246, no. 1, pp. 85–91, 1997.

[42] B. Giese, M. Graber, and M. Cordes, "Electron transfer in peptides and proteins," *Current Opinion in Chemical Biology*, vol. 12, no. 6, pp. 755–759, 2008.

[43] W.-X. Tian and C.-L. Tsou, "Determination of the Rate Constant of Enzyme Modification by Measuring the Substrate Reaction in the Presence of the Modifier," *Biochemistry*, vol. 21, no. 5, pp. 1028–1032, 1982.

[44] Z.-X. Wang and C.-L. Tsou, "Kinetics of substrate reaction during irreversible modification of enzyme activity for enzymes involving two substrates," *Journal of Theoretical Biology*, vol. 127, no. 3, pp. 253–270, 1987.

[45] S. E. Szedlacsek and R. G. Duggleby, "Kinetics of Slow and Tight-Binding Inhibitors," *Methods in Enzymology*, vol. 249, no. C, pp. 144–180, 1995.

Insight into the Mechanistic Basis of the Hysteretic-Like Kinetic Behavior of Thioredoxin-Glutathione...

147

[46] A. Baici and M. Gyger-Marazzi, "The Slow, Tight-Binding Inhibition of Cathepsin B by Leupeptin: A Hysteretic Effect," *European Journal of Biochemistry*, vol. 129, no. 1, pp. 33–41, 1982.

[47] P. Masson, B. N. Goldstein, J.-C. Debouzy, M.-T. Froment, O. Lockridge, and L. M. Schopfer, "Damped oscillatory hysteretic behaviour of butyrylcholinesterase with benzoylcholine as substrate," *European Journal of Biochemistry*, vol. 271, no. 1, pp. 220–234, 2004.

[48] C. Garrido-Del Solo, F. Garcia-Canovas, B. H. Havsteen, E. Valero, and R. Varon, "Kinetics of an enzyme reaction in which both the enzyme-substrate complex and the product are unstable or only the product is unstable," *Biochemical Journal*, vol. 303, no. 2, pp. 435–440, 1994.

[49] J. J. Martínez-González, A. Guevara-Flores, J. L. Rendón, and I. P. D. Arenal, "Auranofin-induced oxidative stress causes redistribution of the glutathione pool in Taenia crassiceps cysticerci," *Molecular and Biochemical Parasitology*, vol. 201, no. 1, pp. 16–25, 2015.

15

Plackett-Burman Design for rGILCC1 Laccase Activity Enhancement in *Pichia pastoris*: Concentrated Enzyme Kinetic Characterization

Edwin D. Morales-Álvarez,[1,2] Claudia M. Rivera-Hoyos,[3] Ángela M. Cardozo-Bernal,[3] Raúl A. Poutou-Piñales,[3] Aura M. Pedroza-Rodríguez,[1] Dennis J. Díaz-Rincón,[4] Alexander Rodríguez-López,[4] Carlos J. Alméciga-Díaz,[4] and Claudia L. Cuervo-Patiño[5]

[1]*Laboratorio de Microbiología Ambiental y de Suelos, Grupo de Biotecnología Ambiental e Industrial (GBAI),*
 Departamento de Microbiología, Facultad de Ciencias, Pontificia Universidad Javeriana (PUJ), Bogotá, Colombia
[2]*Departamento de Química, Facultad de Ciencias Exactas y Naturales, Universidad de Caldas, Manizales, Caldas, Colombia*
[3]*Laboratorio de Biotecnología Molecular, Grupo de Biotecnología Ambiental e Industrial (GBAI), Departamento de Microbiología,*
 Facultad de Ciencias, Pontificia Universidad Javeriana (PUJ), Bogotá, Colombia
[4]*Laboratorio de Expresión de Proteínas, Instituto de Errores Innatos del Metabolismo (IEIM), Facultad de Ciencias,*
 Pontificia Universidad Javeriana (PUJ), Bogotá, Colombia
[5]*Laboratorio de Parasitología Molecular, Grupo de Enfermedades Infecciosas, Facultad de Ciencias,*
 Pontificia Universidad Javeriana (PUJ), Bogotá, Colombia

Correspondence should be addressed to Raúl A. Poutou-Piñales; rpoutou@javeriana.edu.co

Academic Editor: Hartmut Kuhn

Laccases are multicopper oxidases that catalyze aromatic and nonaromatic compounds with concomitant reduction of molecular oxygen to water. They are of great interest due to their potential biotechnological applications. In this work we statistically improved culture media for recombinant GILCC1 (rGILCC1) laccase production at low scale from *Ganoderma lucidum* containing the construct pGAPZαA-*GlucPost*-Stop in *Pichia pastoris*. Temperature, pH stability, and kinetic parameter characterizations were determined by monitoring concentrate enzyme oxidation at different ABTS substrate concentrations. Plackett-Burman Design allowed improving enzyme activity from previous work 36.08-fold, with a laccase activity of 4.69 ± 0.39 UL^{-1} at 168 h of culture in a 500 mL shake-flask. Concentrated rGILCC1 remained stable between 10 and 50°C and retained a residual enzymatic activity greater than 70% at 60°C and 50% at 70°C. In regard to pH stability, concentrated enzyme was more stable at pH 4.0 ± 0.2 with a residual activity greater than 90%. The lowest residual activity greater than 55% was obtained at pH 10.0 ± 0.2. Furthermore, calculated apparent enzyme kinetic parameters were a V_{max} of 6.87×10^{-5} mM s^{-1}, with an apparent K_m of 5.36×10^{-2} mM. Collectively, these important stability findings open possibilities for applications involving a wide pH and temperature ranges.

1. Introduction

Laccases are blue multicopper oxidases (EC 1.10.3.2), catalyzing oxidation reactions for an array of compounds such as diphenols, polyphenols, deamines, aromatic amines, inorganic compounds, and nonphenolic compounds in the presence of redox mediators. During the reaction the substrate is oxidized by donating its electron, where molecular oxygen acts as an electron acceptor and is reduced into water [1–3].

Laccases are enzymes mainly produced in white rot fungi; however they are widely distributed in plants, insects, fungi, and bacteria [1, 4–6]. They have important applications in different industrial settings, as they help to reduce the environmental impact of their waste. These applications include dye bleaching in textile industry, bleaching of cellulose pulp, detoxification of residual waters, toxic compound bioremediation, biosensor construction, fuel cells, fruit juice processing, and synthesis of molecules in the pharmaceutical industry [3, 6–10].

Laccase biotechnological and environmental applications require great enzyme quantities; unfortunately laccases obtained from natural sources are not suitable for long growth periods, low product/biomass ($Y_{p/x}$) or product/substrate ($Y_{p/s}$) yield, and prolonged, complex, and costly isolation procedures [3, 6, 11]. Therefore, heterologous expression is a promising option for greater scale production, using the potential of hosts that are easy to handle and culture, such as bacteria and yeast [12].

Yeast offers fast growth rates, ease of gene manipulation, and posttranslational modification capabilities. *P. pastoris* has been employed for years as an industrial platform for heterologous protein expression. Moreover, it is one of the most effective expression systems to obtain high yield extracellular proteins [14–17]. Additionally, various reports have described better expression and productivity levels thorough culture media optimization [4, 18].

The objective of this work was to increase *Ganoderma lucidum* rGILCC1 laccase activity in *Pichia pastoris* at low scale in the laboratory to determine pH and temperature stability and define its V_{max} and K_m in concentrated supernatant obtained from microbial culture. This was achieved by improving the following factors: nitrogen and source type (organic and inorganic), carbon concentration (glucose), copper concentration, oxygen transfer (media volume/Erlenmeyer flask volume ratio), time of culture, and inoculum percentage.

2. Materials and Methods

2.1. Strain.
The strain was *P. pastoris* X33 containing pGA-PZαA-*LacGluc-Stop* (Clone 1) expression vector with previously optimized synthetic gene *GILCC1* coding for *Ganoderma lucidum* GILCC1 laccase. This strain was kept in 1% YPD (w/v), 2% peptone, 1% yeast extract, and 2% D+ glucose, supplemented with 20% glycerol (w/v), and kept at −80°C [19–21].

2.2. Inoculum Preparation.
Pichia pastoris X33/pGAPZαA-*LaccGluc-Stop* clone 1 Master Cell Bank (MCB) [21] was thawed and used for inoculating 5 mL screw cap tubes with sterile YPD supplemented with 40 μg mL^{-1} zeocin (Z). Tubes were incubated overnight (ON) at 30°C with 180 rpm agitation, followed by inoculation under the same conditions for 12 h in 500 mL Erlenmeyer flasks, containing 100 mL (effective work volume) of fresh YPD-Z. The resulting culture was verified by Gram stain to detect presence of contaminating morphologies and used for factorial design inoculations.

2.3. Plackett-Burman Experimental Design (PBED).
Seven factors were evaluated with two levels each as follows: media volume (150 and 300 mL), CuSO$_4$ concentration (0.1 and

1.0 mM), inoculum percentage (2 and 10% (v/v)), glucose concentration (10 and 30 gL^{-1}), NH$_4$SO$_4$ concentration (5 and 20 mM), peptone concentration (10 and 20 gL^{-1}), and yeast extract concentration (5 and 10 gL^{-1}). The design included a central point evaluated three times; values within central points were 225 mL media, 0.55 mM CuSO$_4$, 6% inoculum (v/v), 20 gL^{-1} glucose, 12.5 mM NH$_4$SO$_4$, 15 gL^{-1} peptone, and 7.5 gL^{-1} yeast extract [22]. For statistical analysis the response variable evaluated was enzyme activity (UL^{-1}). Additionally, to determine if statistically significant differences were observed among treatments (T_1–T_{12}) a one-way ANOVA with Tukey post hoc test was performed, employing a 95% confidence interval (CI, $\alpha = 0.05$). Moreover, Shapiro-Wilk normalization test was applied to verify data quality using SAS V 9.0® 2004 (SAS Institute Inc., Cary, NC, USA).

Note. Enzyme activity (UL^{-1}), protein concentration (mg mL^{-1}), and glucose concentration (gL^{-1}), specific activity (UL^{-1} mg mL^{-1}), and productivity (UL^{-1}h^{-1}) were assayed 0 to 12 h every two hours. The same variables were then evaluated every 24 h up to the end (168 h), since preliminary data revealed that better results were obtained between 156 h and 168 h [19]. Enzyme activity (UL^{-1}) was the response variable utilized for statistical analyses. All improvement assays were carried out in 500 mL Erlenmeyer flasks at 30°C and 180 rpm, employing the same flask for the total 168 h of culture at variable pH starting at 7 ± 0.2. Design Expert V. 9.0 (Stat-Ease, Inc., Minneapolis, MN, USA) software was used to devise experimental design and result analysis. In addition, Sigma Plot V.11.0 software (Systat Software Inc. San José, CA, USA) was employed to graph concentrated enzyme kinetics and results.

2.4. Supernatant Concentrate.
Culture supernatant demonstrating the highest enzymatic activity values with ABTS substrate [2,20-azino-bis(3-ethylbenzothiazoline-6-sulphonic acid)] under conditions previously described [23] (17.9 mg mL^{-1} protein and 4.69 ± 0.39 UL^{-1} enzyme activity, with a specific enzyme activity of 0.26 Umg^{-1} at 168 h of culture) was used. Briefly, culture was centrifuged at 4°C 8,000g and supernatant was filtered in a serial manner through Whatman Number 1 filter paper, followed by 0.45 and 0.22 μm membranes (Pall Corp, Port Washington, NY, USA). The filtrate was concentrated by centrifugation employing a 10 kDa Ultracel regenerated cellulose membrane (Millipore, Billerica, MA, USA) [23]. Approximately 20 mL concentrate was used to perform the enzyme's functional identity by zymography.

2.5. Enzyme Functional Identification.
Zymogram was ran in 12% (w/v) native PAGE under nondenaturing conditions. Activity or functionality was visualized by 0.5 M ABTS stain. BenchMark™ Pre-Stained Protein ladder (Life Technologies™, USA) was used as the molecular weight standard and Lac® (Sigma-Aldrich®, St. Louis, MO, USA) as laccase control.

2.6. Temperature Stability. Concentrated enzyme temperature stability was assayed by incubating for 1 h at the following temperatures: 10, 20, 30, 40, 50, 60, and 70°C; subsequently residual enzyme activity was determined under standard assay conditions [13]. All assays were performed at least three times.

2.7. pH Stability. To establish pH stability supernatant obtained from concentrate was previously incubated for 1 h at 25°C in the absence of substrate using Britton-Robinson buffer [24] with pH values ranging between 2 and 12 ± 0.2, followed by laccase residual enzyme activity determination under standard assay conditions [13]. All assays were performed at least three times.

2.8. Kinetic Constant. Concentrated enzyme kinetic constants were evaluated using ABTS as substrate in a concentration range between 0.05 and 0.5 mM at 0.1 mM intervals in 600 mM sodium acetate buffer at pH 4.5. For all assays 800 μl concentrated enzyme with 4.49 UL^{-1} enzyme activity was employed at 25°C. After hyperbola adjustment using Michaelis-Menten equation $V_{\max}$ and apparent K_m were calculated following Hanes-Woolf linearization method [25], with the aid of SIMFIT software V5.40, 2003 (W.G. Bardsley, University of Manchester, UK) [26]. All kinetic assays were performed at least three times.

2.9. Determination of Total Residual Reducing Sugar Concentration. 3,5-Dinitrosalicylic acid colorimetric method (DNS) was employed to evaluate total residual reducing sugars [27] for each sample (in triplicate). A 0.1 and 2 g L^{-1} D-glucose curve was used as the standard.

2.10. Total Extracellular Protein Concentration Determination. Total extracellular protein concentration was established by Biuret [28] methodology for each sample (in triplicate). A bovine serum albumin (BSA) curve between 0.5 and 5 mg mL^{-1} was used as a standard curve.

2.11. Enzyme Activity Quantification. Enzyme activity was monitored by changes in absorbency at 436 nm (ε_{436} = 29,300 M^{-1}cm^{-1}) as a result of ABTS in a 60 mM (pH 4.5 ± 0.2) sodium acetate buffer. 100 μL 5 mM ABTS as substrate, 800 μL crude extract at room temperature (RT), and 100 μL 600 mM sodium acetate buffer were used. Formation of a green cationic radical was evaluated spectrophotometrically for three minutes. A unit of activity is defined as the quantity of enzyme required to oxidize 1 μmol ABTS in one minute. Blanc solution contained 800 μL distilled water, 100 μL 600 mM sodium acetate buffer, and 100 μL 5 mM ABTS. Enzyme activity was expressed in UL^{-1} [13].

Specific activity was calculated by dividing the obtained enzyme activity for each hour of culture by total protein concentration and expressed in Umg^{-1}:

$$\text{Spec. Act.} \frac{\text{Enz. Act.}}{\text{Prot. Conc.}}, \tag{1}$$

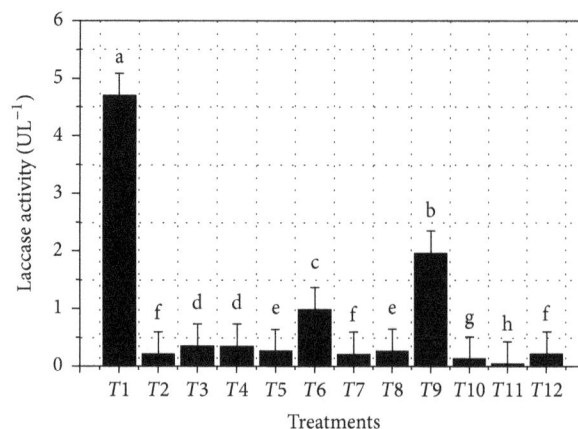

FIGURE 1: PBED mean ± SD treatment results. Each treatment was assayed in triplicate ($n = 3$). Means ± SD were compared among all twelve treatments. $p < 0.05$ was significant.

where enzyme activity (enz. act.) is given in UL^{-1} and protein concentration (prot. conc.) in mg mL^{-1}.

Productivity in function of enzyme activity was expressed as biological activity UL^{-1} h^{-1} (see (2)), calculated in the following manner:

$$P_{\text{Enz.}} = \frac{\text{Enz. Act.}}{\text{Time}}. \tag{2}$$

3. Results

3.1. Plackett-Burman Experimental Design (PBED). ANOVA for a model that was not adjusted for curvature was significant ($p = 0.0034$), allowing for evaluation of the different factors' *"main effect"* on culture extracellular enzyme activity detected at 168 h. Table 1 depicts model's significant values for each factor involved. Polynomial equation (3) represents laccase activity and can be used for predictions based on different evaluated levels for each factor. Additionally, the equation can also result as being useful for factor relative impact identification when comparing obtained coefficients between them

$$\begin{aligned}
\text{Enz. Act.}_{168\,h} = {} & 0.79 - 0.31 \times A + 0.52 \times B + 0.11 \times D \\
& + 0.10 \times E + 0.44 \times F - 0.48 \times G \\
& - 0.84 \times AB - 0.71 \times AD + 0.60 \\
& \times AE + 0.82 \times AG.
\end{aligned} \tag{3}$$

Enzyme activity for each treatment is shown in Table 2 and Figure 1, highlighting the best treatments (T_1 and T_9), with enzyme activity values greater than 1.8 UL^{-1}, as well as mean comparison among the 12 treatments.

For most treatments (Figure 1 and Table 2) enzyme activity exceeded that obtained from previous work at 156 h (>0.13 ± 0.03 UL^{-1}). T_1 and T_9 treatments were the most significant, hence promising, with T_1 attaining after 168 h of culture the highest enzyme activity 4.69 ± 0.39 UL^{-1}.

TABLE 1: Laccase activity ANOVA for a model that was not adjusted for curvature.

Source	Sum of squares	DF	Mean squares	F-value	p value Prob > F
Model	19.57	10	1.96	16.17	**0.0034**
A, culture media volume	1.17	1	1.17	9.71	**0.0264**
B, CuSO$_4$	1.90	1	1.90	15.70	**0.0107**
D, glucose	0.089	1	0.089	0.74	0.4304
E, NH$_4$SO$_4$	0.07	1	0.07	0.58	0.4800
F, peptone	1.17	1	1.17	9.64	**0.0267**
G, yeast extract	1.60	1	1.60	13.24	**0.0149**
Residual	0.61	5	0.12		
Lack of fit	0.05	2	0.025	0.14	0.8777
Error	0.55	3	0.18		
Cor total	20.18	15			
R-square	0.97				
Adjusted R-square	0.91				
Predicted R-square	0.7479				
Adequate precision	16.258				

95% significant values are in bold.

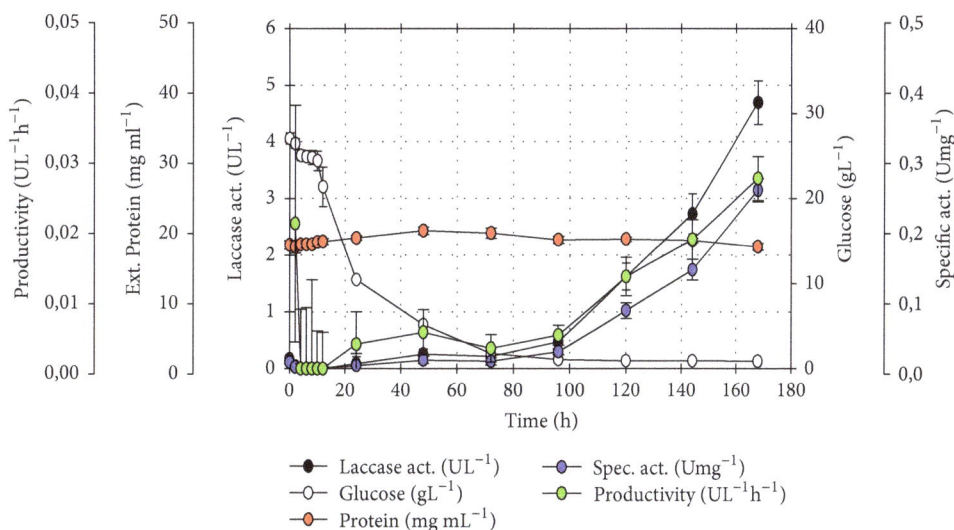

FIGURE 2: Treatment T_1 PBED kinetic follow-up. Treatment 1 [PBED-T_1: 500 mL Erlenmeyer flask containing 150 mL media, 10% inoculum (v/v), 1.0 mM CuSO$_4$, 30 gL^{-1} glucose, 5 mM NH$_4$SO$_4$, 10 gL^{-1} peptone, and 5 gL^{-1} yeast extract], enzy. act. 4.69 ± 0.39 UL^{-1} at 168 h of culture. Assay was carried out in triplicate ($n = 3$).

This represents an approximate 36.08-fold increase in comparison with that obtained from previous work [19].

Laccase activity presented a great variation among the 12 treatments (Table 2), evidencing the relevance and usefulness of culture media improvement. In addition, it is important to note that none of the predictions exceeded the results obtained from treatment T_1.

Positive and negative effects are shown in Table 3 and each factor is involved in PBED percentage contribution on the response variable (enzyme activity UL^{-1}). As can be observed from Table 1 and (3), analysis hierarchy discarded Factor C, inoculum.

PBED kinetic follow-up for T_1 revealed the highest productivity (UL^{-1} h^{-1}) based on enzymatic activity. Enzyme activity (UL^{-1}) and specific activity (Umg^{-1}) were obtained after 168 h of culture (Figure 2).

3.2. Concentrated Enzyme Functional Identification. The functional identification of rGILCC1 enzyme by zymogram (native PAGE) using 0.5 M ABTS in 60 mM sodium acetate buffer stain is shown in Figure 3(a). Commercial laccase presented various active fractions, suggesting the positive control as a possible multimeric laccase (Figure 3(a)).

TABLE 2: PBED observed and predicted values of factors having an effect on laccase activity.

Tx	Factor type	Culture media volume (mL)	$CuSO_4$ (mM)	Inoculum (% v/v)	Glucose (gL^{-1})	NH_4SO_4 (mM)	Peptone (gL^{-1})	Yeast extract (gL^{-1})	Observed enz. activity at 168 h (UL^{-1})	Predicted enz. activity at 168 h (UL^{-1})
T_1	**Factorial**	**150**	**1**	**10**	**30**	**5**	**10**	**5**	**4.6928**	**4.6368**
T_2	Factorial	300	0.1	10	30	20	10	5	0.2133	0.1182
T_3	Factorial	300	1	2	30	20	20	5	0.3413	0.3635
T_4	Factorial	150	1	10	10	20	20	10	0.3413	0.2853
T_5	Factorial	300	0.1	10	30	5	20	10	0.2559	0.2782
T_6	Factorial	150	1	2	30	20	10	10	0.9813	1.0426
T_7	Factorial	150	0.1	10	10	20	20	5	0.2133	0.1573
T_8	Factorial	150	0.1	2	30	5	20	10	0.2559	0.1999
T_9	**Factorial**	**300**	**0.1**	**2**	**10**	**20**	**10**	**10**	**1.9625**	**1.9847**
T_{10}	Factorial	300	1	2	10	5	20	5	0.1279	0.1502
T_{11}	Factorial	300	1	10	10	5	10	10	0.0427	−0.0524
T_{12}	Factorial	150	0.1	2	10	5	10	5	0.2133	0.2746
	Central point	225	0.55	6	20	12.5	15	7.5	0.4693	0.7866
	Central point	225	0.55	6	20	12.5	15	7.5	1.2372	0.7866
	Central point	225	0.55	6	20	12.5	15	7.5	0.9386	0.7866

Best treatments are in bold.

TABLE 3: PBED evaluated factor effect and percentage contribution on laccase activity.

Factor	Effect	Sum of squares	p	% contribution
A, culture media volume	−0.63	1.17	**0.0264**	5.81
B, $CuSO_4$	0.80	1.90	**0.0107**	9.39
D, glucose	0.17	0.09	0.4304	0.44
E, NH_4SO_4	0.15	0.07	0.4800	0.35
F, peptone	0.62	1.17	**0.0267**	5.76
G, yeast extract	−0.73	1.60	**0.0149**	7.92

Significant values are in bold.

3.3. Temperature and pH Stability. rGILCC1 relative enzyme activity (%) temperature and pH stability results obtained from concentrate after incubating the concentrate for 1 h at different temperatures or 25°C at different pHs are shown in Figure 3(b). As was observed enzyme activity between 10 and 60°C was greater than 80%. For pH ranges between pH 2 and pH 11 enzyme activity ranged between 75 and 100%.

3.4. Kinetic Constants. rGILCC1 laccase obtained from concentrate kinetic characteristics for ABTS oxidation is shown in Figure 3(c), where V_{max} values under assay conditions ($V_{max} = 6.87 \times 10^{-5}$ mMs^{-1}) and Michaelis-Menten constant ($K_m = 5.36 \times 10^{-2}$ mM) were also estimated. Hanes-Woolf model best described enzyme concentrate behavior, under assay conditions (temp. 25°C; pH 4.5 ± 0.2).

4. Discussion

As can be seen in Table 1 the hierarchical model was significant, as well as A, B, F, and G factors. In this regard, at 168 h the model's F value was 16.17, implying that the model was significant. In contrast, there was only a 0.34% possibility of a greater F value due to experimental noise. A lack of adjustment F value of 0.14 suggests it was not significant in relation to pure error. In addition, there would be an 87.77% likelihood of a greater lack of adjustment F value as a consequence of noise generated in the experiments. Therefore, a nonsignificant lack of adjustment was positive for the model.

On the other hand, a predicted R^2 of 0.7479 was in agreement with an adjusted R^2 of 0.91, since the difference was less than 0.2. Furthermore, an adequate precision signal/noise ratio presented a 16.258 value, where a ratio greater than 4 is desirable, indicating a suitable signal. In addition, it demonstrated this model can be utilized to navigate through the design's space.

As shown in Figure 1 PBED mean ± SD comparison established treatment results were significantly different, where treatment 1 was the most prominent. Likewise, it is shown in Table 2 that the predicted model did not support obtained results in T_1, despite the minimal difference.

On the other hand, Table 3 result analysis required several considerations. The factor "culture media volume" was significant ($p = 0.0264$), and a negative effect on enzyme activity was observed with a contribution percentage < 10%, implying in an optimization attempt that lower volumes could be tested. It is important to note that Erlenmeyer flaks of the same volume and brand were used; all assays were carried out in the same orbital shaker with the same *setup*, where it is clear that volume was probably associated with oxygen transfer. Decreasing the media volume would represent an increase in oxygen transfer area, which could be favorable. However, contribution percentage was small (5.81) and considered very low to propose new tests.

Furthermore, copper sulfate was also a significant factor ($p = 0.0107$) that had a positive effect on enzyme activity with a contribution percentage of 9.39%, resulting in the highest values among the factors evaluated (Table 3), suggesting that higher concentrations could be tested for optimization. On the contrary, glucose was not significant ($p = 0.4304$), despite its positive effect with a 0.44% contribution. Ammonium sulfate was not a significant factor ($p = 0.4800$) with a positive effect on enzyme activity and 0.35% contribution. Peptone factor was significant ($p = 0.0267$) with a positive effect on enzyme activity and 5.76% contribution, suggesting higher concentrations could be tested to attempt factor optimization. Last, the factor yeast extract was significant ($p = 0.0149$) with a negative effect on enzyme activity and 7.92% contribution. This result suggests lower concentrations could be tested in order to optimize this factor (Table 3).

Furthermore, contribution percentages results did not exceed 10%, and even though A, B, D, E, F, and G obtained factor values suggesting greater or lower concentrations could be assayed depending on response variable effect, carrying-out these tests would not be recommendable. Based on individual contribution percentages the change that could be generated on enzyme activity would not be substantial. Despite a 36.08-fold increase for T_1 in enzyme activity in comparison with our previous work dependent variable values were still low ($T_1 = 4.69 \pm 0.39$ UL^{-1}) with respect to other reported laccase activities. Additionally, Figure 2 T_1 time follow-up highlights that there were no higher values of the variables measured before 168 h. Therefore, the most recommendable option would be to amplify or change the navigation space when studying other conditions that perhaps would increase response variable results. Similarly other values of factors already evaluated would be studied or different factors other than the ones already

(a)

(b)

(c)

Kinetic parameters		Standard error	Confidence limit 95%		p
V_{max} (mM s^{-1})	6.87×10^{-5}	1.16×10^{-5}	3.92×10^{-3}	3.99×10^{-3}	0.000
K_m (mM)	5.36×10^{-2}	5.17×10^{-4}	4.75×10^{-2}	5.03×10^{-2}	0.000

(d)

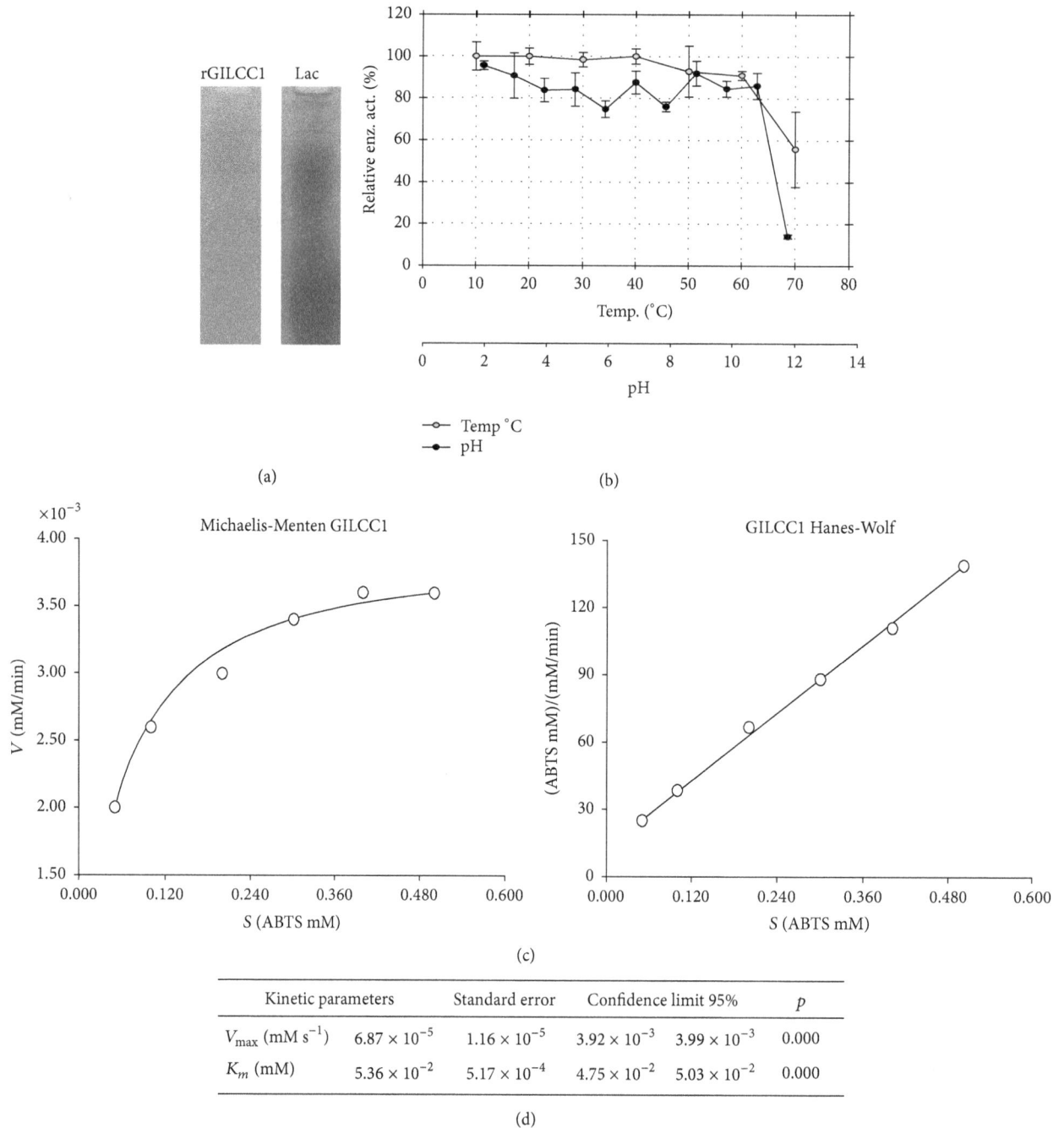

FIGURE 3: rGILCC1 concentrate characterization. (a) Zymogram gel for rGILCC1 enzyme functional identification. As positive control, *Laccase* Lac *(Sigma-Aldrich)*. (b) Relative enzyme activity (%) as a function of rGILCC1 obtained from concentrate after 1 h incubation at different temperatures and pH (assay carried out in triplicate, $n = 3$). Activity was determined as described in Materials and Methods [13]. (c) rGILCC1 obtained from concentrate enzyme kinetics using ABTS as a substrate (assay carried out in triplicate, $n = 3$). (d) K_m (Michaelis-Menten constant), V_{max} (maximal velocity) detailed values.

assayed would be evaluated to achieve greater enzyme activity values.

On the other hand, some aspects of the results drew particular attention. Based on low glucose percentage contribution, the media could require a lower concentration of carbon source. This seems conflicting, since additional carbon would generate increased biomass. Glyceraldehyde-3-phosphate dehydrogenase (GAPDH) is critical in glycolysis. Its promoter P_{GAP} provides constitutive expression on glucose metabolism; therefore it has been widely used for constitutive expression of heterologous proteins. P_{GAP} was governing the expression of *Ganoderma lucidum* optimized laccase synthetic gene GlLCC1 [19]; thus results seem to contradict the fact that enzyme should be produced at the end of the exponential phase (Figure 2).

Our results agree with those by other authors and proposed genes under P_{GAP} expression are not entirely constitutive and could be regulated by additional conditions. Kern et al. (2007) studied an alternate oxidase fused to GFP under P_{GAP} expression. They described GFP fluorescence markedly increased after culture media glucose depletion, while in an intermitted manner small quantities of ethanol were produced, phenomena also described in other investigations [29–31]. No reports support in a detailed manner P_{GAP} constitutive promoter incapability of always producing the metabolite of interest in greater quantities at the end of the exponential phase. Authors argue that it is more reasonable to think of other factors associated with the intrinsic nature of the recombinant protein existing that could influence the velocity, maturation, or enzyme transport [29]. These aspects could be studied in future work in detail, whereas a case in point *Pichia pastoris* rGlLCC1 laccase expression rate and follow-up would be carried out.

Stability study demonstrated that *G. lucidum* rGlLCC1 enzyme expressed in *P. pastoris* X33 was maintained stable at 10 and 60°C and retained over 50% residual enzymatic activity at 70°C (Figure 3(b)). Some authors have performed laccase GlLCC1 enzymatic stability studies, where in contrast to the present study induction of the recombinant enzyme expression in *P. pastoris* used the AOX promoter. You et al. (2014) reported a rapid decrease in enzyme stability at temperatures above 40°C [3]. Likewise, Sun et al. (2012) obtained maximal activity at 55°C; however stability tests revealed that after 20 minutes incubation at 55°C activity decreased [32] considerably [33]. Other authors have expressed the same laccase with some modifications in the N-Terminal sequence, reporting that the enzyme denatured after 10 min at 100°C incubation [34]. Moreover, after one-hour incubation at 50°C, residual activity was below 50% [32]. With respect to obtained stability in other *G. lucidum* laccases results have been variable. Manavalan et al. (2013) reported a maximal laccase-3 activity at 30°C; however, enzyme's half-life was affected even shortly before one-hour incubation at temperatures above 60°C [4]. When comparing previously mentioned pure laccase results with data obtained in this study using enzyme obtained from concentrate, we can argue that results are promising due to the ample range of thermal stability. Our laccase results are comparable or even exceeded GlLCC1 laccase stability or other *G. lucidum* laccases expressed in *P. pastoris*. These are aspects of great importance in terms of considerable potential use in different industrial settings or treatment of contaminated effluents.

rGlLCC1 pH stability presented residual enzyme activities between 80 and 100% for all pHs assayed. Different authors report *P. pastoris* GlLCC1 expressed laccase great pH stability ranging between 2 and 10, obtaining relative enzyme activities above 40% for all cases [3, 4, 32, 33]. However, it is worth noting that this work evaluated up to pH 12, where residual activities of 80% were obtained. This confirms *P. pastoris* rGlLCC1 enzyme stability under the P_{GAP} promoter and its potential use for environmental care.

Different laccase K_m values, whether from fungal or bacterial origin, present an ample substrate range or may even differ from the same substrate. Nonetheless, a higher affinity to ABTS is reported in comparison to other substrates such as syringaldazine or guaiacol among others, with lower oxidation velocity and higher K_m values [4, 5, 35].

Our data revealed a K_m of 5.36×10^{-2} mM; this result is similar to *G. lucidum* laccase-3 [4] with a K_m of 0.047 mM. In contrast other studies have reported discrepant values such as Ko et al. (2001) with a K_m of 0.0037 mM. Other authors using GlLCC1 laccase reported superior K_m values as those obtained in this study with *P. pastoris X33,* namely, You et al. (2014) with 0.521 mM and Sun et al. (2012) with 0.9665 mM [3, 4, 32, 33]. However these differences could be related to laccase purity, since the enzyme used in this study is not in the pure form, hence the need to refer to it as apparent K_m. In addition, the nature of other supernatant components is unknown or could positively or negatively interfere with enzyme activity.

5. Concluding Remarks

In conclusion, culture media was improved for rGlLCC1 laccase production in *Pichia pastoris* [500 mL Erlenmeyer flask containing 150 mL culture media, 10% (v/v) inoculum, 1.0 mM $CuSO_4$, 30 gL^{-1} glucose, 5 mM NH_4SO_4, 10 gL^{-1} peptone, and 5 gL^{-1} yeast extract for 168 h, at 30°C and 180 rpm] with a laccase activity of 4.7 ± 0.4 UL^{-1}. This represents a 36.08-fold increase compared with our previous work. Functionality was identified from enzyme obtained from concentrate through zymogram gel. Additionally, stability at different temperatures increased to a range between 10 and 60°C. Also, at 70°C the enzyme retained 50% residual activity. Moreover, pH stability was observed between 2 and 11 with over 70% residual activity. Apparent kinetic parameters obtained by recombinant laccase demonstrated its affinity for ABTS substrate, as was reported in previous work with molecular docking analysis, and its catalytic efficiency [19], supporting what has been described for other laccases.

Additionally, it is important to note that characterization was performed from concentrated supernatant instead of pure enzyme, since the objective of this group with this and other laccases is to pave the way for liquid residue and contaminated solid treatment. Working for these purposes with pure enzyme is considerably expensive, making it unsustainable. rGlLCC1 laccase can be a promising enzyme for various industries and multiple purposes due to its broad temperature and pH stability. Our next challenge is to increase culture media volume (process scale-up) in addition to augment enzyme activity, since its activity is still considered low.

Authors' Contributions

Edwin D. Morales-Álvarez and Claudia M. Rivera-Hoyos contributed equally to this work.

Acknowledgments

This research was funded by Project ID 00005575 (*Correlación entre la expresión constitutiva la concentración de proteína y la actividad biológica de las lacasas recombinantes POXA1 B de Pleurotus ostreatus y GLlac 1 de Ganoderma lucidum en Pichia pastoris*), Project ID 00006337 (*Optimización del medio de cultivo para producción de la lacasa recombinante POXA 1B de Pleurotus ostreatus en Pichia pastoris*) from *Pontificia Universidad Javeriana*, and Project ID 00006169 (Jóven Investigador Colciencias 2014), Bogotá, Colombia. The authors thank María Lucía Gutiérrez for English editing.

References

[1] J.-A. Majeau, S. K. Brar, and R. D. Tyagi, "Laccases for removal of recalcitrant and emerging pollutants," *Bioresource Technology*, vol. 101, no. 7, pp. 2331–2350, 2010.

[2] S. S. More, P. S. Renuka, K. Pruthvi, M. Swetha, S. Malini, and S. M. Veena, "Isolation, purification, and characterization of fungal laccase from *Pleurotus* sp.," *Enzyme Research*, vol. 2011, Article ID 248735, 7 pages, 2011.

[3] L.-F. You, Z.-M. Liu, J.-F. Lin, L.-Q. Guo, X.-L. Huang, and H.-X. Yang, "Molecular cloning of a laccase gene from *Ganoderma lucidum* and heterologous expression in *Pichia pastoris*," *Journal of Basic Microbiology*, vol. 54, no. 1, pp. S134–S141, 2014.

[4] T. Manavalan, A. Manavalan, K. P. Thangavelu, and K. Heese, "Characterization of optimized production, purification and application of laccase from *Ganoderma lucidum*," *Biochemical Engineering Journal*, vol. 70, pp. 106–114, 2013.

[5] H. Patel, S. Gupte, M. Gahlout, and A. Gupte, "Purification and characterization of an extracellular laccase from solid-state culture of *Pleurotus ostreatus* HP-1," *3 Biotech*, vol. 4, no. 1, pp. 77–84, 2014.

[6] C. M. Rivera-Hoyos, E. D. Morales-Álvarez, R. A. Poutou-Piñales, A. M. Pedroza-Rodríguez, R. Rodríguez-Vázquez, and J. M. Delgado-Boada, "Fungal laccases," *Fungal Biology Reviews*, vol. 27, no. 3-4, pp. 67–82, 2013.

[7] D. Singh Arora and R. Kumar Sharma, "Ligninolytic fungal laccases and their biotechnological applications," *Applied Biochemistry and Biotechnology*, vol. 160, no. 6, pp. 1760–1788, 2010.

[8] W.-T. Huang, R. Tai, R.-S. Hseu, and C.-T. Huang, "Overexpression and characterization of a thermostable, pH-stable and organic solvent-tolerant *Ganoderma fornicatum* laccase in *Pichia pastoris*," *Process Biochemistry*, vol. 46, no. 7, pp. 1469–1474, 2011.

[9] E. D. Morales-Álvarez, C. M. Rivera-Hoyos, L. E. Chaparro-Núñez, C. E. Daza, R. A. Poutou-Piñales, and A. M. Pedroza-Rodríguez, "Decolorization and detoxification of Malachite Green by *Ganoderma lucidum*: key operating parameters and adsorption studies," *Journal of Environmental Engineering*, vol. 143, no. 4, Article ID 04016093, 2017.

[10] E. D. Morales-Álvarez, C. M. Rivera-Hoyos, N. González-Ogliastri et al., "Partial removal and detoxification of Malachite Green and Crystal Violet from laboratory artificially contaminated water by *Pleurotus ostreatus*," *Universitas Scientiarum*, vol. 21, no. 3, pp. 259–285, 2016.

[11] K.-S. Wong, Q. Huang, C.-H. Au, J. Wang, and H.-S. Kwan, "Biodegradation of dyes and polyaromatic hydrocarbons by two allelic forms of *Lentinula edodes* laccase expressed from *Pichia pastoris*," *Bioresource Technology*, vol. 104, pp. 157–164, 2012.

[12] A. Piscitelli, C. Pezzella, P. Giardina, V. Faraco, and G. Sannia, "Heterologous laccase production and its role in industrial applications," *Bioengineered Bugs*, vol. 1, no. 4, pp. 252–262, 2010.

[13] R. Tinoco, M. A. Pickard, and R. Vazquez-Duhalt, "Kinetic differences of purified laccases from six *Pleurotus ostreatus* strains," *Letters in Applied Microbiology*, vol. 32, no. 5, pp. 331–335, 2001.

[14] J. Lin-Cereghino, M. D. Hashimoto, A. Moy et al., "Direct selection of *Pichia pastoris* expression strains using new G418 resistance vectors," *Yeast*, vol. 25, no. 4, pp. 293–299, 2008.

[15] L. M. Damasceno, C.-J. Huang, and C. A. Batt, "Protein secretion in *Pichia pastoris* and advances in protein production," *Applied Microbiology and Biotechnology*, vol. 93, no. 1, pp. 31–39, 2012.

[16] H. A. Córdoba-Ruiz, R. A. Poutou-Piñales, O. Y. Echeverri-Peña et al., "Laboratory scale production of the human recombinant iduronate 2-sulfate sulfatase-Like from *Pichia pastoris*," *African Journal of Biotechnology*, vol. 8, no. 9, pp. 1786–1792, 2009.

[17] R. A. Poutou-Piñales, H. A. Córdoba-Ruiz, L. A. Barrera-Avellaneda, and J. M. Delgado-Boada, "Carbon source feeding strategies for recombinant protein expression in *Pichia pastoris* and *Pichia methanolica*," *African Journal of Biotechnology*, vol. 9, no. 15, pp. 2173–2184, 2010.

[18] D. A. Albarracín Pardo, *Mejoramiento del Medio de Cultivo para Producción de la Lacasa Recombinante POXA 1B de Pleurotus ostreatus en Pichia pastoris*, Microbiology, Pontificia Universidad Javeriana, Bogotá, Colombia, 2016.

[19] C. M. Rivera-Hoyos, E. D. Morales-Álvarez, S. A. Poveda-Cuevas et al., "Computational analysis and low-scale constitutive expression of laccases synthetic genes *GLLCC1* from *Ganoderma lucidum* and *POXA 1B* from *Pleurotus ostreatus* in *Pichia pastoris*," *PLoS ONE*, vol. 10, no. 1, Article ID e0116524, 2015.

[20] H. Sáenz-Suárez, C. Rivera-Hoyos, E. Morales-Álvarez, R. Poutou-Piñales, J. Sáenz-Moreno, and A. Pedroza-Rodríguez, "Modelación computacional preliminar de la estructura 3D de dos lacasas fúngicas," *Salud Arte y Cuidado*, vol. 7, no. 1, pp. 5–16, 2014.

[21] R. A. Poutou, E. Amador, and M. Candelario, "Banco de células primario (BCP): caracterización y papel en la producción de proteínas recombinantes," *Biotecnología Aplicada*, vol. 11, no. 1, pp. 55–59, 1994.

[22] V. Sarria-Alfonso, J. Sánchez-Sierra, M. Aguirre-Morales, I. Gutiérrez-Rojas, N. Moreno-Sarmiento, and R. A. Poutou-Piñales, "Culture media statistical optimization for biomass production of a ligninolytic fungus for future rice straw degradation," *Indian Journal of Microbiology*, vol. 53, no. 2, pp. 199–207, 2013.

[23] J. C. Gonzalez, S. C. Medina, A. Rodriguez, J. F. Osma, C. J. Alméciga-Díaz, and O. F. Sánchez, "Production of *Trametes pubescens* laccase under submerged and semi-solid culture conditions on agro-industrial wastes," *PLoS ONE*, vol. 8, no. 9, Article ID e73721, 2013.

[24] J. E. Reynolds III, M. Josowicz, P. Tyler, R. B. Vegh, and K. M. Solntsev, "Spectral and redox properties of the GFP synthetic chromophores as a function of pH in buffered media," *Chemical Communications*, vol. 49, no. 71, pp. 7788–7790, 2013.

[25] J. B. S. Haldane, "Graphical methods in enzyme chemistry," *Nature*, vol. 179, no. 4564, p. 832, 1957.

[26] F. J. Burquillo, M. Holgado, and W. G. Bardsley, "Uso del paquete estadístico SIMFIT en la enseñanza del análisis de datos en ciencias experimentales," *Journal of Science Education*, vol. 4, no. 1, pp. 8–14, 2003.

[27] G. L. Miller, "Use of dinitrosalicylic acid reagent for determination of reducing sugar," *Analytical Chemistry*, vol. 31, no. 3, pp. 426–428, 1959.

[28] D. T. Plummer, *Introducción a la Bioquímica Práctica*, McGraw-Hill, Bogotá, Colombia, 1981.

[29] A. Kern, F. S. Hartner, M. Freigassner et al., "*Pichia pastoris* 'just in time' alternative respiration," *Microbiology*, vol. 153, no. 4, pp. 1250–1260, 2007.

[30] K. Baumann, M. Maurer, M. Dragosits, O. Cos, P. Ferrer, and D. Mattanovich, "Hypoxic fed-batch cultivation of *Pichia pastoris* increases specific and volumetric productivity of recombinant proteins," *Biotechnology and Bioengineering*, vol. 100, no. 1, pp. 177–183, 2008.

[31] T. Vogl and A. Glieder, "Regulation of *Pichia pastoris* promoters and its consequences for protein production," *New Biotechnology*, vol. 30, no. 4, pp. 385–404, 2013.

[32] E.-M. Ko, Y.-E. Leem, and H. T. Choi, "Purification and characterization of laccase isozymes from the white-rot basidiomycete *Ganoderma lucidum*," *Applied Microbiology and Biotechnology*, vol. 57, no. 1-2, pp. 98–102, 2001.

[33] J. Sun, R.-H. Peng, A.-S. Xiong et al., "Secretory expression and characterization of a soluble laccase from the *Ganoderma lucidum* strain 7071-9 in *Pichia pastoris*," *Molecular Biology Reports*, vol. 39, no. 4, pp. 3807–3814, 2012.

[34] H. X. Wang and T. B. Ng, "Purification of a laccase from fruiting bodies of the mushroom *Pleurotus eryngii*," *Applied Microbiology and Biotechnology*, vol. 69, no. 5, pp. 521–525, 2006.

[35] H. Liu, Y. Cheng, B. Du et al., "Overexpression of a novel thermostable and chloride-tolerant laccase from *Thermus thermophilus* SG0.5JP17-16 in *Pichia pastoris* and its application in synthetic dye decolorization," *PLoS ONE*, vol. 10, no. 3, Article ID e0119833, 2015.

16

Extracellular Pectinase from a Novel Bacterium *Chryseobacterium indologenes* Strain SD and its Application in Fruit Juice Clarification

Karabi Roy ⓘ, Sujan Dey ⓘ, Md. Kamal Uddin ⓘ, Rasel Barua, and Md. Towhid Hossain

Department of Microbiology, University of Chittagong, Chittagong 4331, Bangladesh

Correspondence should be addressed to Sujan Dey; sujan.mbio@cu.ac.bd

Academic Editor: Denise Freire

Pectinase is one of the important enzymes of industrial sectors. Presently, most of the pectinases are of plant origin but there are only a few reports on bacterial pectinases. The aim of the present study was to isolate a novel and potential pectinase producing bacterium as well as optimization of its various parameters for maximum enzyme production. A total of forty bacterial isolates were isolated from vegetable dump waste soil using standard plate count methods. Primary screening was done by hydrolysis of pectin. Pectinase activity was determined by measuring the increase in reducing sugar formed by the enzymatic hydrolysis of pectin. Among the bacterial isolates, the isolate K6 exhibited higher pectinase activity in broth medium and was selected for further studies. The selected bacterial isolate K6 was identified as *Chryseobacterium indologenes* strain SD. The isolate was found to produce maximum pectinase at 37°C with pH 7.5 upon incubation for 72 hours, while cultured in production medium containing citrus pectin and yeast extract as C and N sources, respectively. During enzyme-substrate reaction phase, the enzyme exhibited its best activity at pH of 8.0 and temperature of 40°C using citrus pectin as substrate. The pectinase of the isolate showed potentiality on different types of fruit juice clarification.

1. Introduction

Enzymes are biological molecules that accelerate biochemical reactions. Pectinases are the group of enzymes that prompt the degradation of pectic substances through depolymerization and deesterification reaction [1]. Pectinase is also a well-known term for commercial enzyme preparation during fruit juice clarification. This enzyme disunites polygalacturonic acid into mono-galacturonic acid by opening glycosidic linkages [2].

In the world market, it has been reported that pectinase accounts for 10% of global industrial enzymes produced [3]. Pectinolytic enzymes are produced by many organisms like bacteria, fungi, yeasts, insects, nematodes, protozoan, and plants. Among the various pectinases, bacterial pectinases take more advantages over other pectinases. With the passage of time, many reports have been published on the optimization of different microbiological parameters and fermentation strategies for the production of pectinases [4].

Pectinases have immense applications in fruit juice industries to improve fruit juice clarity and yield [5]. Pectinases also have other various industrial applicationïż£s like scouring of cotton, degumming of plant fibers, waste water treatment, and vegetable oil extraction so used in various industries as pulp industry, textile industry, food industry, and so on. The application of pectinase enzyme to alter the texture or flavor of fruit juice, to increase extraction and clarification, and to reduce viscosity has also been described [6].

Keeping all the above advantages of pectinase enzymes in consideration, the aim of the present study was designed to isolate potential pectinolytic microorganisms, optimize their cultural conditions for maximum pectinase production, and investigate different factors involved in maximum pectinase activity and also to evaluate its potentiality in different fruit juice clarification.

2. Material and Methods

2.1. Sampling and Screening. Pectinase producing bacteria were isolated from vegetable dump waste soil. From the collected samples, 40 bacterial isolates were isolated and purified by following standard plate count techniques described by Dubey and Maheshwari [7]. Among 40 isolates, eight isolates were found to produce pectinase while grown on yeast extract pectin agar (YEP) medium during primary screening. Screening of pectinase producing bacteria was carried out in YEP agar medium containing yeast extract 1%, pectin 1%, agar 1.5%, and NaCl 0.5% (pH 7.0) at 37°C for 48 hours of incubation. After incubation, the colonies showing clear zones upon flooding with iodine-potassium iodide solution (1.0 g iodine, 5.0 g potassium iodide, and 330 mL H_2O) were selected as pectinase producers [8] and the isolate K6 with the maximum zone of diameter was preceded for further studies.

2.2. Identification of Selected Bacterial Strain by Phenotypic and Biochemical Characteristics. The selected isolate was identified by morphological, cultural, and biochemical characteristics. Colony characteristics of the isolate were determined on yeast extract-pectin agar slant and cellular morphology by Gram staining method. For biochemical characteristics, several tests like citrate test, TSI (triple sugar iron) test, indole test, MR-VP, motility, catalase, oxidase, starch hydrolysis, and fermentation tests for various sugars such as glucose, sucrose, lactose, maltose, starch, and mannitol tests were done.

2.3. Molecular Identification of the Bacterial Isolate Using 16S rRNA Sequencing. Simultaneously, this potential isolate was further identified using the molecular tool of 16S rRNA sequencing. In this method, Promega® Wizard® DNA purification kit was used for the extraction of genomic DNA of the selected isolate. 16S rRNA gene region was amplified with the universal primers. The reaction mixtures were 5 μl of template, primers: 1 μl of forward primer: 27F (5′ AGAGTTTGATCCTGGCTCAG 3′), 1 μl of reverse primer: 1492R (5′ TACCTTGTTACGACTT 3′), 6 μl of assay buffer, 2 μl of Taq DNA polymerase, and 5 μl of dNTP mix. PCR products were purified by using the PCR KlenzolTM and it was sequenced with a next-generation DNA sequencing. The sequencing results were then processed using BioEdit software. The nucleotide sequence analysis was done using the Basic Local Alignment Search Tool (BLAST) program on National Center for Biotechnology Information site (https://www.ncbi.nlm.nih.gov). The obtained sequence data were submitted to NCBI GenBank with accession number KY684254. Phylogenetic analysis was conducted in Molecular Evolutionary Genetics Analysis software version 7.0 (MEGA7) [9].

2.4. Preparation of Inoculum. For the production of pectinase enzyme, the inoculum was prepared by inoculating 10 mL of sterilized YEP liquid medium in a test tube with one loop full of pure culture and incubated at 125 rpm at 37°C. The fresh overnight grown pure culture was used as an inoculum for enhanced enzyme production.

2.5. Production of Pectinase Enzyme. The inoculum (5% v/v) was transferred aseptically to 50 mL of production medium (yeast extract pectate broth showing the following composition: yeast extract 1%, pectin 1%, NaCl 0.5%, and distilled water 100 ml) in 250 mL conical flask and incubated at 37°C for 72 hours with 125 rpm in a shaking incubator [10]. After incubation, the production medium was centrifuged at 7000 rpm for 15 min at 4°C to obtain cell-free supernatant. The supernatant was used as the crude enzyme for further studies.

2.6. Pectinase Assay. The pectinase activity was assayed by estimating the amount of reducing sugars released under assay conditions by the enzymatic degradation of citrus pectin. The reaction mixture containing 1.8 ml substrate (citrus pectin) solution and 0.2 ml suitably diluted enzyme solution was incubated at 40°C in water bath for 1 hour. The amount of reducing sugar liberated was quantified by Nelson's modification of Somogyi method [11, 12]. One unit of enzyme activity (U) was defined as the amount of enzyme required to release 1 μmol of reducing sugar per ml per minute under standard assay conditions ($1 U = 1 \mu mol min^{-1} mL^{-1}$) [13]. The color intensity was measured at 500 nm in a colorimeter [SpectroT60 (UV-VIS RS)] and compared with a standard curve prepared with "D-glucose" (25–200 micrograms). Control was maintained with uninoculated media and boiled enzyme. Relative activity of the enzyme was calculated as the percentage by using the following formula:

$$\text{Relative Activity} = \text{Activity of the sample} \times \frac{100}{\text{Maximum activity of the sample}} \quad (1)$$

2.7. Total Protein Estimation. The total protein content was determined by Lowry method [14] measuring the absorbance at 600 nm and compared with a standard curve prepared by bovine serum albumin (BSA).

2.8. Optimization of Different Factors Involved in Maximum Pectinase Production

2.8.1. Effect of Temperature, pH, and Incubation Time. The bacterial isolate was subjected to different culture conditions to derive the optimum conditions for maximum pectinase production. Pectinase production was estimated at different temperatures (27, 30, 37, 40, and 45°C), a wide range of pH (5.0, 5.5, 6.0, 6.5, 7.0, 7.5, 8.0, 8.5, and 9.0), and different incubation periods (24, 48, 72, 96, and 120 hours), and then the enzyme activity was measured.

2.8.2. Effect of Carbon and Nitrogen Sources. To study the effect of carbon and nitrogen sources on pectinase production from the selected isolate, various carbon sources (citrus pectin, glucose, sucrose, and starch) and nitrogen sources (peptone, yeast extract, ammonium chloride, and potassium nitrate) were supplemented into production medium at a concentration of 0.5% w/v and then the pectinase activity was assayed.

2.9. Optimization of Reaction Parameters for the Maximum Pectinase Activity.
Various reaction parameters were studied to determine the optimum conditions of the crude pectinase activity at different temperature, pH, reaction time, and substrate concentration.

2.9.1. Effect of Temperature and pH.
The optimum temperature and pH of the pectinase enzyme were determined by incubating the reaction mixture having pH 7.5 at various temperatures (30, 37, 40, and 45°C), and then a range of pH (6.0, 6.5, 7.0, 7.5, 8.0, 8.5, and 9.0) at 40°C for 1 hour using different buffers, respectively. For this purpose, substrate (1% w/v citrus pectin) was prepared at different pH values (pH 6.0–9.0) using 0.2 M citrate-phosphate buffer (pH 6.0–7.5) and 0.2 M Tris-HCl buffer (pH 8.0–9.0). Then the pectinase activity was assayed by using the standard assay method.

2.9.2. Effect of Reaction Time.
For determination of optimum reaction time, the enzyme stability was studied by incubating the reaction mixture for various time intervals at 10 min, 20 min, 30 min, 40 min, 50 min, and 60 min under optimized temperature and pH and then performing the assay for pectinase activity.

2.9.3. Effect of Substrate Concentration.
To examine the effect of substrate concentration during enzyme-substrate reaction, varying concentrations of citrus pectin (0.5, 1.0, 1.5, 2.0, 2.5, and 3.0%) were prepared and then enzyme activity was measured.

2.10. Crude Pectinase Enzyme in Fruit Juice Clarification.
To observe the effect of crude pectinase on different fruit juice clarification, tubes were labeled as treatment and control. 10 ml of crude pectinase was taken into the test tube (treatment) and 10 ml of distilled water was taken into the control test tube. Then twenty ml of apple/grape juice was added to both tubes. The contents of the tubes were agitated to mix the enzymes throughout the juice. The tubes were kept into the water bath at 40°C. The tubes were observed at 5-minute intervals over one-hour period. After incubation, the solution was filtered [15].

3. Results and Discussion

There are only a few reports on bacterial pectinases till now [16]. In this study, we aimed to find out novel and potential pectinolytic bacteria, to optimize the cultural conditions for enhancing enzyme production and partial characterization of the enzyme. For this purpose, we collected vegetable dump waste soil as the source of potential organisms. Eight isolates were primarily selected from 40 bacterial isolates on the basis of their pectinolytic activity, as shown in Figure 1, during screening and finally, the isolate K6 was selected for further studies.

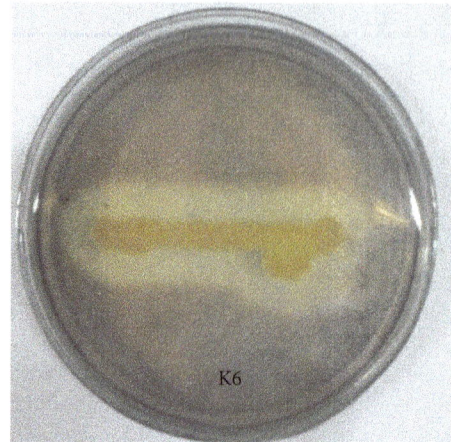

FIGURE 1: Zone of pectin hydrolysis on YEP agar medium of isolate K6 after 48-hour incubation.

3.1. Identification of Selected Bacterial Isolate.
To identify the selected isolate K6, both traditional microbiological methods and modern molecular technologies were considered. On the basis of observed morphological, cultural, and biochemical characteristics, the selected isolate was compared with the standard description in Bergey's Manual of Determinative Bacteriology [17] and the isolate K6 was provisionally identified as *Chryseobacterium indologenes*.

In NCBI database, BLAST showed significant alignments of *Chryseobacterium indologenes* with 95% similarities. Further, the obtained sequence was compared with other related sequences to find the closest homolog at NCBI using BLAST. The construction of a phylogenetic tree (Figure 2) was made on the basis of 16S rRNA gene sequences of isolate *Chryseobacterium indologenes* strain SD and other strains of *Chryseobacterium* species obtained from GenBank database. The related 16 nucleotide sequences were used to construct the phylogenetic tree using the neighbor-joining method [18]. The bootstrap consensus tree inferred from 1000 replicates [19] is taken to represent the evolutionary history of the taxa analyzed [19]. Branches corresponding to partitions reproduced in less than 50% bootstrap replicates are collapsed. The percentage of replicate trees in which the associated taxa clustered together in the bootstrap test (1000 replicates) is shown next to the branches [19]. The evolutionary distances were computed using the Maximum Composite Likelihood method [20] and are in the units of the number of base substitutions per site. The analysis involved 16 nucleotide sequences. All positions containing gaps and missing data were eliminated. There were a total of 1323 positions in the final dataset. Evolutionary analyses were conducted in MEGA7 [9]. Finally, the result attained from the 16S rRNA analyses inferred that the *Chryseobacterium indologenes* strain SD was very close to other *Chryseobacterium indologenes* strains.

3.2. Optimization Studies of the Isolate for Pectinase Production.
To optimize the culture conditions of this potential organism in the production medium different temperature,

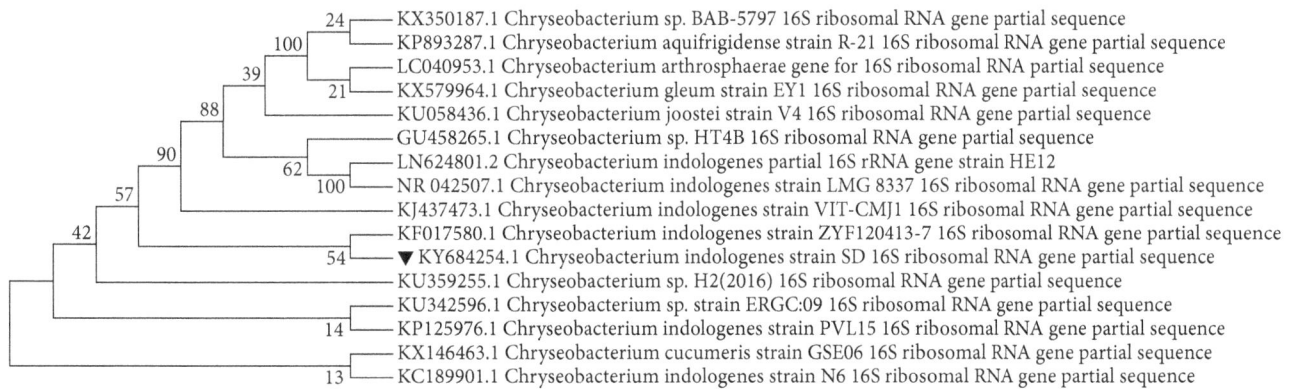

FIGURE 2: Phylogenetic tree constructed on the basis of 16S rRNA gene sequences of *Chryseobacterium indologenes* strain SD with other *Chryseobacterium* sp. obtained from GenBank database. Their names and respective accession numbers are shown on the tree.

TABLE 1: Effect of temperature on pectinase production from *Chryseobacterium indologenes* strain SD.

Incubation temperature (°C)	Enzyme activity (μmolmin^{-1}mL^{-1})	Relative activity (%)
27	0.36 ± 0.050	52.94
30	0.45 ± 0.020	66.18
37	*0.68 ± 0.020*	*100.0*
40	0.24 ± 0.021	35.3
45	0.18 ± 0.026	26.47

Values are mean ± SD of 3 replicates.

TABLE 2: Effect of pH on pectinase production from *Chryseobacterium indologenes* strain SD.

pH	Enzyme activity (μmolmin^{-1}mL^{-1})	Relative activity (%)
5.0	0.37 ± 0.020	66.07
5.5	0.42 ± 0.026	75.00
6.0	0.45 ± 0.026	80.36
6.5	0.47 ± 0.020	83.93
7.0	0.48 ± 0.020	85.71
7.5	*0.56 ± 0.021*	*100.0*
8.0	0.40 ± 0.025	71.43
8.5	0.33 ± 0.020	58.93
9.0	0.21 ± 0.038	37.5

Values are mean ± SD of 3 replicates.

TABLE 3: Effect of incubation period on pectinase production from *Chryseobacterium indologenes* strain SD.

Incubation period (hour)	Enzyme activity (μmolmin^{-1}mL^{-1})	Relative activity (%)
24	0.45 ± 0.020	76.27
48	0.50 ± 0.040	84.75
72	*0.59 ± 0.030*	*100.0*
96	0.52 ± 0.021	88.14
120	0.35 ± 0.032	59.32

Values are mean ± SD of 3 replicates.

pH, incubation period, and various C and N sources were considered. Enzyme production went up with the increase of temperature up to 37°C and then declined (Table 1). The maximum production which occurred at this temperature was 0.679 U/ml (100% of relative activity). This dramatically reduced to nearly 27% at 45°C temperature. In the previous study of Aaisha and Barate (2016) [21], the highest pectinase production was observed from some *Bacillus species* at 37°C which is similar to our current study. The same findings were also reported by the other workers [22] in case of mutagenic strain of *Leuconostoc mesenteroides*.

The same trend was also observed when this organism was allowed to grow in the production medium at varying pH (Table 2). Maximum production (0.564 U/ml) was recorded at pH 7.5. This finding is in accordance with other workers who reported that most of the *Bacillus* sp. produce high amount of pectinase between pH 7.5 and 8 [21, 23]. At highly acidic and alkaline pH, enzyme production decreased by almost 34% and 64%, respectively. From this result, it can be inferred that, at very low and very high pH condition, growth of the organism slows down.

An attempt was made to determine the most favorable time period for enzyme production by the selected isolate and the highest enzyme production (0.594 U/ml) was recorded at 72 hours of incubation (Table 3). The enzyme production gradually decreased to 0.35 μmolmin^{-1}mL^{-1} at 120 hours of incubation which is almost 40% less than that of maximum. This might be due to the accumulation of waste products at prolonged incubation time with limited nutrient sources which consequently suppressed the growth of microorganism. According to Nawawi et al. (2017) [24], maximum pectinase production was determined from the *Bacillus subtilis* ADI1 after 72 hours of incubation which well agreed with our findings.

In this study, we also supplemented different types of carbon and nitrogen sources to find out the suitable production medium for pectinase production by *Chryseobacterium indologenes* strain SD. Among the four C sources, the organism lost almost 50% of production in case of sucrose

TABLE 4: Effect of carbon sources on pectinase production from *Chryseobacterium indologenes* strain SD.

Carbon source	Enzyme activity (μmolmin^{-1}mL^{-1})	Relative activity (%)
Citrus pectin	*0.67 ± 0.020*	*100.0*
Glucose	0.64 ± 0.021	95.52
Sucrose	0.38 ± 0.030	56.72
Starch	0.50 ± 0.020	74.63

Values are mean ± SD of 3 replicates.

TABLE 5: Effect of nitrogen sources on pectinase production from *Chryseobacterium indologenes* strain SD.

Nitrogen source	Enzyme activity (μmolmin^{-1}mL^{-1})	Relative activity (%)
Yeast extract	*0.61 ± 0.020*	*100.0*
Peptone	0.55 ± 0.020	90.16
Potassium nitrate	0.46 ± 0.036	75.41
Ammonium chloride	0.42 ± 0.021	68.85

Values are mean ± SD of 3 replicates.

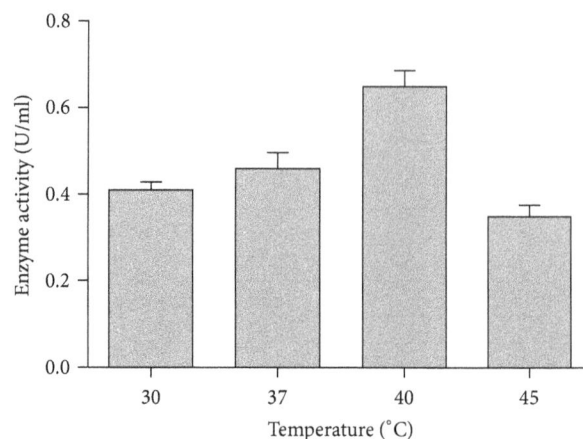

FIGURE 3: Effect of temperature on pectinase activity.

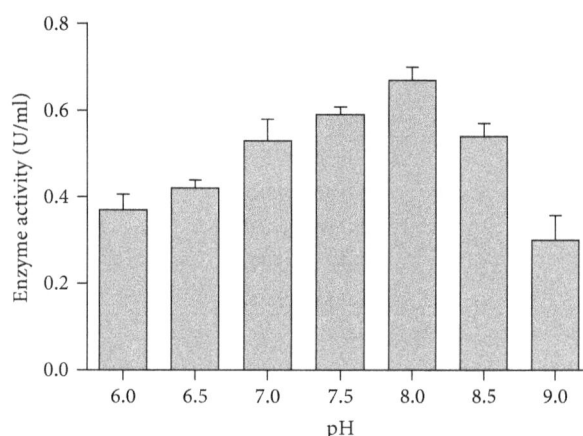

FIGURE 4: Effect of pH on pectinase activity.

utilization. It showed the highest production of 0.671 U/ml, while citrus pectin was used in the medium as C source and almost the same result was found in case of glucose supplement (Table 4) suggesting that the organism exploited citrus pectin more efficiently as compared to other C sources. Prakash et al. (2014) [25] observed the highest production of pectinase with lactose and glucose and Jayani et al. (2010) [26] reported citrus pectin as the best carbon source for pectinase production by *Bacillus sphaericus*. However, some researchers reported the maximum pectinase production from *Bacillus subtilis* ADI1 using rice brain as carbon source [24].

Likewise, the best enzyme production of this isolate was recorded when yeast extract was used as N source in the medium and the organism also produced nearly the same amount of enzyme when peptone was supplemented which indicates that this organism preferred yeast extract as compared to other N sources (Table 5). On top of that, our study revealed that organic nitrogen was used as better N sources by this organism than inorganic sources for enzyme production. These results are completely aligned with the findings of the other workers [25, 26].

Finally, by applying all the optimized parameters, the isolate was allowed to produce the enzyme in the production medium and we observed little increase in pectinase production (0.689 U/ml).

3.3. Total Protein Estimation for Crude Enzyme.
Protein concentration of the crude enzyme was determined by Folin-Lowry method [14]. The total protein content was 1320 μg/ml in the cell-free supernatant of *Chryseobacterium indologenes* strain SD.

3.4. Optimization Studies of Pectinase Activity.
By growing the organism in the production medium under optimized conditions, the crude was collected by centrifugation to determine optimum conditions of the pectinase activity. Then the collected crude enzyme was allowed to react with different substrate concentrations (citrus pectin) at a wide range of temperatures, pH, and reaction time.

In this research work, the crude enzyme obtained from *Chryseobacterium indologenes* strain SD showed maximum activity at 40°C (Figure 3), whereas the organism showed the highest production at 37°C. This result is approximately similar to the result of other studies [27]. However, some studies reported [28] that pectinases from various *Bacillus species* were most active at 50°C and 60°C. From our study, it can be inferred that pectinase enzyme produced by the isolate is a moderately thermophilic enzyme.

Enzyme activity also depends on the pH of the reaction mixture. In our study, crude enzyme showed the highest activity at slightly alkaline pH 8 (Figure 4) and the organism also showed its maximum production at pH 7.5. Therefore, this enzyme can be used for vegetable purees and other preparations which need neutral to slightly alkaline pH [29]. This finding is in accordance with the reports of previous studies [28]. So, the result of our study indicates that this crude enzyme might be alkaline in nature.

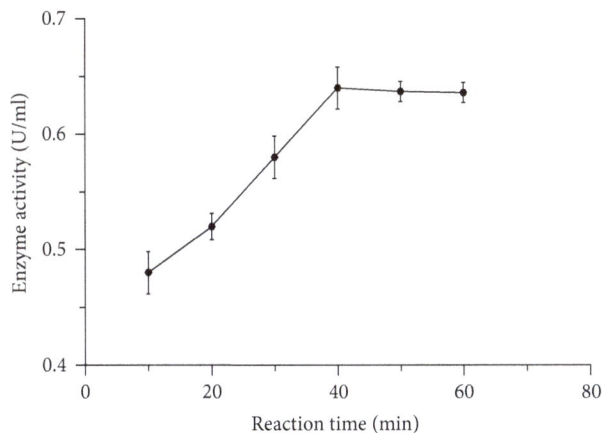

FIGURE 5: Effect of reaction time on pectinase activity.

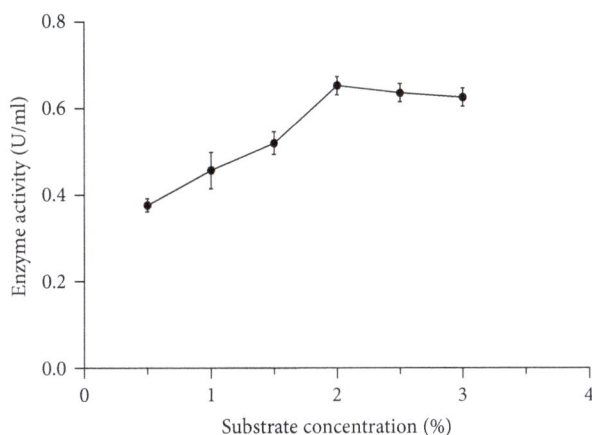

FIGURE 6: Effect of substrate (citrus pectin) concentration on pectinase activity.

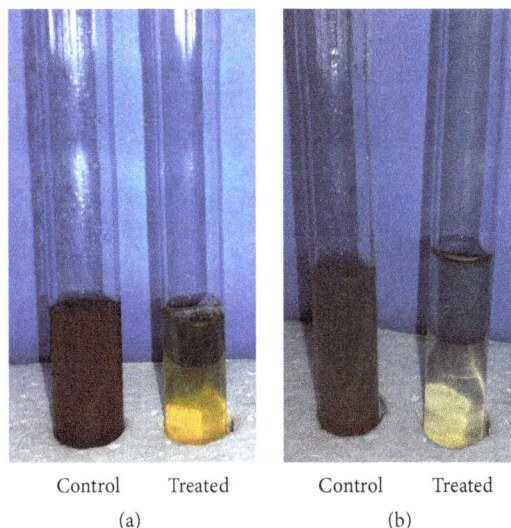

Control   Treated   Control   Treated
         (a)   (b)

FIGURE 7: Application of pectinase in apple juice (a) and grape juice (b) clarification.

4. Conclusion

Nowadays, the need of industrially important enzymes has increased rapidly. Pectinase enzyme has taken great attraction in the field of juice clarification and other commercial applications. In this study, *Chryseobacterium indologenes* strain SD was found as a potential pectinase producer and it is the first report on *Chryseobacterium indologenes* strain SD. In this investigation, the crude enzyme was found to be slightly alkaline in nature and best active at 40°C for 40 minutes of incubation. Further studies can be done for complete characterization and purification of the crude enzyme and purified enzymes can be used in various types of fruit juice clarification.

Acknowledgments

The financial support from the Ministry of Science and Technology and University Grant Commission (UGC), Bangladesh, is thankfully acknowledged.

As incubation time affects the activity of the enzyme, the crude enzyme was allowed to react with 1% of citrus pectin as substrate at optimized pH and temperature for different time intervals to determine its optimum reaction time. In our study, enzyme activity increased with the increase of incubation time up to 40 min and then remained stable in the subsequent incubation period (Figure 5). This result is in a good agreement with other studies [30].

The enzyme assay using different concentrations of substrate (citrus pectin) was observed and found that the enzyme activity augmented up to 2% of pectin and then it showed downward trend before being leveled off at 98% of relative activity in the subsequent increase of substrate concentration (Figure 6). This might be due to complete saturation of enzyme by the substrate.

3.5. Application of Crude Pectinase Enzyme in Fruit Juice Clarification. The effect of crude pectinase of the bacterial isolate *Chryseobacterium indologenes* strain SD was studied for apple/grape juice clarification. The crude enzyme of the selected isolate showed good activities by clarifying the juices as compared to control (Figure 7).

References

[1] D. B. Pedrolli, A. C. Monteiro, E. Gomes, and E. C. Carmona, "Pectin and pectinases: production, characterization and industrial application of microbial pectinolytic enzymes," *The Open Biotechnology Journal*, vol. 3, no. 1, pp. 9–18, 2009.

[2] O. A. Oyewole, S. B. Oyeleke, B. E. N. Dauda, and S. Emiade, "Production of Amylase and Protease Enzymes by Aspergillus niger and Penicillium frequestans Isolated from Abattoir Effluent," *Journal of Microbiology*, vol. 1, no. 5, pp. 174–180, 2011.

[3] F. Stutzenberger, "Pectinase Production," in *Encyclopedia of Microbiology*, J. Lederberg, Ed., vol. 3, pp. 327–337, Academy press, New York, NY, USA, 1992.

[4] S. Kaur, H. P. Kaur, B. Prasad, and T. Bharti, "Production and Optimization of Pectinase by *Bacillus* sp. Isolated From Vegetable Waste Soil," *Indo American Journal of Pharmaceutical Research*, vol. 6, no. 1, pp. 4185–4190, 2016.

[5] I. Alkorta, C. Garbisu, J. M. Llama, and J. L. Serra, "A Review," *ProcessBiochemistry*, vol. 17, pp. 35–41, 1982.

[6] K. Prathyusha and V. Suneetha, "Bacterial pectinases and their potent biotechnological application in fruit processing/Juice production industry: A review," *Journal of Phytology*, vol. 3, no. 6, pp. 16–19, 2011.

[7] R. C. Dubey and D. K. Maheshwari, "Practical Microbiology," in *S. Chand and Co. P Ltd*, p. 413, ISBN, New Delhi, 2nd edition.

[8] L. K. Janani, G. Kumar, and K. V. Bhaskara Rao, "Production and Optimization of Pectinase from *Bacillus* sp. MFW7 using Cassava Waste," *Asian Journal of Biochemical and Pharmaceutical Research*, vol. 1, no. 2, pp. 329–336, 2011.

[9] S. Kumar, G. Stecher, and K. Tamura, "MEGA7: Molecular Evolutionary Genetics Analysis version 7.0 for bigger datasets," *Molecular Biology and Evolution*, vol. 33, no. 7, pp. 1870–1874, 2016.

[10] D. R. Kashyap, S. Chandra, A. Kaul, and R. Tewari, "Production, purification, and characterization of pectinase from a *Bacillus sp*," *World Journal of Microbiology & Biotechnology*, vol. 16, pp. 277–282, 2000.

[11] N. Nelson, "A photometric adaptation of the Somogyi method for the determination of glucose," *The Journal of Biological Chemistry*, vol. 153, pp. 375–380, 1944.

[12] M. Somogyi, "Notes on sugar determination," *The Journal of Biological Chemistry*, vol. 195, no. 1, pp. 19–23, 1952.

[13] D. Silva, E. S. Martins, R. Silva, and E. Gomes, "Pectinase production by Penicillium viridicatum RFC3 by solid state fermentation using agro-industrial by-products," *Brazilian Journal of Microbiology*, vol. 33, pp. 318–324, 2002.

[14] O. H. Lowry, N. J. Rosebrough, A. L. Farr, and R. J. Randall, "Protein measurement with Folin phenol reagent," *The Journal of Biological Chemistry*, vol. 193, pp. 265–275, 1951.

[15] S. A. Mehta, R. Mitali, S. Nilofer, and P. Nimisha, "Optimization of physicological parameters for pectinase production from soil isolates and its applications in fruit juice clarification," *Journal of Environmental Research and Development*, vol. 7, no. 4, pp. 1539–1546, 2013.

[16] R. S. Jayani, S. Saxena, and R. Gupta, "Microbial pectinolytic enzymes: a review," *Process Biochemistry*, vol. 40, no. 9, pp. 2931–2944, 2005.

[17] R. E. Buchannan and N. E. Gibbson, *Bergey's manual of determinative Bacteriology*, Williams and Wilkins Co, Baltimore, Maryland, USA, 8th edition, 1974.

[18] N. Saitou and M. Nei, "The neighbor-joining method: a new method for reconstructing phylogenetic trees," *Molecular Biology and Evolution*, vol. 4, no. 4, pp. 406–425, 1987.

[19] J. Felsenstein, "Confidence limits on phylogenies: an approach using the bootstrap," *Evolution*, vol. 39, pp. 783–791, 1985.

[20] K. Tamura, M. Nei, and S. Kumar, "Prospects for inferring very large phylogenies by using the neighbor-joining method," *Proceedings of the National Acadamy of Sciences of the United States of America*, vol. 101, no. 30, pp. 11030–11035, 2004.

[21] G. Aaisha and D. Barate, "Isolation and Identification of Pectinolytic Bacteria from Soil Samples of Akola Region, India," *International Journal of Current Microbiology and Applied Sciences*, vol. 5, no. 1, pp. 514–521, 2016.

[22] A. B. Saanu, "Pectinolytic Activity of Mutagenic Strain of Leuconostoc Mesenteroides Isolated From Orange and Banana Fruit Waste," *Journal of Applied Microbiology and Biochemistry*, vol. 1, no. 2:7, pp. 1–6, 2017.

[23] O. J. Oumer and D. Abate, "Characterization of Pectinase from Bacillus subtilis Strain Btk 27 and Its Potential Application in Removal of Mucilage from Coffee Beans," *Enzyme Research*, vol. 2017, Article ID 7686904, 7 pages, 2017.

[24] M. H. Nawawi, R. Mohamad, P. M. Tahir, and W. Z. Saad, "Extracellular Xylanopectinolytic Enzymes by *Bacillus subtilis* ADI1 from EFB's Compost," *International Scholarly Research Notices*, vol. 2017, Article ID 7831954, 7 pages, 2017.

[25] S. Prakash, R. Karthik, M. T. Venthan, B. Sridhar, and P. G. Bharath, "Optimization and Production of Pectinase from *Bacillus subtilis* (mtcc 441) by using Orange Peel as a Substrate," *International Journal of Recent Scientific Research*, vol. 5, no. 6, pp. 1177–1179, 2014.

[26] R. S. Jayani, S. K. Shukla, and R. Gupta, "Screening of bacterial strains for polygalacturonase activity: its production by bacillus sphaericus (MTCC 7542)," *Enzyme Research*, vol. 2010, Article ID 306785, 5 pages, 2010.

[27] A. Thakur, R. Pahwa, S. Singh, and R. Gupta, "Production, purification, and characterization of polygalacturonase from mucor circinelloides ITCC 6025," *Enzyme Research*, vol. 2010, Article ID 170549, 7 pages, 2010.

[28] N. Torimiro and R. E. Okonji, "A comparative study of pectinolytic enzyme production by *Bacillus species*," *African Journal of Biotechnology*, vol. 12, no. 46, pp. 6498–6503, 2013.

[29] M. M. C. N. Soares, R. Da Silva, and E. Gomes, "Screening of bacterial strains for pectinolytic activity: Characterization of the polygalacturonase produced by Bacillus sp," *Brazilian Journal of Microbiology*, vol. 30, no. 4, pp. 299–303, 1999.

[30] I. G. Khan and D. L. Barate, "Effect of various parameters on activity of pectinase enzyme," *International Journal of Advanced Research*, vol. 4, no. 1, pp. 853–862, 2016.

17

Characterization of the Catalytic Structure of Plant Phytase, Protein Tyrosine Phosphatase-Like Phytase, and Histidine Acid Phytases and their Biotechnological Applications

Alex Sander Rodrigues Cangussu ⓘ,[1] **Deborah Aires Almeida,**[1]
Raimundo Wagner de Souza Aguiar,[1] **Sidnei Emilio Bordignon-Junior** ⓘ,[2]
Kelvinson Fernandes Viana,[3] **Luiz Carlos Bertucci Barbosa,**[4]
Edson Wagner da Silva Cangussu,[5] **Igor Viana Brandi,**[6] **Augustus Caeser Franke Portella,**[1]
Gil Rodrigues dos Santos,[1] **Eliane Macedo Sobrinho,**[7] **and William James Nogueira Lima**[6]

[1]*Engenharia de Bioprocessos e Biotecnologia e Programa de Pos-Graduação em Biotecnologia,*
 Universidade Federal do Tocantins, Gurupi, TO, Brazil
[2]*Engenharia e Ciências de Alimentos, Universidade Estadual Paulista, São José do Rio Preto, SP, Brazil*
[3]*Laboratório de Biologia Molecular e Bioquímica-ICVN, Universidade Federal da Integração Latino-Americana,*
 Foz do Iguaçu, PR, Brazil
[4]*Engenharia de Bioprocessos e Biotecnologia, Instituto de Recursos Naturais, Universidade Federal de Itajubá, Itajubá, MG, Brazil*
[5]*Faculdade de Medicina, Universidade Estadual de Montes Claros, Montes Claros, MG, Brazil*
[6]*Instituto de Ciências Agrárias, Universidade Federal de Minas Gerais, Montes Claros, MG, Brazil*
[7]*Instituto Federal do Norte Minas Gerais, Araçuaí, MG, Brazil*

Correspondence should be addressed to Alex Sander Rodrigues Cangussu; alexcangussu@uft.edu.br

Academic Editor: Toshihisa Ohshima

Phytase plays a prominent role in monogastric animal nutrition due to its ability to improve phytic acid digestion in the gastrointestinal tract, releasing phosphorus and other micronutrients that are important for animal development. Moreover, phytase decreases the amounts of phytic acid and phosphate excreted in feces. Bioinformatics approaches can contribute to the understanding of the catalytic structure of phytase. Analysis of the catalytic structure can reveal enzymatic stability and the polarization and hydrophobicity of amino acids. One important aspect of this type of analysis is the estimation of the number of β-sheets and α-helices in the enzymatic structure. Fermentative processes or genetic engineering methods are employed for phytase production in transgenic plants or microorganisms. To this end, phytase genes are inserted in transgenic crops to improve the bioavailability of phosphorus. This promising technology aims to improve agricultural efficiency and productivity. Thus, the aim of this review is to present the characterization of the catalytic structure of plant and microbial phytases, phytase genes used in transgenic plants and microorganisms, and their biotechnological applications in animal nutrition, which do not impact negatively on environmental degradation.

1. Introduction

Biocatalysts are important molecules that improve fermentable substrate availability and nutrient absorption, increasing the efficiency of food bioconversion. Phytases, proteases, xylanases, and amylases are typical examples of enzymatic groups that have wide applicability and contribute to increased productivity and profitability [1–3]. In monogastric animals, such as pigs, poultry, and humans, there is little or no phytase activity [4, 5]. However, these animals are fed

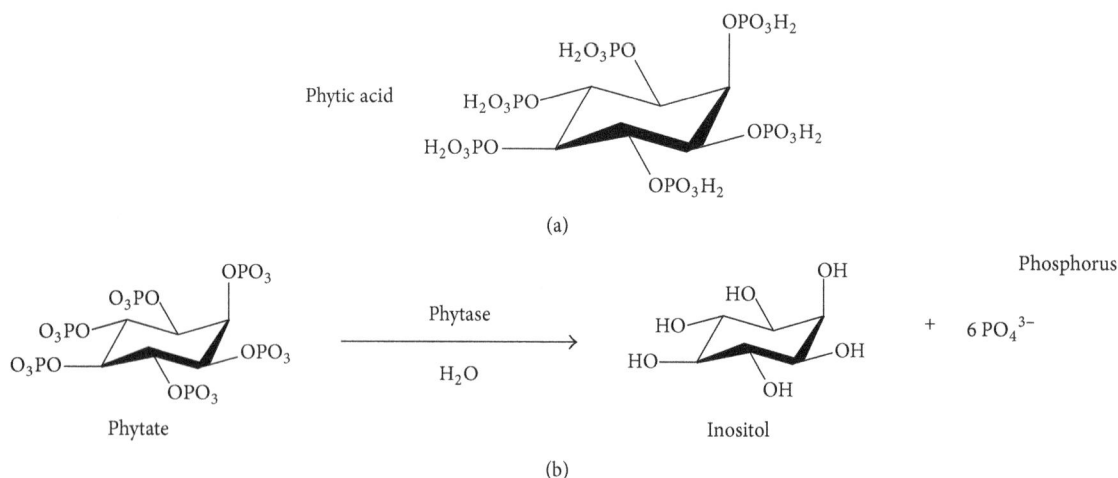

FIGURE 1: Molecular structures. The phytic acid (a) and mechanism of enzyme action (b) developed by ChemBioDraw Ultra version 12 software.

with seeds and grains rich in phytate, which is not completely absorbed. Thus, phytate is excreted in the feces, increasing the environmental pollution by phosphorus [6, 7].

Sustainable agriculture can be achieved by adding phytase to animal feeds. Phytase improves the quality and bioavailability of some important nutrients, reducing the environmental impact of farming at the animal feed stage [8–10]. Thus, several strategies can be explored with the aim of achieving bioprocesses that increase yields and phytase stability by using microorganisms and plants. Among the microorganisms capable of producing phytase are bacteria, such as *Lactobacillus*, *Escherichia*, *Bacillus*, *Xanthomonas*, *Pseudomonas*, and *Klebsiella* spp.; fungi, such as *Penicillium*, *Aspergillus*, and *Rhizopus* spp.; and yeasts, such as *Saccharomyces cerevisiae*, *Schwanniomyces castellii*, *Schizophyllum commune*, *Wickerhamomyces anomalus*, and *Hansenula* spp. [11–13].

Plant growth and productivity can be hindered by low availability of phosphorus (P) in the soil [14]. Therefore, plants have developed mechanisms to improve soil phosphorus availability, with emphasis on the participation of purple acid phosphatases (PAPs). PAPs are acidic metallohydrolases that are involved in plant growth and pathogen defense. PAP-like isoforms have been identified in *Ipomoea batatas*, *Solanum lycopersicum*, and *Hordeum vulgare*. In barley, PAPs are grouped into isogenes HvPAPhy_a, HvPAPhy_b1, and HvPAPhy_b2 [15–17]. Another consideration is the technical compatibility of microbial phytase genes and plants, which enables the generation of transgenic crops. Understanding of the catalytic structure and amino acid sequences of microbial phytases and plants, via analysis of the principal genes, allows for technological advances using phytase without increasing environmental degradation.

Thus, the aim of this review is to present the characterization of the catalytic structure of plants and microbial phytases and their biotechnological applications, while proposing technical alternatives for sustainable agricultural productivity.

2. Phytase and Phosphorus Bioavailability

Phytate ($C_6H_{18}O_{24}P_6$), also known as phytic acid (myo-inositol 1,2,3,4,5,6-hexakisphosphate; PA; IP_6), has a molecular weight of $660\,g\,mol^{-1}$ (Figure 1) and is considered the main source of stored phosphorus in seeds, grains, and vegetables. Phytate is known as a food inhibitor due to its ability to chelate and thus decreases the bioavailability of some important micro- and macronutrients, such as Zn, Mg, Mn, Fe, and Ca, in monogastric animals [18, 19].

Phytate is a phosphorus storage molecule and a constituent of cereals and grains. It has relatively low bioavailability due to its strong adsorption onto soil and unavailability for degradation by soil microorganisms [20, 21]. Plants have developed a variety of mechanisms to overcome this problem, such as upregulation of high-affinity phosphate transporters, improvement of internal phosphatase activity, and secretion of organic acids and phosphatases [22–25]. The latter is an important aspect as it allows mineralized organic phosphorus to be released as inorganic phosphorus into the soil [26]. Phytases are special phosphatase enzymes that catalyze the hydrolysis of phytate into lower inositol phosphates and inorganic phosphorus (Figure 1).

Human beings are limited in their ability to hydrolyze phytate; consequently, phytic acid reduces the amount of minerals required for tissue function and cellular metabolism maintenance [27]. Pigs and poultry are unable to metabolize phytate because they lack the enzyme phytase in their gastrointestinal tract; therefore, it is necessary to add inorganic phosphorus to animal feeds. However, a large amount of phytate and inorganic phosphorus are excreted into the environment because they are not fully absorbed by these animals, causing environmental impacts such as eutrophication of surface waters of lakes and rivers, harmful algal blooms, nitrous oxide production (N_2O; greenhouse gas), growth of toxin-producing microorganisms, and the death of several aquatic species [18, 28]. Thus, it is desirable to use new techniques that aim to increase the quality and bioavailability of phosphorus in animal feed.

TABLE 1: Major genes evaluated in transgenic plants and microorganisms with phytase activity.

Gene	Plant target	Microbial target	Reference
HvPAPhy_a	*Hordeum vulgare* L.	—	Holme et al. [106]
PHY_US417	*Arabidopsis thaliana*	—	Belgaroui et al. [14]
PHYA	*Hordeum vulgare* L.	—	Mohsin et al. [107]
AVP1DOX	*Solanum lycopersicum*	—	Yang et al. [108]
MtPHY1	*Medicago sativa* L.	—	Ma et al. [109]
MtPT1	*Medicago sativa* L.	—	Ma et al. [109]
HvPAPhya	—	*Pichia pastoris*	Dionisio et al. [4]
HvPAPhyb1	—	*Pichia pastoris*	Dionisio et al. [4]
HvPAPhyb2	—	*Pichia pastoris*	Dionisio et al. [4]
sk-57	—	*Pichia pastoris*	Xiong et al. [73]

3. Characterization of Phytase

3.1. Sources of Phytase. Plants, animals, and microorganisms can produce phytase. In plants, it is present in wheat, barley, peas, soybeans, corn, rice, and spinach. The blood of vertebrates such as fish, sea turtles, and reptiles also contains phytase, which is produced by microorganisms such as yeasts, bacteria, and filamentous fungi [18, 19]. Microbial phytase is of great interest to industry due to its high level of production and extracellular activity, for example, in *Aspergillus* spp. [18]. Phytase (myo-inositol hexakisphosphate phosphohydrolase, EC 3.1.3.26 and EC 3.1.3.8) is characterized as a homodimeric enzyme [29, 30]. It is a phosphatase that initiates the sequential release of phytate orthophosphate groups (myo-inositol 1,2,3,4,5,6-hexakisphosphate). It belongs to the class of hydrolases and thereby hydrolyzes phytate (phytic acid) to release inositol phosphates, phosphorus, inositol, and other essential nutrients required for absorption (Figure 1). Moreover, it is an important component of a variety of metabolic processes as released phosphorus favors development, formation and mineralization of animal bone, cellular metabolism, and protein synthesis [31–35].

3.2. Catalytic Structure of Plant Phytase. Many enzymes with phytase activity can be obtained or expressed in plants, animals, and microorganisms [18, 36, 37]. They are classified into the following four groups according to their catalytic mechanisms: histidine acid phosphatases (HAPs), PAPs, Cys or β-helix phosphatases, and protein tyrosine phosphatase-like enzymes (PTP) [38].

Phytases are composed of several phosphatases; however, only some have sufficient phytase activity. The PAPs are acidic metallohydrolases, and their structures are involved in plant growth and pathogen defense. PAPs have a metal center constituting iron, zinc, and manganese [17]. The metal center coordinates the binding site residues with those of other termini. These regions are characterized by seven conserved amino acid residues in five conserved motifs (in italic): *DXG*, *GDXXY*, *GNH(D/E)*, *VXXH*, and *GHXH*. These are involved in the dimetal nuclear center coordination [17, 39–42].

Other PAP-like isoforms have been identified in *I. batatas* and *S. lycopersicum* [15–17]. PAPhy genes in *H. vulgare* are grouped into isogenes: HvPAPhy_a, HvPAPhy_b1, and HvPAPhy_b2 [4]. The isogene HvPAPhy possesses remarkable phytase activities in mature grains and proteins produced by *Pichia pastoris*. Isoform a (HvPAPhy_a) has a molecular mass of 60,29 kDa and contains 544 amino acids, with the ligand-binding sites at amino acid sequence 199, 226, 283, 365, and 402. Isoform_b1 (HvPAPhy_b1) has a molecular mass of 59,51 kDa and contains 536 amino acids, with the ligand-binding sites at amino acid sequence 194, 221, 278, 359, and 396. Isoform b2 (HvPAPhy_b2) has a molecular mass of 59,34 kDa and contains 537 amino acids, with the ligand-binding sites at amino acid sequence 194, 221, 278, 360, and 397 (Figure 2). The genes HvPAPhy_a, HvPAPhy_b1, HvPAPhy_b2, PHY_US417, PHYA, AVPIDOX, MtPHY1, MtPT1, and SK-57 are the major genes evaluated in transgenic plants and microorganisms presenting phytase activity (Table 1).

3.3. Catalytic Structure of Microbial Phytase. Bioinformatics approaches have contributed to increased knowledge about phytase [43]. One approach is comparative sequence analysis of phytase families, which allows for phylogenetic inferences and the prediction of functional sites [4, 34, 44]. Another is molecular modeling, which allows for inferences about enzymatic structure when no three-dimensional structure has been determined [45–47]. Based on the above-mentioned analyses, phytase B from *Aspergillus niger* (histidine acid phosphatases (HAP)) plays an important role in phytate hydrolysis. Its structure includes 460 amino acid residues and is composed of a large α/β-domain with a six-stranded β-sheet surrounded by several α-helices and a small α-domain. It contains five disulfide bonds at positions 52–368, 109–453, 197–422, 206–279, and 394–402, most of which are located in loops next to the surface. These bonds are due to the conformational stability of native phytases A and B and the maintenance of their catalytic activities. Moreover, researchers found that disulfide bonds have an important role in maintaining active site integrity [48]. Oakley [49] presented a structural phytase model of phytase A from *A. niger*, consisting of an α/β-domain, an α-domain, and an N-terminal extension. N-Acetylglucosamine residues are bound to four sites of the phytase structure (N82, N184, N316, and N353) within the active site, which is formed by an α-helix cavity. A structural model of *A. niger* phytase

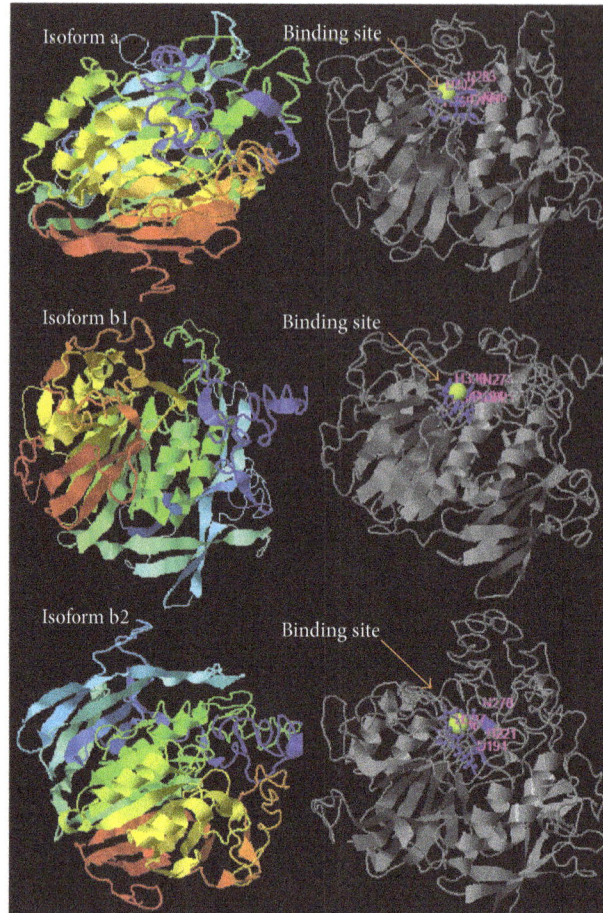

FIGURE 2: Structural model of purple acid phosphatase from barley *(Hordeum vulgare)* proposed by Dionisio et al. [4]. PAPhy genes are grouped in isogenes HvPAPhy_a, HvPAPhy_b1, and HvPAPhy_b2. Isogenes HvPAPhy possess significant phytase activity in the mature grains and proteins already were produced in *P. pastoris*. Structural model used i-TASSER server for protein structure and function prediction (https://zhanglab.ccmb.med.umich.edu/I-TASSER/) [105]. FASTA sequences were obtained from https://www.ncbi.nlm.nih.gov/protein. Isoform a is constituted by 544 amino acids, 60,29 kDa, and Hphob of 49,5%. The ligand-binding site residues from isoform a are represented by amino acid sequence 199, 226, 283, 365, and 402. Isoform b1 is constituted by 536 amino acids and 59,51 kDa with the ligand-binding site residues being represented by amino acid sequence 194, 221, 278, 359, and 396. Isoform b2 is constituted by 537 amino acids and 59,34 kDa. Ligand-binding site residues are constituted by amino acid sequence 194, 221, 278, 360, and 397.

(HAP) proposed by Mishra et al. [50] is represented in Figure 3.

However, PhyA from *Xanthomonas oryzae* pv. *oryzae* (beta propeller phytase (BPHY)) plays a role in the degradation of phytic acid. It encodes a 373-amino acid protein including a 28-amino acid-predicted signal peptide. Its active site is located on the top of the β-propeller having a high conservation of amino acid residues involved in the metal ions binding to a phytase identified from *Bacillus amyloliquefaciens* (BaPhy), which has high- and low-affinity calcium sites responsible for enzymatic activity. PhyA from *X. oryzae* is similar to a phytase identified from *Bacillus amyloliquefaciens* (BaPhy), which has high- and low-affinity calcium sites responsible for enzymatic activity. However, differences in enzymatic activity between phytases can be attributed to differences in phosphate binding affinity [12].

Other types of phytases (HAP) PhyK and AppA are found in plants and microorganisms. PhyK belongs to a group of phytases synthesized by plant-associated bacteria, such as *Xanthomonas campestris*, *Pseudomonas syringae*, and *Erwinia carotovora*. This structural model was proposed by Böhm et al. [51] to explain the binding interaction between an enzyme and a substrate. The structural model observed and proposed by Shivange et al. [52] described the phytase of a mutant *Yersinia mollaretii*. Shivange et al. [52] described five amino acid substitutions, D52N, T77K, K139E, G187S, and V298M, next to the active site loop (S42-T47) and catalytically important residues involved in substrate binding (R37, R41, E241, and D327). Structural analysis of phytase from different environmental sources is important for understanding its specific role in the microenvironment [51]. Figures 4 and 5 represent structural models of *X. oryzae* phytase proposed by Wilkins et al. [53] and *W. anomalus* phytase proposed by Kaur et al. [54].

Wickerhamomyces anomalus has already been described as *Pichia anomala* and *Hansenula anomala*. This species

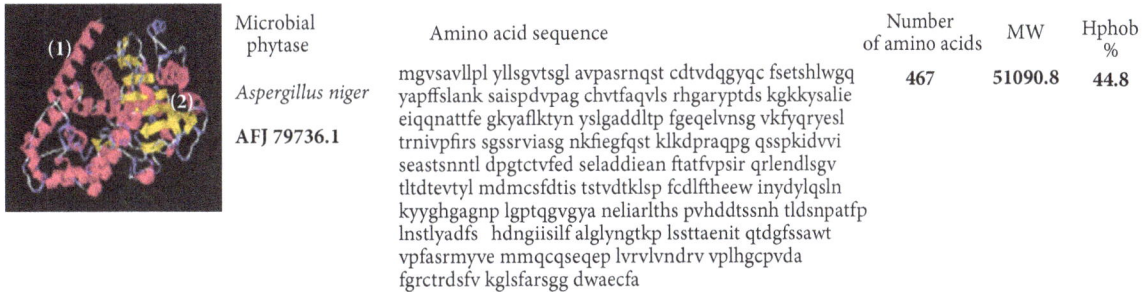

Microbial phytase	Amino acid sequence	Number of amino acids	MW	Hphob %
Aspergillus niger AFJ 79736.1	mgvsavllpl yllsgvtsgl avpasrnqst cdtvdqgyqc fsetshlwgq yapffslank saispdvpag chvtfaqvls rhgaryptds kgkkysalie eiqqnattfe gkyaflktyn yslgaddltp fgeqelvnsg vkfyqryesl trnivpfirs sgssrviasg nkfiegfqst klkdpraqpg qsspkidvvi seastsnntl dpgtctvfed seladdiean ftatfvpsir qrlendlsgv tltdtevtyl mdmcsfdtis tstvdtklsp fcdlftheew inydylqsln kyyghgagnp lgptqgvgya neliarlths pvhddtssnh tldsnpatfp lnstlyadfs hdngiisilf alglyngtkp lssttaenit qtdgfssawt vpfasrmyve mmqcqseqep lvrvlvndrv vplhgcpvda fgrctrdsfv kglsfarsgg dwaecfa	467	51090.8	44.8

FIGURE 3: Structural model of *Aspergillus niger* phytase proposed by Mishra et al. [50] available at https://www.ncbi.nlm.nih.gov/protein and i-TASSER server used for protein structure and function prediction (https://zhanglab.ccmb.med.umich.edu/I-TASSER/) [105]. The domains are identified in the secondary structure elements being alpha-helices (1) and the secondary structure (2).

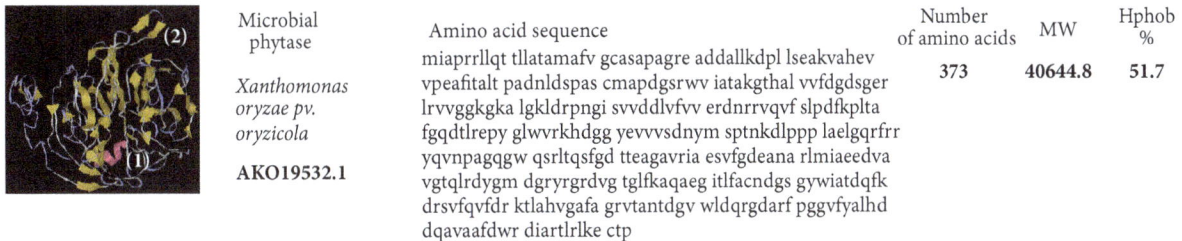

Microbial phytase	Amino acid sequence	Number of amino acids	MW	Hphob %
Xanthomonas oryzae pv. *oryzicola* AKO19532.1	miaprrllqt tllatamafv gcasapagre addallkdpl lseakvahev vpeafitalt padnldspas cmapdgsrwv iatakgthal vvfdgdsger lrvvggkgka lgkldrpngi svvddlvfvv erdnrrvqvf slpdfkplta fgqdtlrepy glwvrkhdgg yevvvsdnym sptnkdlppp laelgqrfrr yqvnpagqgw qsrltqsfgd tteagavria esvfgdeana rlmiaeedva vgtqlrdygm dgryrgrdvg tglfkaqaeg itlfacndgs gywiatdqfk drsvfqvfdr ktlahvgafa grvtantdgv wldqrgdarf pggvfyalhd dqavaafdwr diartlrlke ctp	373	40644.8	51.7

FIGURE 4: Structural model of *Xanthomonas oryzae* phytase proposed by Wilkins et al. [53] available at https://www.ncbi.nlm.nih.gov/protein and i-TASSER server used for protein structure and function prediction (https://zhanglab.ccmb.med.umich.edu/I-TASSER/) [105]. The domains are identified in the secondary structure elements being alpha-helices (1) and the secondary structure (2).

exhibits antimicrobial activities and flavoring features that are responsible for its frequent association with food, beverages, and feed products [13]. Joshi and Satyanarayana [55] described phytase from *P. anomala* as a HAP that is thermostable and used as a feed additive. The PPHY gene of *P. anomala* synthesizes a cell-bound phytase with an ORF of 1389 bp encoding a 462-amino acid protein [53]. Kaur et al. [54] described that the amino acid sequence analysis from *P. anomala* has varying similarity to those of phosphatases from *Pichia stipitis* (62%), *Candida dubliniensis* (51%), *Candida albicans* (51%), and *Arxula adeninivorans* (35%) and phytases from *Debaryomyces castellii* (50%) and *Pichia fabianii* (39%). However, amino acid sequence analysis from *W. anomalus* [54] showed only 5% similarity to phytase from *A. niger* [49].

Structural models of *A. niger*, *W. anomalus*, and *X. oryzae* presented here revealed varying percentages of hydrophobicity between microbial groups. The hydrophobicity of phytase from *A. niger* (44.8%) and that of phytase from *W. anomalus* (48.1%) are similar; however, that of *X. oryzae* is high and different (51.7%) (Figures 3–5). Similarly, the hydrophobicity of HvPAPhy_a is 49.5% (Figure 2). Phytases from *A. niger*, *W. anomalus*, and *H. vulgare* have low hydrophobicity compared with those from *X. oryzae*. The degree of hydrophobicity is an important aspect and reflects the proportions of β-sheets and α-helices in the protein structure. Additionally, the relative amounts of hydrophobic and hydrophilic amino acids partially explain the enzymatic stability. HAP phytase from *Sporotrichum thermophile* showed that surface charge distribution and a high density of hydrophilic amino acids on the surface contribute to the thermal stability of phytase [56].

Moreover, using bioinformatics analysis, Bertrand et al. [57] reported that hydrophobic amino acids are not distributed randomly but form clusters. Thus, it has been suggested that the position of a cluster corresponds to real positions in the secondary structures [57]. Vertical clusters are often associated with β-sheets, whereas horizontal ones often correspond to α-helices [57]. The hydrophobicity of phytases from plants and microorganisms analyzed was represented in this study by Hydrophobicity Clusters Analysis according to Bertrand et al. [57], with the aim of contributing to the analysis of phytase thermostability (Figures 6 and 7).

4. Bioprocess and Strategy of Production

4.1. Transgenic Plants and Phytase. Plant growth and productivity can be hindered by low availability of phosphorus in the soil [14]. Therefore, plants have developed mechanisms that include upregulation of high-affinity phosphate transporters and improvement of internal phosphatase activity [22–24]. Transgenic plants have been used with the aim of increasing the availability of phosphorus in the soil. Genes such as the β-propeller phytases from *Bacillus subtilis* have been inserted in transgenic plants of *Arabidopsis* and tobacco [14, 58].

A recent study revealed overexpression of an extracellular form of the phytase PHY_US417 of *B. subtilis* secreted by transgenic Arabidopsis lines (ePHY) and showed increased phosphate acquisition [14]. Furthermore, phytase genes PhyA and AppA from *A. niger* and *Escherichia coli* have been inserted into *Brassica napus* and increased phosphorus

Microbial phytase	Amino acid sequence	Number of amino acids	MW	Hphob%
Wickerhamomyces anomalus **CBI71332.1**	mvaiqkalvp glylasnyrd vaapelaard qynivkylgg agpyiqlegf gidtkvpdqc tvelvqlymr hgerypglsa gqqqhalvkk lqsynrtitg plsflndyty yvpnedvyel ettpwnsnsp ytgydtavka gaafrakynh lynenktlpv faaasqrvyd tgnffaqgfl gpdylnktvd hvvlseedfl ginslvprwg ckafnsssnd eliaqfptny tqdivkrlte gneglnltts dvsnlfqlca yelsatgysp fcdiftqdel vlhsyasdlq yyytsgpggn ltrtvgavql naslallkqe esdnniwlsf thdtdieifh aalglfdpie plpvnetrfr dmyhhvdvvp mgsrtitekl kcgdetyvrf ivndavvpvp kcqngpgfsc elsdfeayva erlsgidivk dckvpdnvpq eltfywdyqs gqynataeri tr	**462**	**51769.8**	**48.1**

FIGURE 5: Structural model of *Wickerhamomyces anomalus* phytase proposed by Kaur et al. [54] available at https://www.ncbi.nlm.nih.gov/protein and i-TASSER server used for protein structure and function prediction (https://zhanglab.ccmb.med.umich.edu/I-TASSER/) [105]. The domains are identified in the secondary structure elements being alpha-helices (1) and the secondary structure (2).

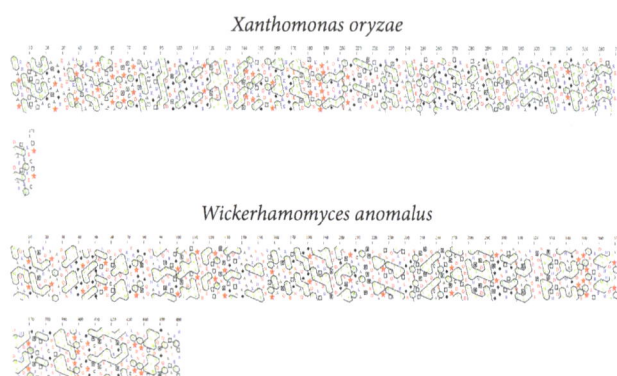

Xanthomonas oryzae

Wickerhamomyces anomalus

Aspergillus niger

Hordeum vulgare L.

FIGURE 6: Secondary structure assignment of microbial phytase by Hydrophobicity Clusters Analysis (HCA) proposed by Bertrand et al. [57]. FASTA data proposed by Kaur et al. [54] are available at https://www.ncbi.nlm.nih.gov/protein. Hydrophobic amino acids are not distributed randomly but form clusters. The clusters are correspondent for the real positions in the regular secondary structures. Vertical clusters are often associated with beta strands and horizontal clusters correspond to alpha-helices. *Xanthomonas oryzae*, Hphob = 51,7%, and *Wickerhamomyces anomalus*, Hphob = 48,1%.

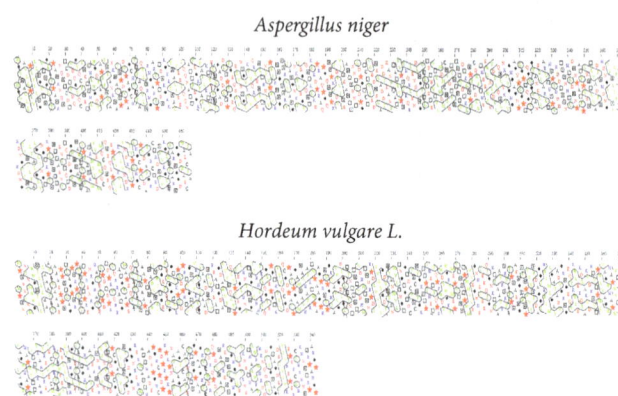

FIGURE 7: Secondary structure assignment of phytase from *Aspergillus niger* and *Hordeum vulgare* L. by Hydrophobicity Clusters Analysis (HCA) proposed by Bertrand et al. [57]. FASTA data proposed by Mishra et al. [50] and Dionisio et al. [4] are available at https://www.ncbi.nlm.nih.gov/protein. Hydrophobic amino acids are not distributed randomly but form clusters. The clusters are correspondent for the real positions in the regular secondary structures. Vertical clusters are often associated with beta strands and horizontal clusters correspond to alpha-helices. *Aspergillus niger*, Hphob = 44,8%, and *Hordeum vulgare* L. isoform *a*, Hphob = 49,5%.

bioavailability in soil and seed phytate [59]. Other reports described a reduction of 50% and 48% in phosphorus secretion by broilers and piglets, respectively, after feeding with transgenic soybean (*Glycine max)* and canola (*B. napus)* seeds with reduced phytase activities [60, 61]. This strategy has also achieved success in decreasing seed phytate levels by endogenous phytase gene expression during the early stages of seed development in transgenic soybean [62]. However, it remains necessary to advance knowledge about the molecular mechanisms regulating phytase formation during grain development and germination [4].

Transgenic wheat and microalgae have been employed for the expression of phytase activity. Transgenic microalgae, such as *Chlamydomonas reinhardtii*, have been revealed as an interesting model system for the production of a fungal phytase (*A. niger* PhyA E228K, mE228K) due to their suitability for genetic transformation and scalability. Lines with high expression levels of in vitro phytase activity were

obtained [63]. Expression of the PhyA gene from *Aspergillus japonicus* in wheat transgenic plants exhibited an increase in phytase activity and iron and zinc contents and a decrease in phytic acid content in seeds. Such reports strengthen an effective method of increasing the bioavailability of iron and zinc and propose a novel way to combat nutritional deficiency [64].

Thus, transgenic plants may be an alternative that warrants further exploration due to the technical compatibility of the genes of plants and microorganisms. It is a promising technology because it promotes efficiency and agricultural productivity, aiming to improve animal nutrition without causing environmental degradation.

4.2. Microbial Phytase. Enzyme production has been observed in various species, especially *A. niger* [20, 65, 66]. It is possible to observe variations in the yield of enzymatic

TABLE 2: Microbial phytase obtained by submerged and solid-state fermentation.

Microorganisms	Process	Yield	Reference
Aspergillus niger mutant	SSF	$154.0 \, \text{U·l}^{-1}$	Bhavsar et al. [79]
Chromobacterium sp.	SF	$7.4–12.4 \, \text{U·ml}^{-1}$	Costa et al. [75]
Schizophyllum sp.	SSF	$55.5 \, \text{U·ml}^{-1}$	Salmon et al. [74]
Bacillus subtilis	SF	$8.5–9.0 \, \text{U·mg}^{-1}$	Fu et al. [69]
Bacillus laevolacticus	SF	$2.957 \, \text{U·ml}^{-1}$	Fu et al. [69]
Aspergillus awamori and fumigatus	SF	$12.86 \text{ and } 20.75 \, \text{U·ml}^{-1}$	Martin et al. [70]
Hansenula polymorpha	SF	$13.5 \, \text{g l}^{-1}$	Mayer et al. [11]

SSF: solid-state fermentation; SF: submerged fermentation; U ml^{-1}: units per milliliter; U mg^{-1}: units per milligram; g l^{-1}: gram per liter.

production due to the physiological peculiarity of each microorganism and varying bioprocess conditions [67]. *A. niger* wild-type strain NRRL3135 produces high amounts of phytase in mineral salts medium, containing glucose and corn starch as the carbon source. However, with the addition of phosphate in the culture medium formulation, there is no evidence of phytase production. *A. niger* can produce two kinds of phytase, PhyA and PhyB, depending only on the pH variation from 2.5 to 5.5, respectively. Moreover, it is possible to use genetic engineering tools to introduce the PhyA gene driven by a specific promoter into another strain of *A. niger*, producing a high yield [68].

Bacillus amyloliquefaciens produces extracellular phytase at 37°C in a medium supplemented with wheat bran and casein hydrolysate by submerged fermentation. Other species also produce high amounts of phytase under specific fermentation conditions. *B. subtilis* can be cultivated at 60°C and pH between 7 and 7.5 to produce high amounts of phytase. Moreover, *B. subtilis* has been used in genetic modifications aimed at the expression of specific phytase genes. Other examples are phytases from *Aspergillus awamori* and *Aspergillus fumigatus* employed by genetic methods modified and evaluated by SDS-PAGE and Western blotting. Thus, these works reveal that the recombinant strain exhibits overexpression, high activity, and thermostability at pH between 4 and 7 at 60°C [69, 70].

Pichia pastoris is a methylotrophic, nonpathogenic yeast and is widely used to produce various heterologous proteins because it has the capacity to perform important and necessary posttranslational modifications of the protein of interest. Transcription of the genes inserted into the vector is driven by the AOX1 promoter mediated by the action of methanol in the medium or in the presence of methanol as a sole carbon source [71, 72]. The phytase gene of 1347 bp from *A. niger* SK-57 can be obtained by PCR and expressed in *P. pastoris* using the AOX1 promoter. Under these conditions, *P. pastoris* produced $6.1 \, \text{g·l}^{-1}$ of purified phytase with an activity of $865 \, \text{U·ml}^{-1}$ and pH optima between 2.5 and 5.5 at 60°C. Thus, *P. pastoris* can be an alternative microorganism for use as a production system for commercial phytase [73]. Another strain of methylotrophic yeast, *Hansenula polymorpha*, can be genetically modified to express phytase effectively under submerged fermentation conditions with the use of the formate dehydrogenase promoter [11].

Moreover, phytase can be obtained from the mushroom *Schizophyllum* sp. Cultivation is usually carried out in solid-state fermentation at 30°C in a culture medium consisting of wheat bran suitable for a few days at pH between 6 and 7 [74]. Gram-negative bacteria of the *Chromobacterium* sp. genus have been isolated from a variety of habitats in Brazilian ecosystems and have already presented phytase activity. *Chromobacterium* sp. usually can be cultivated at 28°C in a specific medium containing phytate. Bacteria of the genera *Escherichia*, *Rahnella*, and *Pseudomonas* also produce phytase [75]. Different bioprocesses for microbial phytase production are described in Table 2.

5. Animal Supplementation and Sustainable Agriculture

In monogastric animals, there is little or no evidence of phytase activity in the digestive tract [4, 5]. In these animals, food is not completely digested, and cellular bioavailability is low. Most of the phytate present in the feed is excreted, increasing the environmental phosphate concentration in areas where livestock grazing is intense [6, 7]. Low-phosphate bioavailability and phosphorus deficiency in animal feed based on dry grain can be supplemented by adding inorganic phosphate. Due to the high market demand for meat, this strategy of phosphorus supplementation is used increasingly in animal feed. However, the practice of adding inorganic phosphorus is unsustainable, because phosphorus is not a renewable resource in nature, and its continued harvesting would result in depletion of this environmental resource in a few decades [6, 9, 67, 76, 77]. Thus, the addition of phytase to animal feed is an interesting and sustainable alternative, due to its ability to increase phosphate bioavailability in diets based on dry grain and to decrease the environmental impacts caused by phosphorus accumulation in the environment [4, 31, 78]. Adding phytase to animal feed also reduces the operational production cost as the inclusion of inorganic phosphate is no longer required. Moreover, phytase also increases the availability of other minerals such as zinc, iron, and manganese, which are important for animal metabolism [7, 67, 79–82]. Animal feed supplementation provides beneficial responses in terms of animal performance such as increased weight of ruminants, broilers, and fishes [82–84]. Other beneficial effects can be achieved by incorporating mixed enzymatic formulations that result in an increase in the coefficients of ileal nutrient digestibility and standardized ileal digestibility by amino acid absorption and energy production [85–88].

Multienzymatic complexes can also provide efficient nutrient absorption. Previous reports described secondary effects associated with the action of multienzymatic complexes added to stimulate hormone synthesis [89, 90]. Bedford and Cowieson [91] analyzed the multifactorial effects of exogenous enzymes on the intestinal microbiota and the improvement of animal performance. Phytase has positive contributions for use in animal feed supplementation; however, it is still necessary to evaluate the specific effects of carbohydrases and proteases on multienzymatic complexes [92–96]. Thus, animal characteristics, such as species or age, are important sources of variation in these studies. However, the main cause is directly associated with characteristics of animal feed in both qualitative and quantitative aspects. The nonstarch portion is diversified and large, which favors increased digestive viscosity and hinders enzymatic action. Studies have reported that this complicates the understanding of the effects of these multienzymatic complexes on animal feeds. Thus, their use and performance can be underestimated mainly due to their variation in the concentrations of enzyme and substrate [97].

Organic acids are other additives that can be supplemented in phytase formulations. Increased animal performance is caused by one or more actions: (1) a decrease in gastric pH favoring pepsin activity in the fed state; (2) immune system improvement due to microbiological selectivity, which reduces Gram-negative bacteria in favor of *Lactobacillus* proliferation in the gastrointestinal tract; and (3) improvement of both mineral and protein digestibility from food [98]. The benefits of organic acids show high interspecific diversity and are largely dependent on the profile of the gastrointestinal tract. In pigs, the gastrointestinal pH is above 6.0, and acidification is beneficial to phytase activity, due to its acidic pH optimum [99]. Thus, enzymatic formulations containing phytase generally include the organic acids, acetic, citric, fumaric, lactic, formic, and propionic acids [98–100], and the following enzymes: xylanases [87, 92–95], cellulase [101], β-glucanase [102], β-mannanase [96, 103], and proteases [104].

Therefore, the addition of supplementary phytase to the feed of monogastric animals shows many advantages, these being associated or not with multienzymatic complexes or with organic acids, providing an adequate supply of phosphorus and other minerals without damaging the environment.

6. Perspectives

Phytase plays an important role by releasing phosphorus and other nutrients from phytate added to monogastric animal feed and decreases the amount of this element in the environment. Its use allows the consumption of phosphorus available in grains used in animal nutrition. The catalytic structure, production, and biotechnological applications of phytase are increasingly evaluated in research to increase the understanding of three-dimensional structures and to optimize bioprocesses with the aim of increasing production and

activity. Plants and different microorganisms can produce phytase, the yields being related to the bioprocess employed. Future perspectives indicate the use of recombinant plants or microorganisms with phytase activity to improve the use of phosphorus and other minerals and to propose a new strategy to improve animal feed. Thus, understanding the catalytic structure of phytase from plants or different microorganisms is an important aspect that allows for sustainable agriculture and attention to the environment.

Acknowledgments

This review was funded by Federal University of Tocantins with financial support of the CNPq.

References

[1] I. Alkorta, C. Garbisu, M. J. Llama, and J. L. Serra, "Industrial applications of pectic enzymes: a review," *Process Biochemistry*, vol. 33, no. 1, pp. 21–28, 1998.

[2] G. S. N. Naidu and T. Panda, "Production of pectolytic enzymes - A review," *Bioprocess Engineering*, vol. 19, no. 5, pp. 355–361, 1998.

[3] L. R. Lynd, P. J. Weimer, W. H. Van Zyl, and I. S. Pretorius, "Microbial cellulose utilization: fundamentals and biotechnology," *Microbiology and Molecular Biology Reviews*, vol. 66, no. 3, pp. 506–577, 2002.

[4] G. Dionisio, C. K. Madsen, P. B. Holm et al., "Cloning and characterization of purple acid phosphatase phytases from wheat, barley, maize, and rice," *Plant Physiology*, vol. 156, no. 3, pp. 1087–1100, 2011.

[5] A. D. Mukhametzianova, A. I. Akhmetova, and M. R. Sharipova, "Microorganisms as phytase producers," *Mikrobiologiia*, vol. 81, no. 3, pp. 291–300, 2012.

[6] B. A. B. Martins, L. M. D. O. Borgatti, L. W. D. O. Souza, S. L. D. A. Robassini, and R. D. Albuquerque, "Bioavailability and poultry fecal excretion of phosphorus from soybeanbased diets supplemented with phytase," *Revista Brasileira de Zootecnia*, vol. 42, no. 3, pp. 174–182, 2013.

[7] E. Humer, C. Schwarz, and K. Schedle, "Phytate in pig and poultry nutrition," *Journal of Animal Physiology and Animal Nutrition*, vol. 99, no. 4, pp. 605–625, 2015.

[8] L. H. S. Guimarães, S. C. Peixoto-Nogueira, M. Michelin et al., "Screening of filamentous fungi for production of enzymes of biotechnological interest," *Brazilian Journal of Microbiology*, vol. 37, no. 4, pp. 474–480, 2006.

[9] G. R. Lelis, L. F. T. Albino, A. A. Calderano et al., "Diet supplementation with phytase on performance of broiler chickens," *Revista Brasileira de Zootecnia*, vol. 41, no. 4, pp. 929–933, 2012.

[10] J. L. Adrio and A. L. Demain, "Microbial enzymes: tools for biotechnological processes," *Biomolecules*, vol. 4, no. 1, pp. 117–139, 2014.

[11] A. F. Mayer, K. Hellmuth, H. Schlieker et al., "An expression system matures: A highly efficient and cost-effective process for phytase production by recombinant strains of Hansenula

polymorpha," *Biotechnology and Bioengineering*, vol. 63, no. 3, pp. 373–381, 1999.

[12] S. Chatterjee, R. Sankaranarayanan, and R. V. Sonti, "PhyA, a Secreted Protein of Xanthomonas oryzae pv. oryzae, Is Required for Optimum Virulence and Growth on Phytic Acid as a Sole Phosphate Source," *Molecular Plant-Microbe Interactions*, vol. 16, no. 11, pp. 973–982, 2003.

[13] J. Schneider, O. Rupp, E. Trost et al., "Genome sequence of Wickerhamomyces anomalus DSM 6766 reveals genetic basis of biotechnologically important antimicrobial activities," *FEMS Yeast Research*, vol. 12, no. 3, pp. 382–386, 2012.

[14] N. Belgaroui, P. Berthomieu, H. Rouached, and M. Hanin, "The secretion of the bacterial phytase PHY-US417 by Arabidopsis roots reveals its potential for increasing phosphate acquisition and biomass production during co-growth," *Plant Biotechnology Journal*, vol. 14, no. 9, pp. 1914–1924, 2016.

[15] A. Durmus, C. Eicken, B. H. Sift et al., "The active site of purple acid phosphatase from sweet potatoes (Ipomoea batatas). Metal content and spectroscopic characterization," *European Journal of Biochemistry*, vol. 260, no. 3, pp. 709–716, 1999.

[16] G. G. Bozzo, K. G. Raghothama, and W. C. Plaxton, "Purification and characterization of two secreted purple acid phosphatase isozymes from phosphate-starved tomato (Lycopersicon esculentum) cell cultures," *European Journal of Biochemistry*, vol. 269, no. 24, pp. 6278–6286, 2002.

[17] S. V. Antonyuk, M. Olczak, T. Olczak, J. Ciuraszkiewicz, and R. W. Strange, "The structure of a purple acid phosphatase involved in plant growth and pathogen defence exhibits a novel immunoglobulin-like fold," *Biology and Medicine*, vol. 1, pp. 101–109, 2014.

[18] R. K. Gupta, S. S. Gangoliya, and N. K. Singh, "Reduction of phytic acid and enhancement of bioavailable micronutrients in food grains," *Journal of Food Science and Technology*, vol. 52, no. 2, pp. 676–684, 2015.

[19] L. Bohn, A. S. Meyer, and S. K. Rasmussen, "Phytate: impact on environment and human nutrition. A challenge for molecular breeding," *Journal of Zhejiang University SCIENCE B*, vol. 9, no. 3, pp. 165–191, 2008.

[20] E. J. Mullaney and A. H. J. Ullah, "The term phytase comprises several different classes of enzymes," *Biochemical and Biophysical Research Communications*, vol. 312, no. 1, pp. 179–184, 2003.

[21] J. Gerke, "Phytate (Inositol Hexakisphosphate) in Soil and Phosphate Acquisition from Inositol Phosphates by Higher Plants. A Review," *Plants*, vol. 4, no. 4, pp. 253–266, 2015.

[22] A. Baker, S. A. Ceasar, A. J. Palmer et al., "Replace, reuse, recycle: Improving the sustainable use of phosphorus by plants," *Journal of Experimental Botany*, vol. 66, no. 12, pp. 3523–3540, 2015.

[23] D. P. Schachtman, R. J. Reid, and S. M. Ayling, "Phosphorus uptake by plants: from soil to cell," *Plant Physiology*, vol. 116, no. 2, pp. 447–453, 1998.

[24] W.-R. Scheible and M. Rojas-Triana, "Sensing, signalling, and control of phosphate starvation in plants: molecular players and applications," *Phosphorus Metabolism in Plants*, vol. 48, pp. 25–64, 2015.

[25] C. P. Vance, C. Uhde-Stone, and D. L. Allan, "Phosphorus acquisition and use: critical adaptations by plants for securing a nonrenewable resource," *New Phytologist*, vol. 157, no. 3, pp. 423–447, 2003.

[26] K. G. Raghothama, "Phosphate acquisition," *Annual Review of Plant Biology*, vol. 50, pp. 665–693, 1999.

[27] G. A. Annor, K. Tano Debrah, and A. Essen, "Mineral and phytate contents of some prepared popular Ghanaian foods," *SpringerPlus*, vol. 5, no. 1, article no. 581, pp. 1–8, 2016.

[28] V. A. McKie and B. V. McCleary, "A novel and rapid colorimetric method for measuring total phosphorus and phytic acid in foods and animal feeds," *Journal of AOAC International*, vol. 99, no. 3, pp. 738–743, 2016.

[29] C. E. Hegeman and E. A. Grabau, "A novel phytase with sequence similarity to purple acid phosphatases is expressed in cotyledons of germinating soybean seedlings," *Plant Physiology*, vol. 126, no. 4, pp. 1598–1608, 2001.

[30] L. H. S. Guimarães, H. F. Terenzi, J. A. Jorge, F. A. Leone, and M. D. L. T. M. Polizeli, "Characterization and properties of acid phosphatases with phytase activity produced by Aspergillus caespitosus," *Biotechnology and Applied Biochemistry*, vol. 40, no. 2, pp. 201–207, 2004.

[31] H. Brinch-Pedersen, L. D. Sørensen, and P. B. Holm, "Engineering crop plants: Getting a handle on phosphate," *Trends in Plant Science*, vol. 7, no. 3, pp. 118–125, 2002.

[32] M.-H. Lim, O.-H. Lee, J.-E. Chin et al., "Simultaneous degradation of phytic acid and starch by an industrial strain of Saccharomyces cerevisiae producing phytase and α-amylase," *Biotechnology Letters*, vol. 30, no. 12, pp. 2125–2130, 2008.

[33] G. R. Lelis, L. F. T. Albino, F. C. Tavernari, and S. Rostagno, "Suplementação dietética de fitase em dietas para frangos de corte," *Revista Eletrônica Nutritime*, vol. 6, no. 2, pp. 875–889, 2009.

[34] V. Kumar, G. Singh, A. K. Verma, and S. Agrawal, "In silico characterization of histidine acid phytase sequences," *Enzyme Research*, vol. 2012, Article ID 845465, 2012.

[35] C. Li, Y. Lin, X. Zheng et al., "Combined strategies for improving expression of Citrobacter amalonaticus phytase in Pichia pastoris," *BMC Biotechnology*, vol. 15, no. 1, article no. 88, 2015.

[36] J. Dvořáková, "Phytase: Sources, Preparation and Exploitation," *Folia Microbiologica*, vol. 43, no. 4, pp. 323–338, 1998.

[37] T.-K. Oh, S. Oh, S. Kim et al., "Expression of Aspergillus nidulans phy gene in Nicotiana benthamiana produces active phytase with broad specificities," *International Journal of Molecular Sciences*, vol. 15, no. 9, pp. 15571–15591, 2014.

[38] E. R. Graminho, N. Takaya, A. Nakamura, and T. Hoshino, "Purification, biochemical characterization, and genetic cloning of the phytase produced by Burkholderia sp. Strain a13," *The Journal of General and Applied Microbiology*, vol. 61, no. 1, pp. 15–23, 2015.

[39] T. Klabunde, N. Sträter, B. Krebs, and H. Witzel, "Structural relationship between the mammalian Fe(III)Fe(II) and the Fe(III)Zn(II) plant purple acid phosphatases," *FEBS Letters*, vol. 367, no. 1, pp. 56–60, 1995.

[40] A. K. Boudalis, R. E. Aston, S. J. Smith et al., "Synthesis and characterization of the tetranuclear iron(iii) complex of a new asymmetric multidentate ligand. A structural model for purple acid phosphatases," *Dalton Transactions*, no. 44, pp. 5132–5139, 2007.

[41] G. Schenk, T. W. Elliott, E. Leung et al., "Crystal structures of a purple acid phosphatase, representing different steps of this enzyme's catalytic cycle," *BMC Structural Biology*, vol. 8, article no. 6, 2008.

[42] R. Kuang, K.-H. Chan, E. Yeung, and B. L. Lim, "Molecular and biochemical characterization of AtPAP15, a purple acid phosphatase with phytase activity, in Arabidopsis," *Plant Physiology*, vol. 151, no. 1, pp. 199–209, 2009.

[43] V. Kumar, G. Singh, P. Sangwan, A. K. Verma, and S. Agrawal, "Cloning, Sequencing, and in silico analysis of β-Propeller Phytase," *Biotechnology Research International*, vol. 2014, Article ID 841353, 11 pages, 2014.

[44] V. Kumar and S. Agrawal, "Short Communication: An insight into protein sequences of PTP-like cysteine phytases," *Nusantara Bioscience*, vol. 6, no. 1, pp. 102–106, 2014.

[45] D. Fu, H. Huang, K. Meng et al., "Improvement of Yersinia frederiksenii phytase performance by a single amino acid substitution," *Biotechnology and Bioengineering*, vol. 103, no. 5, pp. 857–864, 2009.

[46] J. M. Viader-Salvadó, J. A. Gallegos-López, J. G. Carreón-Treviño, M. Castillo-Galván, A. Rojo-Domínguez, and M. Guerrero-Olazarán, "Design of thermostable beta-propeller phytases with activity over a broad range of pHs and their overproduction by pichia pastoris," *Applied and Environmental Microbiology*, vol. 76, no. 19, pp. 6423–6430, 2010.

[47] D. Fu, Z. Li, H. Huang et al., "Catalytic efficiency of HAP phytases is determined by a key residue in close proximity to the active site," *Applied Microbiology and Biotechnology*, vol. 90, no. 4, pp. 1295–1302, 2011.

[48] K. Kumar, M. Dixit, J. M. Khire, and S. Pal, "Atomistic details of effect of disulfide bond reduction on active site of Phytase B from Aspergillus niger: A MD Study," *Bioinformation*, vol. 9, no. 19, pp. 963–967, 2013.

[49] A. J. Oakley, "The structure of *Aspergillus niger* phytase PhyA in complex with a phytate mimetic," *Biochemical and Biophysical Research Communications*, vol. 397, no. 4, pp. 745–749, 2010.

[50] I. G. Mishra, K. B. Bhurat, N. Tripathi, K. Tantwai, and S. Tiwari, "Molecular cloning and characterization of phytase gene from *Aspergillus niger*," in *GenBank AFJ79736.1*, pp. 1–490, 2012, http://www.ncbi.nlm.nih.gov/protein/AFJ79736.

[51] K. Böhm, T. Herter, J. J. Müller, R. Borriss, and U. Heinemann, "Crystal structure of Klebsiella sp. ASR1 phytase suggests substrate binding to a preformed active site that meets the requirements of a plant rhizosphere enzyme," *FEBS Journal*, vol. 277, no. 5, pp. 1284–1296, 2010.

[52] A. V. Shivange, A. Serwe, A. Dennig, D. Roccatano, S. Haefner, and U. Schwaneberg, "Directed evolution of a highly active Yersinia mollaretii phytase," *Applied Microbiology and Biotechnology*, vol. 95, no. 2, pp. 405–418, 2012.

[53] K. E. Wilkins, N. J. Booher, L. Wang, and A. J. Bogdanove, "TAL effectors and activation of predicted host targets distinguish Asian from African strains of the rice pathogen Xanthomonas oryzae pv. oryzicola while strict conservation suggests universal importance of five TAL effectors," *Frontiers in Plant Science*, vol. 6, no. JULY, article no. 536, pp. 1–15, 2015.

[54] P. Kaur, B. Singh, E. Böer et al., "Pphy-A cell-bound phytase from the yeast Pichia anomala: Molecular cloning of the gene PPHY and characterization of the recombinant enzyme," *Journal of Biotechnology*, vol. 149, no. 1-2, pp. 8–15, 2010.

[55] S. Joshi and T. Satyanarayana, "Characteristics and applicability of phytase of the yeast Pichia anomala in synthesizing haloperoxidase," *Applied Biochemistry and Biotechnology*, vol. 176, no. 5, pp. 1351–1369, 2015.

[56] A. K. Maurya, D. Parashar, and T. Satyanarayana, "Bioprocess for the production of recombinant HAP phytase of the thermophilic mold Sporotrichum thermophile and its structural and biochemical characteristics," *International Journal of Biological Macromolecules*, vol. 94, pp. 36–44, 2017.

[57] N. Bertrand, H. Ménager, C. Maufrais et al., "Bioinformatics advance acess," *Bioinformatics*, vol. 25, pp. 3005–3011, 2009.

[58] S.-C. Lung, W.-L. Chan, W. Yip, L. Wang, E. C. Yeung, and B. L. Lim, "Secretion of beta-propeller phytase from tobacco and Arabidopsis roots enhances phosphorus utilization," *Journal of Plant Sciences*, vol. 169, no. 2, pp. 341–349, 2005.

[59] Y. Wang, X. Ye, G. Ding, and F. Xu, "Overexpression of phyA and appA Genes Improves Soil Organic Phosphorus Utilisation and Seed Phytase Activity in Brassica napus," *PLoS ONE*, vol. 8, no. 4, Article ID e60801, 2013.

[60] D. M. Denbow, E. A. Grabau, G. H. Lacy, E. T. Kornegay, D. R. Russell, and P. F. Umbeck, "Soybeans Transformed with a Fungal Phytase Gene Improve Phosphorus Availability for Broilers," *Poultry Science*, vol. 77, no. 6, pp. 878–881, 1998.

[61] Z. B. Zhang, E. T. Kornegay, J. S. Radcliffe, J. H. Wilson, and H. P. Veit, "Comparison of phytase from genetically engineered Aspergillus and canola in weanling pig diets," *Journal of Animal Science*, vol. 78, no. 11, pp. 2868–2878, 2000.

[62] P. Singh, M. Punjabi, M. Jolly, R. D. Rai, and A. Sachdev, "Characterization and expression of codon optimized soybean phytase gene in E. coli," *Indian Journal of Biochemistry and Biophysics*, vol. 50, no. 6, pp. 537–547, 2013.

[63] F. Erpel, F. Restovic, and P. Arce-Johnson, "Development of phytase-expressing chlamydomonas reinhardtii for monogastric animal nutrition," *BMC Biotechnology*, vol. 16, no. 1, article no. 29, 2016.

[64] N. Abid, A. Khatoon, A. Maqbool et al., "Transgenic expression of phytase in wheat endosperm increases bioavailability of iron and zinc in grains," *Transgenic Research*, vol. 26, no. 1, pp. 109–122, 2017.

[65] C. O. Ibrahim, "Development of applications of industrial enzymes from Malaysian indigenous microbial sources," *Bioresource Technology*, vol. 99, no. 11, pp. 4572–4582, 2008.

[66] G. B. Shivanna and G. Venkateswaran, "Phytase production by Aspergillus niger CFR 335 and Aspergillus ficuum SGA 01 through submerged and solid-state fermentation," *The Scientific World Journal*, vol. 2014, Article ID 392615, 2014.

[67] S. Haefner, A. Knietsch, E. Scholten, J. Braun, M. Lohscheidt, and O. Zelder, "Biotechnological production and applications of phytases," *Applied Microbiology and Biotechnology*, vol. 68, no. 5, pp. 588–597, 2005.

[68] R. J. Wodzinski and A. H. J. Ullah, "Phytase," *Advances in Applied Microbiology*, no. 42, pp. 263–302, 1996.

[69] S. Fu, J. Sun, L. Qian, and Z. Li, "Bacillus phytases: Present scenario and future perspectives," *Applied Biochemistry and Biotechnology*, vol. 151, no. 1, pp. 1–8, 2008.

[70] J. A. Martin, R. A. Murphy, and R. F. G. Power, "Purification and physico-chemical characterisation of genetically modified phytases expressed in Aspergillus awamori," *Bioresource Technology*, vol. 97, no. 14, pp. 1703–1708, 2006.

[71] J. L. Cereghino and J. M. Cregg, "Heterologous protein expression in the methylotrophic yeast *Pichia pastoris*," *FEMS Microbiology Reviews*, vol. 24, no. 1, pp. 45–66, 2000.

[72] E. C. Coimbra, F. B. Gomes, J. F. Campos et al., "Production of L1 protein from different types of HPV in *Pichia pastoris* using an integrative vector," *Brazilian Journal of Medical and Biological Research*, vol. 44, no. 12, pp. 1209–1214, 2011.

[73] A.-S. Xiong, Q.-H. Yao, R.-H. Peng, P.-L. Han, Z.-M. Cheng, and Y. Li, "High level expression of a recombinant acid phytase gene in Pichia pastoris," *Journal of Applied Microbiology*, vol. 98, no. 2, pp. 418–428, 2005.

[74] D. N. Salmon, L. C. Piva, R. L. Binati et al., "Formulated products containing a new phytase from Schyzophyllum sp.

phytase for application in feed and food processing," *Brazilian Archives of Biology and Technology*, vol. 54, no. 6, pp. 1069–1074, 2011.

[75] P. S. Costa, A. M. A. Nascimento, C. I. Lima-Bittencourt, E. Chartone-Souza, F. R. Santos, and A. Vilas-Boas, "Chromobacterium sp. from the tropics: Detection and diversity of phytase activity," *Brazilian Journal of Microbiology*, vol. 42, no. 1, pp. 84–88, 2011.

[76] I. Steen, "Phosphorus availability in the 21st century: management of a non-renewable resource," *Phosphorus and Potassium*, vol. 217, pp. 25–31, 1998.

[77] K. R. Jegannathan and P. H. Nielsen, "Environmental assessment of enzyme use in industrial production-a literature review," *Journal of Cleaner Production*, vol. 42, pp. 228–240, 2013.

[78] H. Tan, M. J. Mooij, M. Barret et al., "Identification of novel phytase genes from an agricultural soil-derived metagenome," *Journal of Microbiology and Biotechnology*, vol. 24, no. 1, pp. 113–118, 2014.

[79] K. Bhavsar, V. Ravi Kumar, and J. M. Khire, "High level phytase production by Aspergillus niger NCIM 563 in solid state culture: Response surface optimization, up-scaling, and its partial characterization," *Journal of Industrial Microbiology and Biotechnology*, vol. 38, no. 9, pp. 1407–1417, 2011.

[80] U. Konietzny and R. Greiner, "Bacterial phytase: Potential application, in vivo function and regulation of its synthesis," *Brazilian Journal of Microbiology*, vol. 35, no. 1-2, pp. 11–18, 2004.

[81] E. Casartelli, O. Junqueira, A. Laurentiz, R. Filardi, J. Lucas Júnior, and L. Araujo, "Effect of phytase in laying hen diets with different phosphorus sources," *Revista Brasileira de Ciência Avícola*, vol. 7, no. 2, pp. 93–98, 2005.

[82] B. Troesch, H. Jing, A. Laillou, and A. Fowler, "Absorption studies show that phytase from Aspergillus niger significantly increases iron and zinc bioavailability from phytate-rich foods.," *Food and Nutrition Bulletin*, vol. 34, no. 2, pp. S90–101, 2013.

[83] P. H. Selle and V. Ravindran, "Microbial phytase in poultry nutrition," *Animal Feed Science and Technology*, vol. 135, no. 1-2, pp. 1–41, 2007.

[84] L. Cao, W. Wang, C. Yang et al., "Application of microbial phytase in fish feed," *Enzyme and Microbial Technology*, vol. 40, no. 4, pp. 497–507, 2007.

[85] J. Madrid, S. Martínez, C. López, and F. Hernández, "Effect of phytase on nutrient digestibility, mineral utilization and performance in growing pigs," *Livestock Science*, vol. 154, no. 1-3, pp. 144–151, 2013.

[86] M. R. Bedford, T. A. Scott, F. G. Silversides, H. L. Classen, M. L. Swift, and M. Pack, "The effect of wheat cultivar, growing environment, and enzyme supplementation on digestibility of amino acids by broilers," *Canadian Journal of Animal Science*, vol. 78, no. 3, pp. 335–342, 1998.

[87] P. H. Selle, V. Ravindran, and G. G. Partridge, "Beneficial effects of xylanase and/or phytase inclusions on ileal amino acid digestibility, energy utilisation, mineral retention and growth performance in wheat-based broiler diets," *Animal Feed Science and Technology*, vol. 153, no. 3-4, pp. 303–313, 2009.

[88] E. Kiarie, A. Owusu-Asiedu, P. H. Simmins, and C. M. Nyachoti, "Influence of phytase and carbohydrase enzymes on apparent ileal nutrient and standardized ileal amino acid digestibility in growing pigs fed wheat and barley-based diets," *Livestock Science*, vol. 134, no. 1-3, pp. 85–87, 2010.

[89] J. Singh and P. Kaur, "Optimization of process parameters for cellulase production from Bacillus sp. JS14 in solid substrate fermentation using response surface methodology," *Brazilian Archives of Biology and Technology*, vol. 55, no. 4, pp. 505–512, 2012.

[90] E. G. Jos, C. E. Talita, E. S. Alana et al., "Production, characterization and evaluation of in vitro digestion of phytases, xylanases and cellulases for feed industry," *African Journal of Microbiology Research*, vol. 8, no. 6, pp. 551–558, 2014.

[91] M. R. Bedford and A. J. Cowieson, "Exogenous enzymes and their effects on intestinal microbiology," *Animal Feed Science and Technology*, vol. 173, no. 1-2, pp. 76–85, 2012.

[92] J. E. Lindberg, K. Lyberg, and J. Sands, "Influence of phytase and xylanase supplementation of a wheat-based diet on ileal and total tract digestibility in growing pigs," *Livestock Science*, vol. 109, no. 1-3, pp. 268–270, 2007.

[93] J. C. Kim, J. S. Sands, B. P. Mullan, and J. R. Pluske, "Performance and total-tract digestibility responses to exogenous xylanase and phytase in diets for growing pigs," *Animal Feed Science and Technology*, vol. 142, no. 1-2, pp. 163–172, 2008.

[94] T. A. Woyengo, J. S. Sands, W. Guenter, and C. M. Nyachoti, "Nutrient digestibility and performance responses of growing pigs fed phytase- and xylanase-supplemented wheat-based diets," *Journal of Animal Science*, vol. 86, no. 4, pp. 848–857, 2008.

[95] J. K. A. Atakora, S. Moehn, J. S. Sands, and R. O. Ball, "Effects of dietary crude protein and phytase-xylanase supplementation of wheat grain based diets on energy metabolism and enteric methane in growing finishing pigs," *Animal Feed Science and Technology*, vol. 166-167, pp. 422–429, 2011.

[96] C. H. Mok, J. H. Lee, and B. G. Kim, "Effects of exogenous phytase and β-mannanase on ileal and total tract digestibility of energy and nutrient in palm kernel expeller-containing diets fed to growing pigs," *Animal Feed Science and Technology*, vol. 186, no. 3-4, pp. 209–213, 2013.

[97] V. Ravindran, "Feed enzymes: The science, practice, and metabolic realities," *Journal of Applied Poultry Research*, vol. 22, no. 3, pp. 628–636, 2013.

[98] N. Khodambashi Emami, S. Zafari Naeini, and C. A. Ruiz-Feria, "Growth performance, digestibility, immune response and intestinal morphology of male broilers fed phosphorus deficient diets supplemented with microbial phytase and organic acids," *Livestock Science*, vol. 157, no. 2-3, pp. 506–513, 2013.

[99] A. W. Jongbloed, Z. Mroz, R. Van Der Weij-Jongbloed, and P. A. Kemme, "The effects of microbial phytase, organic acids and their interaction in diets for growing pigs," *Livestock Production Science*, vol. 67, no. 1-2, pp. 113–122, 2000.

[100] Z. Valencia and E. R. Chavez, "Phytase and acetic acid supplementation in the diet of early weaned piglets: Effect on performance and apparent nutrient digestibility," *Nutrition Research*, vol. 22, no. 5, pp. 623–632, 2002.

[101] K. F. Knowlton, M. S. Taylor, S. R. Hill, C. Cobb, and K. F. Wilsont, "Manure nutrient excretion by lactating cows fed exogenous phytase and cellulase," *Journal of Dairy Science*, vol. 90, no. 9, pp. 4356–4360, 2007.

[102] E. Kiarie, A. Owusu-Asiedu, A. Péron, P. H. Simmins, and C. M. Nyachoti, "Efficacy of xylanase and B-glucanase blend in mixed grains and grain co-products-based diets for fattening pigs," *Livestock Science*, vol. 148, no. 1-2, pp. 129–133, 2012.

[103] J. H. Cho and I. H. Kim, "Effects of beta-mannanase supplementation in combination with low and high energy dense diets for growing and finishing broilers," *Livestock Science*, vol. 154, no. 1-3, pp. 137–143, 2013.

[104] M. R. Barekatain, C. Antipatis, M. Choct, and P. A. Iji, "Interaction between protease and xylanase in broiler chicken diets containing sorghum distillers' dried grains with solubles," *Animal Feed Science and Technology*, vol. 182, no. 1-4, pp. 71–81, 2013.

[105] "I-Tasser online protein structure and function predictions," http://zhanglab.ccmb.med.umich.edu/I-TASSER/.

[106] I. B. Holme, G. Dionisio, C. K. Madsen, and H. Brinch-Pedersen, "Barley HvPAPhy_a as transgene provides high and stable phytase activities in mature barley straw and in grains," *Plant Biotechnology Journal*, vol. 15, no. 4, pp. 415–422, 2017.

[107] S. Mohsin, A. Maqbool, M. Ashraf, and K. A. Malik, "Extracellular Secretion of Phytase from Transgenic Wheat Roots Allows Utilization of Phytate for Enhanced Phosphorus Uptake," *Molecular Biotechnology*, vol. 59, no. 8, pp. 334–342, 2017.

[108] H. Yang, X. Zhang, R. A. Gaxiola, G. Xu, W. A. Peer, and A. S. Murphy, "Over-expression of the Arabidopsis proton-pyrophosphatase AVP1 enhances transplant survival, root mass, and fruit development under limiting phosphorus conditions," *Journal of Experimental Botany*, vol. 65, no. 12, pp. 3045–3053, 2014.

[109] X.-F. Ma, S. Tudor, T. Butler et al., "Transgenic expression of phytase and acid phosphatase genes in alfalfa *(Medicago sativa)* leads to improved phosphate uptake in natural soils," *Molecular Breeding*, vol. 30, no. 1, pp. 377–391, 2012.

18

"In Silico" Characterization of 3-Phytase A and 3-Phytase B from *Aspergillus niger*

Doris C. Niño-Gómez,[1] **Claudia M. Rivera-Hoyos,**[1] **Edwin D. Morales-Álvarez,**[2]
Edgar A. Reyes-Montaño,[3] **Nury E. Vargas-Alejo,**[3] **Ingrid N. Ramírez-Casallas,**[1]
Kübra Erkan Türkmen,[4] **Homero Sáenz-Suárez,**[5] **José A. Sáenz-Moreno,**[5]
Raúl A. Poutou-Piñales,[1] **Janneth González-Santos,**[6] **and Azucena Arévalo-Galvis**[7]

[1]*Laboratorio de Biotecnología Molecular, Grupo de Biotecnología Ambiental e Industrial (GBAI),*
 Departamento de Microbiología, Facultad de Ciencias, Pontificia Universidad Javeriana, Bogotá, Colombia
[2]*Departamento de Química, Facultad de Ciencias Exactas y Naturales, Universidad de Caldas, Manizales, Caldas, Colombia*
[3]*Grupo de Investigación en Proteínas, Departamento de Química, Facultad de Ciencias,*
 Universidad Nacional de Colombia (UNAL), Bogotá, Colombia
[4]*Department of Biology, Faculty of Science, Hacettepe University, Beytepe, Ankara, Turkey*
[5]*Unidad de Biología Celular y Microscopía, Decanato de Ciencias de la Salud, Universidad Centroccidental Lisandro Alvarado,*
 Barquisimeto, Venezuela
[6]*Grupo de Bioquímica Computacional y Estructural, Departamento de Bioquímica y Nutrición,*
 Facultad de Ciencias, Pontificia Universidad Javeriana, Bogotá, Colombia
[7]*Laboratorio de Microbiología Especial, Grupo de Enfermedades Infecciosas, Departamento de Microbiología,*
 Facultad de Ciencias, Pontificia Universidad Javeriana, Bogotá, Colombia

Correspondence should be addressed to Edgar A. Reyes-Montaño; eareyesm@unal.edu.co
and Raúl A. Poutou-Piñales; rpoutou@javeriana.edu.co

Academic Editor: Sunney I. Chan

Phytases are used for feeding monogastric animals, because they hydrolyze phytic acid generating inorganic phosphate. *Aspergillus niger* 3-phytase A (PDB: 3K4Q) and 3-phytase B (PDB: 1QFX) were characterized using bioinformatic tools. Results showed that both enzymes have highly conserved catalytic pockets, supporting their classification as histidine acid phosphatases. 2D structures consist of 43% alpha-helix, 12% beta-sheet, and 45% others and 38% alpha-helix, 12% beta-sheet, and 50% others, respectively, and pI 4.94 and 4.60, aliphatic index 72.25 and 70.26 and average hydrophobicity of −0,304 and −0.330, respectively, suggesting aqueous media interaction. Glycosylation and glycation sites allowed detecting zones that can affect folding and biological activity, suggesting fragmentation. Docking showed that H_{59} and H_{63} act as nucleophiles and that D_{339} and D_{319} are proton donor residues. MW of 3K4Q (48.84 kDa) and 1QFX (50.78 kDa) is similar; 1QFX forms homodimers which will originate homotetramers with several catalytic center accessible to the ligand. 3K4Q is less stable (instability index 45.41) than 1QFX (instability index 33.66), but the estimated lifespan for 3K4Q is superior. Van der Waals interactions generate hydrogen bonds between the active center and O_2 or H of the phytic acid phosphate groups, providing greater stability to these temporal molecular interactions.

1. Introduction

Most of the phosphorus (P) present in terrestrial ecosystems is located in the soil. Globally, the terrestrial biota contains 2.6×10^6 g P, which is less than that contained in the soil, which oscillates between 96 and 160×10^6 g P [1]. The highest transference of P from soil to biota occurs through the synthesis of organic compounds containing phosphorus (P) in plants, animals, and microorganisms. The organic compounds containing P are diverse, and their mineralization in the soil allows the P to be recycled back to the biota [1].

Phosphorus is an essential nutrient, which is involved in several biological functions such as regulation of intra- and extracellular pH, accumulation of energy in the form of ATP, lipid transport, and formation of biological membranes [2, 3]. Several compounds with organic phosphorus (oP) have different rates of mineralization. For example, oP from microorganisms (predominantly nucleic acids, 30–50% P in RNA and 5–10% P in DNA) and phospholipids (<10% P) is easily mineralized in soil environments [1]. However, other compounds with oP are not easily mineralized and can accumulate in the soil in substantial amounts. The most significant of these compounds is phytic acid (myo-inositol 1, 2, 3, 4, 5, 6 hexakisphosphate), [1].

Phytic acid is the main form of P storage in cereals, pulses, oilseeds, and nuts and constitutes 1–5% of its dry weight. In forage, one-third of the phosphorus is present as digestible inorganic phosphorus (iP), while the remaining two-thirds are present as oP in the form of phytates [4]. Phytates are a mixture of salts resulting from the union of phytic acid with divalent metal ions such as: Calcium (Ca^{2+}), Copper (Cu^{2+}), Iron (Fe^{2+}), Magnesium (Mg^{2+}), Manganese (Mn^{2+}), and Zinc (Zn^{2+}). Phytic acid can be bound to two different metals such as Calcium (Ca^{2+}) and Magnesium (Mg^{2+}), the resulting mixed salt is called phytin [4].

Phytate constitutes 65–80% of total P in grains and up to 80% of total P in manures of monogastric animals. Due to its negative charge, phytate is strongly adsorbed to various soil components once it is released from plant residues or manure [1].

On the other hand, the accumulation of phytate in the soil is due to the low possibility of being hydrolyzed by the phytase enzymes (E.C. 3.1.3.8), since the phytate dephosphorylation requires the binding of free phytate to the binding pocket of the substrate in the phytase enzyme. Thus, if phytate is tightly bound with soil components, it is not susceptible to be hydrolyzed by enzymes [1].

P from phytate is largely unavailable for monogastric animals, such as pigs and birds, due to the absence or the insufficient amount of phytase enzymes in the gastrointestinal tract to degrade it [5]; in this way it passes without being digested through the gastrointestinal tract. Since phytic acid can not be reabsorbed, feed for pigs and poultry is commonly supplemented with iP in order to meet the requirement of P, which increases production costs [6].

Supplementation with iP, along with the P from phytate excreted by monogastric animals, generates global ecological problems (eutrophication) as the discharge into rivers of wastewater with a high content of phytates results in the proliferation of cyanobacteria, hypoxia, and death of animals from aquatic environments [5]. The P present in phytate that is excreted in the manure of monogastric animals subsequently extends to farmlands, which often contributes to the eutrophication of surface waters, particularly in the areas of intensive livestock of pigs [7].

However, the adverse environmental and nutritional consequences of the presence of phytate in the diet of monogastric animals can be improved by the inclusion of phytases (E.C. 3.1.3.8) in their diet [5]. These enzymes are considered as an environmentally friendly product because (i) they reduce the amount of phosphorus entering the ecosystem, (ii) they reduce the problems caused by eutrophication of water, and (iii) they reduce the constant chelation or sequestration of nutritional factors in the soil, as well as in the digestive tract of poultry and pigs [8]. Phytases are produced by a wide variety of plants, bacteria, fungi, and yeasts. A commercial pair of phytases from the genus Aspergillus (Natuphos® and Ronozyme®) are currently available, as these filamentous fungi are the most prolific extracellular producers of this enzyme [7].

Some studies have shown that microbial sources are more promising for commercial phytase production. Although several strains of bacteria, yeasts, and fungi have been used for production under different conditions, two species, *A. niger* and *A. ficuum*, have been used more frequently for commercial phytase production [9]. Among the best known commercial phytases is found "Natu-phos" (Gist-Brocades NV Company, Netherlands). Natu-phos is a recombinant phytase produced by the expression of the *phy*A gene of *A. ficuum* NRRL 3135 in *A. niger* CBS 513.88, produced in 1994 [4, 7, 9].

In countries like Colombia and Venezuela, there is no legislation regulating the incorporation of phytase enzymes into the feed of monogastric animals, aimed at improving the bioavailability of phosphorus from the diet itself and at the reduction of the amount of phytate excreted in the feces. Therefore, the "in silico" analysis of physicochemical and structural properties, as well as the molecular docking analysis between the enzymes and the ligand, will allow researchers to gather information that is useful for the heterologous expression of the recombinant enzymes.

2. Materials and Methods

2.1. Protein Analysis. The phytases reported until September 13, 2015, were analyzed in the UNIPROT database (The UniProt Consortium) [10]. The PSI-Blast alignment [11] was performed between the amino acid sequences of the phytase reported for *A. niger*, which allowed determining its percentage of similarity and a multiple alignment with the ClustalO programs to identify conserved sites among the selected phytases. The ClustalW alignment allowed comparing the sequences of the two revised phytases: the 3-phytase A and the chains A and B of the 3-phytase B (http://www.ebi.ac.uk) [12].

2.2. Bioinformatic Analysis of the Reported and Revised Phytase from A. Niger. For this analysis, two protein structures resolved by X-ray crystallography (revised proteins) were used for *A. niger*: 3-phytase A and 3-phytase B. The primary sequences of the revised phytases were obtained from UniProtKB, Entry: P34755 and P34752, respectively, while tertiary structures were obtained from Protein Data Bank (PDB) [13], using the ID: 3K4Q and 1QFX, respectively.

2.3. Physicochemical Properties. The physicochemical properties of the amino acid sequences of the revised *A. niger* proteins were evaluated using the following programs: Prot-Param and ProtScale [14] from ExPASy (http://www.expasy.org). The size of the window for the analyses with ProtScale was the basic nine amino acids recommended by the programs to ensure optimum coverage of the sequence when the path is made over it. In the case of the hydrophobicity profile, the Kyte and Doolittle algorithm was used, whose scale considers values between −4.5 and 4.5. The 3D structures of phytases from *A. niger* were visualized using the PyMOL (The PyMOL Molecular Graphics System, Version 1.8 Schrödinger, LLC) program.

2.4. Prediction of N-Glycosylation Sites. For the analysis of potential N-glycosylation sites in both phytases, the NetNG-lyc 1.0 software [15] (http://www.cbs.dtu.dk) was used.

2.5. Prediction of Glycation Sites. For the analysis of the potential glycation sites in both phytases, Netglycate 1.0 software was used [16]. In addition, visualization of the 3D structures and determination of the distances between the ε-NH$_2$ groups of the lysines and the side chains of Glutamate acid residues (E) and Aspartate (D) or basic residues of histidine (H), Arginine (R), and lysine (K) were performed using the SPDB-viewer 4.01 program [17, 18].

2.6. Prediction of Antigenic Peptides. Prediction of antigenic peptides in both proteins was performed by means of the antigenic peptide prediction tool (http://imed.med.ucm.es/Tools/antigenic.pl) [19] of the Complutense University of Madrid.

2.7. Ligand and Molecular Coupling Models (Rigid Docking). The construction of the ligand (phytic acid) was performed using the Spartan version 4.0 molecular modeling program (https://www.wavefun.com/products/spartan.html), which has a graphical interface that allows the construction of the molecule atom by atom, selecting the appropriate hybridization according to the binding site of each element, followed by a minimization of the energy of the generated model.

The molecular coupling models of the reviewed phytases from *A. niger* against the phytic acid as a ligand were performed using the Autodock program [20] and the 3D structure of the ligand (phytic acid) was obtained with the Spartan program (before performing simulations of coupling with Autodock 4.2) [20]. The pocket containing the amino acids that form the highly conserved catalytic center in the revised phytase (histidine acid phosphatases) from *A. niger* (RHGXRXP-D) was identified from reports in the literature [21, 22]. Polar hydrogens were added to each receptor; the grid was located in the pocket of the active site for each model. Phytic acid boxes had the following dimensions and coordinates respectively: for 3-phytase A: X36, Y36 and Z38 (*x*: −6.396, *y*: 8.301 and *z*: 27.885), and for 3-phytase B: X36, Y34 and Z38 (*x*: 23.968, *y*: 71.06 and *z*: 69.576), with a space in both proteins of 0.375 Å. The network parameters and atomic affinity maps were calculated using AutoGrid 4 [20]. Each

coupling simulation was carried out using the Lamarckian genetic algorithm with 2.500.000 energetic evaluations with a population of 150. Finally, the best poses of the ligand were determined based on the results of energy interaction given in kcal/mol.

3. Results

3.1. Computational Characterization. Until September 13, 2015, there were 12651 phytases in UNIPROT of which 115 (0.90%) belonged to *Aspergillus* spp., of which, 33 (28.69%) belonged to *A. niger*. For the genus Aspergillus, 11 phytases in total were reported as revised or cured; the distribution was as follows: one for each species *A. ficuum*, *A. oryzae*, *A. fumigatus,* and *A. nidulans*; two for *A. niger* and *A. awamori*, and three for *A. terreus*. The only two (6.06%) of the revised phytases of *A. niger* were 3-phytase A (E.C. 3.1.3.8, PDB ID: 3K4Q), corresponding to a monomer, and the 3-phytase B (E.C. 3.1.3.8, PDB ID: 1QFX), a homodimer.

The dendrogram between the 33 phytases reported for *A. niger* was obtained using the BLAST tool, and it was found that 3-phytase A (number 17, P34752) is closely related to 78.8% (26/33) of phytases whereas 3-phytase B (number 5, P34754) is only related to 9.1% (3/33) (Figure 1).

The multiple alignment (Clustal O) between the amino acid sequences of the phytases reported for *A. niger* enabled the identification of a highly conserved sequence (**RHGXRXP-HD**) present in the histidine acid phosphatase family, corresponding to the pocket where the active ligand binding site is located (Figure 2).

3.2. Structural Characteristics of the Two Reported and Revised Phytases from A. Niger. The first revised phytase from *A. niger*, the 3-phytase A (PDB ID: 3K4Q), corresponds to a monomer. The second revised phytase, the 3-phytase B (PDB ID: 1QFX), initially corresponds to a homodimer (Chains A and B) which, thanks to its crystallographic symmetry, generates a homotetramer from two dimers.

When performing an alignment (Clustal W) between the sequences of the two chains of the initially dimeric protein (A and B) and the unique sequence of the monomeric protein, it was possible to determine that the chains A and B of the dimer are identical to each other, but different to the monomeric phytase.

3.3. 2D Structures of the Two Reported and Revised Phytases from A. Niger. The 3-phytase A (monomer) (Figure 3) (PDB ID: 3K4Q) has a 2D structure formed by 43% alpha helices, 12% beta sheets, and 45% random coils. A and B chains of 3-phytase B initially form a homodimer (Figure 3) (PDB ID: 1QFX) and its 2D structure corresponds to 38% alpha helices, 12% beta sheets, and 50% random coils. For the 3-phytase A, the active site is conformed by amino acids R_{58}, $\mathbf{H_{59}}$, R_{62}, R_{142}, and $\mathbf{D_{339}}$ [22] (Figure 3(b)), and for the 3-phytase B the active site is conformed by amino acids R_{62}, $\mathbf{H_{63}}$, R_{66}, R_{156}, H_{318}, and $\mathbf{D_{319}}$ [21] (Figures 3(c) and 3(d)).

3.4. Structure of the Homodimer and the Tetramer Formed by Chains A and B of the 3-Phytase B from A. Niger. 3-phytase

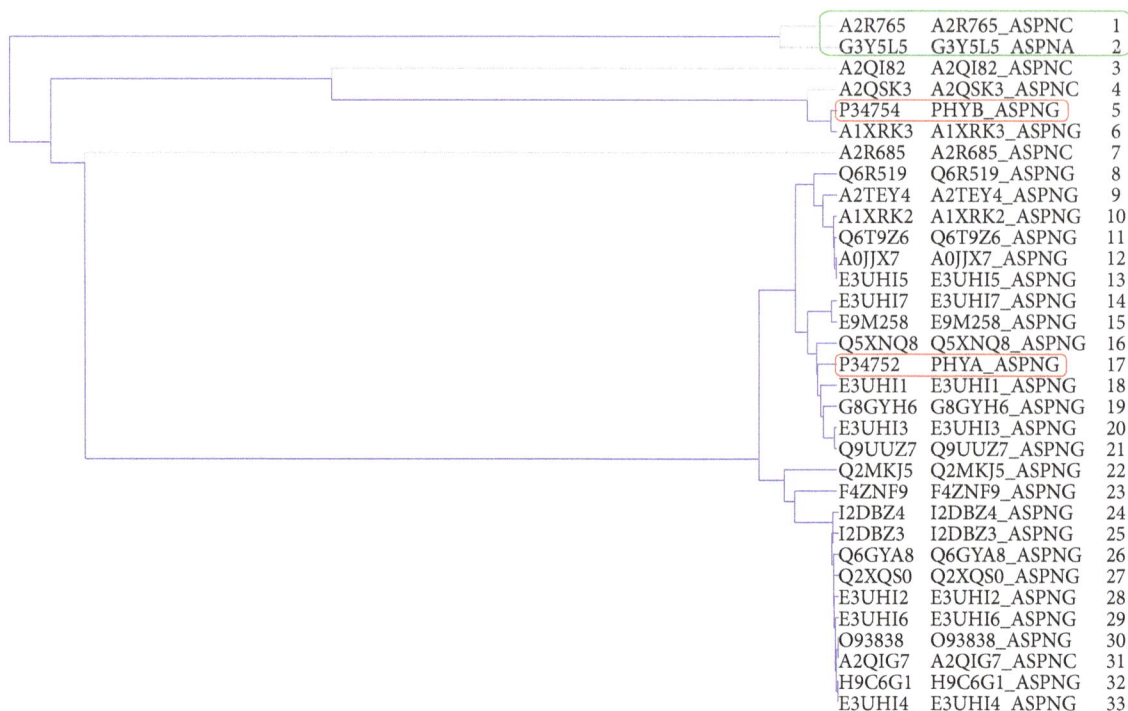

FIGURE 1: *Dendrogram of the 33 phytases reported for A. niger.* The first two phytases that appear are phytases that do not have PDB code or are not characterized within UNIPROT and are far from the two revised phytases. Alignment parameters are predetermined. The default transition matrix is Gonnet; the gap of the opening is 6 bits; the extension interval is of 1 bit. Clustal-Omega uses the HHafign algorithm and its default configuration as its core alignment engine [23].

B is initially a homodimer consisting of two identical chains A and B. The 39 amino acids that allow the interactions that lead to dimerization between the A and B chains are: Lys_{14}-Tyr_{24}, Leu_{27}-His_{29}, Tyr_{36}, Glu_{38}, Ser_{41}-Ala_{45}, Tyr_{120}, Lys_{217}, Leu_{248}, Pro_{252}-Ser_{254}, Gln_{262}-Asp_{263}, Val_{266}-Ser_{267}, Asn_{335}, Arg_{342}, Phe_{345}-Gly_{346}, Ala_{372}, Asp_{393}, Gly_{399}, Tyr_{400}. The crystallographic symmetry generates a tetramer from the two dimers, and 17 amino acids that are involved in the interactions that allow the tetramerization of the protein have been identified: Cys_{109}, Glu_{114}, Thr_{116}, Gly_{118}, Ala_{121}, Leu_{123}-Leu_{124}, Tyr_{127}-Asn_{128}, Asn_{131}, Lys_{163}, Glu_{166}, Tyr_{171}, Arg_{447}, Pro_{450}-Ile_{451}, and Cys_{453} [21] (Figures 3(e) and 3(f)).

In both phytases, the fact that the pocket of the active site is composed mostly of positively charged amino acids (H-histidine and R-Arginine) is highlighted.

3.5. Physicochemical Characterization of the Two Reported and Revised Phytases from A. Niger.
The physicochemical properties of the two phytases reported and reviewed for *A. niger* and obtained by the ProtParam bioinformatics program are detailed in Table 1.

3.6. Hydrophobicity and Accessibility Profiles of the Two Reported and Revised Phytases from A. niger.
Figures 4(a) and 4(b) are the detailed hydrophobicity profiles of the phytases A and B from *A. niger*. The red circles represent the amino acids with a higher hydrophobicity score and the yellow circles represent the amino acids with a lower

hydrophobicity score, according to the values recorded in Table 2.

The accessibility profile of phytases A and B from *A. niger* is shown in Figure 5. Green colored circles represent amino acids with a minimum accessibility value and purple circles represent amino acids with a maximum accessibility value, according to the values recorded in Table 2.

3.7. Prediction of N-Glycosylation Sites in 3-Phytase A and Chain A of 3-Phytase B from A. Niger.
Table 3 details the predictions of possible N-glycosylation sites of phytases A and B from *A. niger* by means of the NetNGlyc 1.0 tool, showing the position of the Asparagines (N), along both phytase chains that are located in an Asn-Xaa-Ser/Thr (where Xaa is any amino acid except proline) and for that reason could be glycosylated.

3.8. Prediction of Glycation Sites in 3-Phytase A and Chain A of 3-Phytase B from A. Niger.
In the case of 3-phytase A, Netglycate 1.0 predicted the glycation potential of seven lysines, whereas the methodology proposed by Sáenz et al., 2016 suggests the glycation of 14 of them. For Lys94, there is no prediction of glycation by either method. Table 4 shows the comparison of the results and distances of the ε-NH2 groups of the lysines and the side chains of acidic or basic residues. Figure 6 shows some distances between lysines and other acidic or basic residues in 3-phytase A.

In the case of 3-phytase A, Netglycate 1.0 allowed predicting the glycation potential of seven lysines along the protein

```
1   P34752   PHYA_ASPNG    63   ------ISPEVPAGCRVTFAQVLSRHGARYPTDSKGKKYSALIEEIQ---QNATTFDGKY   113
2   P34754   PHYB_ASPNG    63   ------IARDPPTGCEVDQVIMVKRHGERYPSPSAGKSIEEALAKVYSI-NTT-EYKGDL   114
3   A2QSK3   A2QSK3_ASPNC  63   ------IARDPPTGCEVDQVIMIKRHGERYPSPSAGKSIEEALAKVYSI-NTT-EYKGDL   114
4   A2QIG7   A2QIG7_ASPNC  63   ------ISPDVPAGCVTFAQVLSRHGARYPTDSKGKKYSALIEEIQ---QNATTFEGKY   113
5   H9C6G1   H9C6G1_ASPNG  63   ------ISPDVPAGCHVTFAQVLSRHGARYPTDSKGKKYSALIEEIQ---QNATTFEGKY   113
6   A1XRK3   A1XRK3_ASPNG  47   ------IARDPPTGCEVDQVIMVKRHGERYPSPSAGKSIEEALAKVYSI-NTT-EYKGDL   98
7   I2DBZ3   I2DBZ3_ASPNG  63   ------ISPDVPAGCHVTFAQVLSRHGARYPTDSKGKKYSALIEEIQ---QNATTFEGKY   113
8   I2DBZ4   I2DBZ4_ASPNG  63   ------ISPDVPAGCHVTFAQVLSRHGARYPTDSKGKKYSALIEEIQ---QNATTFEGKY   113
9   G8GYH6   G8GYH6_ASPNG  63   ------ISPDVPAGKVTFAQVLSRHGARYPTDSKGKKYSALIEEIQ---QNATTFDGKY   113
10  A1XRK2   A1XRK2_ASPNG  44   ------ISPEVPAGCRVTFAQVLSRHGARYPTDSKGKKYSALIEEIQ---QNATTFDGKY   94
11  A0JJX7   A0JJX7_ASPNG  48   ------ISPEVPAGCRVTFAQVLSRHGARYPTDSKGKKYSALIEEIQ---QNATTFDGKY   98
12  Q2XQS0   Q2XQS0_ASPNG  44   ------ISPDVPAGCHVTFAQVLSRHGARYPTDSKGKKYSALIEEIQ---QNATTFEGKY   94
13  Q6GYA8   Q6GYA8_ASPNG  44   ------ISPDVPAGCHVTFAQVLSRHGARYPTDSKGKKYSALIEEIQ---QNATTFEGKY   94
14  Q9UUZ7   Q9UUZ7_ASPNG  63   ------ISPDVPAGCRVTFAQVLSRHGARYPTDSKGKKYSALIEEIQ---QNATTFDGKY   113
15  O93838   O93838_ASPNG  63   ------ISPDVPAGCHVTFAQVLSRHGARYPTDSKGKKYSALIEEIQ---QNATTFEGKY   113
16  Q6T9Z6   Q6T9Z6_ASPNG  63   ------ISPEVPAGCRVTFAQVLSRHGARYPTDSKGKKYSALIEEIQ---QNATTFDGKY   113
17  Q6R519   Q6R519_ASPNG  63   ------ISPDVPAGCRVTFAQVLSRHGARYPTDSKGKKYSALIEEIQ---QNATTFDGKY   113
18  A2TEY4   A2TEY4_ASPNG  63   ------ISPDVPAGCRVTFAQVLSRHGARYPTDSKGKKYSALIEEIQ---QNATTFDGKY   113
19  Q5XNQ8   Q5XNQ8_ASPNG  63   ------ISPDVPAGKVTFAQVLSRHGARYPTDSKGKKYSALIEEIQ---QNATTFDGKY   113
20  F4ZNF9   F4ZNF9_ASPNG  90   ------ISPDVPAGCQVTFAQVLSRHGARYPTDSKGKKYSALIEEIQ---QNATTFKEKY   140
21  A2QI82   A2QI82_ASPNC  59   ------ISRDPPAQCSVDQVIMIKRHGERYPTTGEGTSIEQTLKKINDS-LAGNYSSGDL   111
22  A2R685   A2R685_ASPNC  111  ------DEFPRPKGSNITQMHMLHRHGSRYPNKDEGDDFANWIKAITNATAHGAVFRDEL   164
23  E3UHI1   E3UHI1_ASPNG  44   ------ISPDVPTGCRVTFAQVLSRHGARYPTDSKGKKYSALIEEIQ---QNATTFDGKY   94
24  E3UHI4   E3UHI4_ASPNG  44   ------ISPDVPAGCHVTFAQVLSRHGARYPTDSKGKKYSALIEEIQ---QNATTFEGKY   94
25  E3UHI3   E3UHI3_ASPNG  44   ------ISPDVPAGCRVTFAQVLSRHGARYPTDSKGKKYSALIEEIQ---QNATTFDGKY   94
26  E3UHI2   E3UHI2_ASPNG  44   ------ISPDVPAGCHVTFAQVLSRHGARYPTDSKGKKYSALIEEIQ---QNATTFEGKY   94
27  E3UHI6   E3UHI6_ASPNG  44   ------ISPDVPAGCHVTFAQVLSRHGARYPTDSKGKKYSALIEEIQ---QNATTFEGKY   94
28  E3UHI5   E3UHI5_ASPNG  44   ------ISPEVPAGCRVTFAQVLSRHGARYPTDSKGKKYSALIEEIQ---QNATTFDGKY   94
29  E9M258   E9M258_ASPNG  44   ------ISPDVPAGCRVTFAQVLSRHGARYPTESKGKKYSALIEEIQ---QNVTTFDGKY   94
30  Q2MKJ5   Q2MKJ5_ASPNG  47   ------ISPDVPAGCRVTFAQVLSRHGARYPTESKGKKYSALIEEIQ---QNVTTFDGKY   97
31  E3UHI7   E3UHI7_ASPNG  44   ------ISPDVPAGCRVTFAQVLSRHGARYPTESKGKKYSALIEEIQ---QNVTTFDGKY   94
    A2R765   A2R765_ASPNC  117  LSPNASATHPSPPNIHLHYLDTLLLTGPDN-------APLTGLDAITTTTNTTTT-TTKY   168
    G3Y5L5   G3Y5L5_ASPNA  109  LSPNASATHPSPPNIHLHYLDTLLLTGPDN-------APLTGLDAITTTTNTTTT-TTKY   160
                                        *         :         :    *                    :    :    .
```

```
1   P34752   PHYA_ASPNG    114  AFLKTYNYSLGADDLT-------PFG-EQELVNSGIKFY-QRYESLTRNI--VPFI   158
2   P34754   PHYB_ASPNG    115  AFLNDWTYYVPNECYYNAETTSGPYAG--LLDAYNHGNDYK-ARYGHLWNGET---VVPF   168
3   A2QSK3   A2QSK3_ASPNC  115  AFLNDWTYYVPNECYYNAETTSGPYAG--LLDAYNHGNEYK-ARYGHLWDGET---VVPF   168
4   A2QIG7   A2QIG7_ASPNC  114  AFLKTYNYSLGADDLT-------PFG-EQELVNSGVKFY-QRYESLTRNI----VPFI   158
5   H9C6G1   H9C6G1_ASPNG  114  AFLKTYNYSLGADDLT-------PFG-EQELVNSGVKFY-QRYESLTRNI----VPFI   158
6   A1XRK3   A1XRK3_ASPNG  99   AFLNDWTYYVPNECYYNAETTSGPYAG--LLDAYNHGNDYK-ARYGHLWNGET---VVPF   152
7   I2DBZ3   I2DBZ3_ASPNG  114  AFLKTYNYSLGADDLT-------PFG-EQELVNSGVKFY-QRYESLTRKI----VPFI   158
8   I2DBZ4   I2DBZ4_ASPNG  114  AFLKTYNYSLGADDLT-------PFG-EQELVNSGVKFY-QRYESLTRNI----VPFI   158
9   G8GYH6   G8GYH6_ASPNG  114  AFLKTYNYSLGADDLT-------PFG-EQELVNSGIKFY-QRYESLTRNI----IPFI   158
10  A1XRK2   A1XRK2_ASPNG  95   AFLKTYNYSLGADDLT-------PFG-EQELVNSGIKLY-QRYESLTRNI----IPFI   139
11  A0JJX7   A0JJX7_ASPNG  99   AFLKTYNYSLGADDLT-------PFG-EQELVNSGIKFY-QRYESLTRNI----IPFI   143
12  Q2XQS0   Q2XQS0_ASPNG  95   AFLKTYNYSLGADDLT-------PFG-EQELVNSGKFY-QRYESLTRNI----VPFI   139
13  Q6GYA8   Q6GYA8_ASPNG  95   AFLKTYDYSLGADDLT-------PFG-EQELVNSGKFY-QRYESLTRNI----VPFI   139
14  Q9UUZ7   Q9UUZ7_ASPNG  114  AFLKTYNYSLGADDLT-------PFG-EQELVNSGIKFY-QRYESLTRNI----IPFI   158
15  O93838   O93838_ASPNG  114  AFLKTYNYSLGADDLT-------PFG-EQELVNSGVKFY-QRYESLTRNI----IPFI   158
16  Q6T9Z6   Q6T9Z6_ASPNG  114  AFLKTYNYSLGADDLT-------PFG-EQELVNSGIKFY-QRYESLTRNI----IPFI   158
17  Q6R519   Q6R519_ASPNG  114  AFLKTYNYSLGLDDLT-------PLG-EQELVNSGIKFY-QRYESLTRNI----IPFI   158
18  A2TEY4   A2TEY4_ASPNG  114  AFLKTYNYSLGLDDLT-------PLG-EQELVNSGIKFY-QRYESLTRNI----IPFI   158
19  Q5XNQ8   Q5XNQ8_ASPNG  114  AFLKTYNYSLGADDLT-------PFG-EQELVNSGIKFY-QRYESLTRNI----IPFI   158
20  F4ZNF9   F4ZNF9_ASPNG  141  AFLKTYNYRMGADDLT-------PFG-EQELVNSGIKFY-QRYESLTRNI----VPFI   185
21  A2QI82   A2QI82_ASPNC  112  AFLGNWTYYVPSDCY-EAETSTGPYAG--LNDAYNHGKAYR-QRYGNLYNESS---ILPL   164
22  A2R685   A2R685_ASPNC  165  SFIHDWTYSLGADMLT--------TRG--REDLLESGILNF-YNYGHLYTPGT---KIVA   210
23  E3UHI1   E3UHI1_ASPNG  95   AFLKTYNYSLGADDLT-------PFG-EQELVNSGIKFY-QRYESLTRNI----IPFI   139
24  E3UHI4   E3UHI4_ASPNG  95   AFLKTYNYSLGADDLT-------PFG-EQELVNSGVKFY-QRYESLTRNI----VPFI   139
25  E3UHI3   E3UHI3_ASPNG  95   AFLKTYNYSLGADDLT-------PFG-EQELVNSGIKFY-QRYESLTRNI----IPFI   139
26  E3UHI2   E3UHI2_ASPNG  95   AFLKTYNYSLGADDLT-------PFG-EQELVNSGVKFY-QKSLTRNI----VPFI   139
27  E3UHI6   E3UHI6_ASPNG  95   AFLKTYNYSLGADDLT-------PFG-EQELVNSGIKFY-QRYESLTRNI----VPFI   139
28  E3UHI5   E3UHI5_ASPNG  95   AFLKTYNYSLGADDLT-------PFG-EQELVNSGIKFY-QRYESLTRNI----IPFI   139
29  E9M258   E9M258_ASPNG  95   AFLKTYNYSLGADDLT-------PFG-EQELVNSGIKSY-QRYESLTRNI----IPFI   139
30  Q2MKJ5   Q2MKJ5_ASPNG  98   AFLKTYNYSLGADDLT-------PFG-EQELVNSGVKFY-QRYESLTRNI----VPFI   142
    E3UHI7   E3UHI7_ASPNG  95   AFLKTYNYSLGADDLT-------PFG-EQELVNSGIKFY-QRYESLTRNI----IPFI   139
    A2R765   A2R765_ASPNC  169  PNFP----PLPIATY-PGDGFGGPGPGGHRIPLDAEGLALPAHDPAHIWVSDEYGPYVYK   223
    G3Y5L5   G3Y5L5_ASPNA  161  PNFP----PLPIATY-PGDGFGGPGPGGHRIPLDAEGLALPAHDPAHIWVSDEYGPYVYK   215
                                :       :               *       *        :
```

FIGURE 2: Continued.

				Sequence	
1	P34752	PHYA_ASPNG	354	TLYADFSHDNGIISILFALGLYNGTK-PLSTTTVE------NITQT--------------	392
2	P34754	PHYB_ASPNG	330	PLFFNFAHDTNITPILAALGVLIPNE-DLPLDR------VAFG--------------	365
3	A2QSK3	A2QSK3_ASPNC	330	SLFFNFAHDTNITPILAALGVLIPTE-DLPLDR--------VAFG--------------	365
4	A2QIG7	A2QIG7_ASPNC	354	TLYADFSHDNGIISILFALGLYNGTK-PLSSTTAE------NITQT--------------	392
5	H9C6G1	H9C6G1_ASPNG	354	TLYADFSHDNGIISILFALGLYNGTK-PLSSTTAE------NITQT--------------	392
6	A1XRK3	A1XRK3_ASPNG	314	PLFFNFAHDTNITPILAALGVLIPNE-DLPLDR--------VAFG--------------	349
7	I2DBZ3	I2DBZ3_ASPNG	354	TLYADFSHDNGIISILFALGLYNGTK-PLSSTTAE------NITQT--------------	392
8	I2DBZ4	I2DBZ4_ASPNG	354	TLYADFSHDNGIISILFALGLYNGTK-PLSSTTAE------NITQT--------------	392
9	G8GYH6	G8GYH6_ASPNG	354	TLYADFSHDNGIISILFALGLYNGTK-PLSTTTVE------NITQT--------------	392
10	A1XRK2	A1XRK2_ASPNG	335	TLYADFSHDNGIISILFALGLYNGTK-PLSTTTVE------NITQT--------------	373
11	A0JJX7	A0JJX7_ASPNG	339	TLYADFSHDNGIISILFALGLYNGTK-PLSTTTVE------NITQT--------------	377
12	Q2XQS0	Q2XQS0_ASPNG	335	TLYADFSHDNGIISILFALGLYNGTK-PLSSTTAE------NITQT--------------	373
13	Q6GYA8	Q6GYA8_ASPNG	335	TLYADFSHDNGIISILFALGLYNGTK-PLSSTTAE------NITQT--------------	373
14	Q9UUZ7	Q9UUZ7_ASPNG	354	TLYADFSHDNGIISILFALGLYNGTK-PLSTTTVQ------NITQT--------------	392
15	O93838	O93838_ASPNG	354	TLYADFSHDNGIISILFALGLYNGTK-PLSSTTAE------NITQT--------------	392
16	Q6T9Z6	Q6T9Z6_ASPNG	354	TLYADFSHDNGIISILFALGLYNGTK-PLSTTTVE------NITQT--------------	392
17	Q6R519	Q6R519_ASPNG	354	TLYADFSHDKGIISILFALGLYNGTK-PLSTTTVE------NITQT--------------	392
18	A2TEY4	A2TEY4_ASPNG	354	TLYADFSHDNGIISILFALGLYNGTK-PLSTTTVE------NITQT--------------	392
19	Q5XNQ8	Q5XNQ8_ASPNG	354	TLYADFSHDNGIISILFALGLYNGTK-PLSTTTVQ------NITQT--------------	392
20	F4ZNF9	F4ZNF9_ASPNG	381	TLYADFSHDNGIISILFALGLYNGTK-PLSSTTAE------NITQT--------------	419
21	A2QI82	A2QI82_ASPNC	329	PLFFNFN---ISPIITALGIATPAT-PLNKTR--------IPFPP--------------	361
22	A2R685	A2R685_ASPNC	399	SLYFDFAHDKILLGVLTAFGLRQFADLPFPDYTDQYF-MDVFPPRH--------------	443
23	E3UHI1	E3UHI1_ASPNG	335	TLYADFSHDNGIISILFALGLYNGTK-PLSTTTVE------NITQT--------------	373
24	E3UHI4	E3UHI4_ASPNG	335	TLYADFSHDNGIISILFALGLYNGTK-PLSSTTAE------NITQT--------------	373
25	E3UHI3	E3UHI3_ASPNG	335	TLYADFSHDNGIISILFALGLYNGTK-PLSTTTVQ------NITQT--------------	373
26	E3UHI2	E3UHI2_ASPNG	335	TLYADFSHDNGIISILFALGLYNGTK-PLSSTTAE------NITQT--------------	373
27	E3UHI6	E3UHI6_ASPNG	335	TLYADFSHDNGIISILFALGLYNGTK-PLSSTTAE------NITQT--------------	373
28	E3UHI5	E3UHI5_ASPNG	335	TLYADFSHDNGIISILFALGLYNGTK-PLSTTTVE------NITQT--------------	373
29	E9M258	E9M258_ASPNG	335	TLYADFSHE--------------------------	343
30	Q2MKJ5	Q2MKJ5_ASPNG	338	TLYADFSHDNGIISILFALGLYNGTK-PLSSTTAE------NITQT--------------	376
31	E3UHI7	E3UHI7_ASPNG	335	TLYADFSHDNGIISILFALGLYNGTK-PLSTTTVE------NITQT--------------	373
32	A2R765	A2R765_ASPNC	429	AEYCSF-LDYNVPGQLAKFGLHNGGEQDRWLLNEKWESLALVPVNPDTEDSSG-GNGEKD	486
33	G3Y5L5	G3Y5L5_ASPNA	421	AEYCPF-LDYNVPGQLAKFGLHNGGEQDEWLLNEKWESLALVPVNPDTEDSSTNSGGEKE	479

FIGURE 2: *Multiple alignment (Clustal O) among the 33 phytases reported for A. niger.* The green boxes indicate the two phytases reviewed and the red boxes indicate the amino acids that are highly conserved in the active site. The letters shaded in green correspond to positively charged amino acids.

sequence. When the glycation potential analysis was done by using the methodology proposed by Sáenz et al., 2016, the results suggest the glycation of 14 lysines. For Lys_{217}, there is no prediction of glycation by either method. Table 4 shows the comparison of the results and distances of the ε-NH_2 groups of the lysines and acidic or basic residue side chains. Figure 6 shows some distances between lysines and other acidic or basic residues in 3-phytase B.

3.9. Profile of Antigenicity of 3-Phytase A and Chain A of 3-Phytase B from A. niger. The antigenic propensity or average epitope of 3-phytase A is 1.0304; when the average for the complete protein is greater than 1.0 then all residues having more than 1.0 are potentially antigenic. Figure 7(1A) shows the peaks of antigenicity in green colored circles along the chain of amino acids in 3-phytase A. In Figure 7(1B), the two peaks of antigenicity are located on the surface of the protein with higher score in green color and, additionally, its active center in red color is observed, corresponding to the information registered on the antigenic determinants in Table 5. The amino acid Arg_{58} forms part of the active site of the protein and at the same time is located in an area with a high antigenic propensity, highlighted in underlined font. The letters N highlighted in bold in Table 5 correspond to the positions of the Asparagines (N) which were identified as potential N-glycosylation sites and are also part of some identified antigenic determinant. This finding may be a

contributing factor to the induction of an immune response by 3-phytase A.

The antigenic propensity or average epitope of chain A of 3-phytase B is 1.0234. In Figure 7(2A), the peaks of antigenicity in green circles are observed, along the chain of amino acids in chain A of 3-phytase B. In Figure 7(2B), the peaks of antigenicity in green color are located on the surface of the protein and additionally its active center is observed in red color, corresponding with the information registered on the antigenic determinants in Table 5; the amino acid Arg_{156} is a part of the active site of the protein and, at the same time, is located in an area with a low antigenic propensity, highlighted in underlined font. The letters N highlighted in bold in Table 6 correspond to the positions of the Asparagines (N) which were identified as potential N-glycosylation sites and are also part of some identified antigenic determinant. This finding may be a factor contributing to the induction of an immune response by 3-phytase B.

3.10. Ligand and Molecular Coupling Models (Rigid Docking). The 3D structure of the phytic acid ligand (myo-inositol 1, 2, 3, 4, 5, 6 hexakisphosphate) was generated by the Spartan 4.0 program and is observed in Figure 8. This ligand was used in the molecular coupling models

Molecular coupling models (Rigid Docking) were performed with the two revised phytases from *A. niger* and phytic acid as ligand which was directed to the catalytic

TABLE 1: Physicochemical properties of the two reported and revised phytases from *A. niger*.

Physical and chemical properties (ProtParam)	3-Phytase A (monomer) PDB ID: 3K4Q	3-Phytase B (chain A, Homodimer) PDB ID: 1QFX
aa sequence length	467	479
Signal peptide length	23	19
Mature protein length	444	460
Molecular Weight (kDa)	48,84	58,78
Instability Index	45,41 (Unstable)	33,66 (Stable)
Disulfide bond	5 (Intrachain) in positions: Cys_8-Cys_{17}, $Cys_{48}-Cys_{391}$, $Cys_{192}-Cys_{442}$, $Cys_{241}-Cys_{259}$, $Cys_{413}-Cys_{421}$	5 (Intrachain) in positions: $Cys_{52}-Cys_{368}$, $Cys_{109}-Cys_{453}$, $Cys_{197}-Cys_{422}$, $Cys_{206}-Cys_{279}$, $Cys_{394}-Cys_{402}$
Theoretical Isoelectric Point (iP)	4,94	4,6
Estimated Lifetime	4,4 hours (Reticulocytes of mammals, *in vitro*); >20 hours (yeasts, *in vivo*); >10 hours (*E. coli*, *in vivo*)	1,1 hours (Reticulocytes of mammals, *in vitro*); 3 minutes (yeasts, *in vivo*); 2 minutes (*E. coli*, *in vivo*)
Aliphatic Index	72,25	70,46
Average Hydropathicity (GRAVY)	−0,304	−0,33
Amino acids composition	Ala (A) 29 6.5%	Ala (A) 40 8.7%
	Arg (R) 19 4.3%	Arg (R) 14 3.0%
	Asn (N) 19 4.3%	Asn (N) 38 8.3%
	Asp (D) 29 6.5%	Asp (D) 24 5.2%
	Cys (C) 10 2.3%	Cys (C) 10 2.2%
	Gln (Q) 19 4.3%	Gln (Q) 14 3.0%
	Glu (E) 22 5.0%	Glu (E) 24 5.2%
	Gly (G) 30 6.8%	Gly (G) 37 8.0%
	His (H) 9 2.0%	His (H) 6 1.3%
	Ile (I) 18 4.1%	Ile (I) 19 4.1%
	Leu (L) 36 8.1%	Leu (L) 36 7.8%
	Lys (K) 15 3.4%	Lys (K) 13 2.8%
	Met (M) 4 0.9%	Met (M) 7 1.5%
	Phe (F) 25 5.6%	Phe (F) 20 4.3%
	Pro (P) 22 5.0%	Pro (P) 27 5.9%
	Ser (S) 49 11.0%	Ser (S) 34 7.4%
	Thr (T) 39 8.8%	Thr (T) 32 7.0%
	Trp (W) 4 0.9%	Trp (W) 6 1.3%
	Tyr (Y) 18 4.1%	Tyr (Y) 35 7.6%
	Val (V) 28 6.3%	Val (V) 24 5.2%
	Pyl (O) 0 0%	Pyl (O) 0 0%
	Sec (U) 0 0%	Sec (U) 0 0%
Total number (%) of negatively charged amino acids (Asp + Glu)	51 (11.48%)	48 (10.43%)
Total number (%) of positively charged amino acids (Asp + Glu)	34 (7.65%)	27 (5.86%)

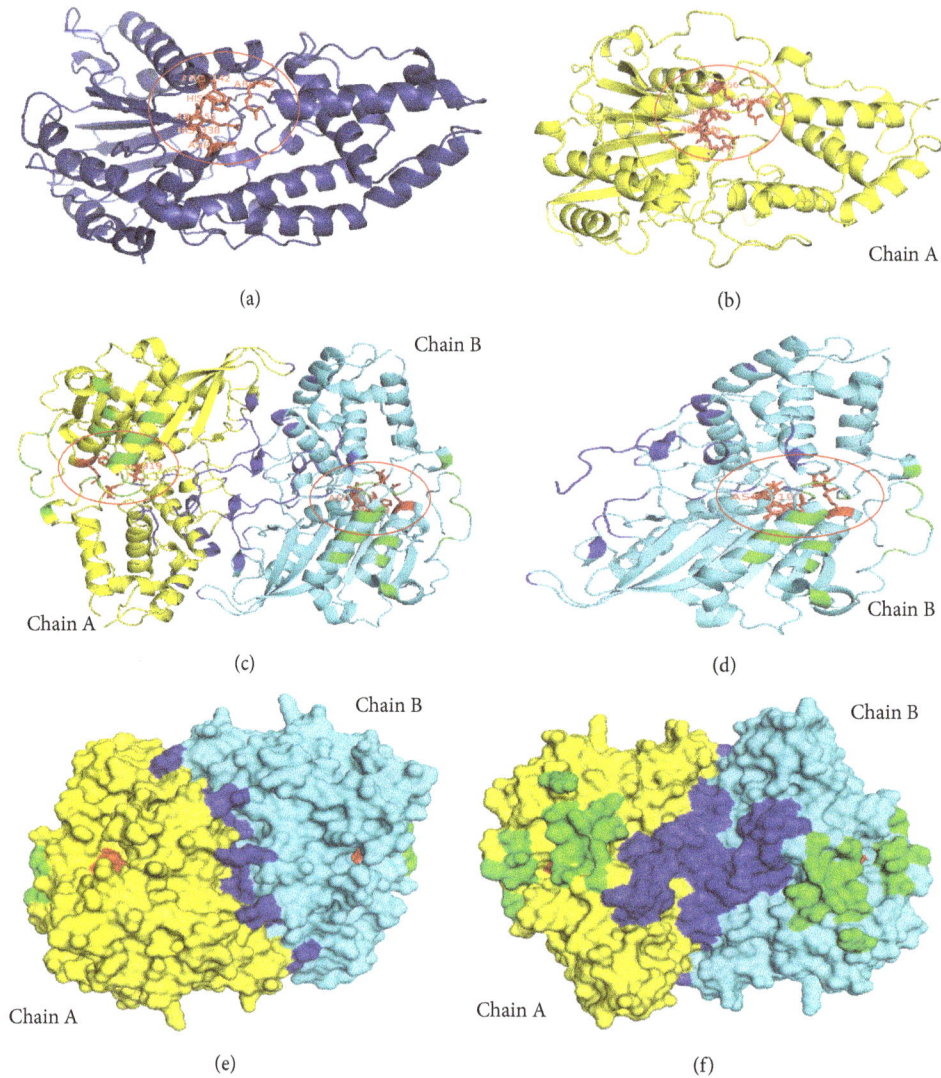

FIGURE 3: (a) *Ribbon diagram of 3-phytase A (PDB ID: 3K4Q)*. The 6 residues marked in red color form part of the pocket of the highly conserved ligand binding active site (R_{58}, $\mathbf{H_{59}}$, G, X(R_{62}), R_{142}, X, P, H_{338}, $\mathbf{D_{339}}$); (b) *ribbon diagram of the 3-phytase B chain A (PDBID: 1QFX)*; the 6 residues indicated in red color form part of the pocket of the highly conserved ligand binding active site (R_{62}, $\mathbf{H_{63}}$, G, X(R_{66}), R_{156}, X, P, H_{318}, and $\mathbf{D_{319}}$); (c) and (d) *active site rear view of the amino acids involved in the interactions that allow dimerization* (king blue color), tetramerization (green color), and active ligand binding sites (red color) in the dimer (left) and in the B chain of the 3-phytase B dimer (right); (e) *surface diagram* of the 3-phytase B top view; (f) *surface diagram* of the 3-phytase B bottom view.

TABLE 2: Hydrophobicity score in the two reported and revised phytases from *A. niger* (ProtScale). Minimum and maximum values of accessibility in the two reported and revised phytases of *A. niger* (ProtScale).

Physicochemical property	3-Phytase A (444 residues)		3-Phytase B (460 residues)	
	Residue position	Score	Residue position	Score
Hydrophobicity	Ala_{164}	*−2.300 (min.)*	Glu_{65}	*−2.633 (min.)*
	Asp_{66}	*−2.289*	Ala_{135}	*−2.200*
	Ser_{314}	*−2.211*	Thr_{204}	*−2.211*
	Ile_{345}	**2.722 (max.)**	Leu_{329}	**3.089 (max.)**
	Leu_{346}	**2.722 (max.)**		
	Residue position	Value	Residue position	Value
Accessibility	Glu_{387}	*3.633 (min.)*	Gln_{56}	*4.089 (min.)*
	Ser_{182}	<u>**7.811**</u>	Ser_{374}	<u>**7.344**</u>
	Gly_{69}	**8.467 (max.)**	Ser_{71}	**7.389 (max.)**

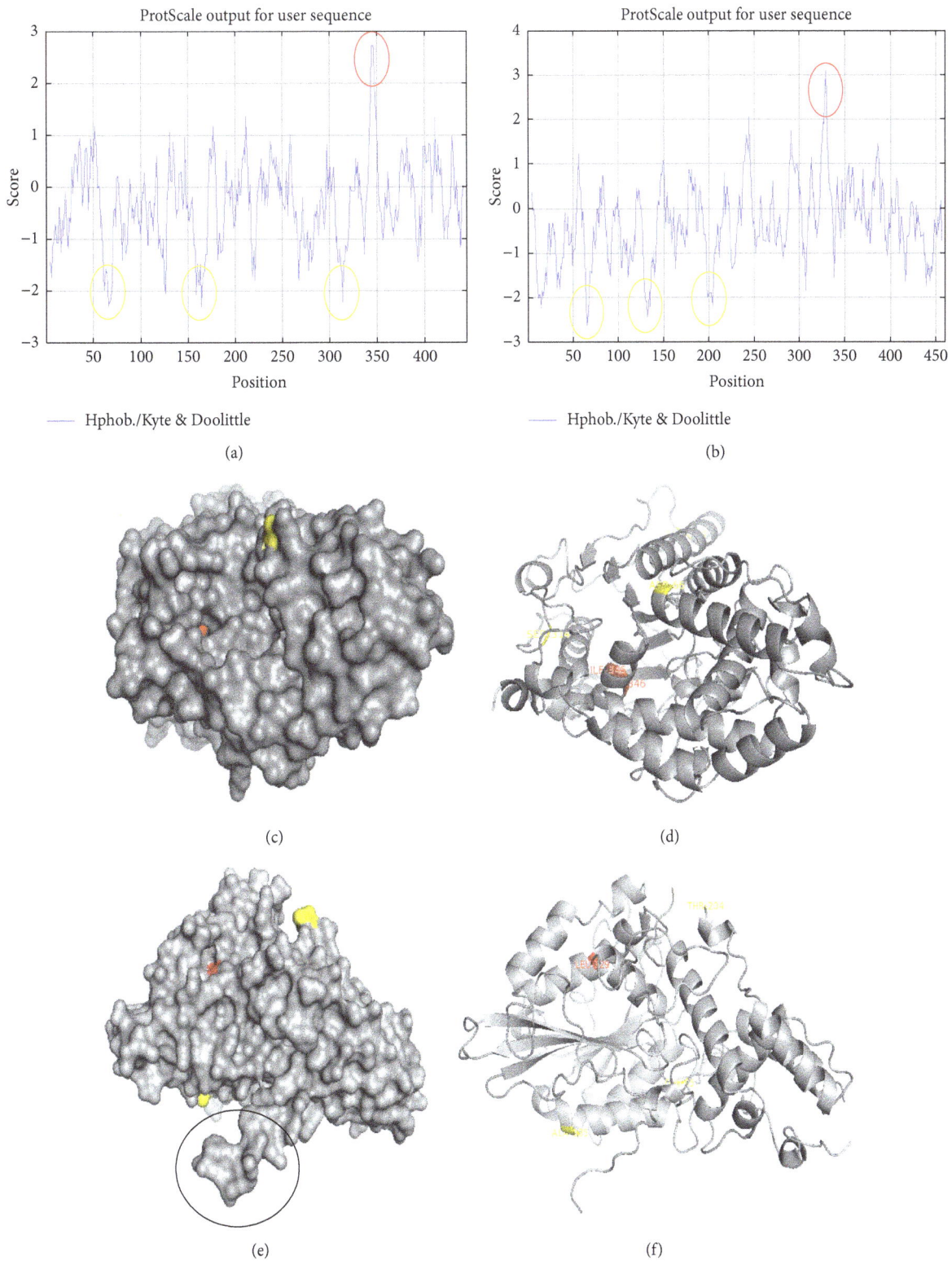

FIGURE 4: *Hydrophobicity profiles* (a and b) of phytases A and B from *A. niger*; (c) and (d) surface and ribbons diagram of the 3-phytase A; (e) and (f) surface and ribbons diagram of the chain A in the 3-phytase B, with the highest scoring amino acids (red color) and lower score (yellow color) of hydrophobicity. The circle in black color refers to the amino acids that allow homodimer formation in the 3-phytase B.

TABLE 3: *Possible N-glycosylation sites. Three phytases A and B from A. niger.*

	aa position	Potential	Jury agreement	N-Glyc
	Prediction of N-glycosylation sites for 3-phytase A (PDB ID: 3K4Q)			
(1)	27 **N**QSS	0.5302	(6/9)	+
(2)	59 **N**ESV	0.6564	(9/9)	++
(3)	105 **N**ATT	0.6414	(7/9)	+
(4)	120 **N**YSL	0.7272	(9/9)	++
(5)	207 **N**NTL	0.5930	(7/9)	+
(6)	230 **N**FTA	0.6720	(8/9)	+
(7)	339 **N**HTL	0.4021	(7/9)	−
(8)	352 **N**STL	0.7211	(9/9)	++
(9)	376 **N**GTK	0.7904	(9/9)	+++
(10)	388 **N**ITQ	0.6418	(8/9)	+
	Prediction of N-glycosylation sites for Chain A of 3-phytase B (PDB ID: 1QFX)			
(1)	87 **N**TTE	0.4822	(4/9)	−
(2)	172 **N**YST	0.6708	(8/9)	+
(3)	208 **N**LTY	0.7573	(9/9)	+++
(4)	231 **N**LTA	0.6870	(9/9)	++
(5)	296 **N**ASL	0.5727	(6/9)	+
(6)	321 **N**ITP	0.1572	(9/9)	− − −
(7)	406 **N**YTS	0.6257	(8/9)	+
(8)	423 **N**VSA	0.5649	(5/9)	+
(9)	439 **N**TTT	0.5323	(7/9)	+

pocket of the enzymes where the active site (RHGXRXP-HD) is located. Figure 9 shows the result of the lower energy docking (−6.3 kcal/mol) between 3-phytase A and phytic acid. Figure 10 provides an overview of the protein and the ligand coupling at the active site.

In Table 6, the results of the five lower docking energies (kcal/mol) obtained with the Autodock program at a distance of 0.375 Å are registered; the amino acids in bold refer to those that make up the highly conserved active site in the histidine acid phosphatases (HAP) and that are involved in the formation of hydrogen bonds with the ligand. In Figure 11, the results recorded in Table 6 are shown. The green dots in Figure 11 refer to the electrostatic interactions involving the formation of hydrogen bonds between the amino acids of the active site and/or the nearest ones to it, and the oxygens or hydrogens of the phytic acid ligand.

Figure 12 shows the docking result between the A chain of the 3-phytase B homodimer and the phytic acid. In Table 6, the results of the first five docking energies (kcal/mol) obtained with the Autodock program are shown. Figure 13 provides an overview of the protein and the ligand coupling at the active site.

In Table 5, the results of the five lower docking energies (kcal/mol) obtained with the Autodock program at a distance of 0.375 Å are recorded; the amino acids in bold refer to those that make up the highly conserved active site in the histidine acid phosphatases (HAP) and that are involved in the formation of hydrogen bonds with the ligand. In Figure 12, the results recorded in Table 2 are evidenced. The green

colored spots in Figure 11 refer to the electrostatic interactions involving the formation of hydrogen bonds between the amino acids of the active site and/or the closest ones to it and the oxygens or hydrogens of the phytic acid ligand.

4. Discussion

The database UNIPROT is characterized for being a nonredundant database, in which, on September 13, 2015, a large number of phytases was reported, 12651 in total, covering different species of microorganisms, plants, and animals. According to Casey and Walsh (2004), 115 (0.9%) phytases belonged to the Aspergillus genus, of which 33 (28.69%) corresponded to the species *A. niger*; they asserted that the genus Aspergillus is one of the most prolific extracellular producers of phytase enzymes [7].

Of these 33 reported phytases, 11 (33.3%) were found as revised for the genus Aspergillus, (phytases that were manually cured by experts in the UNIPROT database). Within this genus, the species *A. terreus* was the one that obtained a greater report of phytases, three in total. However, the three reports for this species corresponded to the same enzyme, 3-phytase A, expressed by the same *phyA* gene. In contrast, the *A. niger* species presented two reports of revised phytases, corresponding to different phytases, the 3-phytase A and the 3-phytase B, expressed by different *phyA* and *phyB* genes, respectively. Both revised phytases from *A. niger* were experimentally obtained by X-ray diffraction [21, 22].

Phytases A2R765 and G3Y5L5 in Figure 1 (green box) were removed from the computational analysis because they

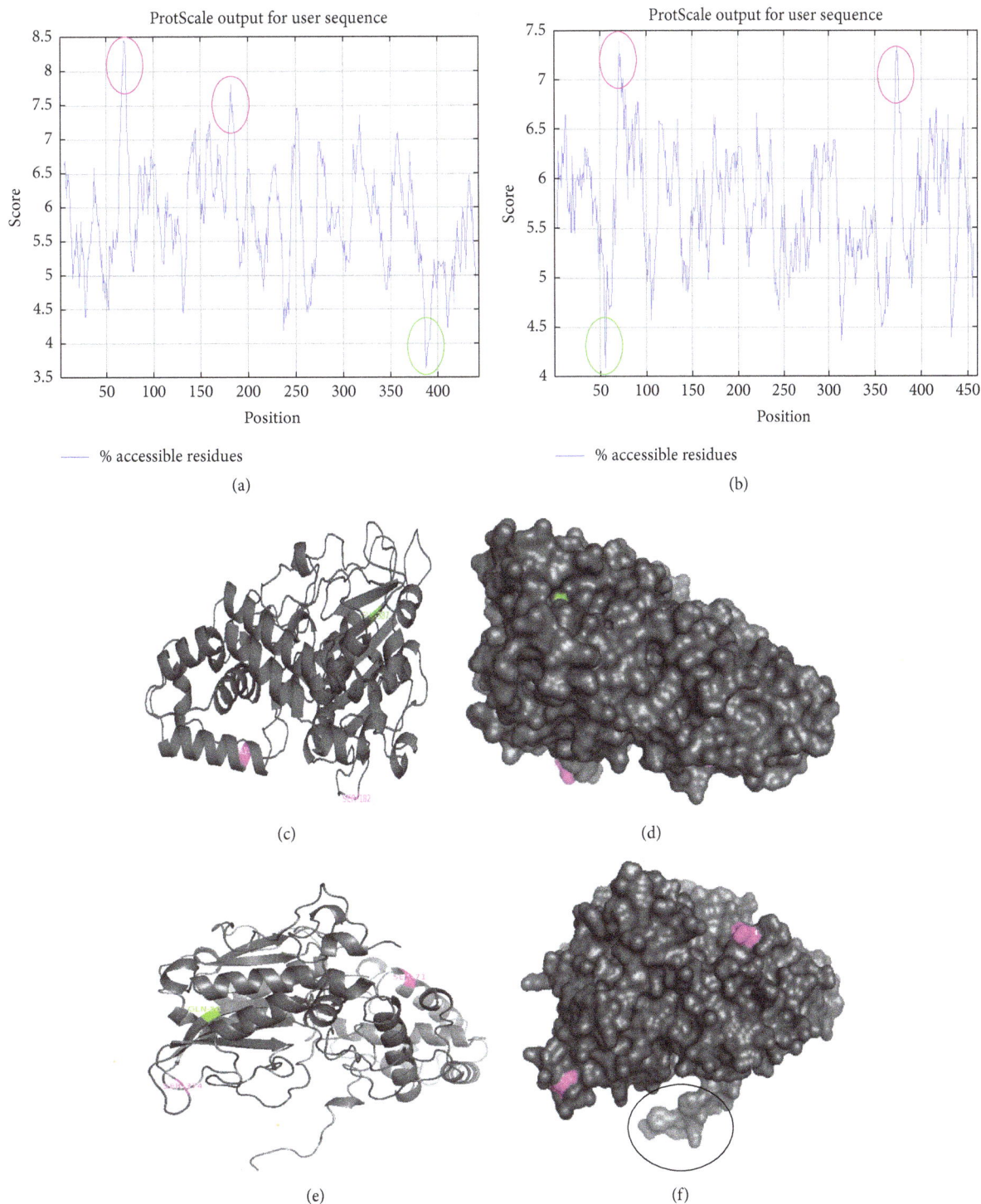

FIGURE 5: *Accessibility profiles* (a and b) of phytases A and B from *A. niger*. (c) and (d) Surface and ribbons diagram of the 3-phytase A. (e) and (f) Surface and ribbons diagram of chain A in 3-phytase B, where the amino acids with the minimum value (green color) and maximum value (purple) of accessibility are observed. The black colored circle refers to the amino acids that allow homodimer formation in 3-phytase B.

do not have PDB code or are not characterized within UNIPROT.

The ClustalW alignment showed that the sequences of these two phytases reviewed, the 3-phytase A with PDB ID: 3K4Q and the 3-phytase B with PDB ID: 1QFX, are different, being from different proteins. However, the sequences of the chains A and B that form the dimer of 3-phytase B are identical (homodimer). The first revised phytase of *A.*

TABLE 4: Distances between lysines and acidic or basic residues in the 3D structure of 3-phytase A and the chain A of 3-phytase B and their relation with the prediction of possible glycation sites.

Lysine position	Acid residue distance: Å	Basic residue distance: Å	Glycation prediction Netglycate 1.0 [16]	[17]
		3-Phytase A (ID: 3K4Q)		
68	Lys_{68}-Glu_{205}: 8.26	Lys_{68}-Lys_{70}: 6.51	X	X
70	Lys_{70}-Asp_{66}: 4.09	Lys_{70}-Lys_{71}: 9.05; Lys_{70}-Lys_{68}: 6.51		X
71	Lys_{71}-Glu_{233}: 9.89		X	X
89	Lys_{89}-Asp_{223}: 5.68		X	X
94	Acid residue location > 13.98	Basic residues location > 13.06	—	—
119	Lys_{119}-Asp_{405}: 4.21; Lys_{119}-Asp_{12}: 4.27			X
148		Lys_{148}-Lys_{149}: 9.73		X
149	Lys_{149}-Glu_{152}: 4.52		X	X
158	Lys_{158}-Asp_{161}: 7.65		X	X
160	Lys_{160}-Asp_{161}: 9.97		X	X
172	Lys_{172}-Asp_{174}: 4.50		X	X
254	Lys_{254}-Asp_{244}: 9.84			X
277	Lys_{277}-Asp_{239}: 7.30; Lys_{277}-Asp_{202}: 9.16; Lys_{277}-Glu_{205}: 4.28	Lys_{277}-Lys_{68}: 7.84; Lys_{277}-Lys_{278}: 6.72		X
278		Lys_{278}-Lys_{277}: 6.72; Lys_{278}-His_{282}: 6.51		X
356	Lys_{356}-Asp_{370}: 8.5; Lys_{356}-Glu_{364}: 9.83			X
	TOTAL		7	14
		Chain A of 3-phytase B (ID: 1QFX)		
14	Lys_{14}-Glu_{19}: 8.26		X	X
28	Lys_{28}-Glu_{38}: 8.55; Lys_{28}-Asp_{22}: 6.95	Lys_{28}-His_{29}: 8.6		X
61	Lys_{61}-Asp_{125}: 6.56	Lys_{61}-His_{360}: 4.63; Lys_{61}-His_{129}: 4.40		X
74	Lys_{74}-Glu_{77}: 9.04; Lys_{74}-Glu_{78}: 4.51; Lys_{74}-Asp_{75}: 5.95			X
82	Lys_{82}-Glu_{78}: 9.46; Lys_{82}-Asp_{236}: 4.15			X
92	Lys_{92}-Glu_{90}: 4.61			X
134		Lys_{134}-His_{139}: 9.21		X
163	Lys_{163}-Glu_{159}: 8.49; Lys_{163}-Glu_{166}: 4.74	Lys_{163}-Arg_{447}: 9.84		X
217	Acid residues location > 11.97	Basic residues location > 13.15	—	—
285	Lys_{285}-Glu_{284}: 7.35			X
307	Lys_{307}-Glu_{308}: 9.81		X	X
413	Acid residues location > 29.31	Basic residues location > 23.84	X	
	Total		3	10

niger, the 3-phytase A, corresponds to a monomer and the second phytase revised, the 3-phytase B, corresponds to a homodimer formed by chains A and B. The two revised phytases from *A. niger* also differ in the amount and type of amino acids that form the signal peptide, being 23 residues for 3-phytase A and 19 for 3-phytase B.

The ClustalO alignment (Figure 2) showed that of the 33 reported phytases, 31 (93.9%) share a highly conserved motif corresponding to the active ligand binding site (RHGXRXP-HD) in the family of the histidine acid phosphatases "HAP" [22, 24, 25], which allows classification within this family. Within these 31 phytases, 3-phytases A and B (Figure 2, green box), revised phytases for *A. niger* species, were found. Phytases that were not characterized within the UNIPROT database (A2R765 and G3Y5L5), which were not assigned PDB code, did not present this highly conserved motif. Initially, it was possible to detect that the majority of the amino acids forming this highly conserved active site correspond to positively charged residues (Figure 2, green shading).

The enzymes 3-phytase A and 3-phytase B, belonging to the histidine acid phosphatases family (HAP), have the same enzymatic code E.C.3.1.3.8 according to the database BRENDA, which corresponds to the enzymes with phosphohydrolase activity. These enzymes catalyze the phosphomonoester bonds of phytic acid, releasing orthophosphate and producing final derivatives such as inositol and inositol monophosphate, which have a lower capacity to bind to metals [25]. This family of phytases share the active site (RHGXRXP-HD), whose catalytic pockets are for 3-phytase A (R_{58}, H_{59}, G, X_{62}, R_{142}, X, P, H_{338}, D_{339}) [22] and for the 3-phytase B (R_{62}, H_{63}, G_{66}, X_{156}, R_{318}, X, P, H_{318}, D_{319}) [21]; both enzymes have activity at acidic pH (2.5–6.0) and at temperatures between 40 and 60°C, have low substrate specificity, and are capable of hydrolyzing phytate up to inositol monophosphate (IP1), [26].

TABLE 5: Antigenic determinants of 3-phytase A and of chain A in 3-phytase B from A. *niger*.

Fragment number	Position Initial	Sequence	Position Final	Total Number a.a.
		Antigenic determinants of 3-phytase A (Long. Total = 444 a.a.)		
		Mean antigenic propensity = 1.0304		
1	23	HLWGQYAPFFSLANESVISPEVPAGCRVTFAQVLSR	58	36
2	374	SAWTVPFASRLYVEMMQCQAEQEPLVRVLVNDRVVPLHGCPVDALGR	420	47
		Antigenic determinants of chain A in 3-phytase B (Long. Total = 460 a.a.)		
		Mean antigenic propensity = 1.0234		
1	322	ITPILAALGVLIPNE	336	15
2	378	TYVRLVLNEAVLPFN	392	15

TABLE 6: Results of lower docking energies between 3-phytase A and the A chain of 3-phytase B versus phytic acid, obtained with the Autodock program.

Classification	Energy (kcal/mol)	Number of hydrogen bonds formed	AA involved
		Docking energy between 3-phytase A and phytic acid	
First	−6.3	8	**Arg$_{58}$**, **His$_{59}$**, **Arg$_{62}$**, **Arg$_{62}$**, **Arg$_{142}$**, **Arg$_{142}$**, **His$_{338}$**, Asp$_{339}$
Second	−6.2	7	Tyr$_{28}$, **Arg$_{62}$**, **Arg$_{62}$**, **Arg$_{142}$**, **Arg$_{142}$**, **His$_{338}$**, Asp$_{339}$
Third	−6.0	7	Tyr$_{28}$, **Arg$_{62}$**, **Arg$_{62}$**, **Arg$_{62}$**, Lys$_{277}$, Lys$_{278}$, Asn$_{340}$
Fourth	−6.0	5	Tyr$_{28}$, **His$_{59}$**, **Arg$_{62}$**, **Arg$_{62}$**, Lys$_{278}$
Fifth	−5.9	8	**Arg$_{62}$**, **Arg$_{142}$**, **Arg$_{142}$**, Lys$_{278}$, His$_{282}$, **His$_{338}$**, Asn$_{340}$, Asn$_{340}$
		Docking energy between 3-phytase B chain A and phytic acid	
First	−6.4	3	**Arg$_{62}$**, **Arg$_{66}$**, Tyr$_{276}$
Second	−6.4	4	**Arg$_{66}$**, **Arg$_{66}$**, Ser$_{69}$, Ser$_{71}$
Third	−6.3	6	**Arg$_{62}$**, Ser$_{71}$, Try$_{154}$, **Arg$_{156}$**, **Arg$_{156}$**, Asn$_{275}$
Fourth	−6.1	3	Asn$_{33}$, Ser$_{69}$, Tyr$_{276}$
Fifth	−5.9	4	Asn$_{33}$, Glu$_{272}$, Asn$_{275}$, Tyr$_{276}$

The 2D structure of the monomer (Figure 3) of the 3-phytase A and the A chain of the 3-phytase B homodimer (Figure 3) have very similar conformations, being in higher proportion random coils (45% and 50%, resp.) and alpha helices (43% and 38% resp.) in both proteins and in a lower proportion (12% in both proteins) (Figure 3) which provides a more compact structure to both proteins. However, 3-phytase A presents a 3D structure formed by a single domain containing 20 alpha helices and only 8 beta sheets, while the A chain of the homodimer has a more complex structure made up of two domains with the active site located in the interface (Figure 3, red color). The largest domain consists of 11 alpha helices and 8 beta sheets, and the smallest consists of only 10 alpha helices.

The 3-phytase B has two identical A and B chains, which have 39 amino acids (Lys$_{14}$-Tyr$_{24}$, Leu$_{27}$-His$_{29}$, Tyr$_{36}$, Glu$_{38}$, Ser$_{41}$-Ala$_{45}$, Tyr$_{120}$, Lys$_{217}$, Leu$_{248}$, Pro$_{252}$-Ser$_{254}$, Gln$_{262}$-Asp$_{263}$, Val$_{266}$-Ser$_{267}$, Asn$_{335}$, Arg$_{342}$, Phe$_{345}$-Gly$_{346}$, Ala$_{372}$, Asp$_{393}$, Gly$_{399}$, Tyr$_{400}$) which allow the interactions that give rise to its dimerization (Figure 3). Kostrewa et al. (1999) obtained a crystal of this protein, formed by a tetramer from two homodimers due to its crystallographic symmetry, and identified 17 amino acids involved in the interactions that facilitate the formation of the tetramer, thus achieving greater stability: Cys$_{109}$, Glu$_{114}$, Thr$_{116}$, Gly$_{118}$, Ala$_{121}$, Leu$_{123}$-Leu$_{124}$, Tyr$_{127}$-Asn$_{128}$, Asn$_{131}$, Lys$_{163}$, Glu$_{166}$, Tyr$_{171}$, Arg$_{447}$, Pro$_{450}$-Ile$_{451}$, Cys$_{453}$ (Figure 3).

Two N-acetylglucosamine residues, NAG472 and NAG473, from a chain of carbohydrates bound to Ans$_{172}$ are involved in the formation of the homotetramer in the crystal. These carbohydrate chains do not represent the complete natural glycosylation, but result from partial deglycosylation. The main tetramerization contacts are located on the opposite side of the entrance to the active site, so that the four active sites of the tetramer are exposed to the solvent and are easily accessible to the substrate [21].

In Figure 3, the amino acids forming the highly conserved ligand binding site (RHGXRXP-HD) in both phytases are detailed, which had initially been determined to be mostly positively charged amino acids (Clustal O alignment), because the ligand "phytic acid" is a very negative organic molecule which, thanks to the presence of the 6 phosphate groups (PO_4^{-3}), each of them located in a carbon atom of the inositol ring, binds to this active site by electrostatic interactions of Van der Waals type that generate temporal molecular couplings [27].

The amino acids indicated in red in both phytases correspond to the amino acids that perform the nucleophilic attack, according to the catalytic mechanism proposed by Oh et al. (2004), who assert that the histidine residue in

(a)

(b)

(c)

(d)

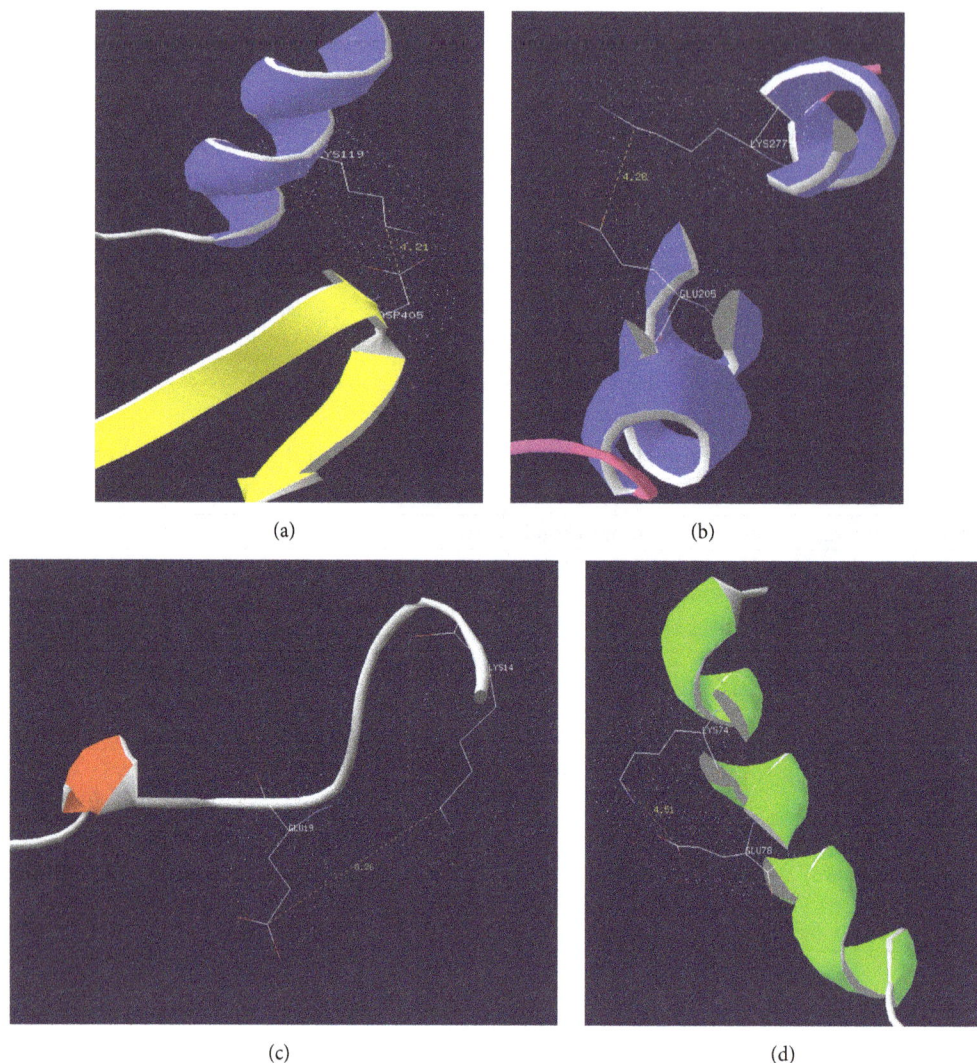

FIGURE 6: *Distances between amino acids involved in glycatio*n, Lysine$_{119}$ (a) and Lysine$_{277}$ (b), and acid residues (Asp$_{405}$ and Glu$_{205}$) in the 3D structure of 3-phytase A. Distances between Lysine$_{14}$ (c) and the Lysine$_{74}$ (d) and acid residues (Glu$_{19}$ and Glu$_{78}$) in the 3D structure of chain A of 3-phytase B.

the highly conserved active site RHGXRXP- serves as a nucleophile in the formation of a covalent phosphohistidine intermediate, while the aspartic acid residue from the C-terminal in the conserved sequence HD serves as a proton donor to the oxygen atom of the cleavable phosphomonoester bond, generating myo-inositol monophosphate as the final product.

The amino acid H$_{338}$ that appears in the legend of Figure 3 between question marks is part of the conserved HD sequence in 3-phytase A that was not reported by Oakley (2010), but that participates in ligand binding; in this research, this amino acid is proposed as a part of the highly conserved active site as observed in molecular docking performed through the Autodock program at a distance of 0.375 Å.

The results obtained by the ProtParam program (Table 1) demonstrated that the 3-phytase A has a length of 444 aa and has a molecular weight of 48.84 kDa. Chain A of 3-phytase B, on the other hand, has a total length of 460 aa in mature state

and a molecular weight of 58.78 kDa. These proteins have a p*I* of 4.94 and 4.6, respectively, which allows them to be classified as acid phytases.

3-phytase A (monomer) has a higher instability index (45.41) compared to chain A of 3-phytase B, whose instability results are inferior (33.66), which allows to catalog the latest as a stable protein (<40 = stable, values >40 = unstable), possibly due to the fact that the initial formation of the homodimer between the identical chains A and B gives it greater stability. Additionally, the formation of five intrachain disulfide bonds in both chains allows them to stabilize their three-dimensional structure.

The half-life of a protein is a prediction of the time it takes half the concentration of a protein in a cell to degrade after its synthesis. The 3-phytase A, being a monomeric protein, has a longer lifespan (>20 hours in yeast, *in vivo*) possibly because the N-terminal group in its sequence is Alanine (Ala1) and the proteins that possess Met, Ser, Ala, Thr, Val, or Gly at the

(1A)

(2A)

(1B)

(2B)

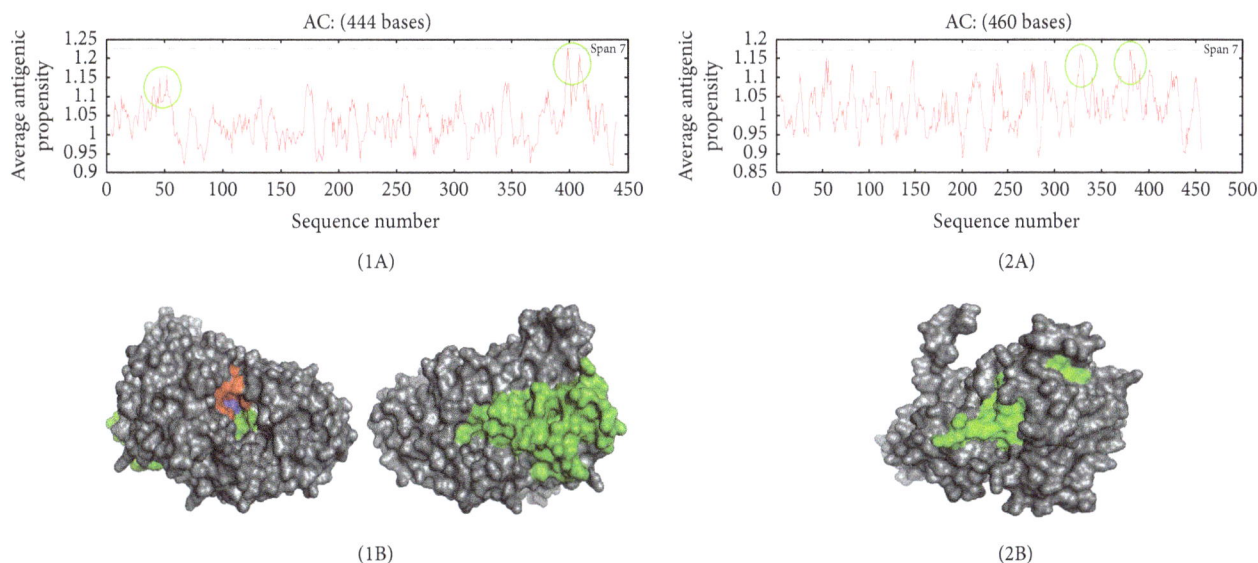

FIGURE 7: *Antigenicity profile.* (1A) 3-Phytase A from *A. niger*; (1B) *location of antigenicity peaks (green color).* 3-Phytase A from *A. niger* and RHGXRXP-HD active site ($\underline{R_{58}}$, $\mathbf{H_{59}}$, R_{62}, R_{142}, H_{338}, and $\mathbf{D_{339}}$); (2A) chain A in 3-phytase B from *A. niger*; (2B) *location of antigenicity peaks (green color).*

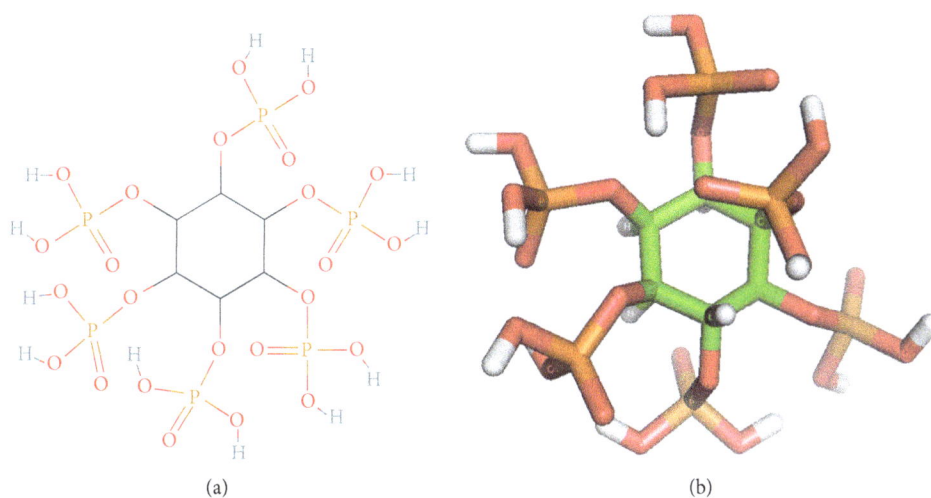

(a)

(b)

FIGURE 8: *3D structure of phytic acid (myo-inositol 1, 2, 3, 4, 5, 6 hexakisphosphate).* (a) Diagram obtained from Pubchem; (b) 3D model generated by the Spartan 4.0 program and visualized with PyMOL.

N-terminal position register lifespans greater than 20 hours [28, 29].

On the contrary, chain A of 3-phytase B yielded a result of 3 minutes of lifespan in yeast, possibly because the N-terminal group in its sequence is phenyl alanine (Phe$_1$) and the proteins that have Phe, Leu, Asp, Lys, or Arg at the N-terminal position register lifespans of less than 3 minutes [28, 29]. It appears that this factor involves the ubiquitin system, which is a small protein (76 amino acids) found in all eukaryotic cells and that undergoes an ATP-dependent reaction with proteins, condensing their C-terminal glycine residues with groups of amino of lysines of the protein to

be labeled. These modified proteins are degraded shortly afterwards by a proteolytic complex which is recognized by the ubiquitin marker and because of this their lifespan is very short [28, 29]. The biological importance of the calculation of this parameter lies in the fact that the production of recombinant enzymes takes into account both the lifespan of the proteins to be expressed as well as their stability, in such way that a reduction of the degradation rate of proteins from heterologous genes is achieved.

The results of the aliphatic index of the two revised phytases from *A. niger*, 3-phytase A and 3-phytase B (72.25 and 70.46, resp.), allow the consideration of the fact that the

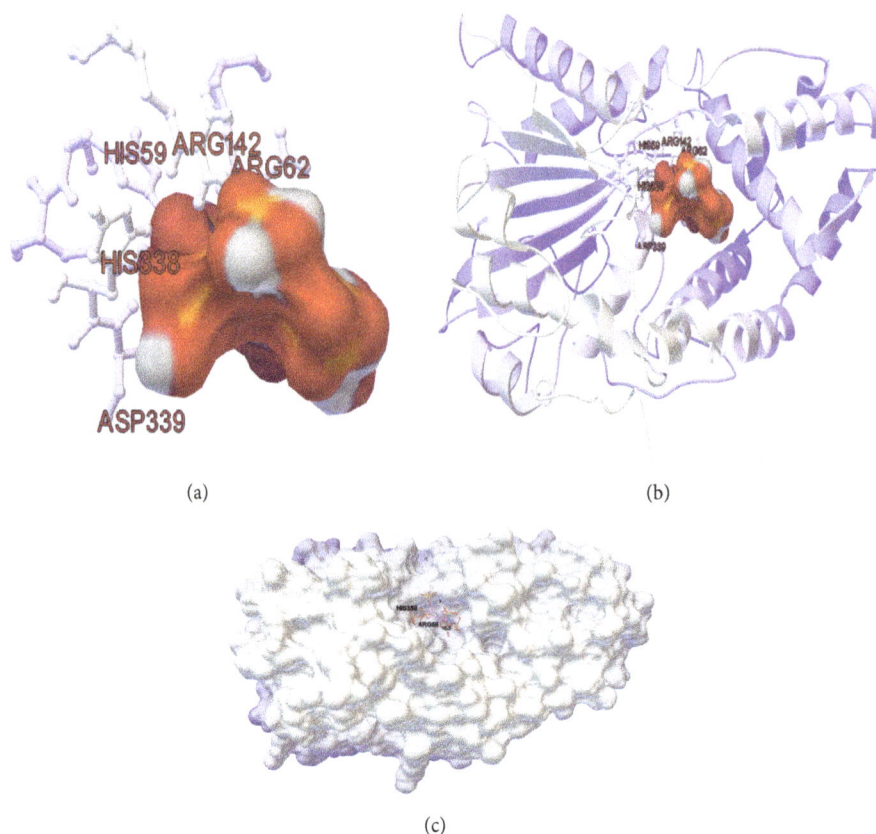

(a)

(b)

(c)

FIGURE 9: *Docking result between 3-phytase A (monomer) and phytic acid.* (a) Active site (RHGXRXP-HD) consisting of residues R_{58}, $\mathbf{H_{59}}$, R_{62}, R_{142}, H_{338}, and $\mathbf{D_{339}}$ versus phytic acid. (b) Ribbons diagram of the amino acids that make up the active site of the protein and (c) surface diagram of the phytic acid ligand attached to the pocket of the active site of the protein, visualized with the program Autodock.

relative volume occupied by its aliphatic side chains Ala, Val, Ile, and Leu increases thermostability in both phytases (both are globular secretory proteins).

The results of GRAVY (grand average of hydropathy) in 3-phytase A and 3-phytase B (-0.304 and -0.33, resp.) are obtained by combining the values of hydrophobicity and hydrophilicity of the side chains in their sequences. These negative values explain the reason why they tend to interact with aqueous media, typical of secretory proteins such as extracellular phytases [27].

The hydrophobicity profile allowed identifying that amino acids with a high score (Ile_{345} y Leu_{346} in 3-phytase A and Leu_{329} in 3-phytase B, Table 2) were found in areas with little exposure (inside alpha helices, Figures 4(c)–4(f)) in both proteins because such amino acids have aliphatic side chains that do not interact easily with aqueous solvents. On the contrary, amino acids with a minimum hydrophobicity score were found in exposed areas of the proteins (Ala_{164} in 3-phytase A and Glu_{65} in 3-phytase B). In the case of Glu_{65}, being an amino acid whose R group does not have positive or negative charges at physiological pH, that is, pH close to 6.5 and 7.0, allows it to be solubilized more easily in aqueous solvents and, in the case of Ala_{164}, to have a short aliphatic side chain that allows it to interact more easily with the aqueous medium. It should be noted that none of these

amino acids were a part of the active site of ligand binding in both phytases. The amino acids indicated within the black circle in Figure 4(e) are involved in the formation of the homodimer in 3-phytase B and therefore are not found within an exposed zone of the protein. In general terms, 3-phytase A and 3-phytase B present few hydrophobic regions, as expected in secretory proteins [27].

The accessibility profile allowed the identification of the amino acids more or less exposed to the solvent, according to the score obtained and recorded in Table 2. For the 3-phytase A and the 3-phytase B, the Gly_{69} and the Ser_{71}, respectively, were located in areas that were very exposed to the solvent and that were not a part of the active ligand binding site (Figures 5(c)–5(f)). The amino acid glycine has a simple structure, is the smallest amino acid, and is the only nonchiral amino acid, characteristics that allow it to acquire special conformations that other amino acids can not, and for this reason obtains a high solvent accessibility score. As for the serine amino acid, although it has an uncharged R polar group (-OH), it is short, very reactive, and hydrophilic, with a tendency to form hydrogen bonds with water.

The amino acids with the lowest accessibility score (Glu_{387} in 3-phytase A and Gln_{56} in 3-phytase B) were located inside the protein as a part of beta sheets (Figures 5(c)–5(f)) because glutamine is a polar amino acid and glutamic acid

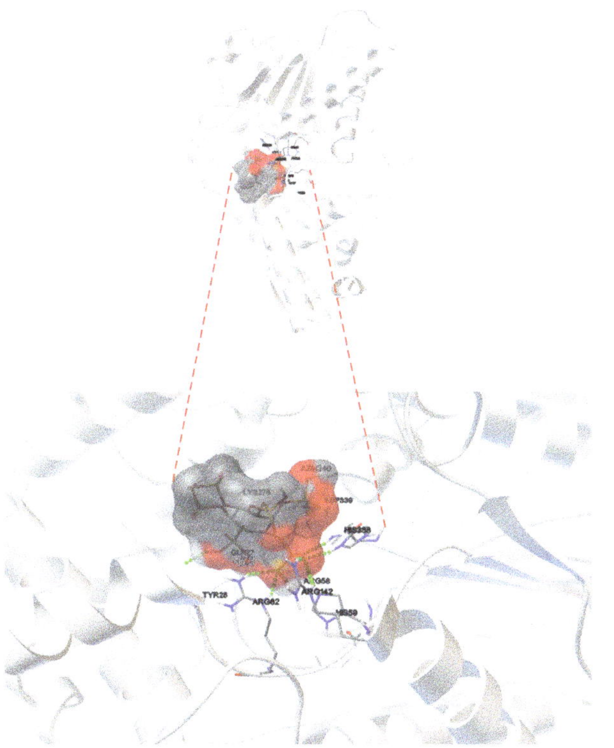

FIGURE 10: *General overview of the docking result between 3-phytase A and phytic acid (lowest energy = −6.3 kcal/mol). Active site (RHGXRXP-HD) consisting of residues* R_{58}, H_{59}, R_{62}, R_{142}, H_{338}, *and* D_{339} *versus phytic acid, visualized with the Autodock program. The red colored areas in the phytic acid correspond to regions with negative charge. Green dots refer to the H Bridges established between the ligand and the amino acids that form the active site in the protein.*

is negatively charged, which reduces its exposure to the solvent. According to the hydrophobicity profile, 3-phytase A and 3-phytase B present a high proportion of zones of easy accessibility to the aqueous medium along their sequences.

Secretory proteins, such as phytase enzymes, have carbohydrate addition or glycosylation sites that allow them to be recognized in the rough endoplasmic reticulum for their future correct folding and secretion [30, 31]. Using the NetNGlyc 1.0 bioinformatics program, 9 N-glycosylation sites for 3-phytase A (monomer) and 7 N-glycosylation sites for 3-phytase B (dimer) chain A were established. The positions of the Asparagines (N) along both phytase chains that were located in an Asn-Xaa-Ser/Thr section (where Xaa is any amino acid except proline) could be glycosylated (Table 3). It stands out that the seven possible N-glycosylation sites predicted for chain A of 3-phytase B must be duplicated because this phytase is formed by two identical chains, A and B, that initially form a homodimer; for that reason, it would have a total of 14 possible N-glycosylation sites for this phytase. The carbohydrate that binds directly to these N-glycosylation sites is normally *N*-acetylglucosamine [21, 22]. These added sugars will promote the correct folding of the phytases, deducing a mechanism of quality control of

synthesis and assembly of the proteins, thus increasing its stability [32].

In glycation, the initial reversible reaction occurs between aldehyde or ketone groups of reducing sugars and $\varepsilon\text{-NH}_2$ groups of lysines or the amino terminal of the protein. Subsequently, there is formation of Amadori products and finally AGEs *(Advanced Glycation End products)* are formed [33]. In general, the amino groups with the lowest pKa value should be more reactive towards glycation due to their nucleophilic capacity [34]. In the case of lysines, it has been suggested that the proximity of nearby residues plays a determining role to be or not glycated. The positively charged amino acids located near the primary structure or the three-dimensional structure decrease the pKa and thereby catalyze the glycation of such lysines [35]. Likewise, it has been suggested that the proximity of an acid residue to a lysine catalyzes the formation of Amadori products, which would make lysine more reactive to be glycated [36]. The results of the comparison of lysine prediction by the Netglycate algorithm [16], which exclusively considers the primary sequence of the protein and the methodology proposed by Sáenz et al. (2016), which uses the 3D structure of the protein, are shown in Table 4 and allow us to point out that, for the case of 3-phytase A, according to the proposal of the spatial relationship between the $\varepsilon\text{-NH}_2$ group and side chains of acidic or basic residues as a requirement for glycation, 14 lysines with distances inferior to 9.89 are considered as potentially glycable. Although these distances exist in the 3D structure, only seven were considered by the algorithm Netglycate 1.0. The only lysine without glycation prediction by the two methodologies presented distances above 13.06 Å between the $\varepsilon\text{-NH}_2$ group and the side chains of basic or acidic residues, a distance that would not allow the chemical interaction between these chemical groups as a requirement for glycation.

For the case of chain A of 3-phytase B, of the 10 lysines susceptible to being glycated as proposed by Sáenz et al. (2016), all with distances lower than 9.48 Å, only two are considered by Netglycate 1.0. The third lysine with prediction of glycation (K_{413}) is not considered by this proposal since the group $\varepsilon\text{-NH}_2$ is separated from the side chains of basic or acidic residues by distances greater than 23.84 Å, a distance that would not allow the chemical interaction between these chemical groups as a requirement for glycation. The only lysine not considered by the two methodologies is K_{217}, whose $\varepsilon\text{-NH}_2$ group is located away from the side chains of basic or acidic residues by distances greater than 13.15 or 11.97 Å, respectively, distances that would not allow the chemical interaction between these chemical groups as a requirement for glycation.

Ninety percent (9/10) of the Netglycate 1.0 predictions, as proposed by Sáenz et al. (2016), can be associated with the spatial relationship between lysines and acidic or basic residues less than 10 Å and the remaining 10% (1/10), the K_{413} of 1QFX, presents distances greater than 23.84 Å to acidic or basic residues even though, according to the data provided by the algorithm Netglycate 1.0, its probability (score) of occurrence of glycation is scarcely 59.8% (data not

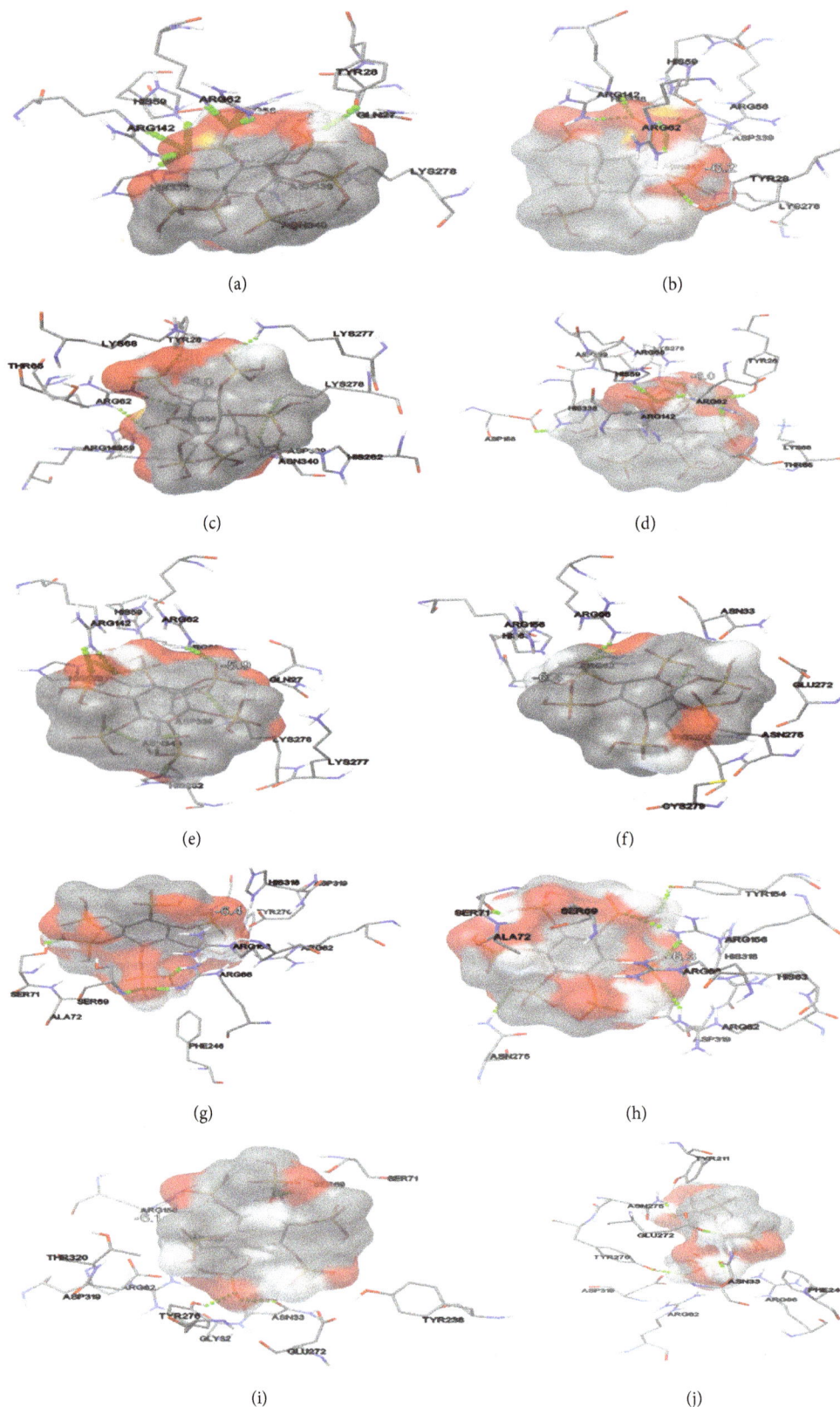

FIGURE 11: *Van der Waals electrostatic interactions involving the formation of hydrogen bonds (green dots)* between the active site amino acids and/or those closest to it in 3-phytase A (a–e), chain A of 3-phytase B (f–j), and the oxygens or hydrogens of the phytic acid ligand. Red colored regions can be seen, corresponding to negative regions in the electrostatic cloud of the ligand that make contact with the amino acids of the protein.

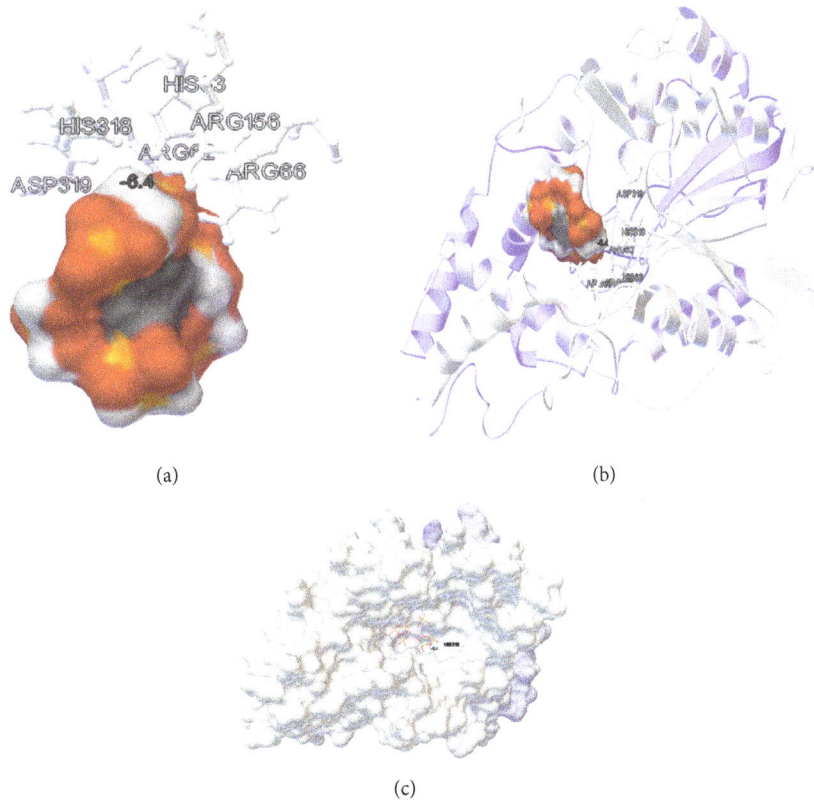

(a)

(b)

(c)

FIGURE 12: *Result of the docking between the A Chain of the homodimer from 3-phytase B and the phytic acid.* (a) Active site (RHGXRXP-HD) consisting of residues R_{62}, **H_{63}**, R_{66}, R_{156}, H_{318}, and **D_{319}** versus phytic acid. (b) Ribbons diagram of the amino acids that compose the active site of the protein and (c) surface diagram of the phytic acid ligand attached to the pocket of the active site of the protein, visualized with the program Autodock.

shown), being the lowest of all; however, it is predicted to be potentially glycable.

These results are in accordance with what Sáenz et al. (2016) proposed, because it is not necessarily the sequence (primary structure) but the spatial relationship in the 3D structure that favors the lysines glycation, provided that the distances of lysines to acidic or basic residues are less than 10 Å (Figure 6). In this sense, this type of enzymes with high percentages of acidic and basic residues and lysines close, in both the primary structure and the 3D structure, to acid residues or other basic residues generates a high probability of chemical interaction between that type of amino acids, required for glycation. Finally, the Netglycate algorithm only predicted as glycable lysines 37.5% of those proposed by Sáenz et al. (2016).

"In vitro" investigations with other proteins that have been brought into contact with reducing sugars show that glycation may affect biological activity [37]. Assays performed with recombinant human interferon-gamma (hIFN-γ) glycoprotein isolates in *E. coli* demonstrated that such purified protein was also prone to progressive proteolysis and covalent dimerization during storage, since late glycation stages cause the cleavage of the peptide bond and the covalent reticulation in lysine and arginine residues (but not of cysteine) [38]; that is to say that glycation promotes protein fragmentation

and is produced in glycated lysines [17]. However, the *"in vitro"* effect caused in proteins should be studied carefully to correctly determine the cause-effect relationship, since the observed phenomena could be the consequence of the glycation of other components that can interact with the proteins or of the reactions between the protein and some by-product generated during glycation [37]. Therefore, the identification of these potential glycation sites in 3-phytase A and 3-phytase B chain A could represent potential sites of fragmentation of the concentrated or purified proteins during prolonged storage times, negatively affecting their biological activity.

In 3-phytase A, the average antigenic propensity was 1.0304, and when the average value is higher than 1.0, the amino acids that are above 1.0 will be potentially antigenic. According to the data recorded in Table 5 and compared to Figure 7(1A) (Green colored circles), two highly antigenic peaks or regions can be identified. The first region groups the amino acids from His_{23}-Arg_{58}, 36 amino acids in total including the amino acid Arg_{58}, which is the first amino acid that forms part of the ligand binding active site and is therefore part of a solvent accessible zone. The second region registers the highest peak of antigenicity and integrates a greater amount of amino acids from Ser_{374}-Arg_{420}, 47 amino acids in total, being located in the opposite side to the active

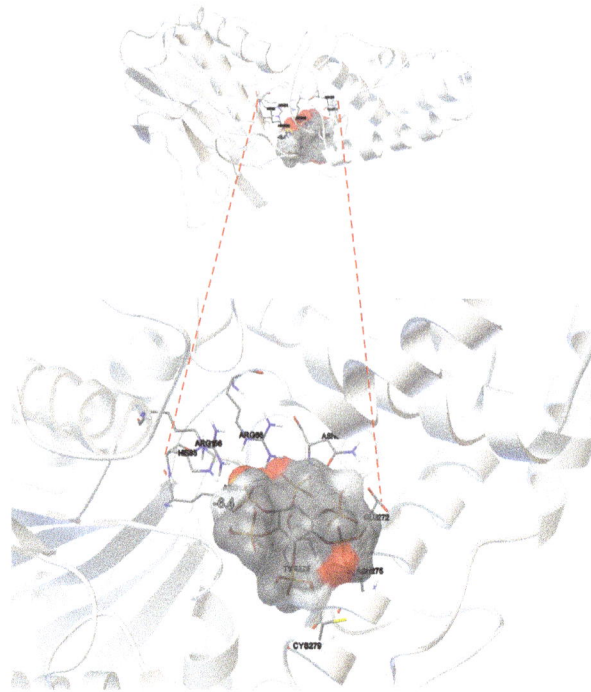

FIGURE 13: *General overview of the docking result between the A chain of the 3-phytase B homodimer and the phytic acid (lowest energy =* *−6.4 kcal/mol).* Active site (R**H**GXRXP-**HD**) consisting of residues R_{62}, $\mathbf{H_{63}}$, R_{66}, R_{156}, H_{318}, and $\mathbf{D_{319}}$ versus phytic acid, visualized with the Autodock program. The red colored areas in the phytic acid correspond to regions with negative charge.

site and in a zone highly exposed to the solvent, as can be observed in Figure 7(1B). The prediction of antigenic peptides takes into account which peptide fragments of a protein are likely to be antigenic. These antigenic fragments should be located in solvent accessible regions and should contain hydrophobic and hydrophilic residues. Therefore, the second region which comprises the highest amount of amino acids and is completely exposed to the solvent would have a higher antigenicity.

In chain A of 3-phytase B, the average antigenic propensity is 1.0234. In contrast to 3-phytase A, the 3-phytase B chain A has more peaks or highly antigenic regions that cluster fewer amino acids; however, according to the data recorded in Table 5 and compared to Figure 7(2A) (green colored circles), two highly antigenic peaks or regions can be identified. The first region groups the amino acids Ile_{322}-Glu_{336}, 15 amino acids in total, located in an area highly exposed to the solvent. The second region contains amino acids Thr_{338}-Asn_{392}, 15 amino acids in total, but not all of them are exposed to the solvent. Five of these 15 are hydrophobic (Val_{380}, Leu_{382}, Val_{383}, Leu_{384}, and Val_{388}) and therefore are located in the interior of the protein (Figure 7(2B)). Therefore, antigenic fragments of the first region, which are all located in solvent accessible regions and contain hydrophobic and hydrophilic residues, would exhibit greater antigenicity.

The positions of the Asparagines (N) that were identified as potential N-glycosylation sites and which are a part of the reported antigenic determinants are highlighted in bold in Table 5 because this is a factor that may contribute to the induction of an immune response by both phytases [39].

Considering the usefulness of phytases as a dietary supplement in monogastric animals, it is also pertinent to consider that the presence of regions with high antigenic propensity along the sequence of these proteins could be translated in the presence of allergens that could trigger an allergic reaction in the host, who ingests them. The process of digestion involves mechanical, chemical, and biochemical processes that allow macronutrients to be transformed into simpler molecules that can be absorbed and used by animals. But despite the fact that these digestive processes take place, significant amounts of protein from diets and that are immunologically active reach the intestinal mucosa of monogastric animals [40]. When there is an actual allergic reaction, the body produces antibodies (proteins that specifically bind to allergens to neutralize and remove them from the body). There are different types of antibodies, but the responsible for allergic reactions to food is known as immunoglobulin E (IgE). The IgE antibody binds to the allergens, triggering an allergic reaction. During this reaction, IgE activates the segregation of signaling molecules in the bloodstream, which simultaneously causes the common symptoms of food allergies such as skin rashes, inflammation, abdominal pain and inflammation, vomiting, and diarrhea [40]. However, in several investigations performed [41–43] in animals, no reports were found on allergic reactions provoked by phytases to the animals involved in the trials.

In the research conducted by Kostrewa et al. (1999) and Oakley (2010), the ligand used to obtain the crystalline structure of 3-phytase A and 3-phytase B, respectively, was myo-inositol-1,2,3,4,5,6-hexakis sulfate (IHS), a potent inhibitor of such enzymes. This chemical compound is isosteric and isoelectric with respect to myo-inositol 1, 2, 3, 4, 5, 6 hexakisphosphate (IHP) and is considered an excellent analogous substrate.

However, the ligand used in this investigation corresponded to the chemical compound myo-inositol 1, 2, 3, 4, 5, 6 hexakisphosphate (IHP), also called phytic acid, the main form of phosphorus storage in the cereals that make up the diet of monogastric animals. This chemical compound is highly negative due to the presence of 6 phosphate groups (PO_4^{-3}) in its inositol ring, which is why the active site of binding to this ligand in phytases is composed mainly of positively charged amino acids (RHGXRXP-HD), [21, 22].

The molecular coupling model (Rigid Docking) directed to the catalytic pocket of 3-phytase A formed by residues Arg_{58}, **His_{59}**, Arg_{62}, Arg_{142}, His_{338}, and **Asp_{339}** [22], and phytic acid as ligand, yielded very interesting results that allowed establishing the formation of Van der Waals electrostatic interactions that generated hydrogen bonds between the amino acids that form the active site of the protein and the oxygen or hydrogen of the phosphate groups of phytic acid (Figure 11). The lowest 5 energies of the molecular coupling result were selected (Table 6) and it was possible to determine which amino acids were forming the hydrogen bonds.

Table 6 shows that in the case of 3-Phytase B there are 6 amino acids (Arg_{58}, **His_{59}**, Arg_{62}, Arg_{142}, His_{338}, and **Asp_{339}**) of the active center involved in the formation of 8 hydrogen bonds with phytic acid, generating the lower energy in docking (−6.3 kcal/mol), which is favorable because the greater number of hydrogen bonds formed between the active site of the enzyme and the ligand favors the stability of this temporary molecular interaction [27].

Research by Oakley (2010) reported that the amino acids that form the active site of the protein and therefore establish electrostatic Van der Waals type interactions with the analogous IHP ligand in the crystal by the X-ray diffraction method at a resolution of 2.20 Å are: Arg_{58}, **His_{59}**, Arg_{62}, Arg_{142} and **Asp_{339}**. However, the active site of the protein in histidine acid phosphatases presents highly conserved residues (RHGXRXP-HD), involving a histidine in the HD segment. By means of the molecular coupling (Rigid Docking) carried out in this investigation at a distance of 0.375 Å, it possible to determine that the His_{338} is involved in the formation of a hydrogen bond with the phytic acid ligand that was not previously reported by Oakley (2010), but was consistent with the information reported in the PDBsum database [22].

For the molecular coupling model (Rigid Docking) directed to the catalytic pocket of chain A of 3-phytase B, formed by residues Arg_{62}, **His_{63}**, Arg_{66}, Arg_{156}, His_{318}, and

Asp_{319} [21] and phytic acid as ligand, the formation of Van der Waals electrostatic interactions that generated hydrogen bonds between the amino acids that form the active site of the protein and the oxygens or hydrogens from the phosphate groups of phytic acid was also detected (Figure 11). The lowest 5 energies of the molecular coupling result were selected (Table 6) and it was possible to determine which amino acids were forming the hydrogen bonds.

Table 6 shows that in the case of 3-Phytase B there are 6 amino acids (Arg_{62}, Ser_{71}, Try_{154}, Arg_{156}, Arg_{156}, and Asn_{275}) of the active center involved in the formation of 4 hydrogen bonds with phytic acid, generating the lower energy in docking (−6.3 kcal/mol).

It is interesting to note that Kostrewa et al., (1999) determined that the active site of 3-phytase B is subdivided into a catalytic center (R_{62}, **H_{63}**, R_{66}, R_{156}, H_{318} y **D_{319}**) and a substrate specificity site (Asp_{75} and Glu_{272}); however, only the amino acid Arg_{66} forms part of this active site and participates in the formation of the hydrogen bonds with the analogous substrate IHS in the crystal [21].

Table 6 shows the energy of docking number 2 (−6.4 Kcal/mol) which involves 3 of these amino acids (Arg_{66}, Ser_{69} and Ser_{71}); although they do not form a part of the active center of the protein, they do form hydrogen bonds with the analog substrate IHS in the crystal. In addition, they are reported in PDBsum. Therefore, taking into account the fact that the stability of this temporal molecular interaction depends mainly on the number of hydrogen bonds formed, the energy that would offer greater stability would be number 3, since it additionally involves a greater number of amino acids than those that are located in the active site of the enzyme [27].

5. Conclusions

The species *Aspergillus niger* expresses two phytases currently reported by the UNIPROT database: 3-phytase A (PDB ID: 3K4Q) corresponding to a monomer and 3-phytase B (PDB ID: 1QFX) corresponding to a homodimer (chains A and B) which, due to its crystallographic symmetry, generates a homotetramer from two dimers. These phytases have been crystallized and the genes encoding them (*phy*A and *phy*B gene, resp.) have been cloned and overexpressed in other microorganisms, which has allowed them to be widely used in the feed industry of monogastric animals. The computational characterization of the two phytases produced by *A. niger*, 3-phytase A and 3-phytase B, made it possible to establish that both phytases belong to the histidine acid phosphatases class, with the active ligand binding site (RHGXRXP-HD) highly conserved. The 3-phytase A and the 3-phytase B chain A possess a molecular length and a molecular weight that do not differ substantially, although the monomer is considered as an unstable protein and the homodimer has a shorter lifespan. The aliphatic index in both phytases allows to conclude that they are thermostable enzymes. The hydrophobicity profiles and accessibility showed that these phytases interact with aqueous media, which is a characteristic of secretory proteins. It was possible to identify possible glycosylation and glycation sites in both phytases, which could affect the correct folding of the proteins and their possible fragmentation during prolonged storage times and therefore their biological activity. Both 3-phytases, A and B, exhibited areas with high antigenic propensity which could affect the immune system of the animal that ingests them. Finally, the molecular coupling models in both phytases allowed verifying the formation of electrostatic interactions of Van der Waals type that generates hydrogen bonds between the amino acids that form the active center of the protein and the oxygens or hydrogens of the phosphate groups of phytic acid, providing greater stability to these temporary molecular interactions.

Authors' Contributions

Doris C. Niño-Gómez and Claudia M. Rivera-Hoyos contributed equally to this work.

Acknowledgments

This work was supported by Pontificia Universidad Javeriana, Bogotá, Colombia (Grant ID 00007390) titled *"Clonación y expresión constitutiva de las Fitasas 3-Fitasa A y 3-Fitasa B de Aspergillus niger bajo el control del pGAP en Pichia pastoris,"* by Universidad Nacional de Colombia, Bogotá, Colombia (Grant DIB 201010020419) and by Universidad Centroccidental Lisandro Alvarado, Barquisimeto, Venezuela (Grant CDCHT ICS-2017-6). The authors thank María Lucía Gutiérrez for English editing.

References

[1] B. L. Lim, P. Yeung, C. Cheng, and J. E. Hill, "Distribution and diversity of phytate-mineralizing bacteria," *The ISME Journal*, vol. 1, no. 4, pp. 321–330, 2007.

[2] A. Acosta and M. Cárdenas, "Enzimas en la alimentación de las aves. Fitasas," *Revista Cubana de Ciencia Agrícola*, vol. 40, no. 4, pp. 377–387, 2006.

[3] Y. Zhu, X. Qiu, Q. Ding, M. Duan, and C. Wang, "Combined effects of dietary phytase and organic acid on growth and phosphorus utilization of juvenile yellow catfish *Pelteobagrus fulvidraco*," *Aquaculture*, vol. 430, pp. 1–8, 2014.

[4] P. Vats and U. C. Banerjee, "Production studies and catalytic properties of phytases (myo-inositolhexakisphosphate phosphohydrolases): an overview," *Enzyme and Microbial Technology*, vol. 35, no. 1, pp. 3–14, 2004.

[5] S. Dahiya and N. Singh, "Isolation and biochemical characterization of a novel phytase producing bacteria *Bacillus cereus* isolate MTCC 10072," *International Journal of Microbial Resource Technology*, vol. 2, no. 2, pp. 1–5, 2014.

[6] M. Lamid, N. N. T. Puspaningsih, and O. Asmarani, "Potential of phytase enzymes as biocatalysts for improved nutritional value of rice bran for broiler feed," *Journal of Applied Environmental and Biological Sciences*, vol. 4, no. 3, pp. 377–380, 2014.

[7] A. Casey and G. Walsh, "Identification and characterization of a phytase of potential commercial interest," *Journal of Biotechnology*, vol. 110, no. 3, pp. 313–322, 2004.

[8] G. E. A. Awad, M. M. I. Helal, E. N. Danial, and M. A. Esawy, "Optimization of phytase production by *Penicillium purpurogenum* GE1 under solid state fermentation by using Box-Behnken design," *Saudi Journal of Biological Sciences*, vol. 21, no. 1, pp. 81–88, 2014.

[9] A. Pandey, G. Szakacs, C. R. Soccol, J. A. Rodriguez-Leon, and V. T. Soccòl, "Production, purification and properties of microbial phytases," *Bioresource Technology*, vol. 77, no. 3, pp. 203–214, 2001.

[10] The UniProt Consortium, "Reorganizing the protein space at the Universal Protein Resource (UniProt)," *Nucleic Acids Research*, vol. 40, no. D1, pp. D71–D75, 2012.

[11] S. F. Altschul, T. L. Madden, A. A. Schäffer et al., "Gapped BLAST and PSI-BLAST: a new generation of protein database search programs," *Nucleic Acids Research*, vol. 25, no. 17, pp. 3389–3402, 1997.

[12] W. Li, A. Cowley, M. Uludag et al., "The EMBL-EBI bioinformatics web and programmatic tools framework," *Nucleic Acids Research*, vol. 43, no. 1, pp. W580–W584, 2015.

[13] H. Berman, K. Henrick, and H. Nakamura, "Announcing the worldwide Protein Data Bank," *Nature Structural Biology*, vol. 10, no. 12, p. 980, 2003.

[14] E. Gasteiger, C. Hoogland, A. Gattiker et al., *Protein Identification and Analysis Tools on the ExPASy Server*, Humana Press, New York, NY, USA, 2005.

[15] G.-Y. Chuang, J. C. Boyington, M. Gordon Joyce et al., "Computational prediction of N-linked glycosylation incorporating structural properties and patterns," *Bioinformatics*, vol. 28, no. 17, Article ID bts426, pp. 2249–2255, 2012.

[16] M. B. Johansen, L. Kiemer, and S. Brunak, "Analysis and prediction of mammalian protein glycation," *Glycobiology*, vol. 16, no. 9, pp. 844–853, 2006.

[17] H. Sáenz-Suárez, R. A. Poutou-Piñales, J. González-Santos et al., "Prediction of glycation sites: new insights from protein structural analysis," *Turkish Journal of Biology*, vol. 40, no. 1, pp. 12–25, 2016.

[18] N. Guex and M. C. Peitsch, "SWISS-MODEL and the Swiss-PdbViewer: an environment for comparative protein modeling," *Electrophoresis*, vol. 18, no. 15, pp. 2714–2723, 1997.

[19] A. S. Kolaskar and P. C. Tongaonkar, "A semi-empirical method for prediction of antigenic determinants on protein antigens," *FEBS Letters*, vol. 276, no. 1-2, pp. 172–174, 1990.

[20] G. M. Morris, H. Ruth, W. Lindstrom et al., "Software news and updates AutoDock4 and AutoDockTools4: automated docking with selective receptor flexibility," *Journal of Computational Chemistry*, vol. 30, no. 16, pp. 2785–2791, 2009.

[21] D. Kostrewa, M. Wyss, A. D'Arcy, and A. P. G. M. Van Loon, "Crystal structure of *Aspergillus niger* pH 2.5 acid phosphatase at 2.4 Å resolution," *Journal of Molecular Biology*, vol. 288, no. 5, pp. 965–974, 1999.

[22] A. J. Oakley, "The structure of *Aspergillus niger* phytase PhyA in complex with a phytate mimetic," *Biochemical and Biophysical Research Communications*, vol. 397, no. 4, pp. 745–749, 2010.

[23] J. Söding, "Protein homology detection by HMM-HMM comparison," *Bioinformatics*, vol. 21, no. 7, pp. 951–960, 2005.

[24] B.-C. Oh, W.-C. Choi, S. Park, Y.-O. Kim, and T.-K. Oh, "Biochemical properties and substrate specificities of alkaline and histidine acid phytases," *Applied Microbiology and Biotechnology*, vol. 63, no. 4, pp. 362–372, 2004.

[25] E. J. Mullaney and A. H. J. Ullah, "The term phytase comprises

several different classes of enzymes," *Biochemical and Biophysical Research Communications*, vol. 312, no. 1, pp. 179–184, 2003.

[26] E. Humer, C. Schwarz, and K. Schedle, "Phytate in pig and poultry nutrition," *Journal of Animal Physiology and Animal Nutrition*, vol. 99, no. 4, pp. 605–625, 2015.

[27] D. L. Nelson and M. M. Cox, *Lehninger, Principles of Biochemistry*, Freeman and Company, New York, NY, USA, 5th edition, 2005.

[28] M. Hochstrasser, "Ubiquitin, proteasomes, and the regulation of intracellular protein degradation," *Current Opinion in Cell Biology*, vol. 7, no. 2, pp. 215–223, 1995.

[29] A. Varshavsky, "The N-end rule: functions, mysteries, uses," *Proceedings of the National Academy of Sciences of the United States of America*, vol. 93, no. 22, pp. 12142–12149, 1996.

[30] A. J. Parodi, "Role of N-oligosaccharide endoplasmic reticulum processing reactions in glycoprotein folding and degradation," *Biochemical Journal*, vol. 348, no. 1, pp. 1–13, 2000.

[31] A. Helenius and M. Aebi, "Intracellular functions of N-linked glycans," *Science*, vol. 291, no. 5512, pp. 2364–2369, 2001.

[32] R. S. López, "El proceso de N-glicosilación de proteínas en *Entamoeba histolytica*," *Revista Latinoamericana de Microbiología*, vol. 48, no. 2, pp. 70–72, 2006.

[33] P. Verbeke, B. F. C. Clark, and S. I. S. Rattan, "Modulating cellular aging in vitro: hormetic effects of repeated mild heat stress on protein oxidation and glycation," *Experimental Gerontology*, vol. 35, no. 6-7, pp. 787–794, 2000.

[34] H. F. Bunn, R. Shapiro, M. McManus et al., "Structural heterogeneity of human hemoglobin A due to nonenzymatic glycosylation," *Journal of Biological Chemistry*, vol. 254, no. 10, pp. 3892–3898, 1979.

[35] N. Iberg and R. Fluckiger, "Nonenzymatic glycosylation of albumin *in vivo*. Identification of multiple glycosylated sites," *Journal of Biological Chemistry*, vol. 261, no. 29, pp. 13542–13545, 1986.

[36] J. W. Baynes, N. G. Watkins, C. I. Fisher et al., "The Amadori product on protein: structure and reactions," *Progress in Clinical and Biological Research*, vol. 304, no. 24, pp. 43–67, 1989.

[37] F. L. González Flecha, P. R. Castello, J. J. Gagliardino, and J. P. F. C. Rossi, "La glucosilación no enzimática de proteínas. Mecanismo y papel de la reacción en la diabetes y el envejecimiento," *Ciencia al Día Internacional*, vol. 3, no. 2, pp. 1–17, 2000.

[38] R. Mironova, T. Niwa, R. Dimitrova, M. Boyanova, and I. Ivanov, "Glycation and post-translational processing of human interferon-γ expressed in *Escherichia coli*," *Journal of Biological Chemistry*, vol. 278, no. 51, pp. 51068–51074, 2003.

[39] M. López-Hoyos, G. Fernández-Fresnedo, H. López-Escribano, and A. de Francisco, "Antigenicidad de las proteínas recombinantes," *Gaceta Médica de Bilbao*, vol. 101, no. 1, pp. 17–21, 2004.

[40] J. L. C. Muñoz, "Antigenicidad de piensos y materias primas proteicas en conejos," in *Departamento de Ciencia Animal*, p. 120, Universidad Politécnica de Valéncia, Valéncia, España, 2016.

[41] Y. B. Wu, V. Ravindran, and W. H. Hendriks, "Effects of microbial phytase, produced by solid-state fermentation, on the performance and nutrient utilisation of broilers fed maize- and wheat-based diets," *British Poultry Science*, vol. 44, no. 5, pp. 710–718, 2003.

[42] J. P. Zhou, Z. B. Yang, W. R. Yang, X. Y. Wang, S. Z. Jiang, and G. G. Zhang, "Effects of a new recombinant phytase on the performance and mineral utilization of broilers fed phosphorus-deficient diets," *Journal of Applied Poultry Research*, vol. 17, no. 3, pp. 331–339, 2008.

[43] D. Šefer, B. Petrujkić, R. Marković et al., "Effect of phytase supplementation on growing pigs performance," *Acta Veterinaria*, vol. 62, no. 5-6, pp. 627–639, 2012.

19

Molecular Analysis of *CYP21A2* Gene Mutations among Iraqi Patients with Congenital Adrenal Hyperplasia

Ruqayah G. Y. Al-Obaidi,[1] **Bassam M. S. Al-Musawi,**[2] **Munib Ahmed K. Al-Zubaidi,**[3] **Christian Oberkanins,**[4] **Stefan Németh,**[4] **and Yusra G. Y. Al-Obaidi**[5]

[1]*Genetic Counseling Clinic and Genetics Laboratory, The Teaching Laboratories, Medical City, Baghdad, Iraq*
[2]*Department of Pathology, College of Medicine, Baghdad University, Baghdad, Iraq*
[3]*Department of Pediatrics, College of Medicine, Baghdad University and Pediatric Endocrine Consultation Clinic,*
 Children Welfare Hospital, Baghdad, Iraq
[4]*ViennaLab Diagnostics GmbH, Gaudenzdorfer Guertel 43-45, 1120 Vienna, Austria*
[5]*National Center of Hematology, Al-Mustansiriya University, Baghdad, Iraq*

Correspondence should be addressed to Bassam M. S. Al-Musawi; abmsadik@yahoo.com

Academic Editor: Hartmut Kuhn

Congenital adrenal hyperplasia is a group of autosomal recessive disorders. The most frequent one is 21-hydroxylase deficiency. Analyzing *CYP21A2* gene mutations was so far not reported in Iraq. This work aims to analyze the spectrum and frequency of *CYP21A2* mutations among Iraqi CAH patients. Sixty-two children were recruited from the Pediatric Endocrine Consultation Clinic, Children Welfare Teaching Hospital, Baghdad, Iraq, from September 2014 till June 2015. Their ages ranged between one day and 15 years. They presented with salt wasting, simple virilization, or pseudoprecocious puberty. Cytogenetic study was performed for cases with ambiguous genitalia. Molecular analysis of *CYP21A2* gene was done using the CAH StripAssay (ViennaLab Diagnostics) for detection of 11 point mutations and >50% of large gene deletions/conversions. Mutations were found in 42 (67.7%) patients; 31 (50%) patients were homozygotes, 9 (14.5%) were heterozygotes, and 2 (3.2%) were compound heterozygotes with 3 mutations, while 20 (32.3%) patients had none of the tested mutations. The most frequently detected mutations were large gene deletions/conversions found in 12 (19.4%) patients, followed by I2Splice and Q318X in 8 (12.9%) patients each, I172N in 5 (8.1%) patients, and V281L in 4 (6.5%) patients. Del 8 bp, P453S, and R483P were each found in one (1.6%) and complex alleles were found in 2 (3.2%). Four point mutations (P30L, Cluster E6, L307 frameshift, and R356W) were not identified in any patient. In conclusion, gene deletions/conversions and 7 point mutations were recorded in varying proportions, the former being the commonest, generally similar to what was reported in regional countries.

1. Introduction

The term congenital adrenal hyperplasia (CAH) comprises a group of autosomal recessive disorders, each due to a deficiency of an enzyme involved in the synthesis of cortisol, aldosterone, or both. The most common form of CAH results from deficiency of the 21-hydroxylase enzyme (aka 21OHD), accounting for about 95% of cases, due to mutations or deletions of *CYP21A2* gene located on 6p21.3 [1].

The condition is usually characterized by either the severe classical form, which includes the salt wasting and simple virilizing forms (manifesting themselves earlier in life), or milder nonclassical or "late-onset" form [2].

21OHD is the most common cause of ambiguous genitalia in female newborns. Affected females are presented with varying degree of genital ambiguity [3]. About 70% of children with classical 21OHD have the salt wasting form, which results primarily from deficient aldosterone synthesis, while nonclassic 21OHD displays symptoms of androgen excess due to mild-to-moderate overproduction of sex hormones that may present at any age [4].

The prevalence of 21OHD as well as its mutation pattern varies among different ethnic populations [5].

The overall worldwide frequency of CAH is estimated to be about 1 per 15,000 live births [6], having higher rates in some Arab countries, for example, 1 : 6400 in Saudi Arabia

TABLE 1: Clinical presentation and age distribution of Iraqi CAH patients with 21-hydroxylase enzyme deficiency.

Clinical presentation	Females number (%)	Males number (%)	Total number (%)	Age range (mean ± SD)
Classic				
Salt wasting form	20 (32.3%)	7 (11.3%)	27 (43.6%)	1 day–3 months (1.01 ± 1.37) months
Simple virilizing form	23 (37.1%)	7 (11.3%)	30 (48.4%)	1 day–14 years (2.61 ± 3.16) years
Nonclassic form (pseudoprecocious puberty)	4 (6.4%)	1 (1.6%)	5 (8%)	8–15 years (9.4 ± 3.13) years
Total	47 (75.8%)	15 (24.2%)	62 (100%)	1 day–15 years (24.69 ± 41.07) months

[7], 1 : 9030 in the United Arab Emirates [8], and 1 : 8000 in the northern part of Palestine [9]. To date there is no report about the incidence or prevalence of CAH among Iraqi people.

Diagnostic challenges arise from similarity of clinical presentations in different enzyme deficiency states causing CAH (21OHD and 11βOHD) in addition to false positive results obtained from neonatal screening tests.

Genetic analysis can be helpful to confirm a diagnosis of CAH, in prenatal diagnosis, in detection of carriers in families having a previously affected child, and to offer better treatment options during pregnancy [10].

2. Materials and Methods

Sixty-two unrelated patients with a clinical diagnosis of 21OHD registered at the Pediatric Endocrine Consultation Clinic/Children Welfare Hospital in Baghdad, Iraq, were recruited for this study.

Patients were enrolled when they presented with ambiguous genitalia with/without dehydration or pseudoprecocious puberty in females or when they had normal male phenotype with dehydration and/or pseudoprecocious puberty.

To establish the clinical diagnosis of 21OHD, initial baseline investigations were performed including measurements of serum electrolytes and 17-hydroxyprogesterone (17OHP), as well as abdominal and pelvic ultrasonography and bone age determination. Cytogenetic evaluation was performed only for sex determination in those with genital ambiguity. Patients who showed 46XY male patterns were excluded from the study.

DNA was extracted from peripheral blood samples. The DNA was amplified in a multiplex polymerase chain reaction (PCR), followed by hybridization to specific wild and mutant oligonucleotide probes designed to detect the 11 most frequent CYP21A2 mutations as well as >50% of large gene deletions/conversions using the CAH StripAssay Kit (ViennaLab Diagnostics, Vienna, Austria). Point mutations covered by the CAH StripAssay are P30L/Exon 1 (c.89C>T), I2Splice/Intron 2 (c.290-13A/C>G), Del 8 bp/Exon 3 (c.329_336 delGAGACTAC), I172N/Exon 4 (c.515T>A), Cluster E6/Exon 6 (c.707T>A, c.710T>A, and c.716T>A), V281L/Exon 7 (c.841G>T), L307 frameshift/Exon 7 (c.920-921insT), Q318X/Exon 8 (c.952C>T), R356W/Exon 8 (c.1066C>T),

TABLE 2: Distribution of Iraqi CAH cases with 21-hydroxylase enzyme deficiency with large deletions/conversions of CYP21A2 gene.

Types of deletions/conversions	Number (%)
P30L, I2Splice, Del 8 bp	3 (4.8)
P30L, I2Splice, Del 8 bp, I172N	2 (3.2)
Cluster E6, V281L, L307 frameshift, Q318X, R356W	5 (8.1)
Complete CYP21A2 gene deletion	2 (3.2)
Total	12 (19.3)

P453S/Exon 10 (c.1357C>T), and R483P/Exon 10 (c.1448G>C). The amplification, hybridization, and detection procedures were performed as reported previously [15].

The study was approved by the ethical committee at the College of Medicine, University of Baghdad, Baghdad, Iraq, and informed consent was obtained from parents of all enrollees.

3. Results

Out of 62 unrelated patients, 47 (75.8%) were females and 15 (24.2%) were males, with a female : male ratio of 3.1 : 1. All patients were Arabs and their ages ranged between 1 day and 15 years [mean ± SD = 24.69 ± 41.07 months].

Fifty-two (82%) cases originated from consanguineous marriages. Fifty-seven (91.9%) patients had the classical form of 21OHD [27 (43.5%) of them had the salt wasting "SW" form and 30 (48.4%) cases had the simple virilizing "SV" form], while the milder nonclassic form was seen in 5 (8.1%) patients and developed later during childhood, Table 1.

Mutations were detected in 42 out of the 62 unrelated patients (67.7%): 31 patients were homozygous for one mutation, 9 patients were heterozygotes, 2 patients were compound heterozygotes with 3 different mutations, and the remaining 20 (32.3%) patients harboured none of the tested mutations.

Mutations were subsequently divided into large gene deletions/conversions and point mutations.

Homozygous large gene deletions/conversions were found in 12 (19.3%) patients (Table 2) and as follows:

(i) Five (8.1%) cases had deletions extending from Cluster E6 to p.R356W.

TABLE 3: Distribution of mutations in the tested alleles and zygosity status among Iraqi CAH cases.

	Homozygotes number (%)	Heterozygotes number (%)	Compound heterozygotes number (%)	Total detected number (%)	Undetected number (%)	Total number (%)
Cases	31 (50) (i) 12 Del (ii) 19 PM	9 (14.51) (i) 0 Del (ii) 9 PM	2 (3.23) (i) 0 Del (ii) 2 PM	42 (67.75) (i) 12 (ii) 30	20 (32.25)	62 (100)
Tested alleles	62 (50)	18 (14.51) (i) 9 detected (7.25) (ii) 9 undetected (7.25)	4 (32.23)	75 (60.48)	49 (39.52)[*]	124 (100)
Mutations	62 (50) (i) 24 Del (ii) 38 PM	9 (7.25) (i) 0 Del (ii) 9 detected PM (iii) 9 undetected[*]	6 (4.83) (i) 0 Del (ii) v6 PM[**]	77 (62.1)[#] (i) 24 Del (ii) 53 PM	49 (39.52)	124 expected mutations[Ψ]

Del = large deletion/conversion. PM = point mutation.

[*] 40 alleles from 20 homozygous cases plus 9 heterozygous cases where only one mutation was found.

[**] Those two cases each carry 3 point mutations in their 2 alleles (I2Splice, Q318X, and Del 8 bp).

[#] 77 mutations were detected in 75 alleles.

[Ψ] 126 total number of mutations (77 were detected and 49 undetected) instead of 124 expected mutations.

TABLE 4: Frequency of detected *CYP21A2* mutations and zygosity status among Iraqi CAH patients.

Type of mutation	Number of cases (number of alleles carrying that mutation)		Number (%) of mutant alleles
	Homozygous	Heterozygous	
P30L	0	0	0
I2Splice	8 (16)	2[*] (2)	18 (14.5%)
Del 8 bp	1 (2)	2[*] (2)	4 (3.2%)
I172N	5 (10)	0	10 (8.1%)
Cluster E6	0	0	0
V281l	1 (2)	3 (3)	5 (4.03%)
L307 frameshift	0	0	0
Q318X	4 (8)	4 + 2[*] (6)	14 (11.3%)
R356W	0	0	0
P453S	0	1 (1)	1 (0.81%)
R483P	0	1 (1)	1 (0.81%)
Large deletions/conversions	12 (24)	0	24 (19.3%)
Total detected mutations	*31 (62)*	*11 (13)[$]*	77 mutations in 75 alleles (60.5%)
Undetected	*20 (40 + 9)[#]*		49 (39.5%)
Total cases	*62 (124)*		

[*] Two of these heterozygous cases were compound, each having 3 mutations (namely, I2Splice, Del 8 bp, and Q318X) in their 2 alleles.

[#] Additional 9 alleles were undetected in the 9 heterozygous cases; only one mutation was detected.

[$] Thirteen mutant alleles were detected in 11 patients (nine of them had a single detected mutation but the remaining two had 3 mutations as mentioned above[*]).

(ii) Three (4.8%) cases had P30L, I2Splice, and Del 8 bp.

(iii) Two (3.2%) cases had a large deletion/conversion ranging from P30L to I172N.

(iv) Two (3.2%) cases had a complete homozygous gene deletion.

Seven out of the 11 point mutations covered by the CAH StripAssay were detected in the enrolled cases; the remaining 4 mutations (P30L, Cluster E6, L307 frameshift, and R356W) were not detected in the studied cases.

Point mutations were detected in 30 (48.4%) out of the total 62 cases (Tables 3 and 4).

The most frequent mutations were I2Splice and Q318X detected in 10 cases each, followed by I172N detected in 5 cases, V281L detected in 4 cases, Del 8 bp detected in 3 cases,

and finally P453S and R483P which were found in one case each.

We subsequently compared genotypes with clinical manifestations; in 27 (43.5%) patients with a classic salt wasting form, we found 22 (81.5%) mutations. The most frequent mutations observed were large deletions/conversions, followed by I2Splice, Q318X, Del 8 bp, and multiple mutations. Five (18.5%) cases showed none of the investigated mutations (Table 5).

Among 30 (48.4%) patients with the simple virilizing form 20 (66.7%) carried a mutation. I172N was the most frequent, followed by Q318X, V281L, I2Splice, deletions/conversions, multiple mutations, P453S, and R483P, while 10 (33.3%) cases showed no mutation.

In 5 (8.1%) cases with nonclassic late-onset form, no mutation was detected (Table 5).

TABLE 5: Frequency of *CYP21A2* mutations according to clinical presentation in 62 Iraqi CAH patients.

Mutation type	Number of patients (%)			
	SW	SV	NC	Total
Large deletions/conversions	11 (40.7)	1 (3.3)	0 (0.0)	12 (19.3)
Q318X	4 (14.8)	4 (13.3)	0 (0.0)	8 (12.9)
I2Splice	5 (18.5)	3 (10.0)	0 (0.0)	8 (12.9)
I172N	0 (0.0)	5 (16.7)	0 (0.0)	5 (8.1)
V281L	0 (0.0)	4 (13.3)	0 (0.0)	4 (6.4)
Del 8 bp	1 (3.7)	0 (0.0)	0 (0.0)	1 (1.6)
Multiple mutations (I2Splice, Del 8 bp, Q318X)	1 (3.7)	1 (3.3)	0 (0.0)	2 (3.2)
P453S	0 (0.0)	1 (3.3)	0 (0.0)	1 (1.6)
R483P	0 (0.0)	1 (3.3)	0 (0.0)	1 (1.6)
Total detected mutations	*22 (81.5)*	*20 (66.7)*	*0 (0.0)*	*42 (67.7)*
Not detected	5 (18.5)	10 (33.3)	5 (100)	20 (32.3)
Total	*27 (100)*	*30 (100)*	*5 (100)*	*62 (100)*

4. Discussion

Molecular analysis of common inherited diseases causing major health problems in Iraq was the center of attention of some recent studies [16–19], aiming to build up a database for disease-causing mutations in this region of the world.

As part of this group of diseases, severe forms of CAH can lead to life-threatening conditions if untreated. Genital ambiguity in affected females is known to have a strong impact on social behaviour and health of patients and also represents a significant burden for health care providers. Further complications include the issue of determining the specific enzyme deficiency in some cases.

Patients were recruited from one of the largest pediatric endocrine clinics in Baghdad, Iraq. Yet, the laboratory diagnostic ability is still limited and discrimination of specific enzyme deficiency remains largely based on clinical criteria and follow-up rather than on solid laboratory confirmation.

For this reason, from all the enrolled 62 unrelated Iraqi CAH patients only the children and not their carrier parents were chosen. The *CYP21A2* mutation analysis was done to confirm the clinical suspicion and to study the spectrum as well as frequency of the mutations along with their clinical impact.

Our findings showed that large deletions/conversions of the *CYP21A2* gene were the most common type of mutation among the group of 42 Iraqi CAH patients where a mutation was detected. In addition, 7 out of 11 screened point mutations were detected in several combinations:

(i) Large deletions/conversions (19.3%) were also described as the most frequent mutation detected in Iran (31.8%) [12].

(ii) The I2Splice mutation (14.5%) was described as the most frequent mutation detected in some regional

countries such as Jordan (35.7%) [11], Iran (28%) [13], and Turkey (22%) [10] and in Western Europe [20].

(iii) The Q318X mutation (11.3%) has been described as the most frequent mutation found in Tunisia (35.3% and 26%) [14, 21].

(iv) The V281L mutation was found in 4.0% of the patients. Interestingly in one case a 2-year-old female with ambiguous genitalia and hypertension only carried a heterozygous V281L mutation. Hypertension and genital ambiguity are typically a sign of 11β-hydroxylase deficiency. The clinical symptoms in conjunction with a V281L mutation suggest a potential *CYP11B2* mutation causing concomitant 11β-hydroxylase deficiency.

(v) Some alleles carried more than one mutation. In this study, two cases with classic presentation had complex alleles comprising 3 mutations (I2Splice, Del 8 bp, and Q318X). This is similar to findings of Baş et al. who found one SW case with a complex allele comprising R356W, V281L, and I172N [10]. Also Kharrat et al. detected one case homozygous for 2 different mutations (I2Splice + Q318X/I2Splice + Q318X) in Tunisia [14]. A German study reported 5 cases with complex alleles, where 3 patients had I172N-F306+t, one case had I172N-R356W, and one case had V281L-R356W [22].

(vi) P453S was found in a single allele (0.8%), which was also detected by Baş et al. in Turkey but in two alleles [10].

(vii) R483P was uniquely found in a single allele (0.8%) but was not detected in the neighboring countries, Table 6.

Four isolated mutations were tested but not identified in any of our cases, namely, P30L, R356W, Cluster E6, and L307 frameshift mutation.

(i) The absence of the P30L mutation is similar to the studies performed in Iran, Turkey, and Tunisia [10, 12–14], while it was reported in Jordan [11], Greece [22], Romania [23], Italy [24], and China [25].

(ii) R356W and Cluster E6 were found in Jordan (3.5%, 1.7%) [11] and Iran in 2 studies (7.95%, 2.27%) [12] and (5%, 4%) [13] and in Turkey (8.8%, 2.2%) [10], but not in our cases.

(iii) A single mutation detected in one (0.8%) case in this study, namely, R483P, was not found in any of the neighboring countries [7, 10, 12].

(iv) The L307 frameshift mutation was not detected in this study, as well as in Iran, Turkey, and Saudi Arabia, and was not tested in Jordan.

(v) All these mutations were detected in the middle European countries [26].

The percentage of alleles with no mutation in this study (39.5%) was high in comparison to other studies [Turkey (15.4%, 22%) [10, 27], Iran (0, 30%) [12, 13], and Tunisia

TABLE 6: Frequency of common allelic mutations in this study compared with previous studies from some of the neighboring and Arab countries.

	Present study, Iraq, 2015	Turkey, 2009 [10]	Jordan, 2011 [11]	Iran, 2011 [12]	Iran, 2008 [13]	Tunisia, 2004 [14]
Total alleles (n)	124	91	92	88	100	102
Not detected (%)	39.5%	15.4%	40%	0	30%	5.9%
Molecular method(s) used	AS-PCR with reverse hybridization for 11 mutations	SB, AS-PCR/ER, and sequencing	ARMS for 8 common mutations & MLPA	AS-PCR for 8 common mutations and sequencing	AS-PCR for 8 mutations	ER + sequencing
P30L	0	0	17.8	0	0	0
I2Splice	14.5	22.0	35.7	14.77	28	17.6
Del 8 bp E3	3.2	4.4	16	0	13	0
I172N	8.1	9.9	16	5.68	9.0	10.8
Cluster E6	0	2.2	1.7	2.27	4.0	0
V281L	4.0	7.6	0	1.14	3.0	0
L307fs	0	0	NT	0	NT	0
Q318X	11.3	3.3	23.2	15.91	9	35.3
R356W	0	8.8	3.5	7.95	5.0	2.0
P453S	0.81	2.2	NT	NR	NT	NR
R483P	0.81	NR	NT	NR	NT	
Deletions/conversions	19.3	8.8/14.3	0	31.8	NT	19.6

NR: not reported; NT: not tested; SB: southern blot; AS-PCR: allele-specific oligonucleotide hybridization by PCR; ER: enzyme restriction; ARMS: amplification refractory mutational screen; MLPA: multiplex ligation dependent probe amplification.

(5.9%)] [14]. Sanger sequencing that has been used in some of the other studies can also detect rare mutations (around 5%) which are not covered by the CAH StripAssay, but it fails to detect larger deletions/conversions. In contrast to Sanger sequencing the StripAssay can detect >50% of large deletions/conversions. However, for the remaining percentage of large aberrations techniques to define the gene copy number need to be applied (MLPA, real-time PCR).

Being an autosomal recessive disorder, 21OHD should affect both sexes equally [28], but the female predominance noted in this study (Table 1) was affected by the type of clinical presentation in this disease causing social problems and impact on health from virilization and dehydration in female patients while males with severe salt loss may die undiagnosed during the neonatal period, whereas moderately affected individuals may stay undetected for several years until symptoms and signs of androgen excess develop [29]. This finding was similarly reported in a Syrian study [30].

Parental consanguinity was found in 51 (82.3%) cases, which reflects the relatively high consanguinity rate among Iraqi population reported to range between 35 and 60% [31]. Similar situations have been seen in other Arab studies (e.g., Saudi Arabia) in which all cases originated from consanguineous marriages.

An interesting observation in the current study is the relatively large number of patients with heterozygous mutations having two related parents [9 (17.7%)]. This finding was similar to studies performed in Tunisia and Turkey, which showed a high frequency of compound heterozygosity (17.6% and 34.8%, resp.) despite a high consanguinity rate [10, 14]

which reflects the high frequency and diversity of mutations in these populations, as well as ours.

Among 11 clinical cases with heterozygous mutations, 9 had only a single detectable mutation. We therefore think that the other allele could host a rare and/or novel mutation or a type of large deletion/conversion not covered by the StripAssay. Baş et al. [10] reported similar findings in 10 cases with one mutant allele (using southern blot (SB) and allele-specific semiquantitative PCR/enzyme restriction and sequencing).

Furthermore, the CYP21A2 promotor region has been reported by several authors to be responsible for contributing to CAH conditions. For instance, in one Chinese study a point mutation in the promoter region of CYP21A2 gene has been shown to reduce the transcription activity by 80% causing CAH [32].

Thus, we are planning to include Sanger sequencing combined with dosage analysis for both, the 9 cases with a single mutation and the 20 undetermined cases. In-depth analysis might extend the spectrum of mutations currently present in our cohort of patients.

Overall, most cases in this study demonstrated good genotype correlation with their expected phenotype.

In the salt wasting and simple virilizing group, genotype-phenotype correlation was as expected, except for two cases: the first carried a homozygous deletion/conversion—supposedly causing severe enzyme deficiency—but it showed a simple virilizing phenotype. This may be explained by improvement of aldosterone biosynthesis in salt wasting form [33, 34].

The second case had homozygous V281L mutation supposedly causing mild nonclassic form but it presented with simple virilization; this may be due to another undetected alteration in the same allele and should be ruled out by gene sequencing [35].

In conclusion, the current study is the first report about the spectrum and frequency of *CYP21A2* mutations among the Arab CAH patients of Iraq showing a high frequency and diversity. Large deletions/conversions were most frequent. Seven out of 11 tested point mutations were also reported, I2Splice and Q318X being the commonest. R483P was uniquely identified in one case only. Female predominance and parental consanguinity were reported at a higher frequency than the general population with a generally good genotype/phenotype correlation.

Competing Interests

The authors report no conflict of interests.

Authors' Contributions

Ruqayah G. Y. Al-Obaidi contributed to collection of data, performing the largest part of the molecular studies, data analysis, and drafting of the manuscript. Bassam M. S. Al-Musawi contributed to the concept and design, part of the molecular studies, data analysis, and drafting of the manuscript; Munib Ahmed K. Al-Zubaidi contributed to patient recruitment, clinical assessment, and diagnosis of enrolled cases; Christian Oberkanins contributed to performing part of the molecular work, data analysis, and drafting of the manuscript. Stefan Németh contributed to performing part of the molecular work, data analysis, and drafting of the manuscript. Yusra G. Y. Al-Obaidi contributed to part of the molecular work. All authors revised and approved the final submitted version of the manuscript.

Acknowledgments

The authors would like to acknowledge the assistance and cooperation of Dr. Ashna Jamal Faik, The Genetics Section, The Central Health Laboratories, Baghdad, Iraq, and those of Dr. Saad A. B. Al-Omer, Dept. of Pathology, College of Medicine, Basrah University.

References

[1] D. P. Merke, "Approach to the adult with congenital adrenal hyperplasia due to 21-hydroxylase deficiency," *Journal of Clinical Endocrinology and Metabolism*, vol. 93, no. 3, pp. 653–660, 2008.

[2] K. Lin-Su, M. C. Macapagal, R. C. Wilson, and M. I. New, "Genetic disorder of the adrenal gland," in *Principles and Practice of Medical Genetics*, D. L. Rimon, J. M. Connor, R. E. Pyeritz, and B. R. Korf, Eds., vol. 2, pp. 2023–2054, Churchill Livingstone, Philadelphia, Pa, USA, 5th edition, 2007.

[3] P. White and T. A. S. S. Bachega, "Congenital adrenal hyperplasia due to 21 hydroxylase deficiency: from birth to adulthood," *Seminars in Reproductive Medicine*, vol. 30, no. 5, pp. 400–409, 2012.

[4] S. F. Witchel and R. Azziz, "Congenital adrenal hyperplasia," *Journal of Pediatric and Adolescent Gynecology*, vol. 24, no. 3, pp. 116–126, 2011.

[5] R. C. Wilson, S. Nimkarn, M. Dumic et al., "Ethnic-specific distribution of mutations in 716 patients with congenital adrenal hyperplasia owing to 21-hydroxylase deficiency," *Molecular Genetics and Metabolism*, vol. 90, no. 4, pp. 414–421, 2007.

[6] S. Pang and A. Clark, "Congenital adrenal hyperplasia due to 21-hydroxylase deficiency: newborn screening and its relationship to the diagnosis and treatment of the disorder," *Screening*, vol. 2, article 105, 1993.

[7] S. Mohamed, S. El-Kholy, N. Al-Juryyan, A. M. Al-Nemri, and K. K. Abu-Amero, "A CYP21A2 gene mutation in patients with congenital adrenal hyperplasia. Molecular genetics report from Saudi Arabia," *Saudi Medical Journal*, vol. 36, no. 1, pp. 113–116, 2015.

[8] H. Al Hosani, M. Salah, H. M. Osman et al., "Expanding the comprehensive national neonatal screening programme in the United Arab Emirates from 1995 to 2011," *Eastern Mediterranean Health Journal*, vol. 20, no. 1, pp. 17–23, 2014.

[9] J. Sack, H. Front, I. Kaiserman, and M. Schreiber, "21-Hydroxylase deficiency: screening and incidence in Israel," *Hormone Research*, vol. 48, no. 3, pp. 115–119, 1997.

[10] F. Baş, H. Kayserili, F. Darendeliler et al., "CYP21A2 gene mutations in congenital adrenal hyperplasia: genotype-phenotype correlation in Turkish children," *Journal of Turkish Pediatric Endocrinology and Diabetes Society*, vol. 1, no. 3, pp. 116–128, 2009.

[11] H. Daggag, D. Hiyesat, N. Khawaja et al., "Prevalence of congenital adrenal hyperplasia (CAH) due to 21-hydroxylase deficiency, in Jordanian sample pool," in *Proceedings of the Endocrine Society's 93rd Annual Meeting & Expo*, National Center for Diabetes Endocrinology and Genetics, Boston, Mass, USA, June 2011.

[12] B. Rabbani, N. Mahdieh, M. T. Ashtiani, M. T. Akbari, and A. Rabbani, "Molecular diagnosis of congenital adrenal hyperplasia in Iran: focusing on CYP21A2 gene," *Iranian Journal of Pediartrics*, vol. 21, no. 2, pp. 139–150, 2011.

[13] A. Ramazani, K. Kahrizi, M. Razaghiazar, N. Mahdieh, and P. Koppens, "The frequency of eight common point mutations in CYP21 gene in Iranian patients with congenital adrenal hyperplasia," *Iranian Biomedical Journal*, vol. 12, no. 1, pp. 49–53, 2008.

[14] M. Kharrat, V. Tardy, R. M'Rad et al., "Molecular genetic analysis of Tunisian patients with a classic form of 21-hydroxylase deficiency: identification of four novel mutations and high prevalence of Q318X mutation," *Journal of Clinical Endocrinology and Metabolism*, vol. 89, no. 1, pp. 368–374, 2004.

[15] S. Németh, S. Riedl, G. Kriegshäuser et al., "Reverse-hybridization assay for rapid detection of common *CYP21A2* mutations in dried blood spots from newborns with elevated 17-OH progesterone," *Clinica Chimica Acta*, vol. 414, pp. 211–214, 2012.

[16] N. A. S. Al-Allawi, B. M. S. Al-Mousawi, A. I. A. Badi, and S. D. Jalal, "The spectrum of β-Thalassemia mutations in Baghdad, Central Iraq," *Hemoglobin*, vol. 37, no. 5, pp. 444–453, 2013.

[17] N. A. S. Al-Allawi, J. M. S. Jubrael, and M. Hughson, "Molecular characterization of β-thalassemia in the Dohuk region of Iraq," *Hemoglobin*, vol. 30, no. 4, pp. 479–486, 2006.

[18] B. M. S. Al-Musawi, N. Al-Allawi, B. A. Abdul-Majeed, A. A. Eissa, J. M. S. Jubrael, and H. Hamamy, "Molecular characterization of glucose-6-phosphate dehydrogenase deficient variants

in Baghdad city—Iraq," *BMC Blood Disorders*, vol. 12, article 4, 2012.

[19] N. A. S. Al-Allawi, S. D. Jalal, A. M. Mohammad, S. Q. Omer, and R. S. D. Markous, "β-Thalassemia intermedia in Northern Iraq: a single center experience," *BioMed Research International*, vol. 2014, Article ID 262853, 9 pages, 2014.

[20] P. C. White and P. W. Speiser, "Congenital adrenal hyperplasia due to 21-hydroxylase deficiency," *Endocrine Reviews*, vol. 21, no. 3, pp. 245–291, 2000.

[21] I. Ben Charfeddine, F. G. Riepe, E. Clauser et al., "Steroid 21-hydroxylase gene mutational spectrum in 50 Tunisian patients: characterization of three novel polymorphisms," *Gene*, vol. 507, no. 1, pp. 20–26, 2012.

[22] M. Dracopoulou-Vabouli, M. Maniati-Christidi, and C. Dacou-Voutetakis, "The spectrum of molecular defects of the CYP21 gene in the Hellenic population: variable concordance between genotype and phenotype in the different forms of congenital adrenal hyperplasia," *The Journal of Clinical Endocrinology and Metabolism*, vol. 86, no. 6, pp. 2845–2848, 2001.

[23] A. G. Sido, M. M. Weber, P. G. Sido, S. Clausmeyer, U. Heinrich, and E. Schulze, "21-Hydroxylase and 11β-hydroxylase mutations in Romanian patients with classic congenital adrenal hyperplasia," *The Journal of Clinical Endocrinology & Metabolism*, vol. 90, no. 10, pp. 5769–5773, 2005.

[24] P. Carrera, L. Bordone, T. Azzani et al., "Point mutations in Italian patients with classic, non-classic, and cryptic forms of steroid 21-hydroxylase deficiency," *Human Genetics*, vol. 98, no. 6, pp. 662–665, 1996.

[25] B. Zhang, Z.-L. Lu, Y. Wang, and H. Tao, "Molecular characterization of mutations and phenotype/genotype correlation in Chinese patients with 21-hydroxylase deficiency," *Acta Genetica Sinica*, vol. 31, no. 9, pp. 950–955, 2004.

[26] V. Dolžan, J. Sólyom, G. Fekete et al., "Mutational spectrum of steroid 21-hydroxylase and the genotype-phenotype association in Middle European patients with congenital adrenal hyperplasia," *European Journal of Endocrinology*, vol. 153, no. 1, pp. 99–106, 2005.

[27] F. Sadeghi, N. Yurur-Kutlay, M. Berberoglu et al., "Identification of frequency and distribution of the nine most frequent mutations among patients with 21-hydroxylase deficiency in Turkey," *Journal of Pediatric Endocrinology and Metabolism*, vol. 21, no. 8, pp. 781–787, 2008.

[28] P. A. Levy and R. W. Marion, "Human genetics and dysmorphology," in *Nelson Essentials of Pediatrics*, K. J. Marcdante and R. M. Kliegman, Eds., p. 150, Elsevier, Philadelphia, Pa, USA, 7th edition, 2015.

[29] R. C. Wilson and M. I. New, "Congenital adrenal hyperplasia," in *Principles of Molecular Medicine*, Humana Press, Totowa, NJ, USA, 1998.

[30] L. I. Alshabab, A. Alebrahem, A. Kaddoura, and S. Al-Fahoum, "Congenital adrenal hyperplasia due to 21-hydroxylase deficiency: a five-year retrospective study in the Children's Hospital of Damascus, Syria," *Qatar Medical Journal*, vol. 2015, no. 1, article 11, 2015.

[31] H. A. Hamamy and Z. S. AL-Hakkak, "Consanguinity and reproductive health in Iraq," *Human Heredity*, vol. 39, no. 5, pp. 271–275, 1989.

[32] K.-K. Chin and S.-F. Chang, "The $^{-104}$G nucleotide of the human CYP21 gene is important for CYP21 transcription activity and protein interaction," *Nucleic Acids Research*, vol. 26, no. 8, pp. 1959–1964, 1998.

[33] P. W. Speiser, L. Agdere, H. Ueshiba, P. C. White, and M. I. New, "Aldosterone synthesis in salt-wasting congenital adrenal hyperplasia with complete absence of adrenal 21-hydroxylase," *The New England Journal of Medicine*, vol. 324, no. 3, pp. 145–149, 1991.

[34] B. Rabbani, M. T. Akbari, N. Mahdieh et al., "Homozygous complete deletion of CYP21A2 causes a simple virilizing phenotype in an Azeri child," *Asian Biomedicine*, vol. 5, no. 6, pp. 889–892, 2011.

[35] R. Marino, P. Ramirez, J. Galeano et al., "Steroid 21-hydroxylase gene mutational spectrum in 454 Argentinean patients: genotype-phenotype correlation in a large cohort of patients with congenital adrenal hyperplasia," *Clinical Endocrinology*, vol. 75, no. 4, pp. 427–435, 2011.

20

Diagnostic Value of Adenosine Deaminase and its Isoforms in Type II Diabetes Mellitus

Bagher Larijani,[1] **Ramin Heshmat,**[2] **Mina Ebrahimi-Rad,**[3]
Shohreh Khatami,[3] **Shirin Valadbeigi,**[3] **and Reza Saghiri**[3]

[1]*Endocrinology and Metabolism Research Center, Endocrinology and Metabolism Clinical Sciences Institute,*
Tehran University of Medical Sciences, Tehran, Iran
[2]*Chronic Diseases Research Center, Endocrinology and Metabolism Population Sciences Institute,*
Tehran University of Medical Sciences, Tehran, Iran
[3]*Department of Biochemistry, Pasteur Institute of Iran, Tehran, Iran*

Correspondence should be addressed to Reza Saghiri; saghiri@pasteur.ac.ir

Academic Editor: Ali-Akbar Saboury

Background and Aims. In the present study, we have investigated the activity of adenosine deaminase (ADA) as a diagnostic marker in type 2 (or II) diabetes mellitus (T2DM). *Design and Methods.* The deaminase activity of ADA1 and ADA2 was determined in serum from 33 patients with type 2 (or II) diabetes mellitus and 35 healthy controls. We also determined the proportion of glycated hemoglobin (HbA1c). *Results.* Our results showed significant differences between total serum ADA (tADA) and ADA2 activities in the diabetic groups with HbA1c < 8 (%) and HbA1c ≥ 8 (%) with respect to the values in healthy individuals ($p < 0.001$). ADA2 activity in patients with high HbA1c was found to be much higher than that in patients with low HbA1c ($p = 0.0001$). In addition, total ADA activity showed a significant correlation with HbA1c ($r = 0.6$, $p < 0.0001$). *Conclusions.* Total serum ADA activity, specially that due to ADA2, could be useful test for the diagnosis of type 2 (or II) diabetes mellitus.

1. Introduction

Diabetes mellitus is a common disorder of glucose homeostasis which grows epidemically. The number of people diagnosed with T2DM is estimated about 380 million by the World Health Organization. This number is expected to increase to 592 million people by the year 2035 [1]. Glycosylated hemoglobin (HbA1c) has been suggested by International Diabetes Federation (IDF) for diagnosis of diabetes. HbA1c is supposed to be the best test for long-term control of blood glucose [2]. Metabolic disturbance and immunological imbalance are two important key factors in type 2 diabetes mellitus. Insulin deficiency and insulin resistance are the most important metabolic factors in type 2 diabetes. Immunological disturbance in type 2 diabetic individuals has been associated with cell-mediated immunity and inappropriate T-lymphocyte function [3]. Adenosine deaminase (adenosine aminohydrolase, EC 3.5.4.4, ADA) is

one of the key enzymes of purine nucleoside metabolism that catalyzes the deamination of adenosine and deoxyadenosine to inosine and deoxyinosine, respectively. ADA, an enzyme essential for the differentiation and proliferation of lymphocytes and monocyte, macrophage system, has been used for monitoring several immune system diseases. This enzyme was considered as a suitable marker of cell-mediate immunity. T cells are the recipient of adenosine signaling produced by B cells. Adenosine decreases the activation of T cells. ADA that exists on the surface of B lymphocytes could inactivate the adenosine signal. ADA is localized on the external surface of erythrocytes this might be in turn associated with inactivation of extracellular adenosine. Binding complexing protein on surface of fibroblasts has been considered to be a surface receptor for ADA [4, 5]. Adenosine deaminase includes two isoforms with unique biochemical properties. ADA1 can exist as a monomer with 30–40 KDa molecular weight or as a dimer with 280 KDa molecular weight while ADA2 is a 110 KDa protein [6, 7]. The ADA1 is found in all cells with the highest activity in lymphocytes and monocytes, whereas ADA2 is the predominant isoform in the serum of normal subjects. The major source of ADA2 is likely to be monocyte-macrophage system that produced it in response to pathogen factors and encoded by CECRI gene. ADA2 is a member of the new family of growth factors called ADCFs (ADA-related growth factors). ADA2 is strongly increased in inflammatory diseases such as rheumatoid arthritis and tuberculosis, AIDS, and diabetes [8–10]. Studies have shown that the protein attached to ADA1 molecule is identified to be CD26/DPPIV [11]. CD26 is a T cell activating antigen and transmembrane glycoprotein. On the surface of T-lymphocytes, ADA binds to CD26 via A2bR (adenosine A2B receptors) [12]. Moreover, CD26/DPPIV inactivates the incretin hormone glucagon-like peptide-1 (GLP-1) and glucose dependent insulin tropic polypeptide (GIP). DPPIV inhibitors stabilize endogenous GLP-1 at physiological concentration, induce insulin secretion in glucose-dependent manner and prevent the degradation of GLP-1 [13]. Recent studies show ADA increase in tissue of diabetic rats. ADA increased activity was reported in the lymph nodes and splenocyte of diabetes-prone BB rats [14]. Genetically, ADA locus in chromosome 20 is associated with locus type 2 diabetes [15]. Adenosine deaminase is an important enzyme for modulating adenosine concentration. Adenosine stimulates insulin activity via several processes such as glucose transport, lipid synthesis, pyruvate dehydrogenase activity, and Leucine oxidation. Adenosine plays an important role in bioactivity of insulin and regulates insulin activity in various tissues such as liver, myocardium white adipose, and skeletal muscles [16, 17]. Adenosine potentiates the action of insulin in myocardium and adipose tissue and inhibits liver and skeletal muscle. Adenosine mimics the action of insulin on glucose and lipid metabolism in adipose tissue [18]. Adenosine has antilipolysis property in adipose tissue and increases insulin sensitivity for glucose transport [19]. Adenosine increases accessibility about 25% of GLUT4 to cell surface for glucose transportation [20]. Adenosine with binding to A1 receptor increases the accumulation of the insulin-induced PIP3 and PKB in postreceptor phase [21]. Adenosine has also been shown to increase gluconeogenesis and glycogenolysis via increasing cyclic AMP (CAMP) by stimulation of hepatic adenylate cyclase through adenosine A2a receptor binding in liver. Both or either of these actions causes the increase of local insulin resistance and glucose output from the liver [22]. In this study we have investigated the alternation of serum ADA activities and its mechanism in type 2 diabetic patients.

2. Material and Method

2.1. Patients. 33 patients with type 2 diabetes who referred to Diabetes Center of Shariati Hospital were diagnosed by the endocrinologist and their blood was drawn during 6 months. No patients displayed any kind of infection or inflammatory diseases which lead to increase in ADA. This group was divided into two groups according to hemoglobin A1c (HbA1c) level, namely, patients whose HbA1c levels were less than 8 and those with HbA1c level more than 8. Blood was also drawn from 35 healthy individuals as the control subjects.

2.2. Blood Sampling Protocol

2.2.1. Serum Samples. Blood samples were obtained from patients and healthy subjects after 12 h fasting. Fasting blood samples were collected to centrifugation at 3,000 rpm for 10 min at 4°C to obtain serum. All samples were stored at −80°C after separation.

2.2.2. Measurement of Adenosine Deaminase. The ADA activity was assayed by the ADA kit of Diazyme Laboratories and with model 912 type Autoanalyzer (Hitachi Co. Ltd., Tokyo, Japan). The ADA assay is based on enzymatic deamination of adenosine to inosine in this kit. Inosine is converted to hypoxanthine by purine nucleoside phosphorylase. Hypoxanthine is then converted to uric acid and hydrogen peroxide (H_2O_2) by xanthine oxidase. H_2O_2 is further reacted with N-ethyl-N-(2-hydroxy-3-sulfopropyl)-3-methylaniline and 4-aminoantipyrine in the presence of peroxidase to generate quinone dye which is monitored in a kinetic manner.

To distinguish between the ADA1 and ADA2 forms, the ADA activity was measured using the same technique with and without EHNA (erythro-9-(2-hydroxy-3-nonyl) adenine), obtained from Sigma-Aldrich (St. Louis, MO, USA). EHNA, a potent selective inhibitor of ADA1, was used at final concentration of 0.1 mmol/L. In its presence, only the ADA2 isoform is active. The ADA1 activity is then calculated by subtracting the ADA2 activity from the total ADA activity.

2.3. Assessment of Glycated Hemoglobin (HbA1c). Diazyme direct enzymatic HbA1c analysis was evaluated on the Hitachi 912 Autoanalyzer (Hitachi Co. Ltd., Tokyo, Japan) using blood samples. We have applied simple enzymatic assay for HbA1C using neutral protease and FPOX. Preparation of samples was carried out by mixing 500 μL of lysis buffer with 25 μL of whole blood and incubating for 10 min at room temperature.

2.4. Statistical Analysis. All results were expressed in terms of mean ± standards deviation (SD). Data were analyzed by Package for Social Sciences (SPSS) version 16. Statistical differences between patient groups and controls were performed by one-way ANOVA test. Correlation between the ADA and HbA1c was measured by means of Pearson's Correlation Coefficient (r). HbA1c was expressed as a percentage. BMI was calculated as weight (kg)/height^2 (m^2). p value < 0.01 was considered statically significant.

3. Results

This study was carried out on 16 patients with HbA1c ≥ 8 who were designated poorly controlled DM and 17 patients with HbA1c < 8 who were well controlled and 35 healthy individuals as control subjects. Mean ± SD values of tADA, ADA1, and ADA2 were found in healthy subjects. Results are shown in Table 1.

TABLE 1: ADA and its isoforms activities in T2DM patients based on low and high HbA1c comparison with healthy controls.

Groups		tADA (IU/L)	ADA2 (IU/L)	ADA1 (IU/L)
Low HbA1c	Mean	16.18	9.41	6.76
	N	17	17	17
	Std. deviation	2.67	3.74	1.67
	Minimum	13	6	2.00
	Maximum	24	22	8.00
	Variance	7.15	14.00	2.81
High HbA1c	Mean	22.44	13.94	8.50
	N	16	16	16
	Std. deviation	7.420	5.19	3.96
	Minimum	16	8	4.00
	Maximum	45	24	22.00
	Variance	55.06	26.99	15.73
Control	Mean	14.00	7.66	6.34
	N	35	35	35
	Std. deviation	1.680	1.73	.48
	Minimum	11	5	6.00
	Maximum	17	11	7.00
	Variance	2.82	2.99	.23

TABLE 2: Anthropometric measurements and clinical parameters of T2DM patients and healthy control.

Characteristics	T2DM ($n = 33$)	Controls ($n = 35$)	p value
Age (years)	60.2 ± 8.7	65.8 ± 9.2	0.0008
BMI (kg/m^2)	29.5 ± 7.00	21.3 ± 6.78	0.005
FPG (mg/dL)	156.7 ± 10.3	92.6 ± 7.58	<0.0001

Data are shown as the means $\pm$ SD. BMI, body mass index; T2D, type 2 diabetes mellitus; FPG, fasting plasma glucose; $p < 0.001$, ANOVA test analysis.

The increasing of tADA, ADA1, and ADA2 activity in subjects with HbA1c < 8 and those with HbA1c $\geq$ 8 is significant in comparison with healthy ones ($p < 0.001$).

Anthropometric measurements and clinical parameters of the study subjects are shown in Table 2. As illustrated in the table, serum ADA activity in T2DM patients with HbA1c high and low was significantly higher than healthy group. These variations specially increased ADA2 in HbA1c $\geq$ 8 in comparison with HbA1c < 8 is investigated to be more.

As shown in Figure 1, we also observed significant positive correlation between serum ADA activities and HbA1c ($r = 0.6$, $p < 0.0001$).

4. Discussion

Type 2 diabetes is accompanied with collection of clinical and biochemical disorders which have been called metabolic syndrome X. These disorders include center obesity, hypertension, atherosclerosis, hypertriglyceridemia, increased cholesterol and LDL, and decreased HDL. The action of cytokines

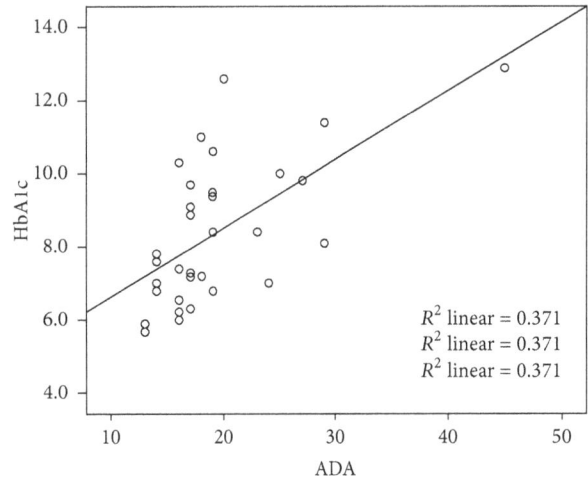

FIGURE 1: The correlation between HbA1c and tADA in diabetic patients.

on the brain, liver, endothelium, and adipose tissue is a major factor of metabolic syndrome X. Cytokines stimulate the acute-phase proteins. In the short term, the acute-phase protein has survival values and regulates homeostasis and in long-term produces diseases [23]. These cytokines such as IL-1, IL-6, and TNF-α are produced from monocytes, macrophage, and adipose tissue. Insulin resistance and hyperglycemia increase the effect of cytokines on the liver and cause the secretion of IL-6 and TNF-α from monocytes and macrophage [24]. On the other hand, production of cytokines from monocytes and macrophage and the increase in acute-phase proteins elevate insulin resistance [25]. Aging, certain dietary components, smoking, and obesity are important factors in cytokine increase and immunity disturbance in type 2 diabetes. Immunity disturbance does not happen in all T2DM individuals and other major factors such as genetics, cytokine sensitivity, and acute-phase response contribute to its existence. Cytokine imbalance effects ADA activity [26]. Variation in cytokines especially cytokines which were produced by Th1 cells is associated with the increase in ADA serum activity. It also activates monocyte-macrophage cell system [27]. Defect in insulin activity required for T-lymphocytes in diabetes leads to abnormal T-lymphocyte proliferation and enhanced ADA activity [28]. The effects of ADA on T-lymphocytes and cytokines represent it as a suitable marker of cell-mediated immunity. The increased ADA levels in inflammatory and autoimmune diseases such as rheumatoid arthritis, tuberculosis, and systemic lupus erythematosus (SLE) make its role more significant. Likewise, immune system and increase and concentration of extracellular adenosine are considered as other major factors in ADA increase. Adenosine is a local hormone which regulates many biological activities. Adenosine causes coronary vasodilatation, bradycardia, inhibition of platelet aggregation, renal vasoconstriction, and regulation of channel ion activity. These processes are carried out via adenosine receptors (A1, A2a, A3, and A2b). In normal conditions, adenosine contains low concentration lesser than $1 \mu m$ and is risen under metabolically conditions such as stress and tissue injuries to $4–10 \mu m$ [29]. Extracellular adenosine concentration is regulated by two mechanisms of transportation of adenosine across the cell membrane and enzymatic regulation of adenosine concentration and can be activate adenosine receptors [9]. There exist two types of nucleoside transporters across the plasma membrane, the equilibrative facilitated-diffusion transport (ENT) and concentrative Na^+ dependent transporters (CNT). Inhibition of these transporters potentiates the action of adenosine. Enzymatic regulation of adenosine concentration in mammals is dependent upon activity of 5-nucleotidase and two utilizing enzymes: adenosine deaminase and adenosine kinase [30]. Investigations on diabetic individuals show the increase in the value of adenosine. Reduction in the activity of adenosine kinase is a major factor in the increase of adenosine in these patients [31]. In addition, variation in adenosine receptors and transporters can change tissue sensitivity to adenosine [32–34]. Studies on adenosine concentration and its effect on cell functions show the important role of adenosine in glucose metabolism. In the range of $0.003–0.5 \mu mol/kg$, adenosine lowers the serum fatty acids and serum insulin. On the other hand, in the range of $0.5–50 \mu mol/kg$, its effect on hepatic A2a receptors stimulates gluconeogenesis and the increase of serum glucose [35]. The increase of adenosine level with decrease in adenosine kinase in diabetes results in deamination of adenosine and the increase in adenosine deaminase. In fact, the increase in ADA activity is a protective mechanism against adenosine elevation [26]. When the adenosine concentration is strongly elevated, ADA enzyme catalyzes deamination of adenosine via ADA1 and ADA2 (ADA-related growth factors) isoforms and decreases its concentration. Since the ADA2 increases CD4-T cells, it can

be useful in stimulating immune system [10]. T-lymphocytes abnormal proliferation, increased secretion cytokines, and increased extracellular adenosine are major factors in the elevation of ADA. The increase of ADA in diabetic patients leads to metabolic changes of insulin especially in adipose tissues. ADA in adipose tissue causes the increase in lipolysis, the augmentation of hyperlipidemia, and the disturbance in antilipolysis activity. High concentration of fatty free acids (FFA) derived from lipolysis elevation causes oxidative phosphorylation and ATP retention in adipocytes [36]. ADA impairs PKB and PI3P production in insulin postreceptor phase and reduces insulin sensitivity in adipocytes [20]. Moreover, endogenic reduction of adenosine by ADA decreases insulin ability for the activation of tyrosine kinase receptor in submaximal concentration of insulin. Therefore, diabetic adipocyte cells require more insulin concentration [21]. ADA reduces GLUT4 accessibility to cell surface for glucose transporters [19]. Adenosine deamination is guided toward hypoxanthine uric acid production and xanthine oxidase enzyme in this process causes superoxides production. The increase membrane peroxidation alters Na^+, K^+, ATPase activity, and transport of metabolites across the membrane. The increase and production of superoxides enhance the risk of cardiovascular diseases in diabetic patients. ADA causes the high rate of uric acid which shows significant correlation between the ADA levels and uric acids in diabetics [21, 37]. On the other hand, high glucose level causes the increase in ADA attached DPPIV protein. Accordingly, the increase in glucose of diabetics leads to high DPPIV-ADA and increase in DPPIV-ADA results in incretin reduction and insulin secretion. Insulin deficiency and irregulation of glucose are major factors in the elevation of ADA in diabetes. Therefore, insulin can be a good quality way of ADA reduction. Insulin reduces ADA activity in diabetic tissues and regulates local concentration of adenosine. Some herbal drugs can reduce ADA and control blood glucose [38]. The ADA-CD26/DPPIV interaction is inhibited by the cell surface glycoprptein (gp120) in HIV1-infected individuals and it might result in increasing levels of serum ADA. ADA-CD26/DPPIV complex plays a main role in the regulation of immune activity and cell adhesion [39, 40]. ADA alongside other immunomodulatory enzymes acts as a deconstructive oxidative marker in diabetes and plays an important role in progression of its complications. ADA increase, especially ADA2, may serve as an immunoenzyme marker in the pathology of type 2 diabetes mellitus.

Abbreviations

HbA1c: Glycated hemoglobin
T2DM: Type II diabetes mellitus
ADA: Adenosine deaminase
DPPIV: Dipeptidyl peptidase 4
PIP3: Phosphatidylinositol-trisphosphate
PKB: Protein kinase B
AMP: Adenosine monophosphate
HDL: High-density lipoprotein
LDL: Low-density lipoprotein
IL: Interleukin

TNF-α: Tumor necrosis factor alpha
FPOX: Fructosyl peptide oxidase
ADGFs: Adenosine-related growth factors.

Competing Interests

The authors declare no conflict of interests.

Acknowledgments

This work was supported by Endocrinology and Metabolism Research Institute. This study represented that ADA could be a useful alternative test to diagnosis of diabetes mellitus.

References

[1] K. M. Goodrich, S. K. Crowley, D.-C. Lee, X. S. Sui, S. P. Hooker, and S. N. Blair, "Associations of cardiorespiratory fitness and parental history of diabetes with risk of type 2 diabetes," *Diabetes Research and Clinical Practice*, vol. 95, no. 3, pp. 425–431, 2012.

[2] L. Czupryniak, "Guidelines for the management of type 2 diabetes: is ADA and EASD consensus more clinically relevant than the IDF recommendations?" *Diabetes Research and Clinical Practice*, vol. 86, no. 1, pp. S22–S25, 2009.

[3] M. S. Prakash, S. Chennaiah, Y. S. R. Murthy, E. Anjaiah, S. Ananda Rao, and C. Suresh, "Altered adenosine deaminase activity in type 2 diabetes mellitus," *Journal, Indian Academy of Clinical Medicine*, vol. 7, no. 2, pp. 114–117, 2006.

[4] J. M. Aran, D. Colomer, E. Matutes, J. L. Vives-Corrons, and R. Franco, "Presence of adenosine deaminase on the surface of mononuclear blood cells: immunochemical localization using light and electron microscopy," *Journal of Histochemistry and Cytochemistry*, vol. 39, no. 8, pp. 1001–1008, 1991.

[5] R. Franco, J. M. Aran, D. Colomer, E. Matutes, and J. L. Vives-Corrons, "Association of adenosine deaminase with erythrocyte and platelet plasma membrane: an immunological study using light and electron microscopy," *Journal of Histochemistry and Cytochemistry*, vol. 38, no. 5, pp. 653–658, 1990.

[6] J. P. J. Ngerer, S. H. Oosthulzeu, S. H. Bissbort, and W. J. Wermaak, "Serum adenosine deaminase: isoenzymes and diagnostic application," *Clinical Chemistry*, vol. 38, no. 7, pp. 1322–1326, 1992.

[7] M. Gupta and V. Nair, "Adenosine deaminase in nucleoside synthesis. A review," *Collection of Czechoslovak Chemical Communications*, vol. 71, no. 6, pp. 769–787, 2006.

[8] A. V. Zavialov, X. Yu, D. Spillmann, G. Lauvau, and A. V. Zavialo, "Structural basis for the growth factor activity of human adenosine deaminase ADA2," *Journal of Biological Chemistry*, vol. 285, no. 16, pp. 12367–12377, 2010.

[9] A. V. Zavialov and Å. Engström, "Human ADA2 belongs to a new family of growth factors with adenosine deaminase activity," *Biochemical Journal*, vol. 391, no. 1, pp. 51–57, 2005.

[10] R. Iijima, T. Kunieda, S. Yamaguchi et al., "The extracellular adenosine deaminase growth factor, ADGF/CECR1, plays a role in *Xenopus* embryogenesis via the adenosine/P1 receptor," *The Journal of Biological Chemistry*, vol. 283, no. 4, pp. 2255–2264, 2008.

[11] J. Kameoka, T. Tanaka, Y. Nojima, S. F. Schlossman, and C. Morimoto, "Direct association of adenosine deaminase with a T cell activation antigen, CD26," *Science*, vol. 261, no. 5120, pp. 466–469, 1993.

[12] E. Gracia, K. Pérez-Capote, E. Moreno et al., "A2A adenosine receptor ligand binding and signalling is allosterically modulated by adenosine deaminase," *Biochemical Journal*, vol. 435, no. 3, pp. 701–709, 2011.

[13] D. J. Drucker, "Dipeptidyl peptidase-4 inhibition and the treatment of type 2 diabetes," *Diabetes Care*, vol. 30, no. 6, pp. 1335–1343, 2007.

[14] G. Wu and E. B. Marliss, "Deficiency of purine nucleoside phosphorylase activity in thymocytes from the immunodeficient diabetic BB rat," *Clinical and Experimental Immunology*, vol. 86, no. 2, pp. 260–265, 1991.

[15] M. M. Amoli, P. Amiri, M. Namakchian et al., "Adenosine deaminase gene polymorphism is associated with obesity in Iranian population," *Obesity Research and Clinical Practice*, vol. 1, no. 3, pp. 173–177, 2007.

[16] J. Espinal, R. A. John Challiss, and E. A. Newsholme, "Effect of adenosine deaminase and an adenosine analogue on insulin sensitivity in soleus muscle of the rat," *FEBS Letters*, vol. 158, no. 1, pp. 103–106, 1983.

[17] M. P. McLane, P. R. Black, W. R. Law, and R. M. Raymond, "Adenosine reversal of in vivo hepatic responsiveness to insulin," *Diabetes*, vol. 39, no. 1, pp. 62–69, 1990.

[18] T. Hoshino, K. Yamada, K. Masuoka et al., "Elevated adenosine deaminase activity in the serum of patients with diabetes mellitus," *Diabetes Research and Clinical Practice*, vol. 25, no. 2, pp. 97–102, 1994.

[19] S. M. Johansson, E. Lindgren, J.-N. Yang, A. W. Herling, and B. B. Fredholm, "Adenosine A1 receptors regulate lipolysis and lipogenesis in mouse adipose tissue—interactions with insulin," *European Journal of Pharmacology*, vol. 597, no. 1–3, pp. 92–101, 2008.

[20] S. J. Vannucci, H. Nishimura, S. Satoh, S. W. Cushman, G. D. Holman, and I. A. Simpson, "Cell surface accessibility of GLUT4 glucose transporters in insulin-stimulated rat adipose cells. Modulation by isoprenaline and adenosine," *Biochemical Journal*, vol. 288, no. 1, pp. 325–330, 1992.

[21] S. Takasuga, T. Katada, M. Ui, and O. Hazeki, "Enhancement by adenosine of insulin-induced activation of phosphoinositide 3-kinase and protein kinase B in rat adipocytes," *The Journal of Biological Chemistry*, vol. 274, no. 28, pp. 19545–19550, 1999.

[22] A. C. Warrier, N. Y. Rao, D. S. Kulpati, T. K. Mishra, and B. C. Kabi, "Evaluation of adenosine deaminase activity and lipid peroxidation levels in diabetes mellitus," *Indian Journal of Clinical Biochemistry*, vol. 10, no. 1, pp. 9–13, 1995.

[23] J. C. Pickup and M. A. Crook, "Is type II diabetes mellitus a disease of the innate immune system?" *Diabetologia*, vol. 41, no. 10, pp. 1241–1248, 1998.

[24] M. Morohoshi, K. Fujisawa, I. Uchimura, and F. Numano, "Glucose-dependent interleukin 6 and tumor necrosis factor production by human peripheral blood monocytes in vitro," *Diabetes*, vol. 45, no. 3, pp. 954–959, 1996.

[25] J. C. Pickup, "Inflammation and activated innate immunity in the pathogenesis of type 2 diabetes," *Diabetes Care*, vol. 27, no. 3, pp. 813–823, 2004.

[26] O. J. Cordero, F. J. Salgado, C. M. Fernández-Alonso et al., "Cytokines regulate membrane adenosine deaminase on human activated lymphocytes," *Journal of Leukocyte Biology*, vol. 70, no. 6, pp. 920–930, 2001.

[27] M. Mokhtari, M. Hashemi, M. Yaghmaei et al., "Serum adenosine deaminase activity in gestational diabetes mellitus and normal pregnancy," *Archives of Gynecology and Obstetrics*, vol. 281, no. 4, pp. 623–626, 2010.

[28] F. Stentz and A. E. Kitabchi, "Activated T lymphocytes in type 2 diabetes: implications from in vitro studies," *Current Drug Targets*, vol. 4, no. 6, pp. 493–503, 2003.

[29] V. Kumar and A. Sharma, "Adenosine: an endogenous modulator of innate immune system with therapeutic potential," *European Journal of Pharmacology*, vol. 616, no. 1-3, pp. 7–15, 2009.

[30] T. Dolzal, "Adenosine deaminase. Review of physiological roles," 2001, http://www.entu.cas.cz/fyziol/seminars/ada/html.

[31] M. Sakowicz-Burkiewicz, K. Kocbuch, M. Grden, A. Szutowicz, and T. Pawelczyk, "Diabetes-induced decrease of adenosine kinase expression impairs the proliferation potential of diabetic rat T lymphocytes," *Immunology*, vol. 118, no. 3, pp. 402–412, 2006.

[32] T. Pawelczyk, M. Sakowicz, M. Szczepanska-Konkel, and S. Angielski, "Decreased expression of adenosine kinase in streptozotocin-induced diabetes mellitus rats," *Archives of Biochemistry and Biophysics*, vol. 375, no. 1, pp. 1–6, 1999.

[33] T. Pawelczyk, M. Grden, R. Rzepko, M. Sakowicz, and A. Szutowicz, "Region-specific alterations of adenosine receptors expression level in kidney of diabetic rat," *The American Journal of Pathology*, vol. 167, no. 2, pp. 315–325, 2005.

[34] M. Sakowicz, A. Szutowicz, and T. Pawelczyk, "Insulin and glucose induced changes in expression level of nucleoside transporters and adenosine transport in rat T lymphocytes," *Biochemical Pharmacology*, vol. 68, no. 7, pp. 1309–1320, 2004.

[35] B. Xu, D. A. Berkich, G. H. Crist, and K. F. LaNoue, "A1 adenosine receptor antagonism improves glucose tolerance in Zucker rats," *The American Journal of Physiology*, vol. 274, no. 2, part 1, pp. E271–E279, 1998.

[36] S. E. Mills, "Regulation of porcine adipocyte metabolism by insulin and adenosine," *Journal of Animal Science*, vol. 77, no. 12, pp. 3201–3207, 1999.

[37] N. Kurtul, S. Pence, E. Akarsu, H. Kocoglu, and Y. Aksoy, "Adenosine deaminase activity in the serum of type 2 diabetic," *Acta Medica*, vol. 47, pp. 33–35, 2004.

[38] T. Pawelczyk, M. Podgorska, and M. Sakowicz, "The effect of insulin on expression level of nucleoside transporters in diabetic rats," *Molecular Pharmacology*, vol. 63, no. 1, pp. 81–88, 2003.

[39] H. Fan, F. L. Tansi, W. A. Weihofen et al., "Molecular mechanism and structural basis of interactions of dipeptidyl peptidase IV with adenosine deaminase and human immunodeficiency virus type-1 transcription transactivator," *European Journal of Cell Biology*, vol. 91, no. 4, pp. 265–273, 2012.

[40] S. Ginés, M. Mariño, J. Mallol et al., "Regulation of epithelial and lymphocyte cell adhesion by adenosine deaminase-CD26 interaction," *Biochemical Journal*, vol. 361, no. 2, pp. 203–209, 2002.

21

A Midgut Digestive Phospholipase A_2 in Larval Mosquitoes, *Aedes albopictus* and *Culex quinquefasciatus*

Nor Aliza Abdul Rahim ⓘ,[1] **Marlini Othman,**[1] **Muna Sabri,**[2] **and David W. Stanley**[3]

[1]*Department of Paraclinical Sciences, Faculty of Medicine and Health Sciences, Universiti Malaysia Sarawak, 94300 Kota Samarahan, Sarawak, Malaysia*
[2]*Department of Basic Medical Sciences, Faculty of Medicine and Health Sciences, Universiti Malaysia Sarawak, 94300 Kota Samarahan, Sarawak, Malaysia*
[3]*United States Department of Agriculture, Agricultural Research Service, Biological Control of Insects Research Laboratory, 1503 S. Providence Road, Columbia, MO 65203, USA*

Correspondence should be addressed to Nor Aliza Abdul Rahim; arnaliza@unimas.my

Academic Editor: Hartmut Kuhn

Phospholipase A_2 (PLA_2) is a secretory digestive enzyme that hydrolyzes ester bond at *sn-2* position of dietary phospholipids, creating free fatty acid and lysophospholipid. The free fatty acids (arachidonic acid) are absorbed into midgut cells. *Aedes albopictus* and *Culex quinquefasciatus* digestive PLA_2 was characterized using a microplate PLA_2 assay. The enzyme showed substantial activities at 6 and 8 $\mu g/\mu l$ of protein concentration with optimal activity at 20 and 25 $\mu g/\mu l$ of substrate concentration in *Aedes albopictus* and *Culex quinquefasciatus*, respectively. PLA_2 activity from both mosquitoes increased in a linear function up to 1 hour of the reaction time. Both enzymes were sensitive to pH and temperature. PLA_2 showed higher enzyme activities in pH 8.0 and pH 9.0 from *Aedes albopictus* and *Culex quinquefasciatus*, respectively, at 40°C of incubation. The PLA_2 activity decreased in the presence of 5 mM *(Aedes albopictus)* and 0.5 mM *(Culex quinquefasciatus)* site specific PLA_2 inhibitor, oleyloxyethylphosphorylcholine. Based on the migration pattern of the partially purified PLA_2 on SDS-PAGE, the protein mass of PLA_2 is approximately 20–25 kDa for both mosquitoes. The information on PLA_2 properties derived from this study may facilitate in devising mosquitoes control strategies especially in the development of inhibitors targeting the enzyme active site.

1. Introduction

Phospholipase A_2 (PLA_2) hydrolyzes the *sn-2* ester bond in phospholipids (PLs) [1]. These enzymes make up a large superfamily of proteins that act in a very wide variety of physiological and pathophysiological actions. PLA_2 actions include digestion of dietary lipids, remodelling cellular membranes, host immune defenses, signal transduction via production of various lipid mediators, and, in the case of platelet activating factor, inactivation of a lipid mediator. Research into noncatalytic PLA_2s and into PLA_2 receptors and binding proteins reveals entirely new biological actions in which PLA_2 acts as a ligand rather than a catalytic enzyme [2, 3]. Here, we focus attention on PLA_2 associated with digestion.

Lipid digestion and absorption take place in the insect midguts. Midgut cells produce and secrete lipases that digest dietary neutral lipids, such as triacylglycerols. PLA_2s are responsible for two separate actions in insect physiology. For one, PLA_2s hydrolyze a fatty acid from the *sn-2* position of dietary PLs. Typically, the fatty acids esterified to the *sn-2* positions are C18 and C20 PUFAs. These fatty acids include linoleic acid, 18:2n-6, and linolenic acid, 18:3n-3, one or the other of which is strictly essential nutritional requirements for most insects and nearly all vertebrates. Hence, midgut PLA_2s are necessary for insects to meet one of their essential nutritional needs. A few insect and invertebrate species express a Δ^{-12} desaturase that inserts a double bond into oleic acid (18:1n-9), yielding 18:2n-6 and obviating the nutritional requirement [4–6]. The desaturation and elongation pathways necessary to convert C18 PUFAs to their C20 counterparts have been documented in several insect species [7, 8], from which we infer insects are able to meet all fatty acid requirements via dietary linoleic and linolenic acids, coupled to the desaturase/elongation pathways. The second important PLA$_2$ action contributes to digestion of dietary neutral lipids. Vertebrates, but not insects, produce bile salts that facilitate lipid digestion by solubilizing neutral lipids. In each PLA$_2$ reaction, hydrolysis of the *sn-2* fatty acid from PL leads to a free fatty acid and to a lysoPLs and these lipids act as the necessary solubilizers that aid lipase digestion of neutral lipids.

PLA$_2$ is the key enzyme responsible in the hydrolysis of arachidonic acid which acts as precursors to lipid mediators such as prostaglandins [9]. PLA$_2$s occur abundantly in venoms [10], in pancreatic juices of mammals, and in synovial fluids [11]. While PLA$_2$s are well characterized in terms of protein and gene structures in mammalian physiology, there is relatively little information on the characteristics of insect digestive PLA$_2$. Nonetheless, these enzymes are very important in insect biology and they may become functional targets in some pest management programs. Seen in this light, there is a real need for new knowledge on insect digestive PLA$_2$s. In this paper, we begin to address that need. Here, we report on the presence and the characteristics of a midgut PLA$_2$ in larvae of the mosquitoes, *Aedes albopictus* and *Culex quinquefasciatus*.

2. Materials and Methods

2.1. Insect. The larvae of *Aedes albopictus* and *Culex quinquefasciatus* were collected from Kampung Semerah Padi, Petra Jaya, Kuching (1°34'59.3''N, 110°19'48.2''E). The larvae were collected with ovitraps filled with a cow-grass infusion solution following methods described by Tang et al. [12].

Ten ovitraps were placed near housing areas at Kampung Semerah Padi, Petra Jaya. The ovitraps were collected and replaced with new ones every week. The larvae collected were pooled together and fed with fish food until the 4th instar. The larvae were identified as *Aedes albopictus* and *Culex quinquefasciatus* [13] and midguts were isolated. A larva to be dissected was placed on a glass plate and the water surrounding the larvae was blotted to dry. The midgut was removed by using a pair of forceps. By holding the thorax with forceps, the 8th abdominal segment was gently pulled using the other forceps so that the entire alimentary canal was drown out. The anal papillae, siphon, and Malphigian tubule attaching to the midgut were removed by pinching with the forceps.

2.2. PLA$_2$ Source Preparation. Midgut samples were homogenized in 200 μl buffer (0.1 M Tris[hydroxymethyl]aminoethane, pH 8; Sigma) mixed with 2 mM phenylthiourea (PTU, Sigma) by using a Bio Masher (Optima, Inc., USA). The homogenates were centrifuged at 735g for 3 minutes, then at 11,750g for 10 minutes. The supernatants were collected and used as the enzyme preparation.

2.3. Phospholipase A$_2$ Assay. PLA$_2$ substrate, 4-nitro-3-(octanoyloxy)benzoic acid (NOB; Enzo Life Sciences, Switzerland), was prepared following Nenad et al. [14] and Beghini et al. [15] with several modifications. The substrate was diluted

with chloroform to 50 mg/ml. 20 μl (1 mg) aliquots were distributed into Eppendorf tubes and all of the moisture was evaporated to dryness. The dry residue was stored at −20°C. Immediately before the assay, the substrate was resuspended in 1 ml of acetonitrile. The suspension was vortexed until all the substrate dissolved.

A standard PLA$_2$ enzyme assay using 96-well plates was conducted following methods by Beghini et al. [15]. The standard assay mixture contains substrate (NOB), enzyme source, and buffer (0.1 M Tris buffer, pH 8) made up of a 200 μl of mixture in each well. After the addition of the enzyme source, the microplate was incubated (Asys Thermostar) for 40 minutes at 40°C. The NOB, through hydrolysis of an ester bond, will convert into a chromophore (4-nitro-3-hydroxybenzoic acid). The absorbance of chromophore concentration produced was quantified using a microplate reader at 405 nm. The effects of Ca^{2+}, substrate and protein concentration, incubation time, pH, and temperature were investigated by varying each parameter.

2.4. Localizing the PLA$_2$ in Larvae. The homogenates were prepared from three different sections of the individual larvae. Three groups, A, B, and C, consisted of gut-free larval bodies, guts and contents, and isolated gut contents. Alimentary canals were removed by pulling out the eighth abdominal segment of the larvae with forceps, while holding its thorax with second forceps. The individual gut was quickly removed and placed into an Eppendorf tube containing 0.1 M Tris buffer and 2 mM PTU (Group B). The remaining bodies were collected into different tubes containing the same buffer (Group A). Isolated gut contents were obtained by separating the gut contents (Group C) from forty individual guts. All samples were prepared for enzyme assay as described in Table 1.

2.5. The Influence of Ca^{2+} on PLA$_2$ Activity. We conducted reactions in the presence of three different buffers: (1) Tris buffer with no additions; (2) Tris buffer amended with 5 mM CaCl$_2$; (3) Tris buffer amended with the Ca^{2+} chelator 5 mM EGTA (ethylene-glycol-bis (β-aminoethyl ether)-N,N,N',N'-tetraacetic acid). Each buffer was used during larvae dissection, homogenization, and enzyme activity assay.

2.6. Characterizing the Mosquito Digestive PLA$_2$. All experiments used midguts plus contents as the enzyme preparations. The influence of substrate and protein concentration, incubation time, pH, temperature, and the effect of site specific inhibitor for PLA$_2$ and OOPC on PLA$_2$ activity were assessed by varying each of the parameters.

2.7. Gel Electrophoresis. Estimation of this digestive PLA$_2$ from *Aedes albopictus* and *Culex quinquefasciatus* was performed according to the tricine-polyacrylamide gel electrophoresis (Tricine-SDS-PAGE) method by Schägger and von Jagow [16].

Migration of digestive PLA$_2$ was compared to the standard protein markers (Sigma) in a range of 26.6 kDa to 1.06 kDa. The protein marker consisted of triosephosphate isomerase from rabbit muscle (26.6 kDa), myoglobin from

TABLE 1: Mosquito larval preparation for localizing PLA$_2$ enzyme experiment.

Groups	Larval sections
A	Gut-free bodies (40 individual larval bodies/pool)
B	Gut and content (40 individual guts/pool)
C	Isolated gut contents (40 individual isolated gut contents/pool)

horse heart (17.0 kDa), α-lactalbumin from bovine milk (14.2 kDa), aprotinin from bovine lung (6.5 kDa), insulin chain B, oxidized, bovine (3.496 kDa), and bradykinin (1.06 kDa). The protein bands on the electrophoresis gel were directly visualized by using silver staining according to the method by Gromova and Celis [17].

2.8. Statistical Analysis. Data were reported as means ± SEM of n experiments as appropriate. The significance of difference between groups was assessed using one-way analysis of variance (ANOVA) followed by Tukey's multiple comparison test to determine the significant group. The confidence limit for significance was $p \leq 0.05$.

3. Results

3.1. Localizing the PLA$_2$ Enzyme in Larvae. To determine the appropriate preparation for characterization of PLA$_2$, three samples were prepared. Substantially high PLA$_2$ enzyme activity from *Aedes albopictus* was recorded in gut plus content preparation ($M = 0.328$, SEM = 0.00) and gut content preparation ($M = 0.3067$, SEM = 0.01). There were no significant differences ($F(2,6) = 5.161$, $p = 0.05$) in PLA$_2$ activity between all the preparations. In contrast, significantly higher PLA$_2$ activity was observed in *Culex quinquefasciatus'* gut content preparation ($M = 0.627$, SEM = 0.01) (Figure 1). Similarly, lower PLA$_2$ activity was recorded in gut-free bodies of *Aedes albopictus* ($M = 0.2733$, SEM = 0.01) and *Culex quinquefasciatus* ($M = 0.220$, SD = 0.02). The gut plus content was used as an enzyme source in all subsequent experiments.

3.2. Calcium Ion (Ca^{2+}) Dependency. *Aedes albopictus* PLA$_2$ activity was significantly higher in the buffer containing EGTA ($M = 0.063$, SEM = 0.01) and Tris buffer ($M = 0.065$, SEM = 0.00), while, for *Culex quinquefasciatus*, the PLA$_2$ activity was significantly high in Tris buffer ($M = 0.089$, SEM = 0.01) compared to enzyme activity in Tris buffer with additional calcium ($M = 0.0087$, SEM = 0.01) (Figure 2). These results suggest that the mosquito larval preparation is independent of Ca^{2+}. Subsequent experiments were conducted in Tris buffer with no added Ca^{2+}.

3.3. Characterization of PLA$_2$ Enzyme. The PLA$_2$ activities from *Aedes albopictus* and *Culex quinquefasciatus* showed a similar trend where the optimum enzyme activities were recorded at 6–8 μg/μL of protein (Figure 3). *Aedes albopictus* PLA$_2$ activity was fairly low at 2 and 4 μg/μl and then increased to a high level at 6 μg/μL ($M = 0.090$, SEM = 0.01) and 8 μg/μL ($M = 0.091$, SEM = 0.01) before it slightly declined at 10 μg/μL ($M = 0.087$, SEM = 0.00). Similarly, the *Culex quinquefasciatus* PLA$_2$ activity increased from 2 μg/μL

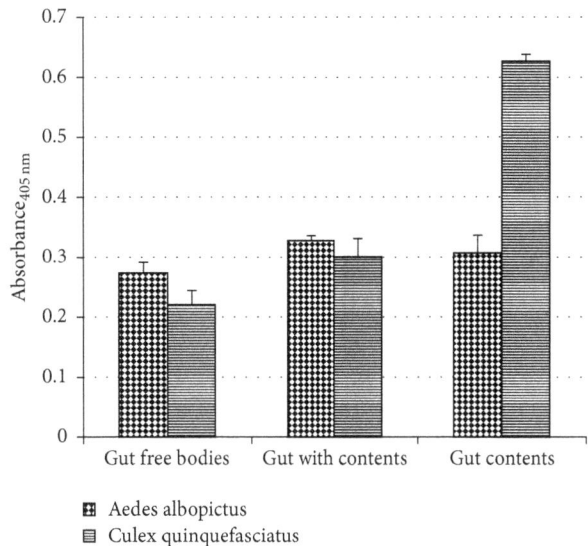

FIGURE 1: The PLA$_2$ activity in different preparations of *Aedes albopictus* and *Culex quinquefasciatus* larvae. 10 μg/μl of protein concentration was reacted with 10 μg/μl of substrate concentration. Each histogram bar shows the mean ± SEM of triplicates from a single experiment representative of at least two experiments.

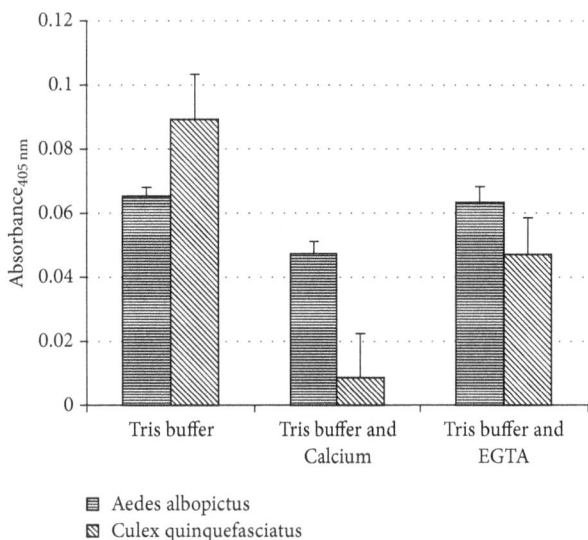

FIGURE 2: The PLA$_2$ activity of *Aedes albopictus* and *Culex quinquefasciatus* in three different buffers. 10 μg/μl of protein concentration was reacted with 10 μg/μl of substrate concentration. Each histogram bar shows the mean ± SEM of triplicates from a single experiment representative of at least two experiments.

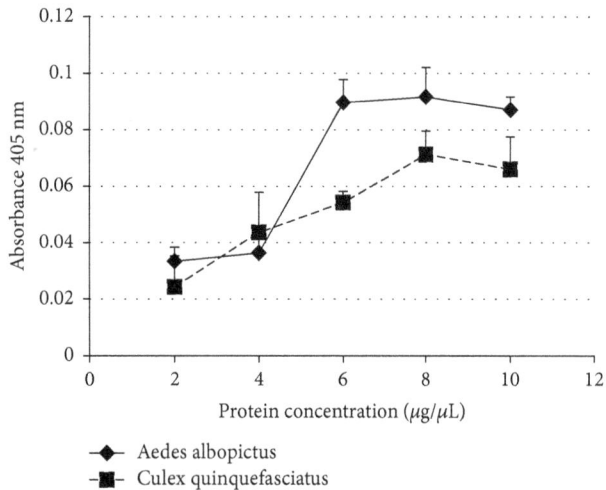

FIGURE 3: The influence of protein concentration on *Aedes albopictus* and *Culex quinquefasciatus* PLA_2 activity. $10\,\mu g/\mu l$ of substrate concentration was used to react with each of the protein concentrations, respectively. Each point represents the mean $\pm$ SEM of triplicates from a single experiment representative of at least two experiments.

($M = 0.0243$, SEM $= 0.01$) of protein until it reached the highest activity at $8\,\mu g/\mu L$ ($M = 0.0713$, SEM $= 0.01$). For subsequent experiment, $6\,\mu g/\mu L$ of protein was used as an enzyme source for PLA_2 assay in *Aedes albopictus* while $8\,\mu g/\mu L$ was used as an enzyme source for PLA_2 assay in *Culex quinquefasciatus*.

The PLA_2 activities from mosquitoes, *Aedes albopictus* and *Culex quinquefasciatus*, were increased with increasing concentration of substrate until the enzyme concentration becomes a limiting factor in the reaction. The *Aedes albopictus* PLA_2 activity increased in a linear manner with increasing substrate concentrations, up to $20\,\mu g/\mu l$ ($M = 0.121$, SEM $= 0.01$) (Figure 4(a)).

PLA_2 activity of *Culex quinquefasciatus* also increased from lower concentration, $5\,\mu g/\mu L$ ($M = 0.103$, SEM $= 0.00$), until it reached its optimal activity at $25\,\mu g/\mu L$ ($M = 0.290$, SEM $= 0.01$) (Figure 4(b)). There was no significant increase of enzyme activity at $30\,\mu g/\mu L$ for both mosquitoes PLA_2. Our standard concentration of substrate was $10\,\mu g/\mu l$ for all subsequent experiments.

The PLA_2 of *Aedes albopictus* was fairly low up to 20-minute incubations ($M = 0.042$, SEM $= 0.01$) and then increased substantially up to 50-minute incubation time ($M = 0.161$, SEM $= 0.02$). There was still another increase at 50 minutes ($M = 0.161$, SEM $= 0.02$) (Figure 5(a)). The enzyme activity remained constant at 60 minutes of incubation time.

On the other hand, there is no PLA_2 activity of *Culex quinquefasciatus* recorded in the first 20 minutes of incubation time. Then, the enzyme activity continued to increase steadily even after an hour ($M = 0.033$, SD $= 0.00$) (Figure 5(b)). We used 40-minute incubations in all experiments.

The digestive PLA_2 was sensitive to temperature. The reaction mixtures were incubated in a range of $24°C$ (room

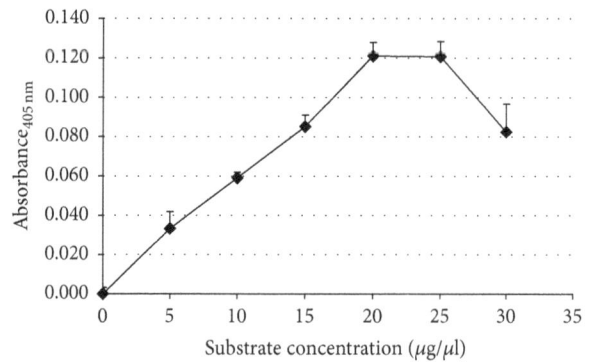

(a) The influence of substrate, NOB, concentration on *Aedes albopictus* PLA_2 activity. $6\,\mu g/\mu l$ of protein concentration was reacted with each concentration of substrate, respectively. Each point represents the mean $\pm$ SEM of triplicates from a single experiment representative of at least two experiments

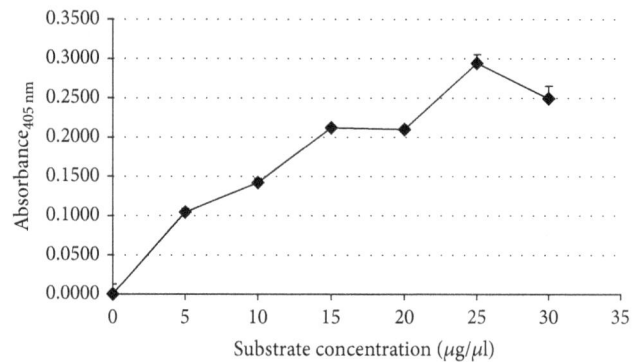

(b) The influence of substrate, NOB, concentration on *Culex quinquefasciatus* PLA_2 activity. $8\,\mu g/\mu l$ of protein concentration was reacted with each substrate concentration, respectively. Each point represents the mean $\pm$ SEM of triplicates from a single experiment representative of at least two experiments

FIGURE 4

temperature) to $70°C$. The *Aedes albopictus* and *Culex quinquefasciatus* PLA_2 activity increased in a linear way from room temperature to a peak at $40°C$ ($M = 0.116$, SEM $= 0.01$; $M = 0.097$, SEM $= 0.01$) (Figures 6(a) and 6(b)). At higher temperatures ($60–70°C$) the enzyme activity declined. Our standard incubation temperature was set at $40°C$.

pH of the reaction mixtures influenced PLA_2 (Figures 7(a) and 7(b)). The PLA_2 activity from *Aedes albopictus* and *Culex quinquefasciatus* increased from acidic condition to a mild alkaline condition. *Aedes albopictus* PLA_2 increased gradually from pH 5.0 ($M = 0.067$, SEM $= 0.01$) until it reached a maximum PLA_2 activity at pH 8.0 ($M = 0.130$, SEM $= 0.01$). At pH 9.0, the enzyme activity slightly decreased, but it is not statistically significant. Similarly, *Culex quinquefasciatus* reached its highest PLA_2 activity at pH 9.0 ($M = 0.1093$, SEM $= 0.01$) before it dropped drastically at pH 10.0 ($M = 0.035$, SEM $= 0.02$). Post hoc multiple comparison (Tukey) analysis showed there was no significant difference in the enzyme activity from pH 6.0 to 9.0. Therefore, pH 8 was selected as a reaction pH for all subsequent reactions.

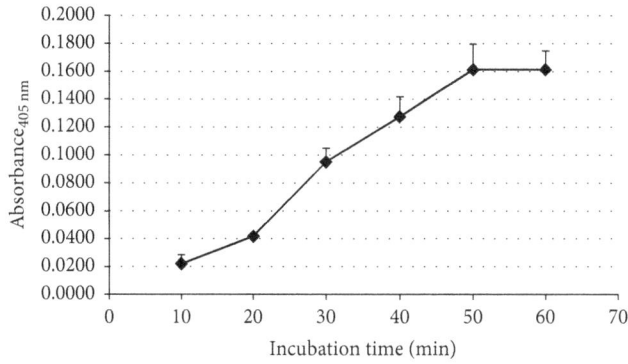

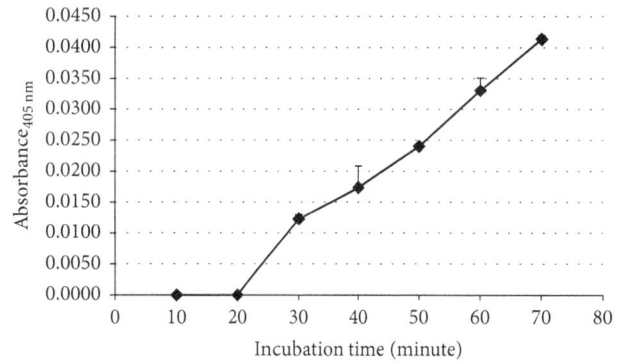

(a) The influence of incubation time on *Aedes albopictus* PLA$_2$ activity. 6 μg/μl of protein concentration was reacted with 10 μg/μl of substrate concentration in each different incubation time. Each point represents the mean ± SEM of triplicates from a single experiment representative of at least two experiments

(b) The influence of incubation time on *Culex quinquefasciatus* PLA$_2$ activity. 8 μg/μl of protein concentration was reacted with 10 μg/μl of substrate concentration in each different incubation time. Each point represents the mean ± SEM of triplicates from a single experiment representative of at least two experiments

FIGURE 5

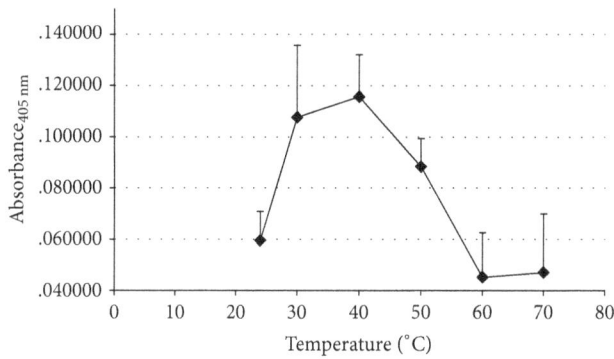

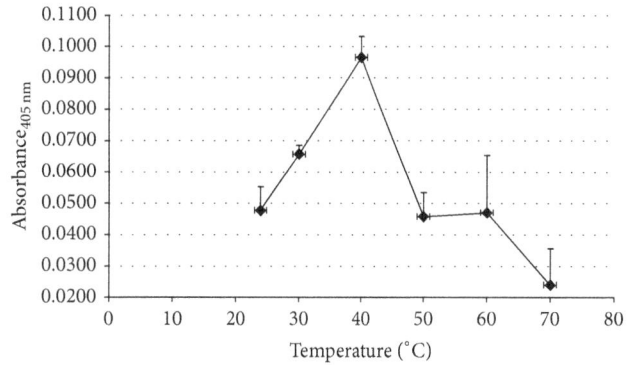

(a) The influence of incubation temperature on *Aedes albopictus* PLA$_2$ activity. 6 μg/μl of protein concentration was reacted with 10 μg/μl of substrate concentration in different incubation temperature, respectively. Each point represents the mean ± SEM of triplicates from a single experiment representative of at least two experiments

(b) The influence of incubation temperature on *Culex quinquefasciatus* PLA$_2$ activity. 8 μg/μl of protein concentration was reacted with 10 μg/μl of substrate concentration in different incubation temperature, respectively. Each point represents the mean ± SEM of triplicates from a single experiment representative of at least two experiments

FIGURE 6

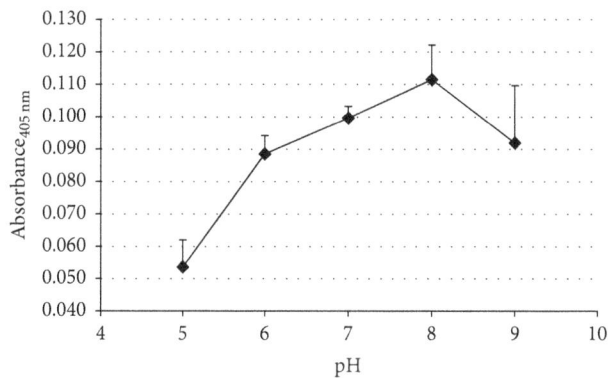

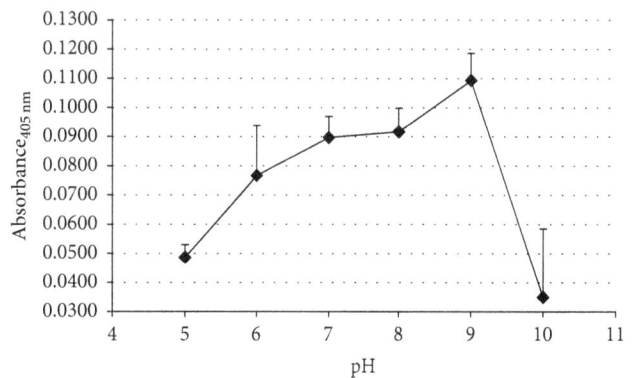

(a) The influence of reaction pH on *Aedes albopictus* PLA$_2$ activity. 6 μg/μl of protein concentration was reacted with 10 μg/μl of substrate concentration in different pH condition, respectively. Each point represents the mean ± SEM of triplicates from a single experiment representative of at least two experiments

(b) The influence of reaction pH on *Culex quinquefasciatus* PLA$_2$ activity. 8 μg/μl of protein concentration was reacted with 10 μg/μl of substrate concentration in different pH condition, respectively. Each point represents the mean ± SEM of triplicates from a single experiment representative of at least two experiments

FIGURE 7

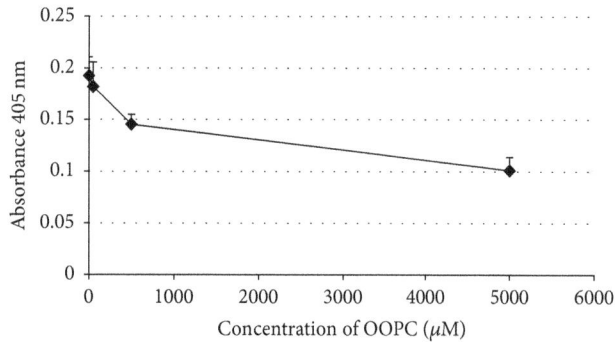

(a) The influence of PLA$_2$ inhibitor, OOPC, against *Aedes albopictus* digestive PLA$_2$ activity. Each histogram bar represents the mean ± SEM of triplicates from a single experiment representative of at least two experiments

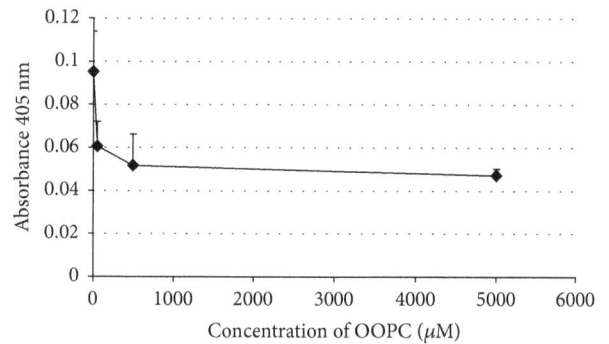

(b) The influence of PLA$_2$ inhibitor, OOPC, against *Culex quinquefasciatus* digestive PLA$_2$ activity. Each histogram bar represents the mean ± SEM of triplicates from a single experiment representative of at least two experiments

FIGURE 8

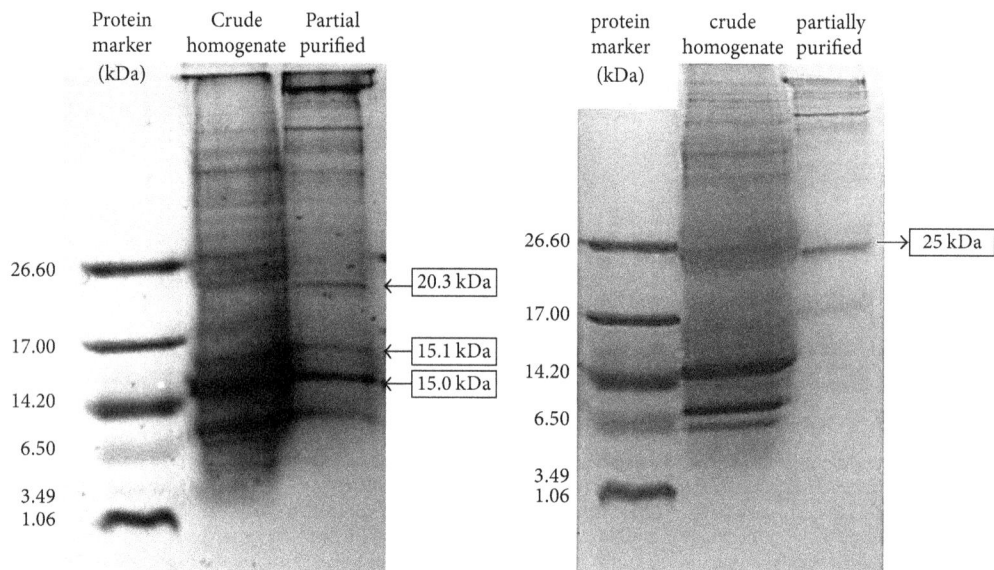

(a) Protein visualization of the crude homogenate and partially purified larval midgut PLA$_2$ of *Aedes albopictus*. Each well was loaded with 120 μg of protein concentration from crude and partially purified sample and 6 μl of protein marker. The gel was stained with silver staining

(b) Protein visualization of the crude homogenate and partially purified larval midgut PLA$_2$ of *Culex quinquefasciatus*. Each well was loaded with 120 μg of protein concentration from crude and partially purified sample and 6 μl of protein marker. The gel was stained with silver staining

FIGURE 9

3.4. The Influence of OOPC on PLA$_2$ Activity. Reactions in the presence of 50 μM to 500 μM of OOPC led to a dose-related decline in PLA$_2$ activity (Figures 8(a) and 8(b)). The PLA$_2$ of *Aedes albopictus* was significantly inhibited in the presence of 5000 μM of OOPC [$F(3, 8) = 5.886$, $p = 0.020$]. However, the PLA$_2$ of *Culex quinquefasciatus* was not statistically inhibited in the presence of PLA$_2$ inhibitor, OOPC [$F(2, 6) = 3.651$, $p = 0.092$].

3.5. Protein Mass Determination of Partially Purified Digestive PLA$_2$. The protein electrophoretic profile of partially purified PLA$_2$ from *Aedes albopictus* showed bands in different sizes which range from 14.6 to 20.3 kDa (Figure 9(a)) while

for partially purified PLA$_2$ from *Culex quinquefasciatus* it showed only one band which was at 25 kDa (Figure 9(b)). The presence of band at a similar size (25 kDa) suggests the presence of similar protein, which is the PLA$_2$.

4. Discussion

In this paper, we report on the characterization of a digestive PLA$_2$ in *Aedes albopictus* and *Culex quinquefasciatus* larvae. During our initial experiment, we compared the PLA$_2$ activities in selected fractions of the alimentary canal. Higher PLA$_2$ enzyme activity was recorded in the midgut plus content and isolated gut content preparations. Similar findings were reported for a related mosquito species, *Aedes aegypti*, where higher PLA_2 enzyme activity was recorded in midgut plus content preparation. These findings suggested that midgut cells secrete more PLA_2 than they store [18].

Secretory PLA_2 is characterized as a low molecular weight molecule (13–55 kDa) [10, 19] that catalyzes substrate in full activity in the presence of calcium. $sPLA_2$ differs from other PLA_2 as it acts extracellularly [20]. The enzyme was secreted from the cells before catalyzing the substrate which usually occurs in the lumen of the insects' midgut [21].

The characterization of PLA_2 in insects was assayed with radioactive substrate in the past. Here, we used a microplate assay using the chromogenic substrate, NOB. NOB is widely used in characterizing PLA_2 from snake venom [22–24] and human serum [14]. Our PLA_2 assays were performed on mosquito samples that have been partially enriched using Heparin column chromatography (1 ml HiTrap Heparin, Sigma, USA). This method successfully enriched the target PLA_2 in primary screwworm preparations [25]. For this work, our simple microplate assay used only a small amount of protein and substrate to obtain an optimal PLA_2 activity. This is an advantage for an investigation with limited amounts of protein sample. Since we collected our larval mosquito from the field, the number of individual larvae in each collection varied depending on their local environmental condition. The microplate assay is a practical but effective method to conduct our experiments with limited enzyme source.

Ca^{2+} is essential for both catalysis and binding of some enzymes to the substrate [26]. A study of Ca^{2+} requirement in primary screwworm PLA_2 preparations showed that the enzyme activity was almost abolished in the presence of calcium chelator, EGTA. The PLA_2 dependency on Ca^{2+} varied across species [21]. Previous studies on PLA_2 from primary screwworm, *C. hominivorax* [25], robber flies, *Asilis* sp. [27], and adult tiger beetles, *Cicindela circumpicta* [28], showed strict Ca^{2+} requirement for catalysis. Several studies on PLA_2 from venom sources such as rattlesnake, *Crotalus durissus cascavella*, venom [15] and sea anemone, *Aiptasia pallid*, nematocyst venom [29] also showed strict Ca^{2+} requirement [30].

PLA_2 activity in the *Aedes albopictus* and *Culex quinquefasciatus* preparations revealed slightly higher enzyme activity in Tris buffer without additional calcium. This is in agreement with PLA_2 from the midgut of tobacco hornworm [31] and *Aedes aegypti* larvae midgut [18].

Generally, enzyme activities are influenced by biophysical parameters, protein and substrate concentrations, pH, temperature, and reaction time. Increasing the amount of either enzyme or substrate generally will increase reaction rates because more active sites are available for reaction and more substrate molecules can bind with the active sites.

The *Aedes albopictus* and *Culex quinquefasciatus* preparations responded to the usual biophysical parameters in a way fairly similar to the *Aedes aegypti* preparations [18] except for its sensitivity to OOPC. PLA_2 activity in *Aedes aegypti* [18] was less sensitive to OOPC than the *Aedes albopictus* and *Culex quinquefasciatus* preparations. *Aedes albopictus*

PLA_2 was inhibited at 5000 μM of OOPC as compared to *Culex quinquefasciatus* PLA_2, which was inhibited at a lower concentration of OOPC (500 μM). In contrast, the PLA_2 activity from *Aedes aegypti* [18] did not show any significant decrease in the presence of OOPC (5–5000 μM).

Similar finding was also recorded for tobacco hornworm, *Manduca sexta* [31], digestive PLA_2 where there was no inhibition in its enzyme activity when exposed to different concentrations of OOPC in a range of 5–500 μM. On the other hand, the primary screwworm PLA_2 preparation was more sensitive to OOPC, 50 μM [25].

The *Aedes albopictus* and *Culex quinquefasciatus* PLA_2 activity increased with time in reactions up to 1 hour, in agreement with PLA_2 enzyme from other insects, *Aedes aegypti* [18] and *C. hominivorax* [25], and PLA_2 from venoms, *Bothrops jararacussu* [22].

PLA_2 from *Aedes albopictus* and *Culex quinquefasciatus* shares some similarity and differences with a related species, *Aedes aegypti*, where the optimal enzyme activity was at 40–50°C. However, the optimum pH condition differs. *Aedes aegypti* PLA_2 activity was optimal at pH 9.0, which is similar with *Culex quinquefasciatus* PLA_2, while for *Aedes albopictus* the enzyme activity declined at pH 9. Although the maximum enzyme activity differs, all mosquitoes PLA_2 reaction studied slowed down in acidic conditions. This is in broad agreement with the pH conditions of insect midguts, which can be very high in lepidoptera and more acidic in mosquitoes [25].

This study has estimated the size of PLA_2 from *Aedes albopictus* and *Culex quinquefasciatus* and provided information on partially purified PLA_2 from crude homogenate of mosquito larval midgut by using Heparin column. In this study, both *Aedes albopictus* and *Culex quinquefasciatus* larval midgut PLA_2 estimated molecular weights were found in a range of secretory insect PLA_2. The molecular weights were estimated at 14.6–25 kDa and 25 kDa for *Aedes albopictus* and *Culex quinquefasciatus*, respectively. Distinct band at 25 kDa was shown in both PLA_2 preparations of *Aedes albopictus* and *Culex quinquefasciatus* which are more likely the protein of interest in this study.

The estimated sizes agreed with the classic characteristic of $sPLA_2$, which consists of small MW enzyme ranging from 13 to 15 kDa which were obtained from various organisms such as snake venoms, porcine pancreas, fungus, and bacteria [1] and for tiger beetle and human PLA_2 which were reported to be 22 kDa and 55 kDa, respectively [19].

However, other investigations such as amino acid sequence and X-ray crystal structures of *Aedes albopictus* and *Culex quinquefasciatus* need to be conducted in order to classify the type of the digestive PLA_2.

5. Conclusions

Aedes albopictus PLA_2 from different sample preparations showed no significant difference in their activity, while for *Culex quinquefasciatus* significantly higher PLA_2 in gut contents was shown if compared to other preparations. PLA_2 from both enzymes did not require calcium (Ca^{2+}) for full enzyme activity and both showed increasing of enzyme activity with increasing concentration of substrate. The PLA_2

enzymatic assay from both mosquitoes showed accumulation of chromogenic substance up to 60 minutes of incubation time at 40°C. Both enzymes reacted in full catalytic activity in alkaline condition. *Aedes albopictus* PLA$_2$ was significantly inhibited by site specific PLA$_2$ inhibitor, OOPC. However, *Culex quinquefasciatus* PLA$_2$ was not significantly inhibited by the same inhibitor. Based on the electrophoretic pattern of the enzyme samples, protein band at 20–25 kDa was observed in both mosquitoes. To conclude, there were no differences between the characteristic of PLA$_2$ from *Aedes albopictus* and *Culex quinquefasciatus* except for its inhibition toward site specific inhibitor PLA$_2$, OOPC, where the inhibitor does not affect the *Culex quinquefasciatus* PLA$_2$ activity.

Disclosure

Mention of trade names or commercial products in this article is solely for the purpose of providing specific information and does not imply recommendation or endorsement by the U.S. Department of Agriculture. All programs and services of the U.S. Department of Agriculture are offered on a nondiscriminatory basis without regard to race, color, national origin, religion, sex, age, marital status, or handicap.

Acknowledgments

The authors would like to thank the Faculty of Medicine and Health Sciences for research laboratory and technical assistances. This work was funded under Fundamental Research Grant Scheme by Ministry of Higher Education Malaysia, Grant no. FRGS/01(24)/835/2012(75).

References

[1] J. E. Burke and E. A. Dennis, "Phospholipase A$_2$ biochemistry," *Cardiovascular Drugs and Therapy*, vol. 23, no. 1, pp. 49–59, 2009.

[2] C. N. Birts, C. H. Barton, and D. C. Wilton, "Catalytic and noncatalytic functions of human IIA phospholipase A$_2$," *Trends in Biochemical Sciences*, vol. 35, no. 1, pp. 28–35, 2010.

[3] E. Hoxha, S. Harendza, G. Zahner et al., "An immunofluorescence test for phospholipase-A$_2$-receptor antibodies and its clinical usefulness in patients with membranous glomerulonephritis," *Nephrology Dialysis Transplantation* , vol. 26, no. 8, pp. 2526–2532, 2011.

[4] T. Aboshi, N. Shimizu, Y. Nakajima et al., "Biosynthesis of linoleic acid in Tyrophagus mites (Acarina: Acaridae)," *Insect Biochemistry and Molecular Biology*, vol. 43, no. 11, pp. 991–996, 2013.

[5] B. Blaul, R. Steinbauer, P. Merkl, R. Merkl, H. Tschochner, and J. Ruther, "Oleic acid is a precursor of linoleic acid and the male sex pheromone in *Nasonia vitripennis*," *Insect Biochemistry and Molecular Biology*, vol. 51, no. 1, pp. 33–40, 2014.

[6] B. Brandstetter and J. Ruther, "An insect with a delta-12 desaturase, the jewel wasp nasonia vitripennis, benefits from nutritional supply with linoleic acid," *Science of Nature*, vol. 103, no. 5, article no. 40, 2016.

[7] D. W. Stanley-Samuelson, R. A. Jurenka, C. Cripps, G. J. Blomquist, and M. de Renobales, "Fatty acids in insects: Composition, metabolism, and biological significance," *Archives of Insect Biochemistry and Physiology*, vol. 9, no. 1, pp. 1–33, 1988.

[8] R. A. Jurenka, D. W. Stanley-Samuelson, W. Loher, and G. J. Blomquist, "De novo biosynthesis of arachidonic acid and 5,11,14-eicosatrienoic acid in the cricket Teleogryllus commodus," *Biochimica et Biophysica Acta (BBA) - Lipids and Lipid Metabolism*, vol. 963, no. 1, pp. 21–27, 1988.

[9] D. Stanley and Y. Kim, "Eicosanoid Signaling in Insects: From Discovery to Plant Protection," *Critical Reviews in Plant Sciences*, vol. 33, no. 1, pp. 20–63, 2014.

[10] D. A. Six and E. A. Dennis, "The expanding superfamily of phospholipase A$_2$ enzymes: classification and characterization," *Biochimica et Biophysica Acta (BBA) - Molecular and Cell Biology of Lipids*, vol. 1488, no. 1-2, pp. 1–19, 2000.

[11] M. Jiménez, J. Cabanes, F. Gandi et al., "A continuous spectrophotometric assay for phospholipase A$_2$ activity," *Analytical Biochemistry*, vol. 319, no. 1, pp. 131–137, 2003.

[12] C. S. Tang, S. G. Lam-Phua, Y. K. Chung, and A. D. Giger, "Evaluation of a grass infusion-baited autocidal ovitrap for the monitoring of Aedes aegypti (L.)," *Dengue Bulletin*, vol. 31, pp. 131–140, 2007.

[13] A. A. Ghani, Medical Entomology. Kuala Lumpur: Institute of Medical Research, Kuala Lumpur, 2006.

[14] P. Nenad, G. Carolyn, E. L. Paul, L. A. Neil, L. A. Misso, and P. J. Thompson, "A simple assay for a human serum phospholipase A2 that is associated with high-density lipoproteins," *Journal of Lipid Research*, vol. 42, no. 10, pp. 1706–1713, 2001.

[15] D. G. Beghini, M. H. Toyama, S. Hyslop, L. Sodek, J. C. Novello, and S. Marangoni, "Enzymatic characterization of a novel phospholipase A$_2$ from Crotalus durissus cascavella rattlesnake (Maracambòia) venom," *The Protein Journal*, vol. 19, no. 7, pp. 603–607, 2000.

[16] H. Schägger and G. von Jagow, "Tricine-sodium dodecyl sulfate-polyacrylamide gel electrophoresis for the separation of proteins in the range from 1 to 100 kDa," *Analytical Biochemistry*, vol. 166, no. 2, pp. 368–379, 1987.

[17] I. Gromova and J. E. Celis, "Protein detection in gels by silver staining: a procedure compatible with mass-spectrometry," *Cell biology: A laboratory Handbook*, vol. 4, pp. 421–429, 2006.

[18] A. R. Nor Aliza and D. W. Stanley, "A digestive phospholipase A$_2$ in larval mosquitoes, Aedes aegypti," *Insect Biochemistry and Molecular Biology*, vol. 28, no. 8, pp. 561–569, 1998.

[19] R. H. Schaloske and E. A. Dennis, "The phospholipase A$_2$ superfamily and its group numbering system," *Biochimica et Biophysica Acta (BBA) - Molecular and Cell Biology of Lipids*, vol. 1761, no. 11, pp. 1246–1259, 2006.

[20] J. D. Bell, S. A. Sanchez, and L. Hazlett theordore, "Liposomes in the study of PLA2 activity," in *In Liposomes Part B*, p. 19, Elsevier Academic Press, Calif, USA, 2003.

[21] D. Stanley, "The non-venom insect phospholipases A$_2$," *Biochimica et Biophysica Acta (BBA) - Molecular and Cell Biology of Lipids*, vol. 1761, no. 11, pp. 1383–1390, 2006.

[22] V. L. Bonfim, M. H. Toyama, J. C. Novello et al., "Isolation and enzymatic characterization of a basic phospholipase A$_2$ from *Bothrops jararacussu* snake venom," *The Protein Journal*, vol. 20, no. 3, pp. 239–245, 2001.

[23] W. Martins, P. A. Baldasso, K. M. Honório et al., "A novel phospholipase A$_2$ (D49) from the venom of the Crotalus oreganus abyssus (North American Grand Canyon rattlesnake)," *BioMed Research International*, vol. 2014, Article ID 654170, 15 pages, 2014.

[24] S. L. Maruňak, L. Leiva, M. E. Garcia Denegri, P. Teibler, and O. Acosta De Pérez, "Isolation and biological characterization of a basic phospholipase A$_2$ from Bothrops jararacussu snake venom," *Biocell*, vol. 31, no. 3, pp. 355–364, 2008.

[25] A. R. Nor Aliza, R. L. Rana, S. R. Skoda, D. R. Berkebile, and D. W. Stanley, "Tissue polyunsaturated fatty acids and a digestive phospholipase A2 in the primary screwworm, Cochliomyia hominivorax," *Insect Biochemistry and Molecular Biology*, vol. 29, no. 11, pp. 1029–1038, 1999.

[26] E. A. M. Fleer, W. C. Puijk, A. J. Slotboom, and G. H. de Haas, "Modification of Arginine Residues in Porcine Pancreatic Phospholipase A2," *European Journal of Biochemistry*, vol. 116, no. 2, pp. 277–284, 1981.

[27] J. M. Uscian, J. S. Miller, R. W. Howard, and D. W. Stanley-Samuelson, "Arachidonic and eicosapentaenoic acids in tissue lipids of two species of predacious insects, Cicindela circumpicta and Asilis sp.," *Comparative Biochemistry and Physiology – Part B: Biochemistry and*, vol. 103, no. 4, pp. 833–838, 1992.

[28] J. M. Uscian, J. S. Miller, G. Sarath, and D. W. Stanley-Samuelson, "A digestive phospholipase A$_2$ in the tiger beetle Cicindella circumpicta," *Journal of Insect Physiology*, vol. 41, no. 2, pp. 135–141, 1995.

[29] G. R. Grotendorst and D. A. Hessinger, "Enzymatic characterization of the major phospholipase A$_2$ component of sea anemone (Aiptasia pallida) nematocyst venom," *Toxicon*, vol. 38, no. 7, pp. 931–943, 2000.

[30] E. A. Dennis, "Diversity of group types, regulation, and function of phospholipase A$_2$," *The Journal of Biological Chemistry*, vol. 269, no. 18, pp. 13057–13060, 1994.

[31] R. L. Rana, G. Sarath, and D. W. Stanley, "A digestive phospholipase A$_2$ in midguts of tobacco hornworms, Manduca sexta L," *Journal of Insect Physiology*, vol. 44, no. 3-4, pp. 297–303, 1998.

22

Production of Thermoalkaliphilic Lipase from *Geobacillus thermoleovorans* DA2 and Application in Leather Industry

Deyaa M. Abol Fotouh,[1] Reda A. Bayoumi,[2,3] and Mohamed A. Hassan[4]

[1] Electronic Materials Research Department, Advanced Technology and New Materials Institute (ATNMRI), City of Scientific Research and Technological Applications (SRTA-City), New Borg El-Arab City, P.O. Box 21934, Alexandria, Egypt
[2] Biology Department, Faculty of Science and Education, Taif University, Khormah Branch, P.O. Box 21974, Taif, Saudi Arabia
[3] Botany and Microbiology Department, Faculty of Science (Boys), Al-Azhar University, P.O. Box 11884, Cairo, Egypt
[4] Protein Research Department, Genetic Engineering and Biotechnology Research Institute (GEBRI), City of Scientific Research and Technological Applications (SRTA-City), New Borg El-Arab City, P.O. Box 21934, Alexandria, Egypt

Correspondence should be addressed to Deyaa M. Abol Fotouh; dabolfotouh@gmail.com
and Mohamed A. Hassan; m_adelmicro@yahoo.com

Academic Editor: Hartmut Kuhn

Thermophilic and alkaliphilic lipases are meeting a growing global attention as their increased importance in several industrial fields. Over 23 bacterial strains, novel strain with high lipolytic activity was isolated from Southern Sinai, Egypt, and it was identified as *Geobacillus thermoleovorans* DA2 using 16S rRNA as well as morphological and biochemical features. The lipase was produced in presence of fatty restaurant wastes as an inducing substrate. The optimized conditions for lipase production were recorded to be temperature 60°C, pH 10, and incubation time for 48 hrs. Enzymatic production increased when the organism was grown in a medium containing galactose as carbon source and ammonium phosphate as nitrogen source at concentrations of 1 and 0.5% (w/v), respectively. Moreover, the optimum conditions for lipase production such as substrate concentration, inoculum size, and agitation rate were found to be 10% (w/v), 4% (v/v), and 120 rpm, respectively. The TA lipase with Triton X-100 had the best degreasing agent by lowering the total lipid content to 2.6% as compared to kerosene (7.5%) or the sole crude enzyme (8.9%). It can be concluded that the chemical leather process can be substituted with TA lipase for boosting the quality of leather and reducing the environmental hazards.

1. Introduction

Lipases (triacylglycerol acylhydrolases, E.C. 3.1.1.3) are ubiquitous enzymes of considerable physiological significance and industrial potential [1]. Lipases catalyze the hydrolysis of triacylglycerols to glycerol and free fatty acids. Today, lipases are the choice of biocatalyst as they show unique chemo-, regio-, enantioselectivities, which enable the production of novel drugs, agrochemicals, and fine products [2].

Due to the ability of many lipases to perform both hydrolytic and synthetic reactions, they find immense applications in industries like foods, detergents, pharmaceuticals, leather, cosmetics, textile, dairy, and even biodiesel [3, 4].

Lipases are widely present in plants and animals, but almost all the commercially available lipases are usually obtained from microorganisms that produce a wide variety of extracellular lipases [5].

Cost of lipase production process was considered as a major obstacle in the industries. Therefore, many efforts are being made to use wastes as raw materials for lipase production. Agricultural residues for lipase production as well as other value added products would hold a prominent position in future biotechnologies, mainly because of its ecofriendliness and flexibility to both developing and developed countries. Several residues such as oil cakes, fibrous residues, and industrial effluent have increasing attention as abundant and cheap renewable feedstock [5, 6]. Enzymes from thermophiles and alkaliphiles have become the subject of special interest for biotechnological applications due to their high stability at adverse operational and/or storage conditions [7].

Many advantages were earned for carrying out biotechnological and industrial processes in high temperatures: high solubility of substrates (in particular for poorly soluble or polymeric molecules) resulting in higher product yield, higher reaction rates, increased availability of substrates, decreased risk of microbial contamination, and lower viscosity of reaction mixtures which in turn reduces the costs related to pumping, filtration, and centrifugation, and saving a great power cost would be exploited for cooling [8, 9].

In the present study, a detailed description of isolation, identification, and optimization of TA lipase production conditions from *G. thermoleovorans* DA2 will be demonstrated. An attempt to utilize restaurant fatty wastes as the main substrate for lipase production was carried out, which may raise the lipolytic activity and decrease the overall cost of the production process. In addition, this approach has a great environmental endeavor through minimizing the ecological hazards accompanied with the accumulation of wastes.

The application of TA lipase from *G. thermoleovorans* DA2 in leather tanning process as a degreasing agent replaced the commonly utilized organic solvent (Kerosene). Kanagaraj et al. reported that it is fundamental to add hydrolytic enzymes such as lipases and proteases in the soaking step for helping the fat degradation and raise the leather quality [10].

Substitution of the traditional chemical tanning processing with more ecofriendly treating procedures, for example, the enzymatic steps, became a necessity because of the recorded environmental hazards resulting from the pollution of water, careless disposal of solid wastes, and gaseous emissions [11].

2. Materials and Methods

2.1. Bacterial Strain.
A wide variety of samples were collected from many localities in Egypt, including desert and hot springs of Southern Sinai, Wadi El-Natron swamps, desert of Qina and Suez governorates, and the soil of El-Basateen slaughter house.

All samples were suspended in sterilized saline solution (0.85% w/v) which were cultured on plates of nutrient agar medium with pH 9 and incubated at 65°C for 48 hrs. The separated colonies had undergone a series of (agar streak method) for purification, and the morphological characteristics of each isolate were investigated. The purpose of this method was to isolate the thermoalkalophilic microorganisms.

2.2. Screening

2.2.1. Qualitative Assay.
All strains were screened for investigating their lipase activity on agar plates containing Rhodamine B 0.001% (w/v), nutrient broth 0.8% (w/v), NaCl 0.4% (w/v), agar 1% (w/v), and olive oil 3%, in distilled water and the pH was adjusted to be 9 [12].

Plates were incubated at 65°C for 18 hrs and the lipase activity was identified as an orange halo zone around colonies under UV light at 350 nm.

2.2.2. Quantitative Assay.
To detect the most potent thermoalkaliphilic lipase producing bacteria, all bacterial isolates were grown on medium composed of yeast extract 1 g; olive oil 10 mL; gum Arabic 10 g; $CaCl_2$ 1 g; and mineral salt solution 1 mL per liter. The medium pH is initially adjusted at 9 by using 6 N NaOH and incubated at 65°C and 100 rpm for 18 hrs. The most potent thermoalkaliphilic lipase producing isolate was purified by "agar streak method" and underwent biochemical investigations and it was identified by 16S rRNA technique.

2.2.3. Bacterial Identification.
The bacterial isolate was identified using 16S rRNA sequence. The genomic DNA was extracted by the following method which was described by Sambrook et al. [13].

The PCR amplification was carried out according to Hassan et al. and Abdou and Hassan. The reaction was performed using forwarded 16S rRNA primer (5′-AAATGG-AGGAAGGTGGGGAT-3′) and reverse 16S rRNA primer (5′-AGGAGGTGATCCAACCGCA-3′). The PCR machine (TECHNE TC-3000, FTC3/02) was programmed as follows: 3 min denaturation at 95°C, followed by 35 cycles that consisted of 1 min at 95°C, 1 min at 58°C, and 1 min at 72°C and the final extension was 10 min at 72°C [14, 15].

The PCR product was cleaned up for sequencing using Qiagen kit for DNA purification from aqueous PCR. DNA sequencing method which was developed by Sanger et al. was carried out using 3130X DNA Sequencer (Genetic Analyzer, Applied Biosystems, Hitachi, Japan) [16].

2.2.4. Sequence Analysis and Phylogenetic Tree Construction.
Similarity of the obtained nucleotide sequence was performed by basic local alignment search tool (BLAST) against reference sequences available in National Center for Biotechnology Information GenBank (NCBI GenBank). The reference sequences were collected from GenBank and the alignments using Clastal W were performed for constructing the phylogenetic tree using MEGA 5 software version 5.1 [17].

2.3. Lipase Production.
G. thermoleovorans DA2 was grown on a liquid production medium containing (%, w/v) yeast extract, 0.1 g; $NaNO_3$, 0.2 g; KH_2PO_4, 0.1 g; $MgSO_4\cdot7H_2O$, 0.05 g; KCl, 0.05 g; and $CaCl_2$, 0.1 g, supplemented with 1 g of fatty restaurant wastes.

The medium was adjusted at pH 9 (pH-Meter Model-420A Orion Co., USA) and 100 mL of medium in 500 mL Erlenmeyer flasks was inoculated with *G. thermoleovorans* DA2 and incubated at 65°C and 100 rpm for 18 hrs on a rotary shaker (Innova J-25 New Brunswick scientific, USA).

2.4. Assay of Thermoalkaliphilic Lipase Activity.
Lipase activity was detected by a spectrophotometric assay using *p*-nitrophenyl laurate (*p*NPL) as a substrate according to Castro-Ochoa et al. and Amara et al. with slight modifications. In brief, the reaction mixture consisted of 0.1 mL enzyme extract, 0.8 mL 0.1 M phosphate buffer (pH 8), and 0.1 mL 0.01 M *p*NPL in isopropanol. The hydrolytic reaction was carried out at 60°C for 30 min and then terminated by 0.25 mL of 0.1 M

Na$_2$CO$_3$ added. The mixture was centrifuged at 10,000 rpm for 15 min and the absorbance was determined at 410 nm using spectrophotometer (spectrophotometer: Lambda EZ 201 Perkin Elmer, USA). One unit of lipase activity was defined as the amount of enzyme that caused the release of 1 μmol of p-nitrophenol (molar absorption coefficient 4.6 mM^{-1} cm^{-1}) from pNP-laurate in 30 min under test conditions [18, 19].

2.5. Determination of Protein Content. Protein content was determined according to Lowry Method [20].

2.6. Optimization of Various Production Conditions of TA Lipase from G. thermoleovorans DA2. Several experiments were conducted to study the effect of physical and nutrients on culture conditions for TA lipase production by *G. thermoleovorans* DA2. All the previously mentioned production conditions were investigated throughout the following experiments.

2.6.1. Effect of Incubation Temperature on Production of TA Lipase from G. thermoleovorans DA2. Most favorable production temperature was studied by incubating the inoculated production medium at varying temperatures (50, 55, 60, 65, 70, 75, 80, 85, and 90°C). The culture filtrate was used for the lipase activity.

2.6.2. Effect of pH of the Medium on the Production of TA Lipase from G. thermoleovorans DA2. For optimization of production pH, the production medium of different pH, namely, 6, 7, 8, 9, 10, 11, and 12, was inoculated with culture and incubated in shaker for 18 hrs at 65°C and lipase activity of culture filtrate was determined.

2.6.3. Effect of Incubation Time on Production of TA Lipase from G. thermoleovorans DA2. To optimize incubation time for the maximum production of lipase, the production medium was incubated at 60°C in the shaker for 12, 18, 24, 36, 48, 56, and 72 hrs.

2.6.4. Effect of Carbon Source on Production of Lipase from G. thermoleovorans DA2. Various carbon sources such as glucose, galactose, xylose, ribose, rhamnose, melezitose, sucrose, lactose, maltose, cellobiose, sorbitol, mannitol, and inulin were used in the production medium at the concentration of 1% (w/v) to check the effect of carbon source on lipase production. The medium without carbon source served as control and the culture filtrate was assayed for enzyme activity.

2.6.5. Effect of Nitrogen Source on Production of TA Lipase from G. thermoleovorans DA2. To study the effect of nitrogen source, various nitrogen sources such as ammonium chloride, ammonium sulphate, ammonium nitrate, ammonium molybdate, ammonium phosphate, sodium nitrite, urea, calcium nitrate, potassium nitrate, and peptone were conducted in the production medium at the concentration of 0.5% (w/v). The main medium included sodium nitrate only served as control. The culture filtrate was assayed for enzyme activity.

2.6.6. Effect of Substrate Concentration on Production of Lipase from G. thermoleovorans DA2. The effect of substrate concentration on the production of TA lipase was determined by varying the concentration of fatty restaurant wastes, that is, 0.25, 0.5, 1, 2, 5, 10, 20, and 40% (w/v).

2.6.7. Effect of Inoculum Size on Production of Lipase from G. thermoleovorans DA2. Heavy cell suspension of *G. thermoleovorans* DA2 was prepared by growing bacterial isolate on nutrient broth and limiting growth level at 3 × 10^7 CFU·mL^{-1}. Different inocula sizes of culture including 0.25, 0.5, 1, 2, 4, 5, and 10% (v/v) were applied.

2.6.8. Effect of Agitation Rate on Production of TA Lipase from G. thermoleovorans DA2. To optimize the agitation rate for maximum lipase production, the inoculated production medium was agitated at different rotations per minute (rpm) such as 40, 80, 120, and 150 rpm. The culture filtrate was used to check the enzyme activity.

2.7. Application of TA Lipase from G. thermoleovorans DA2 in Degreasing of Leather. The crude TA lipase from *G. thermoleovorans* DA2 was used as degreaser agent in leather industry as compared to traditional methods which depends upon solvents and surfactants.

2.7.1. Preparation of Skin Samples. A piece of sheep skin was obtained after it had undergone the common processing steps, that is, liming, dehairing, and bating. Eight skin pieces about 5 × 5 cm dimension were cut and divided into 4 groups (each of 2 pieces):

> Group (A): the skin was soaked in 50 mL (Kerosene) and was incubated at 25°C as a traditional method.
>
> Group (B): the skin was immersed in 50 mL of 10% of the crude TA lipase and was incubated at 60°C.
>
> Group (C): the skin was treated with a mixture of crude TA lipase (10%) + Kerosene in 4 : 1 ratio and was incubated at 60°C.
>
> Group (D): the skin was bathed in a mixture of crude TA lipase (10%) + Triton X-100 in 10 : 1 ratios and was incubated at 60°C.

All groups of skin were treated for 2 hrs then; the skin was collected and dried [21].

2.7.2. Total Lipid Measurement Technique. Total lipid of the treated skin pieces were measured relying on Soxhlet apparatus [22]. The obtained values were then compared with the lipid content of the control.

3. Results

3.1. Bacterial Strain Selection and Identification. Out of seven isolates the most efficient lipase producer was isolated from desert of southern Sinai based on enzyme assay method and it was identified using 16S rRNA and selected for further studies.

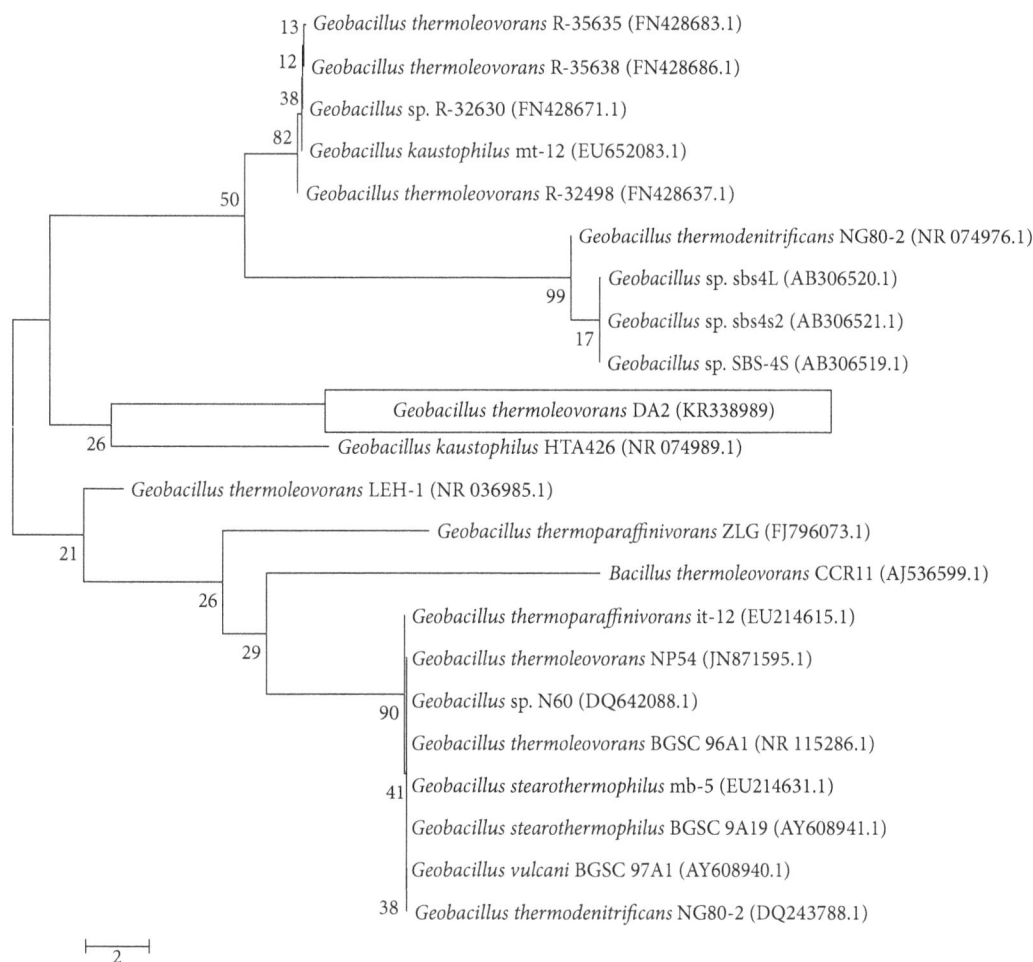

FIGURE 1: Phylogenetic position of *Geobacillus thermoleovorans* DA2 within the genus *Geobacillus*. The branching pattern was generated by neighbor-joining tree method and the GenBank accession numbers of the 16S rRNA nucleotide sequences are indicated in brackets. The bar indicates a Jukes-Cantor distance of 2.

3.2. Bacterial Identification by 16S rRNA.

The amplified 16S rRNA gene fragment was investigated using DNA ladder (Gene ruler 50 bp–1031 bp DNA ladder) and it was 380 bp.

The BLAST algorithm was used to retrieve for homologous sequences in GenBank to the obtained 16S rRNA sequence. The bacterial isolate revealed 99% identity to full genome of *Geobacillus thermoleovorans*, *Geobacillus stearothermophilus*, *G. thermoparaffinivorans*, *G. thermodenitrificans*, and *G. kaustophilus*. Based on the morphological, biochemical, and molecular characteristics, the isolate was identified and released in NCBI GenBank as *Geobacillus thermoleovorans* DA2 under the accession numbers (KR338990), and the branching pattern was analyzed by 500 bootstrap replicates as in Figure 1. *G. thermoleovorans* DA2 is an aerobic, spore-forming, nonmotile rod able to grow at high temperatures (50–80°C) with an optimum growth at 65°C.

3.3. Optimization of Production Conditions of TA Lipase from *G. thermoleovorans* DA2

3.3.1. Effect of Incubation Temperature on Production of TA Lipase from G. thermoleovorans DA2. Most suitable temperature for maximum production of TA lipase (146.85 U/mL) from *G. thermoleovorans* DA2 was found to be 60°C as shown in Figure 2(a).

3.3.2. Effect of pH of the Medium on Production of TA Lipase from G. thermoleovorans DA2. The maximum production of TA lipase from *G. thermoleovorans* DA2 was observed at pH 10 (157.15 U/mL) and after that, the lipase activity was decreased with increasing the pH values as shown in Figure 2(b).

3.3.3. Optimization of Incubation Time for Production of TA Lipase from G. thermoleovorans DA2. The maximum production of lipase was observed at 48 hrs (248.11 U/mL). Optimal incubation time was found to be 48 hrs corresponding to maximum enzyme activity and after this decline in enzyme activity was observed (Figure 2(c)).

3.3.4. Effect of Carbon Source on Production of TA Lipase from G. thermoleovorans DA2. Maximum production of TA lipase was obtained from the medium which was supplemented

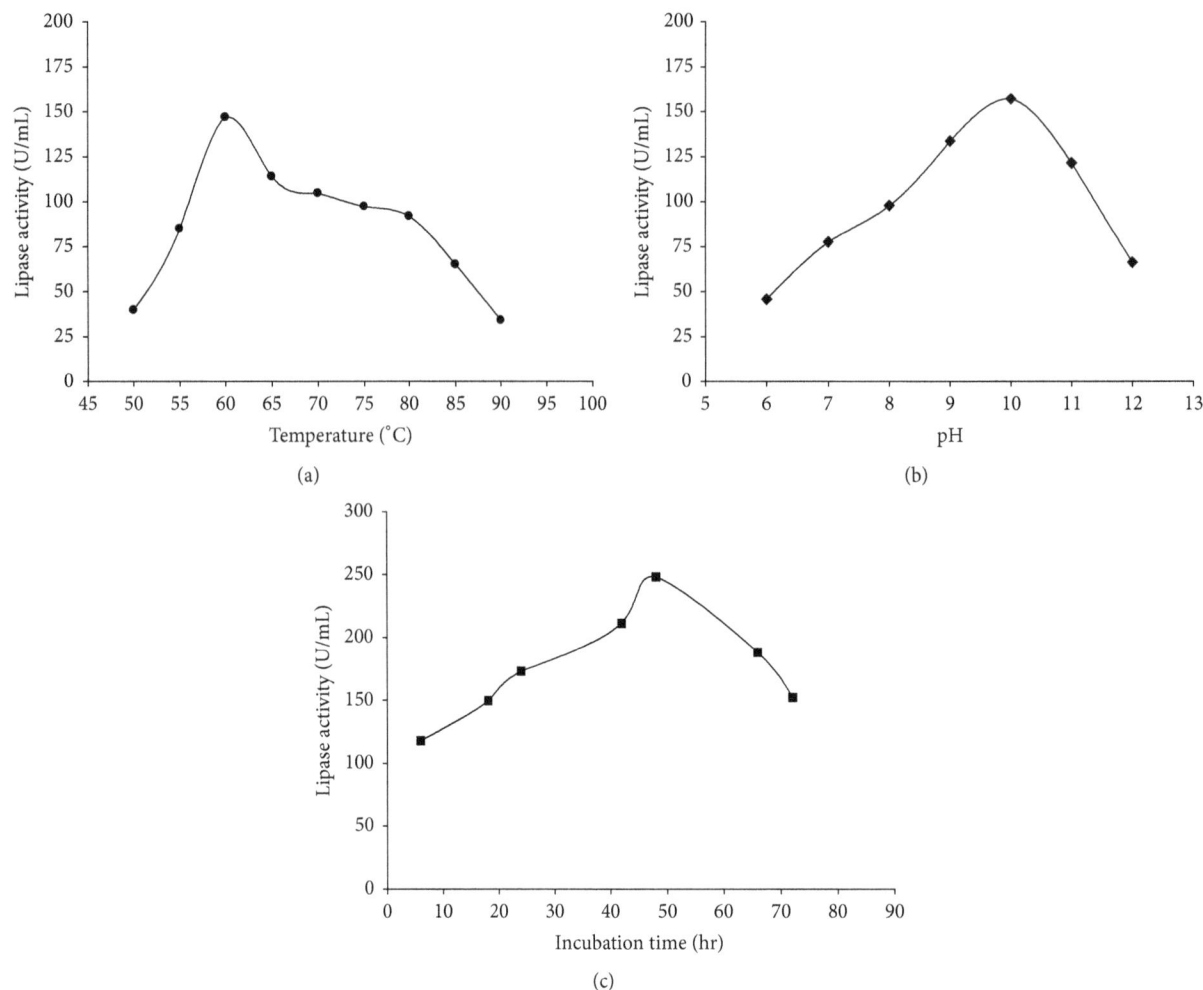

FIGURE 2: (a) Effect of incubation temperature on production of TA lipase from *G. thermoleovorans* DA2. (b) Effect of pH on production of TA lipase from *G. thermoleovorans* DA2. (c) Effect of incubation time on production of TA lipase from *G. thermoleovorans* DA2.

with galactose (1%) as carbon source giving enzyme activity of 701.86 U/mL (Figure 3(a)).

3.3.5. Effect of Nitrogen Source on Production of TA Lipase from G. thermoleovorans DA2. The lipase production was highest (843.04 U/mL) in medium containing ammonium phosphate at a concentration (0.5%) as nitrogen source (Figure 3(b)).

3.3.6. Effect of Substrate Concentration on Production of TA Lipase from G. thermoleovorans DA2. Maximum enzyme activity (892.43 U/mL) was observed with 10% (w/v) fatty restaurant wastes (Figure 3(c)).

3.3.7. Optimization of Inoculum Size for Production of TA Lipase from G. thermoleovorans DA2. Figure 3(d) reveals that the inoculum size (4%, v/v) gave the maximum lipase production with an activity of 917.23 U/mL.

3.3.8. Effect of Agitation Rate on Production of TA Lipase from G. thermoleovorans DA2. The maximum lipase activity (1021.91 U/mL) was observed at 120 rpm as shown in Figure 3(e).

3.4. Application of TA Lipase in Degreasing of Leather. The thermoalkaliphilic lipase from *G. thermoleovorans* DA2 plus Triton X-100 was the most efficient leather degreasing agent where the total lipid content decreased from 17.5% to 2.6% as shown in Table 1.

The mixture of TA lipase plus kerosene decreased the total lipid content to 5.7%; so it came in second level.

4. Discussion

The applications of lipases are constantly increased in industrial and biotechnological fields; therefore, that should be supported by discovering novel lipase types with improved characters. The common about enzymes is their sensitivity to adverse conditions such as high temperature, extreme acidity and/or alkalinity, drought, and high salinity, but extremozymes, enzymes derived from extremophilic microorganisms, are an attractive alternative to tuning a given biocatalyst for a specific industrial application. They are capable of catalyzing their respective reactions in nonaqueous environments, water/solvent mixtures, at extremely high pressures, acidic and alkaline pH, at temperatures up to 140°C, or near

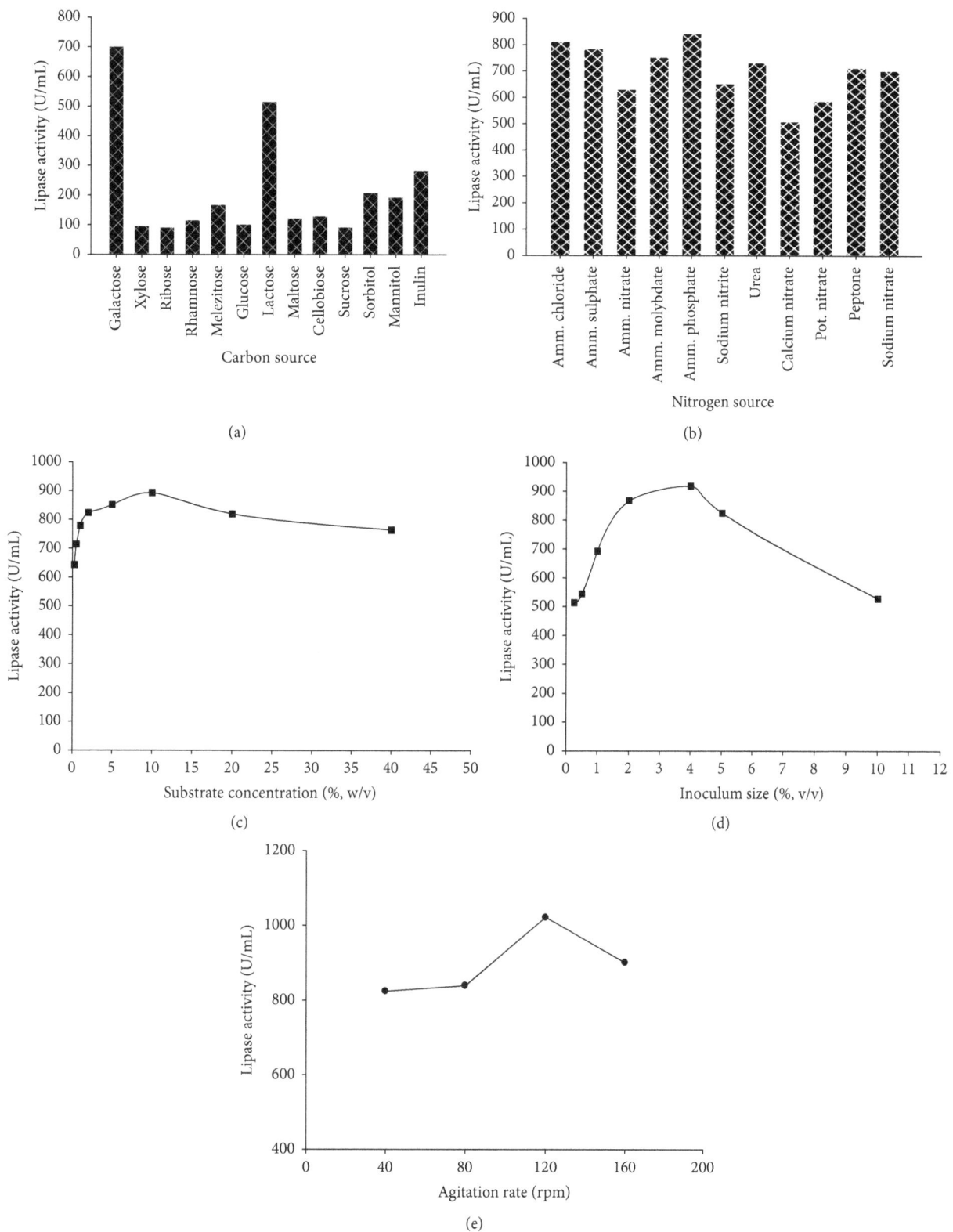

FIGURE 3: (a) Effect of carbon source on production of lipase from *G. thermoleovorans* DA2. (b) Effect of nitrogen source on production of lipase from *G. thermoleovorans* DA2. (c) Effect of substrate concentration on production of TA lipase from *G. thermoleovorans* DA2. (d) Effect of inoculum size on production of TA lipase from *G. thermoleovorans* DA2. (e) Effect of agitation rate on production of lipase from *Geobacillus thermoleovorans* DA2.

TABLE 1: Effect of using the TA lipase from *G. thermoleovorans* DA2 as a degreasing agent and the total lipid content of leather samples.

Group of leather samples	Type of leather treatment	Total lipid content (%)
	Control (without treatment)	17.50
Group (A)	Organic solvent (Kerosene)	7.50
Group (B)	10% crude lipase produced by *G. thermoleovorans* DA2	8.90
Group (C)	10% crude lipase produced by *G. thermoleovorans* DA2 + Kerosene	5.70
Group (D)	10% crude lipase produced by *G. thermoleovorans* DA2 + Triton X-100	2.60

the freezing point of water [23]. The present investigation is an attempt to discover novel bacterial strains with the capability of producing lipases able to tolerate high temperature and high alkalinity as well. Seven of twenty three strains showed different lipolytic activities in incubation temperature 65°C and pH 9; the most potent producing strain is identified as *G. thermoleovorans* DA2 selected to further studies as the most promising source of thermoalkaliphilic lipase.

Optimum temperature in the present investigation came in correspondence with the optimum temperature for lipase produced by *G. stearothermophilus* strain-5 [24]. On the other hand, the temperature for the highest production of lipase by *G. thermodenitrificans* AZ1 was 55°C [25].

Optimum pH value for the present study suggests more alkaliphilic behavior comparing to lipases produced by *Geobacillus thermoleovorans* CCR11 and the alkaliphilic *Bacillus* sp. (KS4) which was found to give their highest lipase productions at pH 8 [26, 27]. Berekaa et al. reported that the optimum pH for lipase production by *G. stearothermophilus* strain-5 was 7 [28].

G. thermoleovorans DA2 was found to consume 48 hrs to give its maximum production of TA lipase comparing to *Bacillus thermoleovorans* CCR11 which is recorded to produce the highest lipase rate after 44 hrs, while the highest production of lipase by the thermophilic strain *Bacillus* sp. strain 42 was recorded after 72 hrs [26, 29].

Galactose was found to be the most supportive carbon source for the TA lipase production by *G. thermoleovorans* DA2, suggesting high ability of the strain to uptake and utilize the monosaccharide galactose in the metabolic processes of lipase production as the lactose disaccharide was the second optimum carbon source. Glycerol was found to be the most inducing carbon source for lipase production by *G. stearothermophilus* strain-5, while it is the highest lipase production by *B. subtilis* ZR-1 supported by the sole olive oil [12, 28].

Peptone was the most supportive nitrogen source for lipase production by *Bacillus* sp. LBN2 [30]. In contrast, mix of yeast extract and peptone was the best for the production of thermoalkaliphilic lipase by *B. coagulans* BTS-3 [31]. Moreover, sole yeast extract was the best for lipase produced by *B. licheniformis* MTCC-10498 [32].

Abd El Rahman found that 12.5% of the slaughter house waste was the optimum substrate concentration for lipase production by *B. stearothermophilus* B-78, while Gayathri et al. have reported that 2% of palm oil was the optimum concentration for lipase production by *B. stearothermophilus* [33, 34].

Ten percent (10%, v/v) was the optimum inoculum for lipase production by *B. licheniformis* MTCC-10498 [32].

However, *Bacillus flexus* XJU-1 gave the best production of lipase depending on 2% (v/v) inoculum size, while 6% (v/v) was the optimum inoculum size for the best lipase production by *Pseudomonas gessardii* [35, 36].

The optimum revolution per minute for the maximum production of TA lipase by *G. thermoleovorans* DA2 was found to be 120 rpm, which came in complete accordance with the optimum agitation rate for highest lipase production by *B. licheniformis* MTCC-10498 [32], while Veerapagu et al. reported that the optimized value for greatest lipase production by *Pseudomonas gessardii* was 160 rpm [36].

Once samples of sheep leather were treated with TA lipase produced by the strain *G. thermoleovorans* DA2 mixed with Triton X-100, fatty content of the leather depleted from 17.5% to 2.5%. This may be due to the elevated processing temperature which takes a hand in destroying the walls of skin fat cells and increase solubility of the fat content very well. Moreover, Triton X-100 had supportive effect in reducing the surface tension of the lipids and facilitated the enzyme efficacy. Saran et al. recorded degreasing rate more than 95% by 5–10% mesophilic lipase enzyme produced by *Bacillus subtilis* solely after treating for 12 hrs [11].

Moreover, their interpretations came in complete accordance with our observations about the higher quality, appearance, and smoothness levels of the enzyme-treated leather compared to chemically processed skin. The obtained results are suggesting that advanced efforts to replace the ecohazardous traditional chemical tanning process with the green enzymatic techniques are worthy to be achieved.

5. Conclusion

In the present study, thermoalkaliphilic lipase was produced from the new bacterial isolate which was identified using the sequence of 16S rRNA gene and biochemical tests as *Geobacillus thermoleovorans* DA2. Wide ranges of growth factors were studied for maximizing TA lipase enzyme. The production of TA lipase with high amount using cheap substrate (fatty restaurant wastes) made the production process cost-effective. The results exhibited the potential application of the TA lipase in leather industry as a degreaser suggesting more ecofriendly tanning processes in the future.

References

[1] R. Sharma, Y. Chisti, and U. C. Banerjee, "Production, purification, characterization, and applications of lipases," *Biotechnol-*

ogy Advances, vol. 19, no. 8, pp. 627–662, 2001.

[2] R. Kaushik, S. Saran, J. Isar, and R. K. Saxena, "Statistical optimization of medium components and growth conditions by response surface methodology to enhance lipase production by *Aspergillus carneus*," *Journal of Molecular Catalysis B: Enzymatic*, vol. 40, no. 3-4, pp. 121–126, 2006.

[3] F. Hasan, A. A. Shah, and A. Hameed, "Industrial applications of microbial lipases," *Enzyme and Microbial Technology*, vol. 39, no. 2, pp. 235–251, 2006.

[4] A. Robles-Medina, P. A. González-Moreno, L. Esteban-Cerdán, and E. Molina-Grima, "Biocatalysis: towards ever greener biodiesel production," *Biotechnology Advances*, vol. 27, no. 4, pp. 398–408, 2009.

[5] A. Salihu, M. Z. Alam, M. I. AbdulKarim, and H. M. Salleh, "Lipase production: an insight in the utilization of renewable agricultural residues," *Resources, Conservation and Recycling*, vol. 58, pp. 36–44, 2012.

[6] J. Sahasrabudhe, S. Palshikar, A. Goja, and C. Kulkarni, "Use of ghee residue as a substrate for microbial lipase production," *International Journal of Scientific & Technology Research*, vol. 1, no. 10, pp. 61–64, 2012.

[7] B. P. Golaki, S. Aminzadeh, A. A. Karkhane et al., "Cloning, expression, purification, and characterization of lipase 3646 from thermophilic indigenous *Cohnella* sp. A01," *Protein Expression and Purification*, vol. 109, pp. 120–126, 2015.

[8] A. Trincone, "Marine biocatalysts: enzymatic features and applications," *Marine Drugs*, vol. 9, no. 4, pp. 478–499, 2011.

[9] S. P. S. Bisht and S. Panda, "Isolation and identification of new lipolytic thermophilic bacteria from an Indian hot spring," *International Journal of Pharma and Bio Sciences*, vol. 2, pp. 229–235, 2011.

[10] J. Kanagaraj, T. Senthilvelan, R. C. Panda, and S. Kavitha, "Eco-friendly waste management strategies for greener environment towards sustainable development in leather industry: a comprehensive review," *Journal of Cleaner Production*, vol. 89, pp. 1–17, 2015.

[11] S. Saran, R. V. Mahajan, R. Kaushik, J. Isar, and R. K. Saxena, "Enzyme mediated beam house operations of leather industry: a needed step towards greener technology," *Journal of Cleaner Production*, vol. 54, pp. 315–322, 2013.

[12] M. Rabbani, M. R. Bagherinejad, H. M. Sadeghi et al., "Isolation and characterization of novel thermophilic lipase-secreting bacteria," *Brazilian Journal of Microbiology*, vol. 44, no. 4, pp. 1113–1119, 2013.

[13] J. Sambrook, E. F. Fritsch, and T. Maniatis, *Molecular Cloning: A Laboratory Manual*, Cold Spring Harbor Laboratory, Cold Spring Harbor, NY, USA, 2nd edition, 1989.

[14] M. A. Hassan, B. M. Haroun, A. A. Amara, and E. A. Serour, "Production and characterization of keratinolytic protease from new wool-degrading *Bacillus* species isolated from Egyptian ecosystem," *BioMed Research International*, vol. 2013, Article ID 175012, 14 pages, 2013.

[15] H. M. Abdou and M. A. Hassan, "Protective role of omega-3 polyunsaturated fatty acid against lead acetate-induced toxicity in liver and kidney of female rats," *BioMed Research International*, vol. 2014, Article ID 435857, 11 pages, 2014.

[16] F. Sanger, S. Nicklen, and A. R. Coulson, "DNA sequencing with chain-terminating inhibitors," *Proceedings of the National Academy of Sciences of the United States of America*, vol. 74, no. 12, pp. 5463–5467, 1977.

[17] K. Tamura, D. Peterson, N. Peterson, G. Stecher, M. Nei, and S. Kumar, "MEGA5: molecular evolutionary genetics analysis using maximum likelihood, evolutionary distance, and maximum parsimony methods," *Molecular Biology and Evolution*, vol. 28, no. 10, pp. 2731–2739, 2011.

[18] L. D. Castro-Ochoa, C. Rodríguez-Gómez, G. Valerio-Alfaro, and R. O. Ros, "Screening, purification and characterization of the thermoalkalophilic lipase produced by *Bacillus thermoleovorans* CCR11," *Enzyme and Microbial Technology*, vol. 37, no. 6, pp. 648–654, 2005.

[19] A. A. Amara, M. A. Hassan, A. T. Abulhamd, and B. M. Haroun, "Non-mucoid *P. aeruginosa* aiming to a safe production of protease and lipase," *International Science and Investigation Journal*, vol. 2, no. 5, pp. 103–113, 2013.

[20] O. H. Lowry, N. J. Rosebrough, A. L. Farr, and R. J. Randall, "Protein measurement with the Folin phenol reagent," *The Journal of Biological Chemistry*, vol. 193, no. 1, pp. 265–275, 1951.

[21] A. Afsar and F. Cetinkaya, "Studies on the degreasing of skin by using enzyme in liming process," *Indian Journal of Chemical Technology*, vol. 15, no. 5, pp. 507–510, 2008.

[22] F. Shahidi, "Extraction and measurement of total lipids," in *Current Protocols in Food Analytical Chemistry*, John Wiley & Sons, 2003.

[23] S. Elleuche, C. Schröder, K. Sahm, and G. Antranikian, "Extremozymes-biocatalysts with unique properties from extremophilic microorganisms," *Current Opinion in Biotechnology*, vol. 29, no. 1, pp. 116–123, 2014.

[24] M. Sifour, T. I. Zaghloul, H. M. Saeed, M. M. Berekaa, and Y. R. Abdel-Fattah, "Enhanced production of lipase by the thermophilic *Geobacillus stearothermophilus* strain-5 using statistical experimental designs," *New Biotechnology*, vol. 27, no. 4, pp. 330–336, 2010.

[25] Y. R. Abdel-Fattah, N. A. Soliman, S. M. Yousef, and E. R. El-Helow, "Application of experimental designs to optimize medium composition for production of thermostable lipase/esterase by *Geobacillus thermodenitrificans* AZ1," *Journal of Genetic Engineering and Biotechnology*, vol. 10, no. 2, pp. 193–200, 2012.

[26] M. G. Sánchez-Otero, I. I. Ruiz-López, D. E. Ávila-Nieto, and R. M. Oliart-Ros, "Significant improvement of *Geobacillus thermoleovorans* CCR11 thermoalkalophilic lipase production using Response Surface Methodology," *New Biotechnology*, vol. 28, no. 6, pp. 761–766, 2011.

[27] D. Sharma, B. K. Kumbhar, A. K. Verma, and L. Tewari, "Optimization of critical growth parameters for enhancing extracellular lipase production by alkalophilic *Bacillus* sp.," *Biocatalysis and Agricultural Biotechnology*, vol. 3, no. 4, pp. 205–211, 2014.

[28] M. M. Berekaa, T. I. Zaghloul, Y. R. Abdel-Fattah, H. M. Saeed, and M. Sifour, "Production of a novel glycerol-inducible lipase from thermophilic *Geobacillus stearothermophilus* strain-5," *World Journal of Microbiology & Biotechnology*, vol. 25, no. 2, pp. 287–294, 2009.

[29] M. A. Eltaweel, R. N. Z. R. A. Rahman, A. B. Salleh, and M. Basri, "An organic solvent-stable lipase from *Bacillus* sp. strain 42," *Annals of Microbiology*, vol. 55, no. 3, pp. 187–192, 2005.

[30] L. Bora and M. Bora, "Optimization of extracellular thermophilic highly alkaline lipase from thermophilic *Bacillus* sp. isolated from hot spring of Arunachal Pradesh, India," *Brazilian Journal of Microbiology*, vol. 43, pp. 30–42, 2012.

[31] S. Kumar, K. Kikon, A. Upadhyay, S. S. Kanwar, and R. Gupta, "Production, purification, and characterization of lipase from thermophilic and alkaliphilic *Bacillus coagulans* BTS-3," *Protein*

Expression and Purification, vol. 41, no. 1, pp. 38–44, 2005.

[32] C. K. Sharma, P. K. Sharma, and S. S. Kanwar, "Optimization of production conditions of lipase from *B. licheniformis* MTCC-10498," *Research Journal of Recent Sciences*, vol. 1, no. 7, pp. 25–32, 2012.

[33] M. A. Abd El Rahman, *Production of thermostable microbial enzymes for application in bio-detergent technology [M.S. dissertation]*, Faculty of Science, Al-Azhar University, Cairo, Egypt, 2006.

[34] V. R. Gayathri, P. Perumal, L. P. Mathew, and B. Prakash, "Screening and molecular characterization of extracellular lipase producing *Bacillus* species from coconut oil mill soil," *International Journal of Science and Technology*, vol. 2, no. 7, pp. 502–509, 2013.

[35] F. N. Niyonzima and S. S. More, "Concomitant production of detergent compatible enzymes by *Bacillus flexus* XJU-1," *Brazilian Journal of Microbiology*, vol. 45, no. 3, pp. 903–910, 2014.

[36] M. Veerapagu, A. S. Narayanan, K. Ponmurgan, and K. R. Jeya, "Screening, selection, identification, production, and optimization of bacterial lipase from oil spilled soil," *Asian Journal of Pharmaceutical and Clinical Research*, vol. 6, supplement 3, pp. 62–67, 2013.

Permissions

List of Contributors

Alex Fernando de Almeida and Kleydiane Braga Dias
Bioprocess Engineering and Biotechnology, Federal University of Tocantins (UFT), Rua Badejós, Chácaras 69/72, Zona Rural, 77402-970 Gurupi, TO, Brazil

Ana Carolina Cerri da Silva and Sâmia Maria Tauk-Tornisielo
Environmental Studies Center (CEA), Universidade Estadual Paulista (UNESP), Avenida 24-A, 1515 Bela Vista, 13506-900 Rio Claro, SP, Brazil

César Rafael Fanchini Terrasan and Eleonora Cano Carmona
Biochemistry and Microbiology Department, Bioscience Institute (IB), Universidade Estadual Paulista (UNESP), Avenida 24-A, 1515 Bela Vista, 13506-900 Rio Claro, SP, Brazil

Luana Cunha, Raquel Martarello, Paula Monteiro de Souza, Marcela Medeiros de Freitas, Kleber Vanio Gomes Barros, Mauricio Homem-de-Mello and Pérola Oliveira Magalhães
Laboratory of Natural Products, School of Health Sciences, University of Brasília, Asa Norte, 70910900 Brasília, DF, Brazil

Edivaldo Ximenes Ferreira Filho
Laboratory of Enzymology, Department of Cell Biology, University of Brasília, Asa Norte, 70910900 Brasília, DF, Brazil

Umi Baroroh
Master of Biotechnology Program, Postgraduate School, Universitas Padjadjaran, Jl. Dipati Ukur 35, Bandung, West Java, Indonesia

Mas Rizky A. A. Syamsunarno
Master of Biotechnology Program, Postgraduate School, Universitas Padjadjaran, Jl. Dipati Ukur 35, Bandung, West Java, Indonesia
Research Center of Molecular Biotechnology and Bioinformatics, Universitas Padjadjaran, Jl. Singaperbangsa 2, Bandung, West Java 40133, Indonesia

Jutti Levita
Master of Biotechnology Program, Postgraduate School, Universitas Padjadjaran, Jl. Dipati Ukur 35, Bandung, West Java, Indonesia
Department of Pharmacology and Clinical Pharmacy, Faculty of Pharmacy, Universitas Padjadjaran, Jl. Raya Bandung-Sumedang Km 21, Jatinangor, Sumedang, West Java 45363, Indonesia

Saadah Diana Rachman and Safri Ishmayana
Department of Chemistry, Faculty of Mathematics and Natural Sciences, Universitas Padjadjaran, Jl. Raya Bandung-Sumedang Km 21, Jatinangor, Sumedang, West Java 45363, Indonesia

Muhammad Yusuf and Toto Subroto
Department of Chemistry, Faculty of Mathematics and Natural Sciences, Universitas Padjadjaran, Jl.Raya Bandung-Sumedang Km 21, Jatinangor, Sumedang, West Java 45363, Indonesia
Research Center of Molecular Biotechnology and Bioinformatics, Universitas Padjadjaran, Jl.Singaperbangsa 2, Bandung, West Java 40133, Indonesia

Preety Vatsyayan
Institute of Analytical Chemistry, Chemo- and Biosensors, University of Regensburg, 93053 Regensburg, Germany

Pranab Goswami
Department of Biosciences and Bioengineering, Indian Institute of Technology Guwahati, Guwahati, Assam 781039, India

Camila Gabriel Kato, Geferson de Almeida Gonçalves, Anacharis Babeto de Sá-Nakanishi, Lívia Bracht, Jurandir Fernando Comar, Adelar Bracht and Rosane Marina Peralta
Postgraduate Program of Food Science, University of Maringá, Avenida Colombo 5790, 87020900 Maringá, PR, Brazil
Department of Biochemistry, University of Maringá, Maringá, PR, Brazil

Flavio Augusto Vicente Seixas
Department of Biochemistry, University of Maringá, Maringá, PR, Brazil

Rosely Aparecida Peralta
Department of Chemistry, Federal University of Santa Catarina, Florianópolis, SC, Brazil

Gastón Ezequiel Ortiz, Diego Gabriel Noseda, María Clara Ponce Mora, Matías Nicolás Recupero, Martín Blasco and Edgardo Albertó
Instituto de Investigaciones Biotecnológicas-Instituto Tecnológico de Chascomús (IIB-INTECH), Universidad Nacional de San Martín (UNSAM) and Consejo Nacional de Investigaciones Científicasy Técnicas (CONICET), San Martín, 1650 Buenos Aires, Argentina

Dennis J. Díaz-Rincón, Ivonne Duque, Erika Osorio, Angela Espejo-Mojica and Carlos J. Alméciga-Díaz
Institute for the Study of Inborn Errors of Metabolism, Facultad de Ciencias, Pontificia Universidad Javeriana, Bogotá, Colombia

Alexander Rodríguez-López
Institute for the Study of Inborn Errors of Metabolism, Facultad de Ciencias, Pontificia Universidad Javeriana, Bogotá, Colombia
Departamento de Química, Facultad de Ciencias, Pontificia Universidad Javeriana, Bogotá, Colombia

Claudia M. Parra-Giraldo
Unidad de Proteómica y Micosis Humanas, Grupo de Enfermedades Infecciosas, Departamento de Microbiología, Facultad de Ciencias, Pontificia Universidad Javeriana, Bogotá, Colombia

Raúl A. Poutou-Piñales and Balkys Quevedo-Hidalgo
Grupo de Biotecnología Ambiental e Industrial (GBAI), Departamento de Microbiología, Facultad de Ciencias, Pontificia Universidad Javeriana, Bogotá, Colombia

Madison A. Smith
MI-SWACO, Shafter, CA 93263, USA

Jesica Gonzalez
University of California, San Francisco, San Francisco, CA 94143, USA

Anjum Hussain, Rachel N. Oldfield, Kathryn A. Johnston and Karlo M. Lopez
California State University, Bakersfield, Bakersfield, CA 93311, USA

Supaporn Klangprapan, Tueanjit Khampitak and Patcharee Boonsiri
Department of Biochemistry, Faculty of Medicine, Khon Kaen University, Khon Kaen 40002, Thailand

Ponlatham Chaiyarit
Department of Oral Diagnosis, Faculty of Dentistry, Khon Kaen University, Khon Kaen 40002,Thailand
Research Group of Chronic Inflammatory Oral Diseases and Systemic Diseases Associated with Oral Health, Khon Kaen University, Khon Kaen 40002, Thailand

Doosadee Hormdee
Research Group of Chronic Inflammatory Oral Diseases and Systemic Diseases Associated with Oral Health, Khon Kaen University, Khon Kaen 40002,Thailand
Department of Periodontology, Faculty of Dentistry, Khon Kaen University, Khon Kaen 40002, Thailand

Amonrujee Kampichai
Department of Periodontology, Faculty of Dentistry, Khon Kaen University, Khon Kaen 40002, Thailand
Dental Department, Fang Hospital, Fang District, Chiangmai 50110, Thailand

Jureerut Daduang
Department of Clinical Chemistry, Faculty of Associated Medical Sciences, Khon Kaen University, Khon Kaen 40002, Thailand
Centre for Research and Development of Medical Diagnostic Laboratories, Khon Kaen University, Khon Kaen 40002, Thailand

Ratree Tavichakorntrakool
Centre for Research and Development of Medical Diagnostic Laboratories, Khon Kaen University, Khon Kaen 40002, Thailand
Department of Clinical Microbiology, Faculty of Associated Medical Sciences, Khon Kaen University, Khon Kaen 40002, Thailand

Bhinyo Panijpan
Faculty of Science, Mahidol University, Bangkok 10400, Thailand

Hemlata Bhosale, Uzma Shaheen and Tukaram Kadam
DST-FIST Sponsored School of Life Sciences, Swami Ramanand Teerth Marathwada University, Nanded 431606, India

Elena N. Levtchenko
Department of Pediatric Nephrology & Growth and Regeneration, University Hospitals Leuven, KU Leuven, UZ Herestraat 49, Leuven, Belgium

Mohamed A. Elmonem
Department of Pediatric Nephrology & Growth and Regeneration, University Hospitals Leuven, KU Leuven, UZ Herestraat 49, Leuven, Belgium
Department of Clinical and Chemical Pathology, Inherited Metabolic Disease Laboratory, Center of Social and Preventive Medicine, Faculty of Medicine, Cairo University, 2 Ali Pasha Ibrahim Street, Room 409, Monira, Cairo, Egypt

Lambertus P. van den Heuvel
Department of Pediatric Nephrology & Growth and Regeneration, University Hospitals Leuven, KU Leuven, UZ Herestraat 49, Leuven, Belgium
Department of Pediatric Nephrology, Radboud University Medical Center, Post 804, Postbus 9101, 6500 HB Nijmegen, Netherlands

Oliyad Jeilu Oumer
Department of Biology, Ambo University, Ambo, Ethiopia

Dawit Abate
College of Natural Science, Addis Ababa University, Addis Ababa, Ethiopia

D. F. Silva, A. F. A. Carvalho, T. Y. Shinya, G. S. Mazali and P. Oliva-Neto
Biological Science Department, Universidade Estadual Paulista (UNESP), Avenida Dom Antônio, 2100 Bairro, Parque Universit´ario, 19806-900 Assis, SP, Brazil

R. D. Herculano
Bioprocess & Biotechnology Department, Universidade Estadual Paulista (UNESP), Rod. Araraquara-Jaú Km 1 Bairro, Machados, 14800-901 Araraquara, SP, Brazil

Juan L. Rendón, Mauricio Miranda-Leyva, Alberto Guevara-Flores, José de Jesús Martínez-González, Irene Patricia del Arenal, Oscar Flores-Herrera and Juan P. Pardo
Departamento de Bioquímica, Facultad de Medicina, Universidad Nacional Autónoma de México, Apartado Postal 70-159, 04510, D.F. México, Mexico

Aura M. Pedroza-Rodríguez
Laboratorio de Microbiología Ambiental y de Suelos, Grupo de Biotecnología Ambiental e Industrial (GBAI), Departamento de Microbiología, Facultad de Ciencias, Pontificia Universidad Javeriana (PUJ), Bogotá, Colombia

Edwin D. Morales-Álvarez
Laboratorio de Microbiología Ambiental y de Suelos, Grupo de Biotecnología Ambiental e Industrial (GBAI), Departamento de Microbiología, Facultad de Ciencias, Pontificia Universidad Javeriana (PUJ), Bogotá, Colombia
Departamento de Química, Facultad de Ciencias Exactas y Naturales, Universidad de Caldas, Manizales, Caldas, Colombia

Claudia M. Rivera-Hoyos, Ángela M. Cardozo-Bernal and Raúl A. Poutou-Piñales
Laboratorio de Biotecnología Molecular, Grupo de Biotecnología Ambiental e Industrial (GBAI), Departamento de Microbiología, Facultad de Ciencias, Pontificia Universidad Javeriana (PUJ), Bogotá, Colombia

Dennis J. Díaz-Rincón, Alexander Rodríguez-López and Carlos J. Alméciga-Díaz
Laboratorio de Expresión de Proteínas, Instituto de Errores Innatos del Metabolismo (IEIM), Facultad de Ciencias, Pontificia Universidad Javeriana (PUJ), Bogotá, Colombia

Claudia L. Cuervo-Patiño
Laboratorio de Parasitología Molecular, Grupo de Enfermedades Infecciosas, Facultad de Ciencias, Pontificia Universidad Javeriana (PUJ), Bogotá, Colombia

Karabi Roy, Sujan Dey , Md. Kamal Uddin, Rasel Barua and Md. Towhid Hossain
Department of Microbiology, University of Chittagong, Chittagong 4331, Bangladesh

Alex Sander Rodrigues Cangussu, Deborah Aires Almeida, Raimundo Wagner de Souza Aguiar, Augustus Caeser Franke Portella and Gil Rodrigues dos Santos
Engenharia de Bioprocessos e Biotecnologia e Programa de Pos-Graduação em Biotecnologia, Universidade Federal do Tocantins, Gurupi, TO, Brazil

Sidnei Emilio Bordignon-Junior
Engenharia e Ciências de Alimentos, Universidade Estadual Paulista, São José do Rio Preto, SP, Brazil

Kelvinson Fernandes Viana
Laboratório de Biologia Molecular e Bioquímica-ICVN, Universidade Federal da Integração Latino-Americana, Foz do Iguaçu, PR, Brazil

Luiz Carlos Bertucci Barbosa
Engenharia de Bioprocessos e Biotecnologia, Instituto de Recursos Naturais, Universidade Federal de Itajubá, Itajubá, MG, Brazil

Edson Wagner da Silva Cangussu
Faculdade de Medicina, Universidade Estadual de Montes Claros, Montes Claros, MG, Brazil

Igor Viana Brandi and William James Nogueira Lima
Instituto de Ciências Agrárias, Universidade Federal de Minas Gerais, Montes Claros, MG, Brazil

Eliane Macedo Sobrinho
Instituto Federal do Norte Minas Gerais, Araçuaí, MG, Brazil

Doris C. Niño-Gómez, Claudia M. Rivera-Hoyos, Ingrid N. Ramírez-Casallas and Raúl A. Poutou-Piñales
Laboratorio de Biotecnología Molecular, Grupo de Biotecnología Ambiental e Industrial (GBAI), Departamento de Microbiología, Facultad de Ciencias, Pontificia Universidad Javeriana, Bogotá, Colombia

Edwin D. Morales-Álvarez
Departamento de Química, Facultad de Ciencias Exactas y Naturales, Universidad de Caldas, Manizales, Caldas, Colombia

Edgar A. Reyes-Montaño and Nury E. Vargas-Alejo
Grupo de Investigación en Proteínas, Departamento de Química, Facultad de Ciencias, Universidad Nacional de Colombia (UNAL), Bogotá, Colombia

Kübra Erkan Türkmen
Department of Biology, Faculty of Science, Hacettepe University, Beytepe, Ankara, Turkey

Homero Sáenz-Suárez and José A. Sáenz-Moreno
Unidad de Biolog´ıa Celular y Microscopía, Decanato de Ciencias de la Salud, Universidad Centroccidental Lisandro Alvarado, Barquisimeto, Venezuela

Janneth González-Santos
Grupo de Bioquímica Computacional y Estructural, Departamento de Bioquímica y Nutrición, Facultad de Ciencias, Pontificia Universidad Javeriana, Bogotá, Colombia

Azucena Arévalo-Galvis
Laboratorio de Microbiología Especial, Grupo de Enfermedades Infecciosas, Departamento de Microbiología, Facultad de Ciencias, Pontificia Universidad Javeriana, Bogotá, Colombia

Ruqayah G. Y. Al-Obaidi
Genetic Counseling Clinic and Genetics Laboratory,The Teaching Laboratories, Medical City, Baghdad, Iraq

Bassam M. S. Al-Musawi
Department of Pathology, College of Medicine, Baghdad University, Baghdad, Iraq

Munib Ahmed K. Al-Zubaidi
Department of Pediatrics, College of Medicine, Baghdad University and Pediatric Endocrine Consultation Clinic, ChildrenWelfare Hospital, Baghdad, Iraq

Christian Oberkanins and Stefan Németh
ViennaLab Diagnostics GmbH, Gaudenzdorfer Guertel 43-45, 1120 Vienna, Austria

Yusra G. Y. Al-Obaidi
National Center of Hematology, Al-Mustansiriya University, Baghdad, Iraq

Bagher Larijani
Endocrinology and Metabolism Research Center, Endocrinology and Metabolism Clinical Sciences Institute, Tehran University of Medical Sciences, Tehran, Iran

Ramin Heshmat
Chronic Diseases Research Center, Endocrinology and Metabolism Population Sciences Institute, Tehran University of Medical Sciences, Tehran, Iran

Mina Ebrahimi-Rad, Shohreh Khatami, Shirin Valadbeigi and Reza Saghiri
Department of Biochemistry, Pasteur Institute of Iran, Tehran, Iran

Nor Aliza Abdul Rahim and Marlini Othman
Department of Paraclinical Sciences, Faculty of Medicine and Health Sciences, Universiti Malaysia Sarawak, 94300 Kota Samarahan, Sarawak, Malaysia

Muna Sabri
Department of Basic Medical Sciences, Faculty of Medicine and Health Sciences, Universiti Malaysia Sarawak, 94300 Kota Samarahan, Sarawak, Malaysia

David W. Stanley
United States Department of Agriculture, Agricultural Research Service, Biological Control of Insects Research Laboratory, 1503 S. Providence Road, Columbia, MO 65203, USA

Deyaa M. Abol Fotouh
Electronic Materials Research Department, Advanced Technology and New Materials Institute (ATNMRI), City of Scientific Research and Technological Applications (SRTA-City), New Borg El-Arab City, Alexandria, Egypt

Reda A. Bayoumi
Biology Department, Faculty of Science and Education, Taif University, Khormah Branch, Taif, Saudi Arabia Botany and Microbiology Department, Faculty of Science (Boys), Al-Azhar University, Cairo, Egypt

Mohamed A. Hassan
Protein Research Department, Genetic Engineering and Biotechnology Research Institute (GEBRI), City of Scientific Research and Technological Applications (SRTA-City), New Borg El-Arab City, Alexandria, Egypt

Index

www.ingramcontent.com/pod-product-compliance
Lightning Source LLC
Chambersburg PA
CBHW080515200326
41458CB00012B/4214